# BICARBONATE, CHLORIDE, AND PROTON TRANSPORT SYSTEMS

ANNALS OF THE NEW YORK ACADEMY OF SCIENCES

Volume 574

# BICARBONATE, CHLORIDE, AND PROTON TRANSPORT SYSTEMS

*Edited by John H. Durham and Marcos A. Hardy*

*The New York Academy of Sciences*
*New York, New York*
*1989*

**Library of Congress Cataloging-in-Publication Data**

Bicarbonate, chloride, and proton transport systems/edited by John
H. Durham and Marcos A. Hardy.
    p.    cm. — (Annals of the New York Academy of Sciences; v.
574)
    Result of a conference held by the New York Academy of Sciences on
Jan. 19–21, 1989 in New York, N.Y.
    Includes bibliographical references.
    ISBN 0-89766-557-0 (alk. paper). — ISBN 0-89766-558-9 (pbk.:
alk. paper)
    1. Biological transport, Active—Congresses.  2. Bicarbonate ions—
Physiological transport—Congresses.  3. Chloride channels—
Congresses.  4. Protons—Physiological transport—Congresses.
I. Durham, John H.  II. Hardy, Marcos A.  III. New York Academy of
Sciences.  IV. Series.
Q11.N5  vol. 574
[QH509]
500 s—dc20
[574.87′5]                           89-13954
                                     CIP

**BiC/PCP**
*Printed in the United States of America*
**ISBN 0-89766-557-0 (cloth)**
**ISBN 0-89766-558-9 (paper)**
**ISSN 0077-8923**

ANNALS OF THE NEW YORK ACADEMY OF SCIENCES

*Volume 574*

*December 29, 1989*

# BICARBONATE, CHLORIDE, AND PROTON TRANSPORT SYSTEMS[a]

*Editors*

JOHN H. DURHAM AND MARCOS A. HARDY

*Conference Organizers*

JOHN H. DURHAM, MARCOS A. HARDY, AND WALTER N. SCOTT

*Advisory Board*

GEORGE A. GERENCSER, ROLF KINNE, JUHA KOKKO, CHRISTOPHER MILLER, REINHART REITHMEIER, ASER ROTHSTEIN, AND CLIFFORD L. SLAYMAN

## CONTENTS

[a] This volume is the result of a conference entitled Bicarbonate, Chloride, and Proton Transport Systems, which was held by the New York Academy of Sciences on January 19–21, 1989 in New York, New York.

## Part 2. Mechanisms of Chloride Cotransport

*Poster Papers*

**Major funding was provided by:**

- CYSTIC FIBROSIS FOUNDATION
- GLAXO GROUP RESEARCH LIMITED
- HASSLE AB
- HOECHST AKTIENGESELLSCHAFT
- HOECHST-ROUSSEL PHARMACEUTICALS INC.
- HOFFMANN-LA ROCHE INC.
- LILLY RESEARCH LABORATORIES
- MERCK SHARP & DOHME RESEARCH LABORATORIES
- NATIONAL HEART, LUNG AND BLOOD INSTITUTE–NIH
- NATIONAL INSTITUTE OF DIABETES AND DIGESTIVE AND KIDNEY DISEASES–NIH
- NATIONAL SCIENCE FOUNDATION
- THE PROCTER & GAMBLE COMPANY
- U.S. ARMY MEDICAL RESEARCH

# Preface

Knowledge of the mechanisms and regulation of chloride, bicarbonate, and proton transport across cells and their membranes has become crucial to our understanding of many physiologic functions at both the cellular and systemic level. Cell pH and volume regulation and gastric, renal, ocular, and airway function are examples of processes that depend on the control and proper operation of anion and proton translocating mechanisms. Osmotic and metabolic functions of many plant and fungal cells also depend on these processes, and such systems have mechanisms in common with those found in animal cells.

Recently there has been an intensification of studies on the mechanisms and especially the regulation of anion and proton transport at both cell and tissue levels. It has become evident not only that these processes participate in the physiologic homeostasis of the living state, but also that dysfunction in these processes can result in significant pathologic conditions. Commensurate with and partially a result of the increased interest in these mechanisms have been the significant strides in knowledge afforded by developments for ion-detecting dyes, signal channel measurements, and the application of genetic and biochemical approaches to the study of anion and proton transport systems. The purpose of this conference was to provide a timely presentation of the current state of knowledge on these important physiologic systems and to provide a forum for investigators examining these processes at the submembrane, membrane, and tissue levels.

We would like to acknowledge our appreciation of the essential guidance of Ellen A. Marks, the assistance of Renée Wilkerson-Brown and Gerri Busacco, and the support provided by the sponsors without which this conference would not have been possible. We thank Dr. Charles Nicholson and the advisory committee at the New York Academy of Sciences, and the members of this conference's scientific committee for participation in the formulation of this conference. We also denote our sincere appreciation for Dr. Walter N. Scott and his critical advice and attention throughout our progress in forming the conference. Finally, we would like to acknowledge the significant assistance and advice freely given by Dr. John Cuppoletti.

JOHN H. DURHAM
MARCOS A. HARDY

WILLIAM A. BRODSKY

# William A. Brodsky: A Tribute

As the seed for the idea for this conference was a New York Academy of Sciences conference formed a decade ago by Bill Brodsky, we thought it appropriate to note that the date of this conference coincides with his seventh decade and at least his fourth decade in science, although some would contend that those decade designations appear to be the other way around.

William A. Brodsky (M.D., Temple University, 1941) completed his clinical training at the Philadelphia General Hospital and the Children's Hospital of Philadelphia. After service in the Medical Corps of the U.S. Army, he started his research career as a Fellow in the laboratories of George M. Guest and Samuel Rapoport at the University of Cincinnati.

Intrigued by the pronounced effects of dehydration and diabetes on children, he and Samuel Rapoport began many insightful and decisive investigations into the effects of hypertonic solute loading on the osmotic functions of the kidney during dehydration and overhydration in normal human subjects as well as those with diabetes mellitus and diabetes insipidus. Their work on solute-induced osmotic diuresis was seminal in our understanding of the handling of water and ions by the human kidney. One of the notable concepts and parameters to arise from that work is that of renal water economy, which is now called free water clearance.

Dr. Brodsky joined the Department of Pediatrics at the University of Louisville in 1951, where he focused his research efforts on the effects of hydration and overhydration on the acid-base excretion by the kidney. During the first few years at Louisville, it was necessary that he share laboratory space with Warren Rehm in the Department of Physiology. This arrangement resulted in an intensely productive friendship and collaboration that enriched the experience of all who worked in their respective laboratories. One of the fruits of that collaboration was the earlier applications of thermodynamics to physiologic processes, namely, that an intracellular osmotic gradient could not account for water transport by the renal tubule.

His interest in the osmotic activity of tissues was extended to show conclusively for the first time that most tissues, and especially the liver, were isoosmotic with plasma. As the countercurrent hypothesis for the osmotic concentration of urine was advanced, his group demonstrated the substantial washout of osmotic activity in the renal medulla during osmotic diuresis, in support of the countercurrent model. Together with Gaspar Carrasquer, he studied the effects of electrolyte loading on acid excretion by the kidney, and first began to examine bicarbonate reabsorption as an underlying process for acidification of the urine.

In 1964 Ted Schilb joined his laboratory and, as the story goes, decided to use some turtles left over from a classroom experiment to examine the transport properties of the urinary bladder. These initial studies founded the use of the turtle and its bladder as a model system for studying basic mechanisms of urinary electrolyte and acid-base excretion. Bill was subsequently invited to join the faculty of the newly formed Mount Sinai School of Medicine where he was appointed as a City University of New York Professor. Since his entire group moved to New York with him, the laboratory remained active, and over the next decade their work expanded into studies of enzymatic mechanisms of Na transport and transport in membrane vesicles. One of those studies, involving the

xiii

effects of stilbenes on ion transport in the bladder, led to the discovery of an electrogenic secretion of alkali and its regulation by *in vivo* conditions as well as extracellular and intracellular messenger systems.

Dr. Brodsky is currently Research Professor of Physiology and Biophysics and Professor of Medicine at Mt. Sinai. He has had a pronounced influence on the careers of those who have had the opportunity to work with him, and his vigorous and innovative research programs will significantly contribute to our knowledge of electrolyte and acid-base balance in the years to come.

JOHN H. DURHAM
WALTER N. SCOTT

# Primary Active Electrogenic Chloride Transport Across the *Aplysia* Gut[a]

GEORGE A. GERENCSER

*Department of Physiology*
*Box J-274, J.H.M.H.C.*
*University of Florida*
*Gainesville, Florida 32610*

As yet, there is no unequivocal, definitive evidence of primary active $Cl^-$ transport ($Cl^-$ pump), that is, an adenosinetriphosphatase (ATPase) that translocates $Cl^-$ up its electrochemical gradient powered by the simultaneous hydrolysis of ATP. In fact, evidence for a $Cl^-$ pump in bacteria[1] and plants[2,3] is more convincing than that demonstrated in animal plasma membranes.[4-6] Also, these studies on the lower life forms have provided the impetus for preliminary and detailed examinations of possible existing $Cl^-$ pumps in animals. The main argument for the anion-stimulated ATPase moving anions up their respective energy gradients in animal plasma membranes is based on parallel observations of anion-stimulated ATPase activity and anion transport phenomena in the same tissue.[4,5] In fact, it was not until the following observation that $HCO_3^-$-stimulated ATPase activity was linked with $Cl^-$ pumping, because no $Cl^-$ activation of this enzyme had been observed. DeRenzis and Bornancin[7] were the first to demonstrate the plasma membrane presence of a ($Cl^- + HCO_3^-$)-stimulated ATPase in goldfish gill epithelium and suggested that the enzyme could participate in the branchial $Cl^-/HCO_3^-$ exchange mechanism.

Gerencser[8,9] has demonstrated that the isolated gut of *Aplysia californica*, bathed in a substrate- and Na-free seawater bathing medium, generates a serosa-negative transepithelial potential difference. The short-circuit current (SCC) was totally accounted for by an active absorptive flux of $Cl^-$ (TABLE 1). The SCC and net $Cl^-$ flux were inhibited by both thiocyanate and acetazolamide but were insensitive to serosally applied ouabain. It was hypothesized that active $Cl^-$ absorption in *Aplysia* enterocytes was mediated by a primary active transport process, for it had been demonstrated that the intracellular $Cl^-$ electrochemical potential was less than that measured in the extracellular medium,[10] even in the absence of extracellular $Na^+$ (TABLE 2).[11] So one need not postulate the existence of an active transport system for $Cl^-$ at the mucosal membrane, for $Cl^-$ movement from the luminal medium into the enterocyte cytoplasm could utilize the energy implicit within its own electrochemical potential gradient directed downhill from the mucosal solution into the cytosol of the *Aplysia* enterocyte. However, the transit of $Cl^-$ across the basolateral membrane from the enterocyte cytoplasm into the serosal solution involves moving up a large electrochemical gradient for $Cl^-$. Thermodynamically, therefore, the site for the active $Cl^-$ transport mechanism has to be the basolateral membrane of the *Aplysia* enterocyte.

[a] These studies were supported in part by National Institutes of Health Grants 1-T01-AM-05697-02 and -03, AM-17361, and AM-00367, Division of Sponsored Research (DSR, University of Florida) Seed Award 229K15, and Award 122101010, Whitehall Foundation Grant 78-156 ck-1, and the Eppley Foundation for Research.

**TABLE 1.** Chloride Fluxes in Tris-Cl Seawater Media

| | $J_{ms}$ | $n$ | $J_{sm}$ | $n$ | $J_{ms}^{Net}$ | SCC | $n$ |
|---|---|---|---|---|---|---|---|
| Before thiocyanate addition | $216.7 \pm 14.2$ | 6 | $180.1 \pm 13.2$ | 6 | $36.6 \pm 6.8$ | $30.3 \pm 7.9$ | 6 |
| After thiocyanate addition | $175.7 \pm 8.1$ | 6 | $173.9 \pm 8.6$ | 6 | $1.8 \pm 1.3$ | $0.8 \pm 0.7$ | 6 |

Values are means $\pm$ SE in neq $\cdot$ cm$^{-2}$ $\cdot$ min$^{-1}$; $n$ = no. of experiments; $J$ = flux; SCC = short-circuit current. Table is taken from Gerencser,[9] with permission.

Presently, without question, the greatest argument regarding Cl$^-$-stimulated ATPase activity is its localization within the subarchitecture of cells. It seems that Cl$^-$-stimulated ATPase activity resides in both microsomal and mitochondrial fractions of cell homogenates.[12] However, Schuurmans Stekhoven and Bonting[12] have stated that microsomal or plasma membrane localization of this enzyme is entirely due to mitochondrial contamination, hence the dispute. If Cl$^-$-stimulated ATPase activity is exclusively of mitochondrial origin, it is very difficult to conceive a mechanism which is the ATPase that can drive net Cl$^-$ transport across plasma membranes. Conversely, if the Cl$^-$-stimulated ATPase is located in the plasma membrane, it would not be difficult to envision primary active Cl$^-$ transport by this enzyme which could possibly act as the prime mover of net Cl$^-$ transport between the intracellular and extracellular space, analogous to the (Na$^+$ + K$^+$)-stimulated ATPase and its role in the net transport of Na$^+$ and K$^+$ across plasma membranes.[12]

Gerencser and Lee[5,13] demonstrated the existence of a Cl$^-$-stimulated ATPase activity in *Aplysia* enterocyte basolateral membranes, proving an extramitochondrial origin of this enzyme. Their finding that the subcellular membrane fraction having a high specific activity in Na$^+$-K$^+$-ATPase (a basolateral membrane marker enzyme), but having no perceptible cytochrome $c$ oxidase activity and a significantly reduced succinic dehydrogenase activity (mitochondrial marker enzymes), supports the notion of the nonmitochondrial origin of this enzyme (TABLE 3). The observation that there was very little NADPH-cytochrome $c$ reductase activity in the basolateral membrane fraction (TABLE 3) also suggests that the membranes in this fraction were relatively free from endoplasmic reticulum and Golgi body membrane contamination.[14] The failure of oligomycin to inhibit Cl$^-$-ATPase activity in the basolateral membrane fraction is also consistent with the nonmitochondrial origin of the anion-ATPase (TABLE 4). Supporting this contention is the corollary finding that oligomycin inhibited mitochondrial anion-stimulated ATPase activity. The finding that efrapeptin, a direct inhibitor of mitochondrial F$_1$-ATPase activity,[15] significantly inhibited Mg$^{2+}$-ATPase activity in the

**TABLE 2.** Intracellular Cl$^-$ Activities and Mucosal Membrane Potentials in NaCl and Tris-Cl Seawater Media

| | $a_{Cl}^i$ (mM) | $a_{Cl}^{eq}$ (mM) | $\psi_m$ (mV) | $n$ |
|---|---|---|---|---|
| NaCl | $13.9 \pm 0.5$ (30) | 28.6 | $-64.6 \pm 1.3$ (48) | 7 |
| Tris-Cl | $9.1 \pm 0.3$ (32) | 21.4 | $-72.2 \pm 1.4$ (51) | 7 |
| | $p < 0.01$ | | $p < 0.01$ | |

Values are means $\pm$ SE; numbers in parentheses are number of observations; $n$ is number of animals. Polarity of $\psi_m$ is relative to mucosal solution. $a_{Cl}^i$, intracellular Cl activity. Table is taken from Gerencser,[11] with permission.

mitochondrial and not in the basolateral membrane fraction also supports the notion that the basolateral membrane fraction is of extramitochondrial origin.[13]

Biochemical properties of this enterocyte basolateral membrane-localized $Cl^-$-stimulated ATPase include the following: (1) pH optimum = 7.8; (2) ATP being the most effective nucleotide hydrolyzed; (3) also stimulated by $HCO_3^-$, $SO_3^=$, and $S_2O_3^=$ but inhibited by $NO_2^-$ and $NO_3^-$; (4) apparent $K_m$ for $Cl^-$ = 10.3 mM (FIG. 1) while the apparent $K_m$ for ATP = 2.6 mM (FIG. 2); and (5) a requirement for $Mg^{2+}$ which has an optimal concentration of 3 mM.[13]

Additionally, the ATPase activity stimulated by $Cl^-$ was strongly inhibited by thiocyanate, vanadate, and acetazolamide but not inhibited by ouabain (TABLE 5). These results with inhibitors strongly suggest a possible participation by the $Cl^-$-stimulated ATPase in net chloride absorption by the *Aplysia* gut.[13] The finding that anion-stimulated ATPase is inhibited by thiocyanate but not by ouabain has

TABLE 3. Distribution of Marker Enzymes and Anion-Stimulated ATPase during Preparation of Plasma Membranes from *Aplysia* Enterocytes

| Enzyme | Homogenate | $P_2$ (Mitochondria) | S-III (Basolateral Membranes) |
|---|---|---|---|
| Total protein (mg) | 208.59 ± 24.95 | 15.13 ± 1.95 | 5.47 ± 1.09 |
| $Na^+$-$K^+$-ATPase | 0.85 ± 0.11 | 0.25 ± 0.23 | 5.26 ± 1.96 |
| 5'-Nucleotide | 0.41 ± 0.17 | 0.37 ± 0.11 | 0.89 ± 0.33 |
| Cytochrome *c* oxidase | 0.50 ± 0.12 | 0.97 ± 0.23 | ND |
| Succinic dehydrogenase | 22.30 ± 4.90 | 83.40 ± 27.60 | 4.17 ± 2.40 |
| NADPH-cytochrome *c* reductase | 3.73 ± 0.28 | 7.44 ± 1.25 | 1.32 ± 0.18 |
| $Mg^{2+}$-ATPase | 2.69 ± 0.57 | 5.10 ± 0.84 | 9.16 ± 1.82 |
| $HCO_3^-$-ATPase | 3.85 ± 0.83 | 7.26 ± 0.91 | 14.12 ± 2.02 |
| $Cl^-$-ATPase | 2.51 ± 0.59 | 2.50 ± 0.79 | 5.77 ± 2.26 |

Values are means ± SE from 9–11 different preparations. Enzyme activity is expressed as $\mu mol \cdot h^{-1} \cdot mg\ protein^{-1}$ for $Na^+$-$K^+$-ATPase and 5'-nucleotidase; $\Delta log$ (ferrocytochrome *c*) $\cdot min^{-1} \cdot mg\ protein^{-1}$ for cytochrome *c* oxidase; $\mu mol \cdot min^{-1} \cdot mg\ protein^{-1}$ for succinic dehydrogenase; $nmol \cdot min^{-1} \cdot mg\ protein^{-1}$ for NADPH-cytochrome-*c* reductase; $\mu mol \cdot 15\ min^{-1} \cdot mg\ protein^{-1}$ for $Mg^{2+}$-, $HCO_3^-$-, and $Cl^-$-ATPase. $P_2$, pellets from 9,500-*g* centrifugation; S-III, 40–50% sucrose interface; ND = not detectable. Starting gut mucosa was ~1.5 g. Conditions for enzyme assays were as described in text. Table from Gerencser and Lee,[13] with permission.

also been demonstrated in many tissues known to perform active anion transport and to contain anion-stimulated ATPase activity.[16,17] Additionally, thiocyanate inhibits anion transport across a variety of epithelial systems.[16,18] The parallel between ouabain insensitivity and thiocyanate sensitivity to $Cl^-$-stimulated ATPase activity and net active $Cl^-$ absorption in the *Aplysia* gut warrants conjecture that the active $Cl^-$ absorptive mechanism could be driven by a $Cl^-$-stimulated ATPase found in the enterocyte plasma membrane. Additional support for this contention rests with the present finding that $Cl^-$-stimulated ATPase activity of the basolateral membranes is inhibited by vanadate.[13] Vanadate has been shown to inhibit ATPases, which form high-energy phosphorylated intermediates while having no effect on the mitochondrial anion-sensitive ATPase.[19] These results strongly suggest that the $Cl^-$-stimulated ATPase is an ion-transporting ATPase of the $E_1$-$E_2$ variety rather than the $F_0$-$F_1$ type.

**TABLE 4.** Effect of Oligomycin on $Cl^-$-$HCO_3^-$-ATPase Activity

| Oligomycin $10^{-9}$ M | $Mg^{2+}$-ATPase Specific Activity ($\mu$mol · 15 min$^{-1}$ · mg$^{-1}$) | % Inhibition | $HCO_3^-$-ATPase Specific Activity ($\mu$mol · 15 min$^{-1}$ · mg$^{-1}$) | % Inhibition | $Cl^-$-ATPase Specific Activity ($\mu$mol · 15 min$^{-1}$ · mg$^{-1}$) | % Inhibition |
|---|---|---|---|---|---|---|
| **H (Homogenate)** | | | | | | |
| 0 | 4.38 ± 0.16 | 0 | 4.84 ± 2.09 | 0 | 1.46 ± 0.10 | 0 |
| 6.8 | 2.80 ± 0.49 | 36 | 3.36 ± 1.81 | 31 | 1.28 ± 0.47 | 12 |
| 13.6 | 2.10 ± 0.07 | 52 | 2.77 ± 0.86 | 43 | 0.71 ± 0.04 | 51 |
| **P₂ (Mitochondrial Pellet)** | | | | | | |
| 0 | 4.24 ± 1.41 | 0 | 4.00 ± 1.68 | 0 | 1.06 ± 0.26 | 0 |
| 6.8 | 2.90 ± 0.41 | 32 | 4.14 ± 2.05 | 0 | 1.13 ± 0.27 | 0 |
| 13.6 | 1.77 ± 1.33 | 58 | 2.45 ± 0.17 | 39 | 0.68 ± 0.02 | 36 |
| **S-III (Basolateral Membranes)** | | | | | | |
| 0 | 14.65 ± 3.21 | 0 | 14.63 ± 3.41 | 0 | 2.97 ± 0.43 | 0 |
| 6.8 | 12.65 ± 2.85 | 14 | 14.79 ± 4.71 | 0 | 2.37 ± 0.47 | 20 |
| 13.6 | 9.47 ± 2.26 | 35 | 14.04 ± 3.31 | 4 | 3.16 ± 0.91 | 0 |

Values are means ± SE from three different experiments. Oligomycin was preincubated with enzymes in reaction mixture at indicated concentrations for 10 minutes at 30°C. Table taken from Gerencser and Lee,[13] with permission.

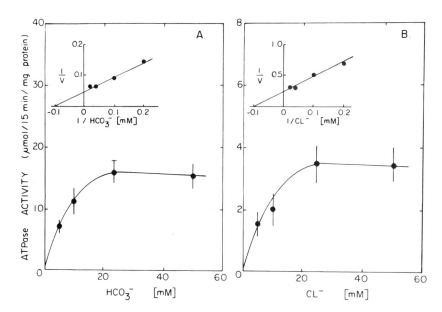

**FIGURE 1.** Effect of Cl⁻ concentration on ATPase activity. Values are means ± SE from three different experiments. *Insets,* Lineweaver-Burk plots for ATPase activity; line was determined by linear regression analysis ($r^2 = 0.984$). Figure is taken from Gerencser and Lee,[13] with permission.

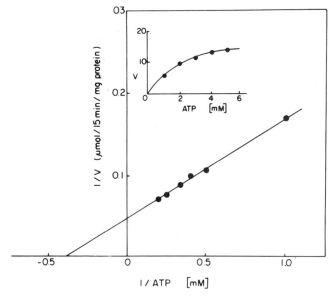

**FIGURE 2.** Lineweaver-Burk plots for Cl⁻-ATPase activity. Concentrations of ATP were varied as indicated. Values are means from five different experiments, and line was determined by linear regression analysis ($r^2 = 0.992$). *Inset,* enzyme activity with different concentrations of ATP. Figure is taken from Gerencser and Lee,[13] with permission.

**TABLE 5.** Effects of Inhibitors on $Cl^-$-ATPase Activity

| Inhibitor | Concentration (mM) | Specific Activity | | |
|---|---|---|---|---|
| | | $Mg^{2+} + Cl^-$-ATPase | $Cl^-$-ATPase | % Inhibition |
| Control | | $16.8 \pm 0.3$ | 6.7 | 0 |
| Thiocyanate | 10.1 | $7.7 \pm 0.6$ | 4.5 | 33 |
| Acetazolamide | 1.0 | $11.1 \pm 1.0$ | 1.4 | 79 |
| Acetazolamide | 2.0 | $7.7 \pm 2.0$ | 0.4 | 94 |
| Ouabain | 1.0 | $16.5 \pm 0.5$ | 7.0 | 0 |
| Vanadate | 0.5 | $14.0 \pm 1.4$ | 4.4 | 34 |
| Vanadate | 1.0 | $10.5 \pm 1.1$ | 2.5 | 63 |

Values are means $\pm$ SE from 3–5 different experiments. Specific activity is expressed as $\mu mol \cdot 15\ min^{-1} \cdot mg\ protein^{-1}$. Inhibitors were preincubated with enzymes in reaction mixture at indicated concentrations for 10 minutes at 30°C. Conditions for enzyme assay were as described in text. Table from Gerencser and Lee,[13] with permission.

Acetazolamide inhibited $Cl^-$-stimulated ATPase activity in the *Aplysia* gut.[13] This finding has also been demonstrated in blue crab gill $HCO_3^-$-ATPase.[20] Although acetazolamide has been shown to be a specific inhibitor of carbonic anhydrase,[21] it has also been demonstrated to be a $Cl^-$ transport inhibitor.[22] Thus the data further strengthen the notion that the $Cl^-$-stimulated ATPase, which is inhibited by acetazolamide, may be involved in $Cl^-$ transport across the molluscan gut.

Furthermore, Gerencser and Lee[23] have demonstrated an ATP-dependent $Cl^-$ uptake in *Aplysia* inside-out enterocyte basolateral membrane vesicles that was inhibitable by thiocyanate, vanadate, and also by acetazolamide.[24] The ATP-driven $Cl^-$ uptake was obtained in the absence of $Na^+$, $K^+$, $HCO_3^-$, or a pH gradient between the intra- and extravesicular space,[23,25] which is strong suggestive evidence that the $Na^+$-$K^+$-ATPase enzyme,[12] $Na^+/Cl^-$ symport,[5,26] $K^+/Cl^-$ symport,[27] $Na^+/K^+/Cl^-$ symport,[27] $Cl^-/HCO_3^-$ or $Cl^-/OH^-$ antiport,[5] and $K^+/H^+$ antiport[12] are not mechanisms that are involved in the accumulation of $Cl^-$ within the vesicles in the presence of ATP.

To further elucidate the electrogenic nature of the ATP-dependent $Cl^-$ transport process, several experimental maneuvers were performed by Gerencser[25] as follows. First, an inwardly directed valinomycin-induced $K^+$ diffusion potential, making the vesicle interior electrically positive, enhanced ATP-driven $Cl^-$ uptake compared with vesicles lacking the ionophore (FIG. 3). Second, an inwardly di-

**TABLE 6.** Effect of ATP on Transport Parameters in Basolateral Membrane Vesicles

| Experimental Condition | $Cl^-$ Uptake (nmol/mg protein) | n | Vesicular Membrane Potential Difference (mV) | n |
|---|---|---|---|---|
| +ATP | $102.7 \pm 7.9$ | 3 | $-34.9 \pm 2.5$ | 12 |
| −ATP | $49.7 \pm 5.9$ | 3 | $0.0 \pm 5.2$ | 12 |
| +Nonhydrolyzable ATP analog (5'-adenylyl-imidodiphosphate) | $59.6 \pm 8.3$ | 3 | $-1.3 \pm 0.9$ | 12 |
| $NO_3^-$ for $Cl^-$ (mole for mole) | | | $+3.0 \pm 4.6$ | 3 |

Values are means $\pm$ SE; $n$ = number of experiments. Table is taken from Gerencser *et al.*,[6] with permission.

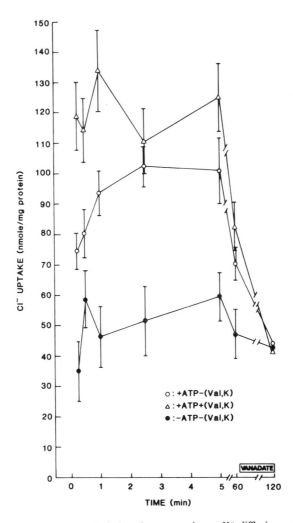

**FIGURE 3.** Effect of valinomycin-induced transmembrane K$^+$ diffusion potential on time course of ATP-dependent Cl$^-$ uptake by *Aplysia* enterocyte plasma membrane vesicles. $\Delta$ = vesicles whose reaction mixture contained 40 mM K$^+$ and 0.18 mM valinomycin but were otherwise under isosmotic equilibrium tracer exchange conditions; O = same as above except that no valinomycin or K$^+$ was present on either side of the vesicular membrane; ● = same as above except that no valinomycin, K$^+$, or ATP was present on either side of the vesicular membrane. pH 7.8 on both sides of the vesicular membrane was used throughout these experiments. Extra- and intravesicular medium composition was as follows: 10 mM Tris-HEPES (pH 7.8), 250 mM sucrose, 3 mM MgSO$_4$, and 25 mM choline chloride. Values reported are means of nine different experiments (40–45 animals). Figure is taken from Gerencser,[25] with permission.

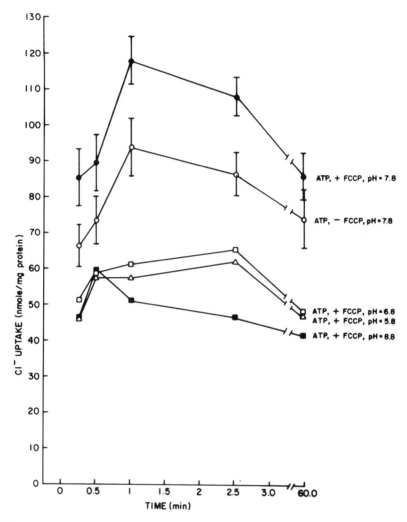

**FIGURE 4.** Effect of FCCP-induced transmembrane $H^+$ conductance on time course of ATP-dependent $Cl^-$ uptake by various populations of enterocyte plasma membrane vesicles bathed in reaction mixtures with different pHs. ● = vesicles whose reaction mixture contained 0.12 mM FCCP at pH 7.8 and 5 mM ATP but otherwise were under isosmotic equilibrium tracer exchange conditions. ○ = same as above except FCCP was not present on either side of the vesicular membrane. □ = same as above except FCCP was present in the reaction mixture and the extravesicular pH was 6.8. ■ = same as above except extravesicular pH was 8.8. △ = same as above except extravesicular pH was 5.8. Initial intravesicular pHs for all vesicle populations were 7.8. Extra- and intravesicular medium composition was as follows: 10 mM Tris-HEPES, 250 mM sucrose, 3 mM $MgSO_4$, and 25 mM choline chloride. Values reported are means of nine different experiments (40–45 animals) or means ± SE from nine different experiments (40–45 animals). Figure is taken from Gerencser,[25] with permission.

TABLE 7. Effect of Bilateral iso-pH, FCCP, and ATP on Total Steady-state $Cl^-$ Uptake in EPMV

| pH | $Cl^-$ Uptake (nmol/mg protein) |
|---|---|
| 5.8 | 56.9 ± 8.7 |
| 6.8 | 53.5 ± 6.5 |
| 7.8 | 96.8 ± 8.1 |
| 8.8 | 50.9 ± 9.2 |

Values are means ± SE from 3 different experiments (12–18 animals). FCCP (0.12 mM) was preincubated with membrane vesicles in the reaction mixture (50 μl containing 10 mM Tris-HEPES, 250 mM sucrose, 3 mM $MgSO_4$, and 25 mM choline chloride) at the indicated pH for 10 minutes at 25°C. Intravesicular pH matched extravesicular pH, with intravesicular medium composition as follows: 10 mM Tris-HEPES, 250 mM sucrose, 3 mM $MgSO_4$, and 25 mM choline chloride. The uptake of $^{36}Cl^-$ was measured for 2.5 minutes at 25°C. EPMV = enterocyte plasma membrane vesicles. Table is taken from Gerencser,[25] with permission.

rected FCCP-induced $H^+$ electrodiffusion potential, making the vesicle interior less negative, increased ATP-dependent $Cl^-$ uptake compared to control (FIG. 4). Third, ATP plus $Cl^-$ increased intravesicular negativity measured by lipophilic $TPMP^+$ distribution across the vesicular membrane. Both ATP and $Cl^-$ appeared to be necessary for generating the negative intravesicular membrane potential, because substituting a nonhydrolyzable ATP analog for ATP, in the presence of $Cl^-$ in the extravesicular medium, did not generate a potential above that of control (TABLE 6). Likewise, substituting $NO_3^-$ for $Cl^-$ in the extra- and intravesicular media, in the presence of extravesicular ATP, caused no change in potential difference above that of control (TABLE 6).[25] These results also suggest that hydrolysis of ATP is necessary for the accumulation of $Cl^-$ in the vesicles. Furthermore, vanadate, acetazolamide, and thiocyanate inhibited the (ATP + $Cl^-$)-dependent intravesicular negativity.[4,24,25] In addition, it has been demonstrated that the pH optimum of the $Cl^-$-stimulated ATPase[13] coincides exactly with the pH optimum of 7.8 of the ATP-dependent $Cl^-$ transport in the same fraction of *Aplysia* enterocyte basolateral membrane vesicles (TABLE 7).[25] These results, combined with the aforementioned ones, are strongly consistent with the hypothesis that the active electrogenic $Cl^-$ transport mechanism in *Aplysia* gut is an electrogenic, $Cl^-$-transporting, $Cl^-$-stimulated ATPase or a primary $Cl^-$ pump found in the enterocyte-basolateral membrane.

## ACKNOWLEDGMENTS

The author acknowledges his gratitude to his technologists, students, and collaborators for their able contributions to the studies reviewed and performed herein.

## REFERENCES

1. LANYI, J. K. 1986. Halorhodopsin: A light-driven chloride ion pump. Ann. Rev. Biophys. Chem. **15:** 11–28.
2. GRADMANN, D. 1984. Electrogenic $Cl^-$ pump in the marine alga *Acetabularia*. *In*

Chloride Transport Coupling in Biological Membranes and Epithelia. G. A. Gerencser, ed., : 13–62. Elsevier, Amsterdam.

3. HILL, B. S. 1984. Metabolic coupling of chloride transport in higher plant cells. *In* Chloride Transport Coupling in Biological Membranes and Epithelia. G. A. Gerencser, ed., : 1–11. Elsevier, Amsterdam.

4. GERENCSER, G. A. 1986. Properties and functions of $Cl^-$-stimulated ATPase. Trends Life Sci. **1:** 1–18.

5. GERENCSER, G. A. & S. H. LEE. 1983. $Cl^-$-stimulated adenosine triphosphatase: Existence, location and function. J. Exp. Biol. **106:** 143–161.

6. GERENCSER, G. A., J. F. WHITE, D. GRADMANN & S. L. BONTING. 1988. Is there a $Cl^-$ pump? Am. J. Physiol. **255:** R677–R692.

7. DeRENZIS, G. & M. BORNANCIN. 1977. $Cl^-/HCO_3^-$ ATPase in the gills of *Carassius auratus:* Its inhibition by thiocyanate. Biochim. Biophys. Acta **467:** 192–207.

8. GERENCSER, G. A. 1981. Effects of amino acids on chloride transport in *Aplysia* intestine. Am. J. Physiol. **240:** R61–R69.

9. GERENCSER, G. A. 1984. Thiocyanate inhibition of active chloride absorption in *Aplysia* gut. Biochim. Biophys. Acta **775:** 389–394.

10. GERENCSER, G. A. & J. F. WHITE. 1980. Membrane potentials and chloride activities in epithelial cells of *Aplysia* gut. Am. J. Physiol. **239:** R445–R449.

11. GERENCSER, G. A. 1983. Electrophysiology of chloride transport in *Aplysia* (mollusk) intestine. Am. J. Physiol. **244:** R143–R149.

12. SCHUURMANS STEKHOVEN, F. & S. L. BONTING. 1981. Transport adenosine triphosphatases: Properties and functions. Physiol. Rev. **61:** 1–76.

13. GERENCSER, G. A. & S. H. LEE. 1985. $Cl^-/HCO_3^-$-stimulated ATPase in intestinal mucosa of *Aplysia*. Am. J. Physiol. **248:** R241–R248.

14. MIRCHEFF, A. K., G. SACHS, S. D. HANNA, C. S. LABINER, E. RABON, A. P. DOUGLAS, M. W. WALLING & E. M. WRIGHT. 1979. Highly purified basal lateral plasma membranes from rat duodenum: Physical criteria for purity. J. Membr. Biol. **50:** 343–363.

15. BULLOUGH, D. A., C. G. JACKSON, P. J. F. HENDERSON, R. B. BEECHEY & P. E. LINNETT. 1982. The isolation and purification of the elvapeptins: A family of peptide inhibitors of mitochondrial ATPase activity. FEBS Lett. **145:** 258–262.

16. KATZ, A. I. & F. H. EPSTEIN. 1971. Effect of anions on adenosine triphosphatase of kidney tissue. Enzyme **12:** 499–507.

17. SACHS, G., W. E. MITCH & B. I. HIRSCHOWITZ. 1965. Frog gastric mucosal ATPase. Proc. Soc. Exp. Biol. Med. **119:** 1023–1027.

18. EPSTEIN, F. H., J. MAETZ & G. DeRENZIS. 1973. Active transport of chloride by the teleost gill: Inhibition by thiocyanate. Am. J. Physiol. **224:** 1295–1299.

19. AKERA, T., K. TEMMA & K. TAKEDA. 1983. Cardiac actions of vanadium. Fed. Proc. **42:** 2984–2988.

20. LEE, S. H. 1982. Salinity adaptation of $HCO_3^-$-dependent ATPase activity in the gills of blue crab (*Callinectes sapidus*). Biochim. Biophys. Acta **689:** 143–154.

21. MAREN, T. H. 1977. Use of inhibitors in physiological studies of carbonic anhydrase. Am. J. Physiol. **232:** F291–F297.

22. WHITE, J. F. 1980. Bicarbonate-dependent chloride absorption in small gut: Ion fluxes and intracellular chloride activities. J. Membr. Biol. **53:** 95–107.

23. GERENCSER, G. A. & S. H. LEE. 1985. ATP-dependent chloride transport in plasma membrane vesicles from *Aplysia* gut. Biochim. Biophys. Acta **816:** 415–417.

24. GERENCSER, G. A. 1986. Inhibition of a chloride pump by acetazolamide in the gut of *Aplysia californica*. J. Exp. Biol. **125:** 391–393.

25. GERENCSER, G. A. 1988. Electrogenic ATP-dependent $Cl^-$ transport by plasma membrane vesicles from *Aplysia* intestine. Am. J. Physiol. **254:** R127–R133.

26. FRIZZELL, R. A., M. FIELD & S. G. SCHULTZ. 1979. Sodium-coupled chloride transport by epithelial tissues. Am. J. Physiol. **236:** F1–F8.

27. GREGER, R. & E. SCHLATTER. 1984. Mechanisms of chloride transport in vertebrate renal tubule. *In* Chloride Transport Coupling in Biological Membranes and Epithelia. G. A. Gerencser, ed. : 271–346. Elsevier, Amsterdam.

# Halorhodopsin: A Light-Driven Active Chloride Transport System[a]

LÁSZLÓ ZIMÁNYI[b] AND JANOS K. LANYI

*Department of Physiology and Biophysics*
*University of California*
*Irvine, California 92717*

Halorhodopsin,[1,2] one of the retinal proteins in the cytoplasmic membrane of halobacteria, is an inward-directed light-driven electrogenic pump for chloride ions that generates an inside-negative membrane potential similar to that of bacteriorhodopsin, which transports protons out of the cell interior. However, the physiologic role of halorhodopsin might be not only to generate a transient proton-motive force on illumination, but also to maintain cell volume.[3] This is because in these organisms the high (several molar) external NaCl concentration in the medium is balanced mostly by intracellular KCl, and although the replacement of $Na^+$ with $K^+$ can be accomplished, as in many other systems, by a combination of a $H^+/Na^+$ antiporter[4,5] and electrogenic $K^+$ uptake,[6,7] the net uptake of $Cl^-$ requires an active accumulation system. Indeed, as with protons in the case of bacteriorhodopsin, a second transport pathway for active chloride transport exists in the dark,[8] apparently driven independently, by protonmotive force.

The elements of the foregoing hypothesis can be observed in vesicles prepared from *Halobacterium halobium* cell envelopes containing halorhodopsin.[9] Thus, in the absence of $K^+$ (e.g., in 3 M NaCl), illumination causes the inward flow of chloride ions, which is detectable by direct determination of the accumulated chloride in vesicles equilibrated first with $Na_2SO_4$ or phosphate. When the illumination is started, there is an initial passive influx of protons, which slows as a concentration gradient for protons (inside acid) develops. During this time, $Na^+$ takes over as the main counterion to the $Cl^-$ movement. Once the protonmotive force approaches zero, the net proton flux ceases, and light will drive the continued uptake of NaCl instead. Indeed, illumination is seen to cause swelling of the vesicles, particularly when gramicidin is added to increase the electrical potential-driven secondary $Na^+$ uptake.

Halorhodopsin, as bacteriorhodopsin, requires no other component than the opsin, a small (MW about 26,000) integral membrane protein, and the retinal, for the light-driven transport. In both proteins the retinal is attached to a lysine via a protonated Schiff base. The intimate association of the retinal with various amino acid residues in halo-opsin is indicated by the fact that the wavelength maximum of the pigment is shifted from 440 nm, that of a protonated retinal Schiff base in solution, to 578 nm. Thus, halorhodopsin, as bacteriorhodopsin, is a purple protein. Absorption of a photon causes the isomerization of the retinal from all-*trans* to 13-*cis*; this initiates a sequence of thermally driven reactions which lead back to the parent pigment in a few tens of milliseconds (the "photocycle"). Understanding the relation between these reactions, detected by either transient or low temperature spectroscopy of the retinal chromophore in the visible and infrared

a This work was supported by a grant (GM 29498) from the National Institutes of Health.
b Permanent address: Institute of Biophysics, Biological Research Center of the Hungarian Academy of Sciences, Szeged, Hungary.

regions, and the binding and translocation of the chloride ion will provide important clues to the chloride transport mechanism.

## THE HALORHODOPSIN PHOTOCYCLE

Measurement of transient changes in the absorbance of the pigment amounts, in essence, to following directly the course of a single turn-over reaction (i.e., transport) cycle at ambient temperature,[10–16] after it is initiated by light. A somewhat different measurement can be done at low temperatures[17] (e.g., 120 K), where photostationary states are established. These arise because here, depending on the temperature, various intermediates of the photocycle are frozen in, and they are interconvertible with others through slow (minutes or hours) thermal processes and photoreactions of their own. However, both kinds of investigative strategies strive to identify the species that form the sequence of reactions leading back to the parent pigment. As might be expected, events in the photocycle depend on whether or not the transported anion is provided: with chloride present, at least some of the species that arise and decay will reflect steps in the transprotein (and therefore transmembrane) movement of the anion; with it absent a futile cycle will result, which dissipates energy in ways other than anion translocation.

In the presence of chloride the first stable species, which arises within picoseconds, absorbs near 600 nm,[14,18] and we refer to it as $HR_K$ (in analogy with a similar intermediate in bacteriorhodopsin, which had been named K). FTIR spectra identify the retinal configuration in $HR_K$ as distorted 13-*cis*.[19] $HR_K$ decays to another species named $HR_{KL}$, which absorbs near 580 nm, but with lower amplitude and a broader absorption band than that of halorhodopsin, in less than several tens of nanoseconds. FIGURE 1 shows difference spectra, measured with gated optical multichannel spectroscopy, at increasing delay times beginning 60 ns after a flash. The first of these spectra (60 ns) represents a time where virtually all of the HR that entered the photocycle exists as $HR_{KL}$. The last of the spectra (20 ms) shown represents a time range where the recovery of HR takes place. Between these time domains, the spectral transformations evident in FIGURE 1 can be resolved into two transitions: one near 1 $\mu$s and another near 1 ms. These are visualized by subtracting an initial difference spectrum from each of the following ones, as in FIGURE 2. The resulting net difference spectra in FIGURE 2A indicate the appearance of a blue-shifted species at the expense of $HR_{KL}$. We refer to the product of this transformation as $HR_L$, whose absorption maximum is at 526 nm. At later times, concurrently with the beginning of the recovery of HR, part of $HR_L$ is converted into a red-shifted species, as shown in FIGURE 2B. The latter is $HR_O$, whose absorption maximum is at 635 nm. Kinetic modeling[12,13,16] indicates that $HR_O$ is in a chloride-dependent equilibrium with $HR_L$, and it lies between $HR_L$ and HR in the photocycle sequence.

The spectra of the HR species could be calculated exactly from the difference spectra, if the amount of HR that enters the photocycle were known. The latter is estimated, however, by trial and error, that is, by generating spectra that have no negative absorption, agree with the shape of known rhodopsin spectra, and the like. Fortunately, the spectrum of $HR_L$ is rather well described because circumstances can be created in which it is virtually the only photoproduct,[17] and all of our spectral reconstructions rely on obtaining good spectra for this intermediate. FIGURE 3 shows spectra for $HR_{KL}$, $HR_L$, and $HR_O$ obtained in this way, scaled to the spectrum of HR. The transient spectroscopy[15,16] and other previous results[11-14] indicate that the HR photocycle in the presence of chloride contains the sequence, HR —hv $\to$ $HR_K$ $\to$ $HR_{KL}$ $\to$ $HR_L$ $\leftrightarrow$ $HR_O$ $\to$ HR.

Similar experiments were performed[16] in the presence of nitrate, which was thought to be a nontransported anion. However, at high concentrations of nitrate (e.g., 1 M), a chloride-like photocycle is observed, parallel with a second photocycle which consists of the sequence, HR —hv $\to$ $HR_K'$ $\to$ $HR_{KO}$ $\to$ $HR_O$ $\to$ HR. The amount of chloride contamination in the nitrate is much too small to account for

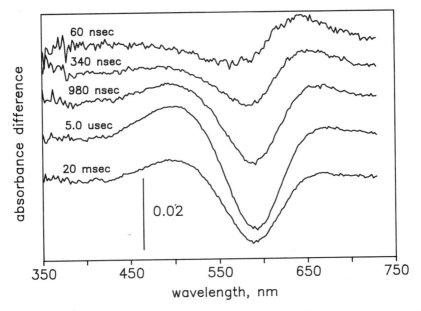

**FIGURE 1.** Difference spectra determined by optical multichannel spectroscopy (method described elsewhere[15,16]) at various delay times after a brief (<1 ns) laser flash. The traces represent differences between the absorption spectrum of a purified halorhodopsin sample at the indicated delay times and the same sample once the photocycle had decayed. The exposure window was much shorter than the delay times. The buffer contained 400 mM $Na_2SO_4$, 100 mM NaCl, 0.5% Lubrol PX, and 25 mM MES, pH 6.0; the temperature was 3°C.

this. It therefore appears that nitrate can substitute for chloride in generating an $HR_L$-type photocycle, but much less effectively. $HR_O$, which appears in both photocycles, is the same species, because both $HR_{KO}$ $\to$ $HR_O$ $\to$ $HR_L$ and $HR_{KL}$ $\to$ $HR_L$ $\to$ $HR_O$ reaction sequences can be observed under some circumstances.[16] The entire photocycle reaction is shown in FIGURE 4 (but without additional known light-dependent effects in this system, such as the slow deprotonation of the Schiff base, isomerization of the retinal to the 9-*cis* form, and their reversal by blue light).

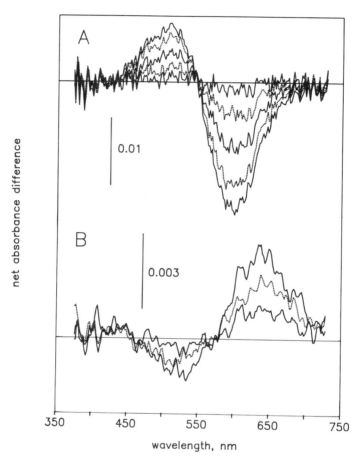

**FIGURE 2.** Net difference spectra in two time-domains of the halorhodopsin photocycle. The traces were calculated from difference spectra, such as in FIGURE 1, by subtracting an initial difference spectrum from each of subsequent difference spectra. **(A)** Difference spectra (in the order of increasing magnitude) for 0.34, 0.55, 0.98, 1.83, and 5.0 $\mu$s *minus* an 0.233-$\mu$s difference spectrum. The change is the $HR_{KL} \rightarrow HR_L$ reaction. **(B)** Difference spectra (in the order of increasing magnitude) for 0.47, 1.35, and 3.8 ms *minus* an 0.135-ms difference spectrum. Because at these delay times the recovery of HR is already significant, each of the spectra was normalized before the subtraction, to the full extent of the photoconversion. The change is the $HR_L \rightarrow HR_O$ reaction.

## ANION-BINDING SITES IN HALORHODOPSIN

Halorhodopsin was shown to transport chloride, bromide, and to a limited extent iodide,[9,20,21] but not other anions. Half-maximal transport is at 40 mM chloride. Possible transport of nitrate at much higher concentrations is difficult to determine, because this is a permeant anion and will have chaotropic effects on the membrane as well. The effects of anions on the halorhodopsin chromophore and its photoreactions identified two distinct binding sites[22]: (a) site I, which is relatively nonspecific and causes a small blue-shift and a rise in the pK of the Schiff base, and (b) site II, which has the same specificity as the transport (although it now seems to bind nitrate with low affinity) and causes a small red-shift and the appearance of $HR_{KL}$ and $HR_L$ in the photocycle (cf. above). A third, nonspecific surface binding site is identified from $^{35}Cl$ NMR line-broadening.[23] The relation of these binding sites to specific steps of the anion transport is not yet known, but it is likely that site II participates in the uptake and site III in the release of the transported anion. If so, the photocycle scheme readily provides a transport scheme as well: HR releases the anion in the $HR_L \leftrightarrow HR_O$ step, and regains it after its recovery but before the absorption of the next photon, as shown in FIGURE 4. The uptake of chloride is driven by the high chloride concentration in the growth medium of the halobacteria. Although release of the chloride in the scheme is an equilibrium reaction, it is driven in the forward direction by the irreversible step that follows it. Thus, there is cyclic uptake and release as halorhodopsin passes though the photocycle. In the absence of the transported anion (i.e., when site II is not occupied), a truncated photocycle results (FIG. 4), which lacks, most conspicuously, the blue-shifted species, $HR_L$.

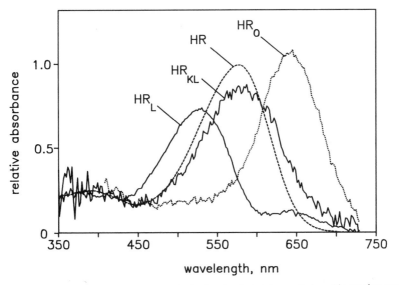

**FIGURE 3.** Absolute spectra for the halorhodopsin photointermediates. Absorption spectra were calculated from data, such as shown in FIGURES 1 and 2, and scaled to the halorhodopsin spectrum. The absorption maxima are: $HR_{KL}$, 578 nm; $HR_L$, 526 nm; and $HR_O$, 635 nm. Because $HR_O$ is produced in very small amounts in the presence of 100 mM chloride, its spectrum was calculated from difference spectra measured in buffer containing 100 mM $NaNO_3$. The spectrum of $HR_L$ contains a small absorption band near 650 nm; it is likely to be an additional intermediate at 20 $\mu$s, not accounted for by the proposed scheme. The spectrum of $HR_{KL}$ was calculated by extrapolation of measured difference spectra to zero time.

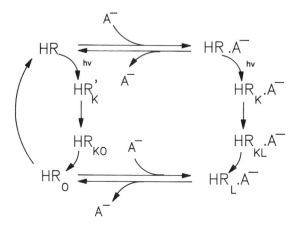

**FIGURE 4.** Proposed scheme for the halorhodopsin photocycles. $A^-$ stands for chloride or bromide, but it can also be nitrate when its concentration is high.

## THE HALORHODOPSIN STRUCTURE

Halo-opsin must accommodate the retinal chromophore and its photoreactions, and provide chloride binding sites as well as possible conformational changes during the photocycle, to accomplish the unidirectional chloride translocation. The primary structure of halorhodopsin in *H. halobium* is known from the sequence of its structural gene[24]; a second halorhodopsin from *Natronobacterium pharaonis*,[25] with 65% sequence identity to the first one, was recently purified and its gene cloned and sequenced.[26] The halo-opsins are quite similar to bacterioopsin: they also can be readily arranged into seven transmembrane helical segments (helices A–G), oriented in the same way in the membrane,[27] and many residues are conserved in the three proteins. Most interesting is the fact that the halo-opsins contain very few intramembranous positively charged residues, which would qualify as anion-binding sites, and these are all present in bacterioopsin as well. It is as expected, however, that several of the acidic residues thought to be involved in proton transfer in bacteriorhodopsin are missing in the halorhodopsins.

Resonance Raman spectra of halorhodopsin[28-31] clearly indicate that the environment of the retinal in this pigment is similar to that in bacteriorhodopsin. This environment is thought to consist, at least partly, of several charged residues, which influence the electronic configuration of the retinal. Replacement of the retinal in bacteriorhodopsin with selected analogs[32] provided artificial pigments whose absorption maxima could be interpreted in terms of the external point-charge model of rhodopsin chromophores,[33] where the removal of a negative group from the immediate vicinity of the Schiff base and the addition of an ion-pair near the ionone ring generated all of the spectroscopic difference between a free and a protein-bound retinal (opsin shift). These experiments were repeated with halorhodopsin,[34] and the close similarity between the results with this pigment and bacteriorhodopsin suggested that the arrangement of charges around the retinal must be essentially the same in the two proteins. Examination of conserved and nonconserved residues in these two proteins (as well as in the third, pharaonis halorhodopsin) readily provides candidates for the proposed charged residues. FIGURE 5 shows a view of halorhodopsin, making use of the analogy with bacteriorhodopsin, with helices F and G removed for clarity. (Residues on these helices are shown with bold lettering.) Retinal is bound to the single lysine in the protein via the positively charged Schiff base on helix G, and asp-238 (also on helix G) is its likely counterion. The negative charge near the ring is almost certainly asp-141 (on helix C). The positive charge near the ring may be either arg-200 (on helix F) or arg-108 (on helix C). According to the external point charge theory, the two negative and one positive residues will determine the color of the pigment.

## TRANSPORT MECHANISM

If this structural scheme is taken as correct, the residues just described must be considered also in any proposed chloride transport mechanism, because there are no positively charged residues inside the protein other than those shown in FIGURE 5. Site I, which is localized near the Schiff base, may involve the two arginines, at the base of helix B and the middle of helix C, whereas site II, which is localized closer to the ring, may involve arg-161 and arg-200, the two arginines on helices E and F.[34] Site III may be the cluster of arginines on the loop between helices A and B. A minimal mechanistic scheme for chloride transport would identify the location of the transported anion at these (or perhaps other) places in the protein, for each of those intermediates in the photocycle that are now given in the general form, $HR_X.A^-$ (FIG. 4). Such a scheme is not yet available, although the small number of positive residues in the protein allows relatively few alternatives.

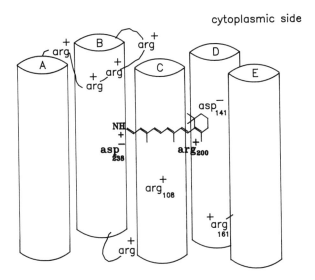

**FIGURE 5.** Proposed structure for halorhodopsin based on its similarity[24,34] to bacteriorhodopsin. Helices F and G (which would continue the clockwise sequence of helices in front of the others) have been removed for the sake of clarity. Residues shown in bold characters are on helix F or G. A few charged residues are not shown; these are either negatively charged or located on interhelical loops.

In a general sense, there are two kinds of chloride transport mechanisms: (1) one where the anion is translocated more by virtue of the retinal isomerization-dependent spatial displacement of the positively charged Schiff base,[35] in a manner analogous to proton transport in bacteriorhodopsin, and (2) one where the anion is moved more by a protein-conformation dependent spatial rearrangement of arginine residues. The first of these mechanisms emphasizes the role of the Schiff base, whereas the second the participation of the protein in the translocation. A combination of the two would reconcile the elegance of a single kind of mechanism for the two pigments with observations suggesting that it is site II, nearer to the ionone ring, rather than site I, whose specificity bears resemblance to the specificity of the transport.

## REFERENCES

1. LANYI, J. K. 1986. Halorhodopsin: A light-driven chloride pump. Ann. Rev. Biophys. Biophys. Chem. **15:** 11–28.
2. HEGEMANN, P., J. TITTOR, A. BLANCK & D. OESTERHELT. 1987. Progress in halorhodopsin. *In* Retinal Proteins. Yu.A. Ovchinnikov, ed.: 333–352. VNU Science Press, Utrecht.
3. MEHLHORN, R. J., B. SCHOBERT, L. PACKER & J. K. LANYI. 1985. ESR studies of light-dependent volume changes in cell envelope vesicles from Halobacterium halobium. Biochim. Biophys. Acta **809:** 66–73.
4. LANYI, J. K. & R. E. MACDONALD. 1976. Existence of electrogenic hydrogen ion/sodium ion antiport in Halobacterium halobium cell envelope vesicles. Biochemistry **15:** 4608–4614.
5. EISENBACH, M., S. COOPER, H. GARTY, R. M. JOHNSTONE, H. ROTTENBERG & S. R. CAPLAN. 1977. Light-driven sodium transport in sub-bacterial particles of Halobacterium halobium. Biochim. Biophys. Acta **465:** 599–613.
6. GARTY, H. & S. R. CAPLAN. 1977. Light-dependent rubidium transport in intact Halobacterium halobium cells. Biochim. Biophys. Acta **459:** 532–545.
7. WAGNER, G., R. HARTMANN & D. OESTERHELT. 1978. Potassium uniport and ATP synthesis in Halobacterium halobium. Eur. J. Biochem. **89:** 169–179.
8. DUSCHL, A. & G. WAGNER. 1986. Primary and secondary chloride transport in Halobacterium halobium. J. Bacteriol. **168:** 548–552.
9. SCHOBERT, B. & J. K. LANYI. 1982. Halorhodopsin is a light-driven chloride pump. J. Biol. Chem. **257:** 10306–10313.
10. WEBER, H. J. & R. A. BOGOMOLNI. 1981. P588, a second retinal-containing pigment in Halobacterium halobium. Photochem. Photobiol. **33:** 601–608.
11. HEGEMANN, P., D. OESTERHELT & M. STEINER. 1985. The photocycle of the chloride pump halorhodopsin. I. Azide catalyzed deprotonation of the chromophore is a side reaction of photocycle intermediates inactivating the pump. EMBO J. **4:** 2347–2350.
12. OESTERHELT, D., P. HEGEMANN & J. TITTOR. 1985. The photocycle of the chloride pump halorhodopsin. II. Quantum yields and a kinetic model. EMBO J. **4:** 2351–2356.
13. LANYI, J. K. & V. VODYANOY. 1986. Flash spectroscopic studies of the kinetics of the halorhodopsin photocycle. Biochemistry **25:** 1465–1470.
14. TITTOR, J., D. OESTERHELT, R. MAURER, H. DESEL & R. UHL. 1987. The photochemical cycle of halorhodopsin: Absolute spectra of intermediates obtained by flash photolysis and fast difference spectra measurements. Biophys. J. **52:** 999–1006.
15. ZIMÁNYI, L., L. KESZTHELYI & J. K. LANYI. 1989. Transient spectroscopy of bacterial rhodopsins with optical multichannel spectroscopy. 1. Comparison of the photocycles of bacteriorhodopsin and halorhodopsin. Biochemistry **28:** 5165–5172.
16. ZIMÁNYI, L. & J. K. LANYI. 1989. Transient spectroscopy of bacterial rhodopsins with optical multichannel analyser. 2. Effects of anions on the halorhodopsin photocycle. Biochemistry **28:** 5172–5178.

17. ZIMÁNYI, L. & J. K. LANYI. 1989. Low temperature photoreactions of halorhodopsin. 2. Description of the photocycle and its intermediates. Biochemistry **28:** 1662–1666.
18. POLLAND, H. J., M. A. FRANZ, W. ZINTH, W. KAISER, P. HEGEMANN & D. OESTERHELT. 1985. Picosecond events in the photochemical cycle of the light-driven chloride pump halorhodopsin. Biophys. J. **47:** 55–59.
19. ROTHSCHILD, K. J., O. BOUSCHE, M. S. BRAIMAN, C. A. HASSELBACHER & J. L. SPUDICH. 1988. Fourier transform infrared study of the halorhodopsin chloride pump. Biochemistry **27:** 2420–2424.
20. HAZEMOTO, N., N. KAMO, Y. KOBATAKE, M. TSUDA & Y. TERAYAMA. 1984. Effect of salt on photocycle and ion pumping of halorhodopsin and third rhodopsinlike pigment of Halobacterium halobium. Biophys. J. **45:** 1073–1077.
21. BAMBERG, E., P. HEGEMANN & D. OESTERHELT. 1984. Reconstitution of the light-driven electrogenic ion pump halorhodopsin into black lipid membranes. Biochim. Biophys. Acta **773:** 53–60.
22. SCHOBERT, B., J. K. LANYI & D. OESTERHELT. 1986. Effects of anion binding on the deprotonation reactions of halorhodopsin. J. Biol. Chem. **261:** 2690–2696.
23. FALKE, J. J., S. I. CHAN, M. STEINER, D. OESTERHELT, P. TOWNER & J. K. LANYI. 1984. Halide binding by the purified halorhodopsin chromoprotein. II. New chloride binding sites revealed by 35-Cl NMR. J. Biol. Chem. **259:** 2185–2189.
24. BLANCK, A. & D. OESTERHELT. 1987. The halorhodopsin gene. II. Sequence, primary structure of halorhodopsin, and comparison with bacteriorhodopsin. EMBO J. **6:** 265–273.
25. BIVIN, D. B. & W. STOECKENIUS. 1986. Photoactive retinal pigments in haloalkalophilic bacteria. J. Gen. Microbiol. **132:** 2167–2177.
26. LANYI, J. K., A. DUSCHL, G. W. HATFIELD, K. MAY & D. OESTERHELT. 1989. The derived primary structure of a halorhodopsin from Natronobacterium pharaonis: Structural, functional and evolutionary implications for bacterial rhodopsins and halorhodopsins. J. Biol. Chem., in press.
27. SCHOBERT, B., J. K. LANYI & D. OESTERHELT. 1988. Structure and orientation of halorhodopsin in the membrane: A proteolytic fragmentation study. EMBO J. **7:** 905–911.
28. ALSHUTH, T., M. STOCKBURGER, P. HEGEMANN & D. OESTERHELT. 1985. Structure of the retinal chromophore in halorhodopsin. A resonance Raman study. FEBS Lett. **179:** 55–59.
29. SMITH, S. O., M. J. MARVIN, R. A. BOGOMOLNI & R. A. MATHIES. 1984. Structure of the retinal chromophore in the hR578 form of halorhodopsin. J. Biol. Chem. **259:** 12326–12329.
30. MAEDA, A., T. OGURUSU, T. YOSHIZAWA & T. KITAGAWA. 1985. Resonance Raman study on binding of chloride to the chromophore of halorhodopsin. Biochemistry **24:** 2517–2521.
31. FODOR, S. P., R. A. BOGOMOLNI & R. A. MATHIES. 1987. Structure of the retinal chromophore in the hRL intermediate of halorhodopsin from resonance Raman spectroscopy. Biochemistry **26:** 6775–6778.
32. SPUDICH, J. L., D. A. MCCAIN, K. NAKANISHI, M. OKABE, N. SHIMIZU, H. RODMAN, B. HONIG & R. A. BOGOMOLNI. 1986. Chromophore/protein interaction in bacterial sensory rhodopsin and bacteriorhodopsin. Biophys. J. **49:** 479–483.
33. KAKITANI, H., T. KAKITANI, H. RODMAN & B. HONIG. 1985. On the mechanism of wavelength regulation in visual pigments. Photochem. Photobiol. **41:** 471–479.
34. LANYI, J. K., L. ZIMÁNYI, K. NAKANISHI, F. DERGUINI, M. OKABE & B. HONIG. 1988. Chromophore/protein and chromophore/anion interactions in halorhodopsin. Biophys. J. **53:** 185–191.
35. OESTERHELT, D., P. HEGEMANN, P. TAVAN & K. SCHULTEN. 1986. Trans-cis isomerization of retinal and a mechanism for ion translocation in halorhodopsin. Eur. Biophys. J. **14:** 123–129.

# Electrogenic Chloride Transport in Algae[a]

D. GRADMANN

*Pflanzenphysiologisches Institut der Universität*
*D-3400 Göttingen, West Germany*

In the outer plasma membrane of animal cells the $Na^+/K^+$ ATPase is the predominant device of active ion transport. It is widely accepted that the corresponding system in fungi and plants is an $H^+$ ATPase. This knowledge is based on investigations of cells of fresh-water algae (mostly the giant internodial cells of *Characean* species) or of higher plant cells that actually live in an equivalent environment of low salinity. The question now for marine algae is: do they have an $H^+$ ATPase like their green relatives, or do they operate a $Na^+/K^+$ ATPase type of pump in their Ringer-like environment? This question has been investigated in some detail on the giant cells of the unicellular marine alga *Acetabularia*. The answer was not as clearcut as the question, but it provided evidence for the operation of an electrogenic $Cl^-$ pump.

This report essentially follows the line of a contribution in a recent review on $Cl^-$ pumps.[1] It starts with a brief introduction of the object and a qualitative documentation of an electrogenic $Cl^-$ pump. The following topics are treated in more detail: (1) current-voltage relationships of the pump, their sensitivity to the external $Cl^-$ concentration, $[Cl^-]_o$, and their reaction-kinetic analysis; (2) reaction-kinetic prediction and experimental verification of voltage-dependent $^{36}Cl^-$ efflux through the pump; and (3) ATP synthesis by reversal of the pump.

Biochemical identification, characterization, purification, and reconstitution of this $Cl^-$ ATPase are important goals. However, to date there is only one report on these subjects[2] and this attempt is not very conclusive. Therefore, these issues are not treated here. This contribution concludes with some remarks on the existence of $Cl^-$ ATPases in other organisms and on the physiologic meaning of a $Cl^-$ pump.

## THE OBJECT

Adult specimens of the most familiar species, *A. acetabulum* (former *A. mediterranea*), are approximately cylindrical "stalks" about 50 mm in length and about 0.3 mm in diameter with one nucleus in the basal "rhizoid" and an apical "cap" (about 10 mm in diameter) that is organized in radial chambers that contain reproductive organs at the end of the lifetime of a specimen. During past decades, the culturing conditions have changed a little, without apparent effects on the ion transport properties. At present, the cells are cultured in a synthetic medium.[3]

For electrophysiologic experiments, young stalks (about 30 mm in length and 0.2 mm in diameter, with undeveloped cap) have been used preferentially because their relatively tender cell wall is easy to puncture and the geometry is simpler to calculate. In more recent experiments, the cytoplasm (about 10% of the cell volume) of little older cells (about 40 mm in length) is gently centrifuged into the still capless apex; approximately spherical segments of these vacuole-free prepa-

[a] This work was supported by the Deutsche Forschungsgemeinschaft.

rations are tied off and cut from the cytoplasm-depleted part of the stalk. Use of the spherical, devacuolated segments bypasses some difficulties that otherwise arise in analysis of the data from compartmentation and from cable properties of the intact cells. Some comparative experiments on normal cells and on centrifuged preparations (both devacuolated and cytoplasm-depleted) have demonstrated qualitative identity of the electrical properties of the plasmalemma of these three types of preparations.[4,5]

## PRIMARY EVIDENCE FOR AN ELECTROGENIC Cl⁻ PUMP

The observations and conclusions in this section originate from independent and equivalent studies.[6-8] The results are illustrated schematically in FIGURE 1.

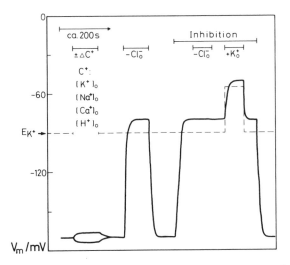

**FIGURE 1.** Paradigmatic scheme of evidence for an electrogenic Cl⁻ pump in *Acetabularia* by recordings of transmembrane voltage ($V_m$) and its particular response on changes in experimental conditions as indicated; $E_K$ = Nernst equilibrium voltage for K⁺.

The approximate distribution of the major ions on the two sides of the plasmalemma (cytoplasm/seawater) of normal *Acetabularia* cells (in mM) is as follows: 400/10 K⁺, 70/460 Na⁺, 0.001/10 Ca²⁺, 500/500 Cl⁻; pH 7/8. Thus the steady-state Goldman voltage for passive ion diffusion cannot be more negative than the Nernst equilibrium voltage for K⁺ ($E_K$) of about −90 mV. However, the normal transmembrane voltage ($V_m$) of *Acetabularia* (about −170 mV) is far more negative, which unambiguously points to the operation of an electrogenic pump. Variations in the concentrations of various cations (such as K⁺, Na⁺, or Ca²⁺) in the external medium have little effect on the resting $V_m$. However, when the external Cl⁻ concentration, $[Cl^-]_o$, is lowered (e.g., to 10% of seawater), there is a (hyper-Nernstian, i.e., more than 58 mV per decade of substrate concentration) shift of $V_m$ to about −80 mV, that is, to a voltage that is compatible with the Goldman equation for the ionic relations just given. As judged by the positive sign of this change in $V_m$ on a decrease, this effect *per se* does not prove $Cl^-$ to be the substrate of the electrogenic pump. This effect could also be caused by enhanced outward diffusion of $Cl^-$.

For crucial evidence, another experiment is necessary. An equivalent change in $V_m$ occurs when the pump is inhibited (by cold or by inhibitors of energy metabolism). In this state, $V_m$ is insensitive to $[Cl^-]_o$, but now behaves as a $K^+$ electrode. The results of these simple experiments demonstrate (a) the predominance of a primary $Cl^-$ importing pump and (b) a small permeability for passive $Cl^-$ diffusion, which is, in fact, an essential precondition to the avoidance of futile cycling.

## CURRENT-VOLTAGE RELATIONSHIP OF THE PUMP

The following section is mainly based on voltage-clamp experiments,[9–11] on tracer-flux experiments,[5,12] and on theoretical work.[13,14] By current subtraction between pairs of membrane current-voltage curves, $i_m(V_m)$, measured with the pump either stimulated or inhibited compared with the control state, the current-voltage relationship of the pump, $i_p(V_m)$, can be determined under favorable circumstances. In the case of inhibition without a thermodynamic shift (e.g., by cold, but not by inhibition of the power supply[14,15]), such pairs of $i_m(V_m)$ curves intersect at the equilibrium voltage of the pump, $E_p$. Several attempts consistently provided a value of $E_p$ of about $-190$ mV.[8,9,16] Because the phosphate potential in *Acetabularia* is about $-450$ mV,[17] the pump is expected to transport two $Cl^-$ per hydrolysis of one ATP (if it is an ATPase, as is demonstrated below).

Depending on the experimental conditions, three different types of $i_p(V_m)$ curves can be determined. On a rectangular voltage step, the "early" current (immediately after recharging of the membrane capacitance) yields a straight line, $i_o(V_m)$. After relaxation, the "late" current yields another curve, $i_\infty(V_m)$, with a region of negative slope conductance (approximately between $-150$ mV and $E_K$).[b] Current responses $\Delta i$ to small and short voltage pulses ($\Delta V_m$) that are superimposed on the voltage step provide an early conductance ($g_0 = \Delta i_m/\Delta V_m$) for each steady-state voltage. This $g_0$ reflects a property of the pump because it vanishes when the pump is inhibited. Integration of this (bell-shaped) function $g_0(V_m)$ over $V_m$ yields a family of $i_m(V_m)$ curves. Choosing the integration constant to yield zero current at $E_p$, provides the third type of $i(V_m)$ curve of the pump. These $i_p(V_m)$ curves have been analyzed in some detail[18,19] and are focused upon here.

An example of $i_p(V_m)$ is given in FIGURE 2. The points are measured data; the curve is the result of fitting the four rate constants $\alpha^0$, $\beta^0$, $\gamma$, and $\delta$ of Equation 1 to the data, where the short-circuit rate constants $\alpha^0$ and $\beta^0$ are the apparent rate constants $\alpha$ and $\beta$ at zero voltage. This equation,

---

[b] In other words, this property means that at $V_m$ more positive than about $-80$ mV, the pump virtually ceases to operate. Under this condition, difficulties are expected in attempts to isolate the pump when the enzyme is short-circuited in free solution and therefore cannot be identified by its activity. On the other hand, the pathway for $K^+$ diffusion and its gating properties have also been demonstrated to generate the phenomenon of negative slope conductance for $V_m$ more negative than $E_K$.[23] However, this slow voltage-gating of the $K^+$ channels with its inductive behavior for positive going voltages (and capacitive behavior for negative going voltages)[24] can be distinguished from the slow electrical response of the pump which displays only capacitive characteristics in the voltage range between $E_p$ and $E_K$.[9]

$$i_p(V_m) = zFN \frac{\alpha\gamma - \beta\delta}{\alpha + \beta + \gamma + \delta}, \tag{1}$$

describes the steady-state $i(V_m)$ of all cyclic reaction systems with one $V_m$-sensitive reaction step (Class I systems[13]). For this description, these reaction systems can be reduced to a two-state model. (For symbols see inset of FIGURE 2.) Voltage sensitivity enters the system by $\alpha = \alpha^0 \cdot U$ and $\beta = \beta^0/U$ with $U = exp(V_m zF/2RT)$, where $z$ (here $z = -2$ for two $Cl^-$ per reaction cycle), $F$, $R$, and $T$ have their usual thermodynamic meanings, and the factor 2 in the denominator of the exponent stays for the assumption of a symmetric Eyring barrier. $N$ is the density of the pumps (in moles per membrane area). The value of the scaling factor $N$ (about 50 nmol m$^{-2}$) is obtained from the electrodynamic properties of the pump,[20] which may only be mentioned here.

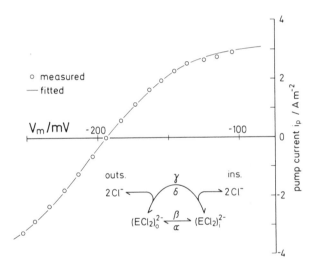

**FIGURE 2.** Example of current-voltage relationship of the pump, $i_p(V_m)$, in artificial seawater; circles measured, line fitted by inset model (Equation 1), with $\alpha^0 = 9.6 \cdot 10^{-2}$ s$^{-1}$, $\beta^0 = 6.8 \cdot 10^5$ s$^{-1}$, $\gamma = 3.2 \cdot 10^2$ s$^{-1}$, $\delta = 4.2 \cdot 10^2$ s$^{-1}$, $z = -2$, and $N = 50$ nmol m$^{-2}$.

Numerical values for the fitted four two-state rate constants of the pump are given in the legend to FIGURE 2. With these parameters, the unidirectional efflux of $Cl^-$ through the pump and its voltage sensitivity, $\Phi_{io}(V_m)$ can be calculated by Equation 2, which holds for Class I systems with fast equilibration of the substrate with the transporter:[12]

$$\Phi_{io}(V_m) = nN \frac{\alpha(\beta + \gamma)}{\alpha + \beta + \gamma + \delta} \tag{2}$$

where n is a stoichiometry factor (here n = 2). This predicted relationship (Equation 12 with the empirical parameters as listed in the legend to FIGURE 2) is given by the curve in FIGURE 3. Voltage-dependent efflux $\Phi_{io}(V_m)$ of $Cl^-$ in *Acetabularia* can be measured by tracer $^{36}Cl^-$ efflux analysis from individual cells with two impaled electrodes (one for current injection and another one for voltage record-ing). Subtracting $\Phi_{io}(V_m)$ under conditions in which the pump is inhibited from $\Phi_{io}(V_m)$ when the pump is operating yields $\Phi_{io,p}(V_m)$, the $Cl^-$ efflux-voltage rela-tionship of the pump. Such experimental data are given by the points and error bars in FIGURE 3. The quantitative coincidence with the predicted curve strongly confirms that $Cl^-$ is the substrate of the electrogenic pump in *Acetabularia*.

### EFFECT OF $[Cl^-]_o$ ON $i_p(V_m)$

This section is based on steady-state voltage-clamp experiments.[10,11] For reac-tion-kinetic modeling of the effect of the substrate concentration on the $i(V_m)$ relationship of a Class I transporter, a three-state model (with three reversible

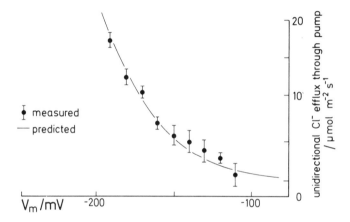

**FIGURE 3.** Voltage-dependent (unidirectional) $Cl^-$ efflux through pump; dots and bars measured; line predicted by two-state model (*inset*, FIG. 2, Equation 2) with data in legend to FIGURE 2 and $n = 2$.

reaction steps) is necessary to describe each of the three essential reaction steps explicitly.[13] These are: substrate-dependent binding/debinding, voltage-depen-dent reorientation of the loaded binding site, and the unaffected rest of the reac-tion cycle. This model with its symbols is given in FIGURE 4 (*inset*) for the electrogenic $Cl^-$ pump in *Acetabularia*. The corresponding $i(V_m)$ equation is:

$$i_p(V_m) = zFN \frac{ace - bdf}{a(c + d + e) + b(d + e + f) + c(e + f) + df} \qquad (3)$$

with $a = a^0 \cdot U$, $b = b^0/U$, $d = \bar{d}[Cl^-]_o^2$ for a stoichiometry of 2 $Cl^-$ per cycle, where the short-circuit rate constants $a^0 (\neq \alpha^0)$ and $b^0(\neq \beta^0)$ are the apparent rate constants $a$ and $b$ at zero voltage and the fundamental rate constant $\bar{d}$ is the apparent rate constant $d$ at the standard concentration 1 M $[Cl^-]_o$.

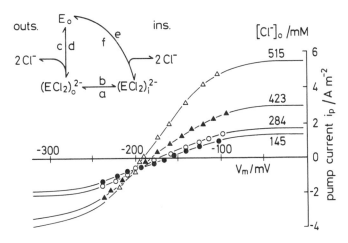

**FIGURE 4.** Example of $[Cl^-]_o$ dependence of steady-state current-voltage curve of the pump (differences in corresponding data in FIGURE 2 are due to a different culture); variation of $[Cl^-]_o$ in artificial seawater by partial substitution of $Cl^-$ by $I^-$; measured data (triangles and circles) and curves fitted by Equation 3 (symbols in *inset*), the parameter being $a = a^0 \cdot U$ with $a^0 = 0.36$ s$^{-1}$, $b = b^0/U$ with $b^0 = 1.8 \cdot 10^6$ s$^{-1}$, $c = 2.0 \cdot 10^6$ s$^{-1}$, $d = \bar{d}[Cl^-]_o^2$ with $\bar{d} = 5.8 \cdot 10^6$ s$^{-1}$ M$^{-2}$, $e = 1.04 \cdot 10^3$ s$^{-1}$, $f = 0.67 \cdot 10^3$ s$^{-1}$, $z = -2$, and $N = 50$ nmol m$^{-2}$.

FIGURE 4 shows examples of $i_p(V_m)$ as measured under various $[Cl^-]_o$. The curves are fits of Equation 3 to the plotted data points. These fits comprise seven parameters: the six rate constants and $N$ as a scaling factor (see Equation 3). Five parameters were forced to be in common for all four data sets, and two parameters were free for particular adjustment to each of the four sets of experimental data. For the best fits (curves in FIG. 4) these two parameters were $N$ (corre-

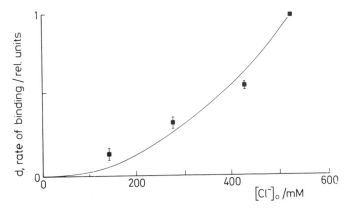

**FIGURE 5.** Apparent rate constant for association of external $Cl^-$ to transporter molecule, $d$, as function of external $Cl^-$ concentration; dots and bars: resulting data from experiments and fits in FIGURE 4, scaled to unity at normal $[Cl^-]_o$ in seawater; line: theoretic expectation (parabola) for a stoichiometry of 2 $Cl^-$ per reaction cycle.

sponding to inactivation of the pump by alternate anionic substrates that bind to the pump but are not translocated) and the rate constant $d$ for binding of $Cl^-$ outside. The resulting relationship of $d$ as a function of $[Cl^-]_o$ in FIGURE 5 fits the theoretical parabola for a $2:1$ stoichiometry. The example in FIGURES 4 and 5 holds for substitution of external $Cl^-$ by the alternate halogenide anion $I^-$. Equivalent results were obtained by substitution of $Cl^-$ by the large, organic benzene-

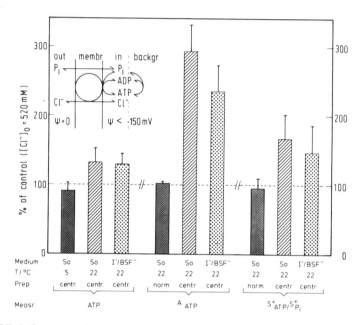

**FIGURE 6.** Summary of measurements of level (ATP), $^{32}$P-labeling ($A_{ATP}$), and specific radioactivity of ATP related to specific radioactivity of cellular $P_i(S^*_{ATP}/S^*_{Pi})$ after 10 seconds of exposure to $^{32}$P containing medium; results are from normal (norm.) or cytoplasm-depleted, centrifuged (centr.) stalk segments at normal temperature (22°C) when the pump is working, or in the cold (5°C) when the pump is inhibited; percentages of results obtained under a strong outward electrochemical gradient for $Cl^-$ (zero $Cl^-$ outside) compared with results from normal conditions (about 500 mM $Cl^-$ outside), set to 100% are shown; zero $Cl^-$ outside was achieved either by replacement of seawater by isotonic sorbitol solution (*So*) or by substitution of $Cl^-$ by $I^-$ or by benzenesulfonate$^-$ ($I^-/BSF^-$); *inset:* scheme of working hypothesis.

sulfonate$^-$ anion (BSF$^-$), and by dilution of the salt medium by isotonic sorbitol solutions (the latter causes an increase of $N$ by about 20%).

So far, the essential criteria of primary $Cl^-$ pumping (electrogenicity and substrate identification) are fulfilled. Moreover, essential properties of the pump, such as stoichiometry, transport rates, and voltage-sensitivity, are known and consistent with a simple reaction-kinetic model, as given by the inset in FIGURE 4 with the numerical data in the legend to this figure.

## ATP SYNTHESIS BY REVERSAL OF THE PUMP

The following section refers to two coherent studies.[17,21] As a first guess, ATP was assumed to drive the pump. Conclusive evidence for this hypothesis would be the demonstration of ATP synthesis by conditions in which the pump is thermodynamically forced to operate in reverse. Because $i_p(V_m)$, under normal conditions, shows appreciable negative currents at $V_m < E_p$ (FIGS. 2 and 4), reversibility of the pump is anticipated. Alternatively, the pump can be reversed electrochemically by depletion of $[Cl^-]_o$, when $Cl^-$ is forced to leave the cell. Because the effects of this treatment on the ATP relations are expected to be membrane related, they should be more pronounced when the cytoplasmic background of ATP metabolism is reduced. This has been achieved in *Acetabularia* by investigating stalk segments that are depleted of cytoplasm by gentle centrifugation.

For indication of ATP synthesis, the ATP level was measured as well as the radioactivity, A, in the ATP fraction ($A_{ATP}$) and the intracellular ratio of the specific radioactivities of ATP and of inorganic phosphate ($S^*_{ATP}/S^*_{Pi}$) under normal conditions and in the absence of external $Cl^-$. Further controls were measurements in the cold (5°C), when the pump is known to be inhibited,[7,8] and measurements on normal (uncentrifuged) stalk segments with a regular, high cytoplasm/membrane ratio. This should obscure the membrane-related portion of the ATP relations. Depletion of external $Cl^-$ is accomplished again either by replacing $Cl^-$ by $I^-$ or $BSF^-$ or by reducing the entire salt content by isotonic sorbitol solution. Because under these sorbitol conditions (*So*) external $K^+$ is diluted as well, $V_m$ can be expected to stay rather negative (about $-150$ mV are measured, in fact, under *So* conditions) and to cause enhanced effects of an outward-directed electrochemical gradient of $Cl^-$ compared with $I^-$ or $BSF^-$ conditions.

FIGURE 6 (*inset*) is a scheme of the working hypothesis. The results of the performed experiments and controls, as summarized in FIGURE 6, provide strong and consistent evidence for membrane-located ATP synthesis driven by an outward gradient of $Cl^-$ when active electrogenic pumps are present (at normal temperatures but not in the cold). In conclusion, the electrogenic pump in the plasmalemma of *Acetabularia* is a primary $Cl^-$ translocating ATPase.

## REMARKS

Independent studies[6-8] have revealed the operation of an electrogenic $Cl^-$ pump in the plasmalemma of the marine alga *Acetabularia*. In addition, there is electrophysiologic evidence for an electrogenic $Cl^-$ pump (also with a negative slope conductance in a certain voltage range) in the plasmalemma of *Halicystis*,[22] another marine alga that is related to *Acetabularia* (*Dasycladaceae*). In search of further examples of electrogenic $Cl^-$ pumps in marine algae, Ulrike Homann has measured $V_m$ of *Bryopsis plumosa* (a *Dasycladacean* species as well) under various conditions. The basic (unpublished) results are: a resting voltage of about $-80$ mV, which is apparently insensitive to external $Cl^-$ and $H^+$ but is reversibly increased (by about $-10$ mV with a time constant of about 10 seconds) by some 10 W m$^{-2}$ white light. With regard to footnote *b*, these findings are still compatible with the idea of an electrogenic $Cl^-$ pump that is active only at voltages more negative than $E_K$. In this case, elimination of depolarizing artifacts and recordings of $i_m(V_m)$ curves are necessary before this matter can seriously be discussed.

Electrogenic pumps of other marine algae have not been investigated. If the $Cl^-$ pumps were common in (primary) marine algae, one might speculate that the ancient plant cell was capable of operating a $H^+$ pump as well as a $Cl^-$ pump. The capability for $Cl^-$ pumps could have degenerated in fresh-water algae and in higher plant cells (with their low-salt environment), and $H^+$ pumps may have vanished in marine cells, just as a matter of substrate availability. Conversely, it cannot be obligatory for marine algae to run a $Cl^-$ pump, because there are salt-tolerant plant cells that operate a $H^+$ pump and no $Cl^-$ pump, such as *Lamprothamnium*, a *Characean* species that is considered to be secondary salt tolerant.

For the question of the physiologic significance of the electrogenic $Cl^-$ pump in *Acetabularia*, there is a standard "Mitchellian" answer: the primary, electrogenic ion pump creates an electrochemical driving force to fuel secondary, electrophoretic (or electroneutral) transport processes, such as the uptake of nutrients. However, there are no reports of corresponding cotransport systems in *Acetabularia*. (Our own results on this subject are negative and we have not attempted to have them published, therefore.)

Another possible physiologic role of this $Cl^-$ pump is direct acquisition of $HCO_3^-$ for photosynthesis, which is a problem for giant cells with their small surface/volume ratio. In fact, the number of active pumps is increased in photosynthetically active cells,[19] and the ATPase activity of the presumptive pump in isolated membrane fractions is about sevenfold in $HCO_3^-$ compared to $Cl^{-2}$. An electrical effect of the absence or presence of external $HCO_3^-$ has not been found,[8] which might be due to the small $HCO_3^-$ availability (about 2 mM) in normal seawater. Conversely, a $HCO_3^-$ pump with such poor selectivity against the overwhelming $Cl^-$ supply seems very unlikely and ineffective, especially at the apparent stoichiometry (2), which intrinsically favors the more concentrated substrate.

## REFERENCES

1. GERENCSER, G. A., J. F. WHITE, D. GRADMANN & S. BONTING. 1988. Is there a $Cl^-$ pump? Am. J. Physiol. 255: R677–R692.
2. GOLFARB, V. & D. GRADMANN. 1983. ATPase activities in partially purified membranes of *Acetabularia*. Plant Cell Rep 2: 152–155.
3. SCHWEIGER, H. G., P. DEHM & S. BERGER. 1977. Culture conditions for *Acetabularia*. *In* Progress in Acetabularia Research. C.L.F. Woodcock, ed. : 319–330. Academic Press, New York, NY.
4. FREUDLING, C. & D. GRADMANN. 1979. Cable properties and compartmentation in *Acetabularia*. Biochim. Biophys. Acta 552: 358–365.
5. MUMMERT, H. 1979. Transportmechanismen für $K^+$, $Na^+$ and $Cl^-$ in stationären und dynamischen Zuständen bei Acetabularia. Dissertation, University of Tübingen.
6. SADDLER, H. D. W. 1970a. The ionic relations of *Acetabularia mediterranea*. J. Exp. Bot. 21: 345–359.
7. SADDLER, H. D. W. 1970b. The membrane potential of *Acetabularia mediterranea*. J. Gen. Physiol. 55: 802–821.
8. GRADMANN, D. 1970. Einfluss von Licht, Temperatur und Aussenmedium auf das elektrische Verhalten von *Acetabularia*. Planta 93: 323–353.
9. GRADMANN, D. 1975. Analog circuit of the *Acetabularia* membrane. J. Membr. Biol. 25: 183–208.
10. HANSEN, U.-P., D. GRADMANN, J. TITTOR, D. SANDERS & C. L. SLAYMAN. 1982. Kinetic analysis of active transport: Reduction models. *In* Plasmalemma and Tonoplast: Their Functions in the Plant Cell. D Marmé, E. Marrè & R. Hertel, eds. : 77–84. Elsevier, Amsterdam.
11. GRADMANN, D., J. TITTOR & V. GOLDFARB. 1982. Electrogenic $Cl^-$ pump in *Acetabularia* Phil. Trans. R. Soc. Lond. B Biol. Sci. 299: 447–457.

12. MUMMERT, H., U.-P. HANSEN & D. GRADMANN. 1981. Current-voltage curve of electrogenic Cl⁻ pump predicts voltage-dependent Cl⁻ efflux in *Acetabularia*. J. Membr. Biol. **62:** 139–148.

13. HANSEN, U.-P., D. GRADMANN, D. SANDERS & C. L. SLAYMAN. 1981. Interpretation of current-voltage relationships of "active" ion transport systems. I. Steady state reaction-kinetic analysis of Class-I mechanisms. J. Membr. Biol. **58:** 139–148.

14. GRADMANN, D., H. G. KLIEBER & U.-P. HANSEN. 1987. Reaction-kinetic parameters for ion transport from steady-state current-voltage curves. Biophys. J. **51:** 569–585.

15. BLATT, M. 1986. Interpretation of steady-state current-voltage curves: Consequences and implications of current subtraction in transport studies. J. Membr. Biol. **92:** 91–110.

16. GRADMANN, D. & W. KLEMKE. 1974. Current-voltage relationship of the electrogenic pump in *Acetabularia mediterranea*. *In* Membrane Transport in Plants. U. Zimmermann & J. Dainty, eds. : 131–138. Springer-Verlag, Berlin.

17. GOLDFARB, V., D. SANDERS & D. GRADMANN. 1984a. Phosphate relations of *Acetabularia:* Phosphate pool, adenylate phosphates and ³²P influx kinetics. J. Exp. Bot. **35:** 626–644.

18. GRADMANN, D. 1984. Electrogenic Cl⁻ pump in the marine alga *Acetabularia*. *In* Chloride Transport Coupling in Biological Membranes and Epithelia. G. A. Gerencser, ed. : 13–61. Elsevier, Amsterdam.

19. GRADMANN, D., U.-P. HANSEN & C. L. SLAYMAN. 1982. Reaction-kinetic analysis of current-voltage relationships for electrogenic pumps in *Neurospora* and *Acetabularia*. Curr. Top. Membr. Transp. **16:** 257–276.

20. TITTOR, J., U.-P. HANSEN & D. GRADMANN. 1983. Impedance of the electrogenic Cl⁻ pump in *Acetabularia:* Electrical frequency entrainments, voltage-sensitivity and reaction kinetic interpretation. J. Membr. Biol. **75:** 129–139.

21. GOLDFARB, V., D. SANDERS & D. GRADMANN. 1984b. Reversal of electrogenic Cl⁻ pump in *Acetabularia* increases level and ³²P labelling of ATP. J. Exp. Bot. **35:** 645–658.

22. GRAVES, J. S. & J. GUTKNECHT. 1977. Current-voltage relationships and voltage sensitivity of the Cl⁻ pump in *Halicystis parvula*. J. Membr. Biol. **36:** 83–95.

23. BERTL, A. & D. GRADMANN. 1987. Current-voltage relationships of potassium channels in the plasmalemma of *Acetabularia*. J. Membr. Biol. **99:** 41–49.

24. BERTL, A., H.-G. KLIEBER & D. GRADMANN. 1988. Slow kinetics of a potassium channel in *Acetbularia*. J. Membr. Biol. **102:** 141–152.

# Possible Role of Outwardly Rectifying Anion Channels in Epithelial Transport[a]

JOHN W. HANRAHAN[b] AND JOSEPH A. TABCHARANI

*Department of Physiology*
*McGill University*
*Montreal, Quebec, Canada*

Electrogenic chloride transport provides the driving force for fluid secretion by many epithelial tissues.[1] Chloride enters the cell by electroneutral cotransport of NaCl or NaK2Cl from the serosal side, and exits by electrodiffusion through an apical membrane conductance.[2] Passive exit is the rate-limiting step for transepithelial Cl secretion; therefore, it is not surprising that apical Cl conductance is a site of regulation by secretagogues.[3]

It is now clear that many epithelia secrete bicarbonate when stimulated by cAMP. Bicarbonate secretion has a variety of functions. In the duodenum, cAMP-stimulated secretion of $HCO_3$-rich fluid insulates the mucosa from acidic gastric chyme.[4,5] In the choroid plexus, cAMP-stimulated bicarbonate transport drives the secretion of cerebrospinal fluid.[6] In the pancreas, transport of bicarbonate-rich fluid helps to clear the ducts of enzymes secreted by the acini and neutralizes acid in the duodenum so that enzymes can operate near optimal pH.[7] In the skate, alkaline fluid produced by Marshall's gland is thought to maintain sperm motility by raising the pH of the urine, whereas in the turtle, cAMP-stimulated $HCO_3$ secretion by the bladder contributes to acid-base balance.[9,10] Despite the prevalence of cAMP-activated, electrogenic $HCO_3$ secretion in epithelia, the mechanisms underlying $HCO_3$ secretion remain to be established.

One model for electrogenic $HCO_3$ secretion across the turtle urinary bladder was proposed by Stetson *et al.*[11] According to this scheme, the apical membrane contains both $Cl/HCO_3$ exchangers and a cAMP-activated anion conductance (FIG. 1). Some $HCO_3$ exits by anion exchange, driven by the inward [Cl] gradient. Chloride that enters the cell by this exchange process leaks back to the lumen through a parallel conductance. Activation of the apical Cl conductance by cAMP would cause intracellular Cl activity to fall, and this would increase the inward Cl gradient driving anion exchange, thereby stimulating $HCO_3$ secretion. In the absence of selectivity data, Stetson *et al.*[11] proposed that the conductance mediates efflux of both Cl and $HCO_3$. Variations of this model can account for $HCO_3$ secretion across many epithelia, although recent data from rat pancreatic duct are more consistent with $HCO_3$ exit predominantly through electroneutral anion exchange.[12]

[a] This work was supported by the U.S. and Canadian C.F. Foundations and by the Medical Research Council of Canada. J. W. H. is the recipient of an MRC Research Scholarship.

[b] Address for correspondence: John W. Hanrahan, Department of Physiology, McGill University, McIntyre Med. Sci. Bldg., 3655 Drummond St., Montreal, Quebec, Canada H3G 1Y6.

Several important properties of transepithelial $HCO_3$ secretion parallel those of Cl secretion. Both transport systems are active, electrogenic, and stimulated by maneuvers that raise intracellular [cAMP]. Also, $HCO_3$- and Cl-secreting tissues are both postulated to have high apical membrane Cl conductance. These similarities raise interesting questions regarding the nature of apical Cl channels and their role in $HCO_3$ secretion. Is the same channel responsible for apical Cl conductance in $HCO_3$- and Cl-secreting tissues? A novel, cAMP-stimulated Cl channel has been described in the rat pancreatic duct, suggesting a different Cl channel may be involved in $HCO_3$-secretion.[13] Another question concerns the mechanism of $HCO_3$ exit through the apical membrane: Does $HCO_3$ efflux occur entirely by apical anion exchange, or could anion channels also mediate a significant portion of the $HCO_3$ flux?

In the first part of this paper we review recent studies of Cl channels in a human pancreatic ductal cell line and compare their properties with channels in

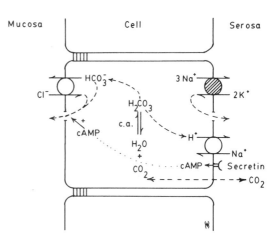

**FIGURE 1.** Cellular model proposed for electrogenic bicarbonate secretion across turtle urinary bladder. (Stetson et al.,[11] 1985.)

two Cl-transporting epithelia, $T_{84}$, which secretes Cl actively,[14] and the sweat duct, which absorbs Cl passively.[15] In the second part of the paper we describe recent studies of Cl channel permeability to bicarbonate.

## IDENTIFICATION OF OUTWARDLY RECTIFYING Cl CHANNELS IN $HCO_3$- AND Cl-SECRETING EPITHELIA

### *Conductance*

The PANC-1 cell line was studied as a model for the human pancreatic duct. PANC-1 cells form polarized monolayers and have many ultrastructural features characteristic of differentiated pancreatic ducts, including microvilli, apically localized cytoplasmic vesicles, epithelial intermediate filaments, and complete tight junctional complexes.[16] Biochemical evidence for PANC-1 differentiation is provided by observations that (1) the Na/K ATPase is localized in the basolateral membrane, (2) PANC-1 cells contain gamma-glutamyltranspeptidase and carbonic anhydrase, but lack the acinar enzyme amylase, and (3) PANC-1 cells secrete large, sulfated proteins resembling those produced by principal cells of rat and human pancreatic duct.

FIGURE 2 shows the channel that was observed most frequently in apical membrane patches from PANC-1 cells. The amplitude and kinetics of channels in $T_{84}$ and reabsorptive sweat duct cells were qualitatively similar. In each case, currents at $+50$ mV were larger than those at $-50$ mV, when the patches were bathed symmetrically with 150 mM NaCl solutions. The current-voltage relationship for a channel from a PANC-1 cell is shown in FIGURE 3A. In symmetrical 150 mM NaCl, the slope conductance was 31 pS at the reversal potential, 23 pS at negative potentials (between $-20$ and $-30$ mV), and 70 pS at large positive voltages (between $+60$ and $+80$ mV). Also, rectification could be described by a simple model comprised of a single dominant energy barrier of $\sim5.3$ kCal/mole that senses 38% of the membrane field, starting from the cytoplasmic side of the membrane.

Estimates of conductance and barrier parameters for PANC-1 channels were identical to those in $T_{84}$ and reabsorptive sweat ducts (FIG. 3B). The channel was easily recognized in patches by its rectification and by its kinetics at several holding potentials.

### Selectivity

Selectivity of the PANC-1 channel for anions over cations was demonstrated in excised, inside-out patches by partially replacing 66% of the NaCl in the bath with sufficient sucrose to keep the osmotic pressure constant. This caused the reversal potential to shift by $-25.8$ mV, suggesting a mean permeability ratio ($P_{Na}/P_{Cl}$) of 0.033. The channel was not highly selective among anions; for example, when bath Cl was replaced by iodide the reversal potential shifted by $+15.8$ mV, indicating higher permeability to iodide at zero net current ($P_I/P_{Cl} = 1.86$). Replacing bath Cl with acetate shifted the reversal by $-10$ mV, indicating that acetate permeability is lower than Cl permeability but is still significant ($P_{acetate}/P_{Cl} = 0.67$; FIG. 4). Identical iodide and acetate permeability ratios were obtained when Cl channels from sweat gland and $T_{84}$ cells were used, and the estimates are in close agreement with those reported previously in Cl-transporting epithelia.[17]

In the pancreatic duct, weak channel selectivity among anions might explain why fluid secretion persists after $HCO_3$ has been replaced with other anions. Depending on the species, fluid secretion occurs at 42%–70% of the control rate when $HCO_3$ is completely replaced by acetate.[18,19] Formic, proprionic, butyric, and other organic acids that can enter the cell by nonionic diffusion also support fluid secretion to varying degrees in the absence of $HCO_3$. (See ref. 20 for review.) The secretion rate with each substitute is proportional to its concentration and depends on pH. Organic anions produced intracellularly through the action of basolateral Na/H exchange (see FIG. 1) may leak out through apical anion channels, leading to net Na and fluid secretion. The alternative hypothesis, that apical $HCO_3$ efflux is mediated by anion exchange, would require several anion exchangers in parallel or a single exchanger with unusually wide substrate specificity. Of the anions tested to date, those capable of partially supporting fluid secretion can also permeate the outwardly rectifying anion channel.

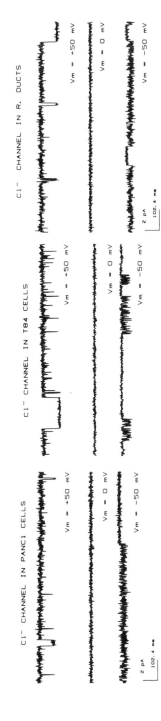

**FIGURE 2.** Single channels recorded using inside-out membrane patches from PANC-1, $T_{84}$, and cultured sweat duct cells. The pipette solution contained (mM): 150 NaCl, 1 EGTA, 10 HEPES, 2 CaCl$_2$, pH 7.3. The bath solution was identical except [CaCl$_2$] was lowered to 1 mM. The outwardly rectifying channel was observed in these three culture systems representing HCO$_3$-secreting, Cl-secreting, and Cl-absorbing epithelia.

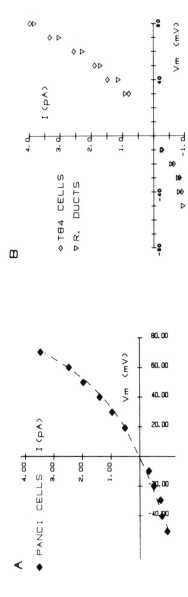

**FIGURE 3.** **(A)** Current-voltage relationship of a PANC-1 Cl channel bathed symmetrically with 150 mM NaCl solution. The I/V relationship was fitted (*dashed line*) with the simplest model consisting of a single barrier of 5.3 kcal/mol and an electrical distance from the inside of 0.349. **(B)** I/V relationship of Cl channels from $T_{84}$ and cultured sweat duct cells bathed on both sides with 150 mM NaCl solution.

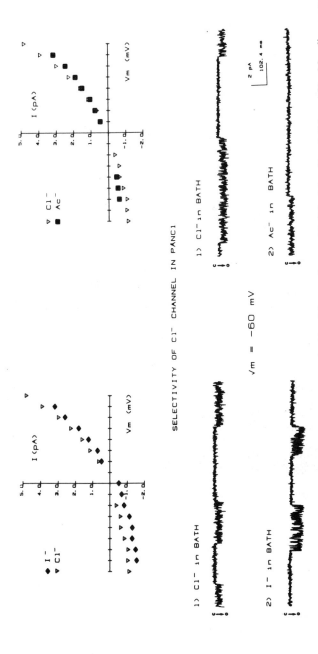

**FIGURE 4.** Effect of replacing 150 mM NaCl on the cytoplasmic side of the membrane with 150 mM sodium iodide or 150 mM sodium acetate. Reversal potentials were interpolated after fitting the curve with a fourth-degree polynomial and then added to the liquid junction potential. The results indicate $P_I/P_{Cl} = 1.86$ and $P_{acetate}/P_{Cl} = 0.67$.

### Voltage Dependence

The anion channel in PANC-1 cells had three types of voltage dependence: (1) Cl channels that were not spontaneously active after excision could be activated by clamping the membrane potential to large positive voltages (greater than +80 mV);[21,22] (2) Once a channel was activated, its probability of being in the open state immediately after stepping to a particular potential increased with depolarization.[17,23,24] (3) FIGURE 5 shows a third type of voltage dependence, inactivation. This occurred when the membrane potential was held at voltages greater than +50 mV. The rate of inactivation increased with increasingly positive voltage and could be removed by reversing the polarity of the membrane potential or by partial depolarization. We observed voltage-dependent inactivation using anion channels from all three cell types, although this phenomenon has not been reported in previous patch clamp studies.

The voltage dependence of the outwardly rectifying anion channel has not been studied in great detail because it has no obvious physiologic significance. Nevertheless, voltage dependence provides a biophysical signature of the channel and supports the idea that the same anion channel molecule is present in $HCO_3$ and Cl transporting tissues.

### Activation by db-cAMP

Little spontaneous channel activity was observed in cell-attached patches of PANC-1 monolayers. When the pipette contained 150 mM NaCl, the single channel currents that were occasionally observed reversed polarity when the patch was depolarized by 22–37 mV relative to the membrane potential. The reversal potentials ($E_{rev}$) are within the range of Cl electrochemical potentials measured in secretory epithelia (e.g., −22 to −35 mV in trachea;[25] −34 mV in shark rectal gland[26]).

FIGURE 6 shows recordings obtained from cells in normal bathing saline solution (NBS; mEq/L) 144 Na, 4 K, 2 Ca, 1 Mg, 150 Cl, 10 HEPES, and 5 mM glucose, pH 7.4 at 37° C ± 0.2. The membrane patch was hyperpolarized −40 mV

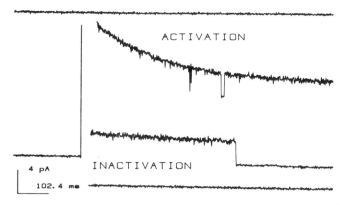

**FIGURE 5.** Inactivation of a PANC-1 anion channel after the membrane potential is stepped from 0 to +80 mV.

**FIGURE 6. (A)** Effect of adding $5 \times 10^{-4}$ M db-cAMP on channel activity in a PANC-1 cell. Channel activity was recorded in the cell-attached configuration with the membrane patch continuously hyperpolarized by 40 mV. Channel activity increased dramatically after the addition of db-cAMP (compare noise at i and ii). **(B)** Same patch as in **A** but in the inside-out configuration. At least six outwardly rectifying anion channels were spontaneously active when the patch was initially excised, but these could be inactivated by stepping the voltage from 0 mV to +80 or +100 mV.

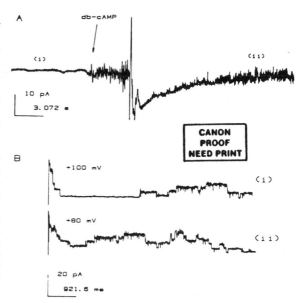

relative to the membrane potential throughout the experiment. Open probability was negligible initially (at "i") but increased dramatically after adding cAMP ($5 \times 10^{-4}$ M). Channel activity recorded after cAMP addition was too intense to resolve single channel openings (see "ii"); however, outwardly rectifying anion channels were spontaneously active upon excision. When the same patch was excised and bathed with symmetrical NaCl solutions, inactivation of at least six outwardly rectifying anion channels was observed after stepping the potential from 0 mV to +80 or +100 mV (FIG. 6B). In summary, the PANC-1 anion channel is responsive to cAMP, as expected for a channel involved in secretin-stimulated $HCO_3$ secretion. Effects of the "first messengers" secretin and VIP have not been tested using these cell-attached patches, but might be useful for studies of receptor differentiation in the PANC-1 cell line.

## PERMEABILITY TO HCO₃

In this section we consider whether the outwardly rectifying anion channel could mediate a component of the transepithelial $HCO_3$ flux. Bicarbonate permeability was studied in symmetrical $HCO_3$ solutions, under biionic conditions with outwardly and inwardly directed $HCO_3$ gradients, and with mixtures of Cl and $HCO_3$.

### Symmetrical HCO₃

Replacing Cl with $HCO_3$ on both sides of the patch reduced the channel conductances by approximately half at all potentials but did not affect the kinetics or voltage dependence noticeably. Moreover, the degree of outward rectification was not altered, and the bicarbonate current-voltage relationship was again well described using a single barrier model in which the barrier was located at an electrical distance of 38% of the field from the inside. However, the calculated barrier for $HCO_3$ permeation was 0.6 kcal/mol higher than that for Cl permeation.

## $HCO_3$ Influx

$HCO_3$ influx was studied under two sets of biionic conditions, first using a large pH gradient (so that only a small transmembrane $pCO_2$ gradient would be needed to maintain external $[HCO_3]$ constant) and then with a large $pCO_2$ gradient (so that $HCO_3$ permeability could be measured in the presence of a small pH gradient and near absence of $CO_3^=$). The bath was gassed continuously with 5% $CO_2$ when it contained $HCO_3$. It was important to correct apparent reversal potentials for the large $Cl$-$HCO_3$ liquid junctions at the reference electrode ($-7.35$ mV).

In the first series of experiments, positive current (carried by a flow of anions from the pipette to bath) was measured when the bath contained 30 mM Cl and the pipette contained either 30 mM ionic $CO_2$ ($HCO_3^-$ + $CO_3^=$) or 30 mM Cl. Pipette solutions containing $HCO_3$ or Cl had the same pH (9.7), but the former was equilibrated with air (0.03% $CO_2$) and the latter with 100% nitrogen. Surprisingly, the mean slope conductance at $\sim +90$ mV was 35.5 $\pm$ 1.16 pS with $HCO_3$ in the pipette, as compared to 31.1 $\pm$ 0.87 pS with Cl in the pipette suggesting $HCO_3$ (and possibly $CO_3^=$) are carried more effectively by the channel. On the other hand, permeability ratios calculated from reversal potentials supports the opposite conclusion ($P_{HCO_3}/P_{Cl}$ = 0.63 $\pm$ 0.03). Conductance ratios are probably more accurate under these conditions because reversal potentials had to be extrapolated and were thus more susceptible to error. Nevertheless, the high $HCO_3$ : Cl conductance ratio obtained with a pH gradient is puzzling and probably not caused by carbonate permeation because sulfate, which has a smaller hydrated radius than does $CO_3^=$ (1.84° A $vs$ 2.12° A), permeates poorly compared to chloride ($G_{SO_4}/G_{Cl}$ = 0.19; Tabcharani and Hanrahan, unpublished observations).

In the second series of experiments, $HCO_3$ influx was studied using inside-out patches exposed to a smaller pH gradient but a substantial transmembrane $pCO_2$ gradient. Both the pipette and the bath solutions initially contained 150 mM $NaHCO_3$ and were preequilibrated with 5% $CO_2$ at pH 8.27. After recording the current-voltage relationship, the bath was flushed with 150 mM NaCl ($N_2$ equilibrated), and a second I/V curve was obtained. Replacing bath $HCO_3$ with Cl shifted the (corrected) reversal potential from 0 to 17.3 $\pm$ 1.05, indicating a $HCO_3$ : Cl permeability ratio of 0.48 $\pm$ 0.03. Interestingly, currents measured at depolarizing potentials (carried by $HCO_3$ flowing from the pipette to the bath) were consistently larger when the bath contained some Cl. In other words, Cl had a stimulatory effect on $HCO_3$ permeation from the opposite side.

## $HCO_3$ Efflux

To study $HCO_3$ permeation through the channel, I/V curves were obtained using inside-out patches exposed to $HCO_3$ in the pipette and 150 mM NaCl or 150 mM $NaHCO_3$ in the bath. FIGURE 7A shows I/V curves obtained from a PANC-1 cell in symmetrical 150 mM Cl and after replacing bath Cl by $HCO_3$. The mean

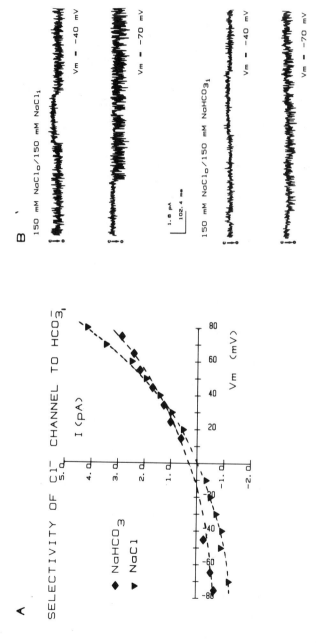

**FIGURE 7.** **(A)** Effect of replacing 150 mM bath (cytoplasmic) Cl with 150 mM $HCO_3$ on the current-voltage relationship. The bicarbonate solution was maintained at pH 8.27 by gassing the bath with 5% $CO_2$. The reversal potential indicates a selectivity $P_{HCO_3}/P_{Cl} = 0.521$. **(B)** Traces showing Cl and $HCO_3$ flow through the anion channel when the cytoplasmic side of the membrane is exposed to 150 mM NaCl or 150 mM $HCO_3$.

reversal potential measured immediately after replacing bath Cl and corrected for liquid junctions was $-15.35 \pm 0.69$ mV, indicating a permeability ratio of $0.52 \pm 0.01$. A very similar ratio was obtained for slope conductances when currents were carried by $HCO_3$ and Cl at physiologic potentials ($G_{HCO_3}/G_{Cl} = 0.55$ between $-30$ and $-50$ mV). $HCO_3$ currents were clearly resolved (FIG. 7B). To test for possible artifacts resulting from asymmetrical $CO_2$ exposure, I/V curves during the first 45 seconds were compared with those determined after 25–30 minutes of exposure to 5% $CO_2$ in the bath. The I/V curves at early and late times were identical. This finding suggests that $HCO_3$ did not accumulate significantly in the pipette tip during experiments, because, as just shown, the channel is permeable to inward $HCO_3$ flow and significant elevation of pipette [$HCO_3$] should have shifted the reversal potential in the negative direction.

### HCO₃ Permeability in Mixed Solutions

The preceding results show that $HCO_3$ permeation of the Cl channel can occur in both directions from pure $HCO_3$ solutions; however, this may not be the case *in situ* where other anions, particularly Cl, could influence $HCO_3$ permeability. To test this we compared I/V curves under biionic conditions (Cl in the pipette, $HCO_3$ in the bath) before and after adding 10 mM Cl to the $HCO_3$ solution. In a second series of experiments, 150 mM bath $HCO_3$ was replaced with a solution containing 25 mM $HCO_3$ and 125 mM Cl. The permeability ratios calculated under these conditions were $P_{HCO_3}/P_{Cl} = 0.70$ and $0.50$, respectively. This indicates that Cl does not decrease the permeability of the outwardly rectifying channel to other anions.

### DISCUSSION

Our results indicate that the outwardly rectifying anion channel, which has been described previously in various Cl-transporting epithelia (cultured cells from airway and colon) is also abundant in cells derived from a bicarbonate-secreting tissue, the pancreatic duct. Transepithelial $HCO_3$ secretion by PANC-1 has not yet been demonstrated; nevertheless, it expresses many ultrastructural and biochemical properties consistent with differentiation. The high density of outwardly rectifying anion channels in PANC-1 cells and their activation by cAMP suggest some role in $HCO_3$ secretion. As shown in FIG. 1, the first major function of outwardly rectifying anion channels during cAMP-stimulated $HCO_3$ secretion may be to allow Cl efflux at the apical membrane, thereby enhancing the inward Cl gradient that drives anion exchange. As already discussed, this Cl efflux would also resupply the anion exchanger with luminal Cl.[12,13]

Another function of the outwardly rectifying anion channel in $HCO_3$-secreting epithelia may be to mediate some of the $HCO_3$ efflux, particularly when the inward Cl gradient no longer provides sufficient energy to drive the anion exchanger. Several observations suggest that this might be the case. First, if one assumes typical intracellular Cl and $HCO_3$ activities and high luminal [Cl] (resembling acinar fluid; >100 mM), measured luminal [$HCO_3$] in rat pancreatic duct approaches the maximum (~70 mM) that could be achieved by Cl-$HCO_3$ exchange. However, it is not clear if such high [$HCO_3$] could be attained *in vivo* because the luminal [Cl] concentration changes reciprocally with [$HCO_3$] as fluid moves down the ductal system.[7,20] That is, luminal [Cl] falls in concert with $HCO_3$

secretion, which implies that the inward Cl gradient would be partially dissipated before the [$HCO_3$] reached its theoretical maximum. Also, some species have luminal $HCO_3$ concentrations that are probably too high to be generated by anion exchange: Luminal [$HCO_3$] reaches 120–140 mM in humans, and values of 142 and 160 mM have been reported in dogs and pigs, respectively. (See reviews 7, 20, and 29.) Finally, it is well established that anion gradients decline along the main duct. This reequilibration is saturable and may be mediated by Cl/$HCO_3$ exchange.[28] Dissipation of the $HCO_3$ and Cl gradients by anion exchange itself suggests that an additional mechanism is involved in generating them.

Could bicarbonate permeation through the outwardly rectifying anion channel contribute significantly to apical membrane conductance? In the smallest ducts where [Cl] is high and [$HCO_3$] low, $HCO_3$ probably contributes little conductance when compared to the contribution of Cl. However, as luminal [$HCO_3$] increases to 70–160 mM and [Cl] declines to 20–40 mM $HCO_3$ would probably contribute a large fraction of the total apical membrane conductance, despite being less permeant than Cl through the channel.

The presence of the outwardly rectifying anion channel in PANC-1 certainly does not rule out the possibility that other anion-selective channels are involved in cAMP-stimulated $HCO_3$ secretion. A novel Cl channel was recently described in the apical membrane of rat pancreatic duct by Gray et al.[30] The conductance of that channel is ~4 pS, and it has a linear current-voltage relationship and voltage-independent gating. Most importantly, in cell-attached patches, the open probability of the 4-pS channel increased approximately fivefold when cells were exposed to 10 nM secretin or 1 mM dibutyryl-cAMP. It will be important to identify all the anion channels that participate in pancreatic $HCO_3$ secretion and to establish their roles. The outwardly rectifying channel is implicated by the fact that its modulation and pancreatic $HCO_3$ secretion are both disrupted in cystic fibrosis (CF). Defective modulation of this channel would also explain the similarity of transport abnormalities in airway and pancreatic cells. The severity of the CF defect in pancreatic duct is more variable than in airway cells; some CF patients have almost normal pancreatic function. The presence of more than one cAMP-regulated Cl channel in pancreatic duct (i.e., the 4-pS channel[30] and the outward rectifier) might explain this variability if, for example, the low-conductance Cl channel could in some instances compensate for abnormal regulation of the outward rectifier.

In summary, this paper showed the existence of cAMP-activated, outwardly rectifying anion channels in a pancreatic ductal cell line. The results further suggest that $HCO_3$ secretion may be partially mediated by this anion-selective channel, in addition to Cl/$HCO_3$ exchange. Both exit mechanisms would require current flow at the apical membrane and would therefore appear as electrogenic transport when studied transepithelially. The relative contribution of channel-mediated flux to transepithelial $HCO_3$ secretion has not been determined and probably varies with species and position along the ductal tree. For example, anion exchange may mediate the largest fraction of the $HCO_3$ flux within the intralobular and small interlobular ducts, whereas the channel-mediated component accounts for the large $HCO_3$ gradients achieved by some species (including humans) and for the Cl-independent component of secretin-stimulated secretion. Clearly, simultaneous measurements of net $HCO_3$ flux and apical net electrochemical gradients for Cl and $HCO_3$ would be useful in defining the fluxes. It will also be interesting to learn if the outwardly rectifying anion channel is involved in electrogenic, cAMP-stimulated $HCO_3$ secretion across other model preparations, such as the turtle bladder, frog choroid plexus and duodenum.

## ACKNOWLEDGMENTS

We thank Drs. I. Novak, M. Gray, and W. Marshall for useful discussions, Dr. J. Riordan and T. Jensen for providing sweat gland cultures, and R. Wolanski and D. Elie for help with data collection and cell culture.

[Note added in proof: The functions of outwardly rectifying anion channels are not yet established, but several recent observations are incongruous with a major contribution to apical membrane conductance in human pancreatic and sweat duct epithelium. First, we have found activation of the outward rectifier by cAMP to be inconsistent in PANC-1 cells, and modulation of this channel was not observed in a recent study of cultured human fetal pancreas.[31] Second, our estimate of the selectivity sequence of the outward rectifier in sweat duct cells ($1.62\ NO_3 > 1.56\ I > 1.17\ Br > 1\ Cl > 0.54\ HCO_3$) is clearly different from the one observed in native sweat duct epithelium (Quinton, this volume). Together these observations suggest that another channel, such as the one reported by Gray et al.,[13] may account for the largest fraction of apical anion conductance. A role in cell volume regulation has recently been proposed for outwardly rectifying channels in $T_{84}$ cells.[32]]

## REFERENCES

1. SILVA, P., J. STOFF, M. FIELD, L. FINE, J. FORREST & F. H. EPSTEIN. 1977. Mechanisms of active chloride secretion by shark rectal gland: Role of Na-K-ATPase in chloride transport. Am. J. Physiol. **233:** F298–F306.

2. FRIZZELL, R. A. & S. G. SCHULTZ. 1979. Sodium-coupled chloride transport by epithelial tissues. Am. J. Physiol. **236:** F1–F8.

3. KLYCE, S. D. & R. K. S. WONG. 1977. Site and mode of adrenalin action on chloride transport across the rabbit corneal epithelium. J. Physiol. **266:** 777–799.

4. SIMSON, J. N. L., A. MERHAV & W. SILEN. 1981. Alkaline secretion by amphibian duodenum. III. Effects of dbcAMP, theophylline, and prostaglandins. Am. J. Physiol. **241:** G529–536.

5. FLEMSTROM, G. 1987. Gastric and duodenal mucosal bicarbonate secretion. In Physiology of the Gastrointestinal Tract, 2nd ed. L. R. Johnson, ed.: 1011–1029. Raven Press, NY.

6. SAITO, Y. & E. M. WRIGHT. 1983. Bicarbonate transport across the frog choroid plexus and its control by cyclic nucleotides. J. Physiol. **336:** 635–648.

7. NOVAK, I. 1988. Pancreatic bicarbonate secretion. In pH Homeostasis. Mechanism and control. D. Haussinger, ed.: 447–470. Academic Press, London.

8. MAREN, T. H., J. A. RAWLS, J. W. BURGER & A. C. MYERS. 1963. The alkaline (Marshall's) gland of the skate. Comp. Biochem. Physiol. **10:** 1–16.

9. SATAKI, N., J. H. DURHAM, G. EHRENSPECK & W. A. BRODSKY. 1983. Active electrogenic mechanisms for alkali and acid transport in turtle bladders. Am. J. Physiol. **244:** C259–C269.

10. DURHAM, J. H., C. MATONS & W. A. BRODSKY. 1987. Vasoactive intestinal peptide stimulates alkali excretion in turtle urinary bladder. Am. J. Physiol. **252:** C428–C435.

11. STETSON, D. L., R. BEAUWENS, J. PALMISANO, P. P. MITCHELL & P. R. STEINMETZ. 1985. A double-membrane model for urinary bicarbonate secretion. Am. J. Physiol. **249:** F546–F552.

12. NOVAK, I. & R. GREGER. 1988. Properties of the luminal membrane of isolated perfused rat pancreatic ducts. Effects of cyclic AMP and blockers of chloride transport. Pflügers Arch. **411:** 546–553.

13. GRAY, M. A., J. R. GREENWELL, & B. E. ARGENT. 1988. Secretin-regulated chloride channel on the apical plasma membrane of pancreatic duct cells. J. Memb. Biol. **105:** 131–142.

14. DHARMSATHAPHORN, K., K. G. MANDEL, H. MASUI & J. A. MCROBERTS. 1985. Vasoactive intestinal polypeptide-induced chloride secretion by a colonic epithelial cell line. J. Clin. Invest. **75:** 462–471.
15. QUINTON, P. M. 1983. Chloride impermeability in cystic fibrosis. Nature **301:** 421–422.
16. MADDEN, M. E. & M. P. SARRAS, JR. 1988. Morphological and biochemical characterization of a human pancreatic ductal cell (PANC-1). Pancreas **3:** 512–528.
17. HALM, D. R., G. R. RECHKEMMER, R. A. SCHOUMACHER & R. A. FRIZZELL. 1988. Apical membrane chloride channels in a colonic cell line activated by secretory agonists. Am. J. Physiol. **254:** C505–C511.
18. CASE, R. M., J. HOLZ, D. HUTSON, T. SCRATCHERD & R. D. A. WYNNE. 1979. Electrolyte secretion by the isolated cat pancreas during replacement of extracellular bicarbonate by organic anions and chloride by inorganic anions. J. Physiol. **286:** 563–576.
19. SEOW, F. & J. A. YOUNG. 1984. Anionic dependency of secretin-stimulated secretion by the isolated perfused rat pancreas. In Secretion: Mechanism and Control. R. M. Case, J. M. Lingard & J. A. Young, eds.: 97–102. Manchester Univ. Press, Manchester.
20. CASE, R. M. & B. E. ARGENT. 1986. Bicarbonate secretion by pancreatic duct cells: Mechanisms and control. In The Exocrine Pancreas: Biology, Pathobiology, and Diseases. V. L. W. Go, J. D. Gardner, F. P. Brooks, E. Lebenthal, E. P. DiMagno & G. A. Scheele, eds.: 213–243. Raven Press, New York.
21. SHOUMACHER, R. A., R. L. SHOEMAKER, D. R. HALM, E. A. TALLANT, R. A. WALLACE & R. A. FRIZZELL. 1987. Nature **330:** 752–754.
22. LI, M., J. D. MCCANN, C. M. LIEDTKE, A. C. NAIRN, P. GREENGARD & M. J. WELSH. 1988. Nature **331:** 358–360.
23. FRIZZELL, R. A., D. R. HALM, G. RECHKEMMER & R. L. SHOEMAKER. 1986. Chloride channel regulation in secretory epithelia. Fed. Proc. **45:** 2727–2731.
24. HAYSLETT, J. P., H. GÖGELEIN, K. KUNZELMANN & R. GREGER. 1987. Characteristics of apical chloride channels in human colon cells ($HT_{29}$). Pflügers Arch **410:** 487–494.
25. WELSH, M. J. 1983. Intracellular chloride activities in canine tracheal epithelium: Direct evidence for sodium-coupled intracellular chloride accumulation in a chloride secreting epithelium. J. Clin. Invest. **71:** 1392–1401.
26. FRIZZELL, R. A. & M. E. DUFFEY. 1980. Chloride activities in epithelia. Fed. Proc. **39:** 2860–2864.
27. REINHARDT, R., R. J. BRIDGES, W. RUMMEL & B. LINDEMANN. 1987. Properties of an anion-selective channel from rat colonic enterocyte plasma membranes reconstituted into planar phospholipid bilayers. J. Membrane Biol. **95:** 47–54.
28. GREENWELL, J. R. 1977. The selective permeability of the pancreatic duct of the cat to monovalent ions. Pflügers Arch. **367:** 265–270.
29. KUIJPERS, G. A. & J. J. H. H. M. DEPONT. 1987. Role of proton and bicarbonate transport in pancreatic cell function. Ann. Rev. Physiol. **49:** 87–103.
30. CASE, R. M., A. A. HARPER & T. SCRATCHERD. 1969. The secretion of electrolytes and enzymes by the pancreas of the anaesthetized cat. J. Physiol. **201:** 335–348.
31. GRAY, M. A. et al. 1989. Am. J. Physiol. **257:** C240–C251.
32. WORRELL, R. T. et al. 1989. Am. J. Physiol. **256:** C1111–C1119.

# Phosphorylation-Dependent Regulation of Apical Membrane Chloride Channels in Normal and Cystic Fibrosis Airway Epithelium[a]

MICHAEL J. WELSH, MING LI, JOHN D. McCANN,
JOHN P. CLANCY, AND MATTHEW P. ANDERSON

*Howard Hughes Medical Institute*
*Department of Internal Medicine and*
*Physiology and Biophysics*
*University of Iowa College of Medicine*
*Iowa City, Iowa 52242*

Electrolyte transport by airway epithelia controls the quantity and composition of the respiratory tract fluid, thereby contributing to normal mucociliary clearance.[1] Airway epithelia have the capacity both for active sodium ($Na^+$) absorption from the mucosal surface to the submucosal surface and for active chloride ($Cl^-$) secretion from the submucosal surface to the mucosal surface. The relative contributions of $Cl^-$ secretion and $Na^+$ absorption to overall transport vary, depending on the neurohumoral environment, the airway region, and the species. Here we focus on $Cl^-$ secretion, specifically the apical membrane $Cl^-$ channel.

## Cl⁻ TRANSPORT BY AIRWAY EPITHELIA

Transepithelial $Cl^-$ secretion is a two-step process. $Cl^-$ enters the cell across the basolateral membrane and is accumulated at a value greater than that predicted for electrochemical equilibrium. $Cl^-$ then exits the cell onto the mucosal surface via apical membrane $Cl^-$ channels. Regulation of apical $Cl^-$ conductance controls, in part, the rate of transepithelial $Cl^-$ secretion: the addition of hormones and neurotransmitters increases apical $Cl^-$ conductance and the rate of transepithelial $Cl^-$ secretion.[1]

In cystic fibrosis (CF) airway epithelia, regulation of the apical $Cl^-$ conductance is defective; hormones and neurotransmitters that increase cellular levels of cAMP fail to stimulate $Cl^-$ secretion.[2,3] Moreover, the addition of membrane permeant analogs of cAMP fails to stimulate $Cl^-$ secretion or to increase apical $Cl^-$ permeability. As a result, CF airway epithelia are $Cl^-$ impermeable, a defect that may contribute to abnormal respiratory tract fluid, impaired mucociliary clearance, and the pathophysiology of the disease.[4] $Cl^-$ impermeability is also observed in several other epithelia affected by CF, including the sweat gland duct, the sweat gland secretory coil, the intestine, and probably the pancreas.

[a] Portions of the work from the author's laboratory were supported by grants from the National Institutes of Health (HL29851 and HL42385) and the National Cystic Fibrosis Foundation. J. P. Clancy is an American Heart Association Student Fellow. J. D. McCann is supported by the March of Dimes Birth Defects Foundation.

## REGULATION OF Cl⁻ CHANNELS ON THE CELL

To better understand the mechanism of Cl⁻ secretion and regulation of apical Cl⁻ conductance, we and others used the patch-clamp technique to study airway epithelial cells.[5-7] In cell-attached recordings, the addition of secretagogues, such as isoproterenol, activated[b] Cl⁻ channels. As shown in FIGURE 1, a characteristic feature of these channels was their outwardly rectifying current-voltage relationship in the presence of symmetrical Cl⁻ concentrations. Evidence that this outwardly rectifying Cl⁻ channel is responsible for apical membrane Cl⁻ conductance is that: (a) it was activated by hormones and neurotransmitters that increase apical Cl⁻ conductance;[5-7] (b) it has an anion selectivity sequence of SCN > I > Cl = Br > F, which is the same as the anion permeability sequence of the apical membrane (unpublished observation); (c) it was blocked by carboxylic acid analogs which also block apical Cl⁻ conductance;[5] (d) it was found in patches of membrane obtained from confluent sheets of cells where only the apical membrane was accessible to the recording pipet (in contrast, the basolateral K⁺ channel was not observed in the apical membrane of confluent sheets of cells);[5-7] and (e) as indicated below, its regulation was altered in CF.

When isoproterenol, prostaglandin E, or 8-Br-cAMP was added to CF cells, we never saw activation of Cl⁻ channels in cell-attached patches. This observation is consistent with previous transepithelial studies showing that the CF apical membrane is Cl⁻ impermeable.

## ACTIVATION OF Cl- CHANNELS IN EXCISED, CELL-FREE MEMBRANE PATCHES

Although secretagogues failed to activate Cl⁻ channels in CF cells, Cl⁻ channels were present in the membrane; after membrane patches were ripped off the cell, Cl⁻ channels activated in patches from both normal and CF cells.[6,7] Once they were excised and activated, Cl⁻ channels from CF cells had the same conductive and kinetic properties as did Cl⁻ channels from normal cells. This observation suggested that the Cl⁻ impermeability of CF apical membranes resulted not from the absence of the channel protein, but rather from defective regulation of the Cl⁻ channel.

On some occasions, channels activated immediately after excision of the patch, independent of what voltage was maintained across the patch. Most often, however, channels remained in an inactivated state until the membrane was depolarized to relatively large voltages (+80 to +140 mV).[8,9] Once activated by depolarization, channels remained in the activated state even when membrane voltage was returned to less depolarizing values. We have observed a similar phenomenon when we increase the bath temperature from 23°C to 37°C.

The mechanisms responsible for Cl⁻ channel activation by patch excision, by depolarization, and by an increase in bath temperature are uncertain. These phenomena are only observed in excised patches; similar maneuvers do not activate channels in cell-attached patches. These results led us to speculate that the chan-

---

[b] The Cl⁻ channel functions in at least two modes. We refer to an "inactivated" channel as one that is unstimulated or quiescent and always in the closed state. An "activated" channel is one that has been stimulated and spontaneously flickers back and forth between the open and closed state.

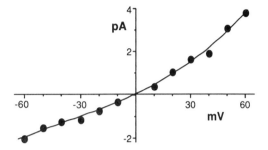

**FIGURE 1.** Current-voltage relationship of the apical membrane Cl⁻ channel in an excised, inside-out patch bathed in symmetrical 140 mM NaCl solutions.

nel might be regulated by an inhibitor that maintained it in an inactivated state; excision of the membrane patch could relieve the inhibition by altering the interactions between the channel and the inhibitor or by allowing an inhibitor to diffuse away from the channel. An alternative possibility is that these maneuvers may somehow physically alter channel conformation, thereby causing it to remain in an activated state. In any case, the observation that large membrane depolarization activated normal and CF Cl⁻ channels required that we avoid such large depolarizing voltages when investigating other modes of channel activation. More importantly, it allowed us to use depolarization as a "tool" to determine if a Cl⁻ channel was present in a membrane patch.

## ACTIVATION OF Cl⁻ CHANNELS BY cAMP-DEPENDENT PROTEIN KINASE

The studies showing that Cl⁻ channels were regulated by some intracellular second messenger and that regulation of CF Cl⁻ channels was defective raised the question: What second messenger system regulates the Cl⁻ channel? There was substantial circumstantial evidence that cAMP was involved: many agonists that stimulate Cl⁻ secretion increase cellular levels of cAMP, and the addition of poorly metabolized, membrane permeant analogs of cAMP stimulate Cl⁻ secretion. In addition, the observation that β-adrenergic agonists caused a normal accumulation of cAMP in CF cells and that cAMP analogs failed to activate CF Cl⁻ channels suggested that the defect in CF lay distal to cAMP accumulation.

In many cells, the biologic effects of cAMP result from activation of cAMP-dependent protein kinase (PKA), resulting in phosphorylation of target proteins. To determine if PKA regulates Cl⁻ channels, we used excised, inside-out patches of membrane and added purified catalytic subunit of PKA plus ATP to the internal (cytosolic) surface.

FIGURE 2 shows a representative example. In the top traces, the channel was in the inactive state; no channel was observed to open. Subsequent addition of PKA plus ATP activated the channel. Activation required the addition of both ATP and PKA; neither added alone was sufficient. Moreover, the addition of boiled catalytic subunit failed to activate Cl⁻ channels. In some patches, PKA did not activate Cl⁻ channels; to determine if those patches contained Cl⁻ channels, we depolarized the membrane. However, in patches in which PKA failed to activate channels, no further channels were activated by depolarization, suggesting that no channel was present in the patch of membrane, that is, the patch was blank. Those results showed that the addition of PKA plus ATP mimics the effect of secretagogues such as the $\beta$-adrenergic agonist isoproterenol and exogenous cAMP in regulating Cl⁻ channels. Thus, it appears that the Cl⁻ channel or a regulatory protein closely associated with the channel is phosphorylated, resulting in channel activation.[8,9]

As just indicated, in CF airway cells, Cl⁻ channels are not activated by secretagogues such as isoproterenol, even though cellular levels of cAMP increase appropriately. To further localize the defect in CF we added PKA plus ATP to excised patches from CF cells, but despite this maneuver, no channels were activated. However, channels were present in the CF patches, because they could be activated by subsequent membrane depolarization. Once they were activated, CF Cl⁻ channels showed the same single-channel conductive properties as did Cl⁻ channels from normal cells. These results suggest that in CF epithelial cells the effector system distal to PKA is abnormal.[8,9]

## ACTIVATION OF Cl⁻ CHANNELS BY PKC

Protein kinase C (PKC) may play an important role in regulating ion channels in several types of cells. Previous studies in canine airway epithelium used the membrane permeant activator of PKC, phorbol 12-myristate 13-acetate (PMA), to suggest that PKC might regulate secretion.[10,11] The addition of PMA to cell monolayers caused complex effects on Cl⁻ secretion; PMA caused a transient stimulation of secretion, but then inhibited the secretory response to subsequent addition of a membrane-permeant cAMP analog. Because those studies did not allow the transport effects to be localized to the Cl⁻ channel, and because PMA may have effects on secretion other than those mediated by PKC, we examined the effect of PKC on Cl⁻ channels in excised patches.[12] Phosphorylation requires the presence of PKC (we used highly purified PKC from rat or mouse brain or a partially purified preparation from canine tracheal epithelium with similar results), phos-

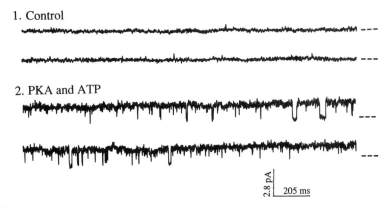

**FIGURE 2.** Activation of a Cl⁻ channel in an excised, inside-out patch by addition of catalytic subunit of cAMP-dependent protein kinase (PKA) and ATP (1 mM). Tracings were obtained at +40 mV, under control conditions and 98 seconds after addition of the phosphorylation solution.

1. Control, $[Ca^{2+}] < 10$ nM

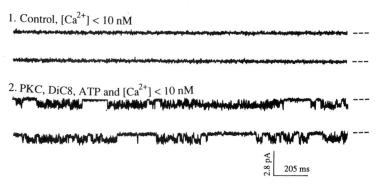

2. PKC, DiC8, ATP and $[Ca^{2+}] < 10$ nM

**FIGURE 3.** Activation of a $Cl^-$ channel in an excised, inside-out patch by addition of PKC, DiC8 (1 $\mu$g/ml), and ATP (1 mM) at a low $[Ca^{2+}]$ (<10 nM). Tracings were obtained at −40 mV, under control conditions and 132 seconds after addition of the phosphorylation solution. *Dashed line* shows the zero-current level. (Adapted from ref. 12 with permission.)

phatidylserine (the membrane patch served as a source of phospholipid), and a diacylglycerol or tumor-promoting phorbol ester (we used dioctanoylglycerol (DiC8), or diolein, or PMA with similar results). Some forms of PKC also require $Ca^{2+}$; therefore, we performed experiments at either a low (<10 nM) or a high (1 $\mu$M) $[Ca^{2+}]$.

FIGURE 3 shows the effect of adding PKC at a low $[Ca^{2+}]$. Under baseline conditions: (A) no channels were activated in the patch. (The dashed line indicates the zero-current level.) When PKC, PMA, and ATP were added to the solution bathing the internal surface of the patch, a channel activated and flickered back and forth between the open and closed state (B). Activation by PKC required the presence of PKC, ATP, and a diacylglycerol or PMA; the addition of any two alone was insufficient to activate the channel. Activation occurred an average of 126 seconds after the addition of all three agents. In contrast, channels did not activate during 480 seconds of observation in paired patches not exposed to PKC.

When we repeated the same experiment using CF cells, we were never able to activate $Cl^-$ channels with PKC, even though in many cases the patches contained $Cl^-$ channels as evidenced by activation with membrane depolarization.

These observations indicate that PKC phosphorylates and activates the channel at a low $[Ca^{2+}]$.

## INACTIVATION OF $Cl^-$ CHANNELS BY PKC

We also examined the effect of PKC at a high $[Ca^{2+}]$ (1 $\mu$M). Initially, we added PKC, ATP, and either PMA or a diacylglycerol to the internal surface of an excised membrane patch to determine if the channel activated. However, no channels activated under these conditions during a 6-minute observation period. Then, to determine if a channel was present in the patch, we depolarized the membrane; however, no channels opened with depolarization. In contrast, in paired experiments not exposed to PKC, depolarization activated channels in the majority of patches. These results suggested that PKC at a high $[Ca^{2+}]$ may prevent channel activation by large membrane depolarization.

To test this notion more directly, we activated channels by membrane depolarization, as shown in FIGURE 4. After depolarization the channel was in an activated state, spontaneously flickering between the open and closed states (1). We next added ATP (2) and then PMA (3); neither altered channel kinetics. However, on addition of PKC to the ATP- and PMA-containing solution (4), the channel inactivated. In control patches not exposed to PKC, Cl⁻ channels remained in the activated state for at least 10 minutes, indicating that inactivation was not a

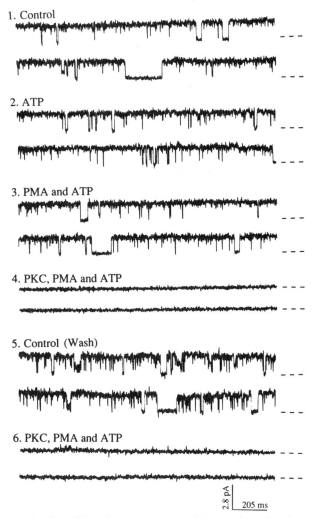

**FIGURE 4.** Inactivation of a Cl⁻ channel in an excised, inside-out patch by addition of PKC, PMA (100 nM), and ATP (1 mM) at a high [Ca$^{2+}$] (1 $\mu$M). Tracings were obtained at +40 mV with the indicated agents added to the internal solution. (From ref. 12 with permission.)

random event. Inactivation required the addition of PKC, a diacylglycerol or PMA, ATP, and a high $[Ca^{2+}]$ (1 $\mu$M). These results suggest that a high $[Ca^{2+}]$, PKC phosphorylates the channel, causing it to inactivate.

In some cases inactivation was reversible (FIG. 4–5). Removal of the phosphorylation solution resulted in reactivation. These results suggest that membrane-associated phosphatase has access to the channel in the membrane patch.

When we activated CF Cl$^-$ channels by excision and depolarization, we were able to inactivate them by the addition of PKC and its cofactors at 1 $\mu$M Ca$^{2+}$. In CF cells the effect of PKC was also reversible. These results indicate that PKC-dependent inactivation of Cl$^-$ channels is normal in CF and show that at least one phosphorylation-dependent function of the channel is intact in CF.

## SUMMARY

The observations described herein allow us to make several inferences about PKC and regulation of normal and CF Cl$^-$ channels. FIGURE 5 shows a model that summarizes these observations. In this model, for the sake of clarity, we refer to

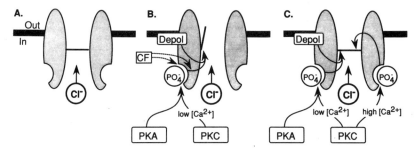

**FIGURE 5.** Model of the mechanism of Cl$^-$ channel regulation by phosphorylation. See text for details. (From ref. 12 with permission.)

the channel as a single entity, but note that it may consist of multiple subunits and associated proteins.

FIGURE 5A shows the channel in an inactivated state following excision from the cell. The channel can be activated by strong membrane depolarization, via an unknown mechanism, or by phosphorylation with PKA or PKC at a low $[Ca^{2+}]$ (FIG. 5B). We speculate that PKA and PKC may phosphorylate and activate the channel at the same site, or region of the channel, because phosphorylation-dependent activation by both is defective in CF. This result suggests that the CF defect might lie in a defective phosphorylation site on the channel, or associated protein, or in the mechanism that converts phosphorylation into a change in channel conformation, such as activation.

Activated channels can be inactivated by PKC at a high $[Ca^{2+}]$ (FIG. 5C). At high $[Ca^{2+}]$, PKC maintains the channel in an inactivated state and it inactivates channels that have been activated by PKC at low $[Ca^{2+}]$, by depolarization, or by PKA. Both activation and inactivation appear to result from phosphorylation; neither can be explained by down-regulation of the channel.

There are several possible ways to explain the two opposite effects of PKC on the Cl$^-$ channel: different responses may be due to an effect of Ca$^{2+}$ on the channel, on PKC, or on the interaction between the two. The channel apparently has two different phosphorylation sites (one activating and one inactivating), and $Ca^{2+}$ might determine which site is phosphorylated, in any of several ways. First, $Ca^{2+}$ might change channel conformation, making different sites accessible to PKC. Second, the $Ca^{2+}$ dependence of PKC might be influenced by the nature of the substrate, so that one phosphorylation site might not require $Ca^{2+}$ for an effect. Third, the interaction of PKC with the membrane might be $Ca^{2+}$ dependent; in the absence of $Ca^{2+}$, PKC might phosphorylate an extrinsic site on the channel, and in the presence of $Ca^{2+}$ PKC might phosphorylate a site on the channel associated with the membrane. Alternatively, different effects of PKC could be caused by isozymes that phosphorylate different sites: a $Ca^{2+}$-independent form that activates the channel and a $Ca^{2+}$-dependent form that inactivates the channel. This is possible because the purified PKC preparations that we use probably contain more than one isozyme. In any case, each of these alternatives requires that PKC show substrate specificity for two different phosphorylation sites on the channel.

## REFERENCES

1. WELSH, M. J. 1987. Electrolyte transport by airway epithelia. Physiol. Rev. **67:** 1143–1184.
2. WIDDICOMBE, J. H., M. J. WELSH & W. E. FINKENBEINER. 1985. Cystic fibrosis decreases the apical membrane chloride permeability of monolayers from cells of tracheal epithelium. Proc. Natl. Acad. Sci. USA **82:** 6167–6171.
3. COTTON, C. U., M. J. STUTTS, M. R. KNOWLES, J. T. GATZY & R. C. BOUCHER. 1987. Abnormal apical cell membrane in cystic fibrosis respiratory epithelium. An in vitro electrophysiologic analysis. J. Clin. Invest. **79:** 30–85.
4. WELSH, M. J. & R. B. FICK. 1987. Perspective: Cystic fibrosis. J. Clin. Invest. **80:** 1523–1526.
5. WELSH, M. J. 1986. An apical-membrane chloride channel in human tracheal epithelium. Science **232:** 1648–1650.
6. WELSH, M. J. & C. M. LIEDTKE. 1986. Chloride and potassium channels in cystic fibrosis airway epithelia. Nature **322:** 467–470.
7. FRIZZELL, R. A., G. RECHKEMMER & R. L. SHOEMAKER. 1986. Altered regulation of airway epithelial cell chloride channels in cystic fibrosis. Science **233:** 558–560.
8. LI, M., J. D. McCANN, C. M. LIEDTKE, A. C. NAIRN, P. GREENGARD & M. J. WELSH. 1988. Cyclic AMP-dependent protein kinase opens chloride channels in normal but not cystic fibrosis airway epithelium. Nature **331:** 358–360.
9. SCHOUMACHER, R. A., R. L. SHOEMAKER, D. R. HALM, E. A. TALLANT, R. W. WALLACE & R. A. FRIZZELL. 1987. Phosphorylation fails to activate chloride channels from cystic fibrosis airway cells. Nature **330:** 752–754.
10. WELSH, M. J. 1987. Effect of phorbol ester and calcium ionophore on chloride secretion in canine tracheal epithelium. Am. J. Physiol. **253** (Cell Physiol. **22**): C828–C834.
11. BARTHELSON, R A., D. B. JACOBY & J. H. WIDDICOMBE. 1987. Regulation of chloride secretion in dog tracheal epithelium by protein kinase C. Am. J. Physiol. **253**(Cell Physiol **22**): C802–C808.
12. LI, M., J. D. McCANN, M. A. ANDERSON, J. P. CLANCY, C. M. LIEDTKE, A. C. NAIRN, P. GREENGARD & M. J. WELSH. Regulation of chloride channels by protein kinase C in normal and cystic fibrosis airway epithelia. Science (in press).

# Cl⁻ Efflux in Brown Adipocytes

## A Possible Mechanism for α-Adrenergic Plasma Membrane Depolarization

LEONARDO DASSO,[a] EAMONN CONNOLLY,[b]
AND JAN NEDERGAARD[c]

*The Wenner-Gren Institute*
*The Arrhenius Laboratories F3*
*University of Stockholm*
*S-106 91 Stockholm, Sweden*

Physiologic stimulation of brown-fat cells occurs via the release of norepinephrine from the sympathetic nervous system. Norepinephrine elicits an increased rate of oxygen consumption (thermogenesis), and the biochemical processes behind this increase are functionally distinct for the $\beta$- and the $\alpha_1$-receptors found on the cells.[1,2]

Norepinephrine also causes depolarization of the brown adipocyte plasma membrane.[3] First a rapid, predominantly $\alpha$-adrenergically mediated depolarization occurs, which is followed by repolarization before the onset of a second, slow depolarization which depends on $\beta$-adrenergic stimulation and seems to be secondary to the increased metabolism of the cells.[4]

To determine the molecular mechanisms underlying $\alpha$-adrenergic depolarization we have investigated ionic movements across the plasma membrane of isolated brown adipocytes. Alpha$_1$-adrenergic stimulation leads to $Ca^{2+}$ mobilization from intracellular stores and to activation of an apamin-sensitive $Ca^{2+}$-dependent $K^+$ channel, but this mediates efflux and cannot cause depolarization.[5] $Ca^{2+}$ entry is not stimulated $\alpha_1$-adrenergically.[6] Norepinephrine induces increased $Na^+$ influx, but this is $\beta$-adrenergic.[7] Thus, $Ca^{2+}$, $Na^+$, and $K^+$ fluxes have all been found unlikely candidates for causing $\alpha$-adrenergic membrane depolarization.

$Cl^-$, which earlier was believed to move passively across plasma membranes, has been found to deviate from this behavior in many cell types. We therefore investigated whether an adrenergically stimulated $Cl^-$ efflux pathway could be observed in isolated brown-fat cells.[8,9] Brown-fat cells were isolated from hamsters and preincubated in $^{36}Cl^-$, and the efflux was followed as the loss of radioactivity when the cells were transferred to an unlabeled medium.

We found a rapid basal efflux of $^{36}Cl^-$, but as this efflux was apparently electroneutral (not influenced by chemical depolarization with KCl), it may represent a $Cl^-/Cl^-$ exchange or a $Cl^-/cation$ symport.

However, the addition of norepinephrine to the cells during efflux led to significant stimulation of the rate of $Cl^-$ efflux (FIG. 1). This efflux was inhibited by the $\alpha_1$-adrenergic antagonist prazosin (FIG. 1A), whereas the $\beta$-antagonist proprano-

[a] Present address: Cátedra de Bioquímica, Gral. Flores 2124, Montevideo, Uruguay.
[b] Present address: KabiVitrum AB, Stockholm, Sweden.
[c] Correspondence address: Jan Nedergaard, Arrhenius Laboratories F3, Stockholms Universitet, S-106 91 Stockholm, Sweden.

lol was without effect (not shown). It was thus mediated via $\alpha_1$ receptors and therefore was expected to be secondary to either an increase in cytosolic $Ca^{2+}$ levels or stimulation of protein kinase C via the diacylglycerol released from $PIP_2$. We found that the $Ca^{2+}$ ionophore A23187 could not evoke Cl⁻ efflux from the cells; thus, the efflux is probably not $Ca^{2+}$ stimulated. One consequence of protein kinase C activation is stimulation of phospholipase $A_2$ activity. This enzyme can be inhibited by mepacrine. The addition of mepacrine inhibited norepinephrine-stimulated Cl⁻ efflux (FIG. 1B), indicating that Cl⁻ efflux may be mediated by the action of phospholipase $A_2$ and the ensuing release of arachidonic acid. This can be transformed into eicosanoids which may be responsible for activation of the Cl⁻ efflux pathway (FIG. 2).

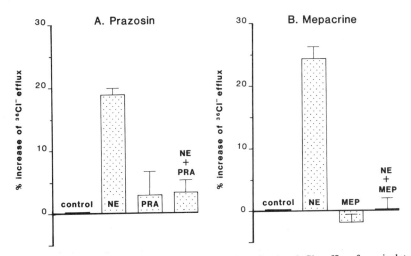

**FIGURE 1.** Effects of inhibitors of norepinephrine-stimulated Cl⁻ efflux from isolated brown adipocytes. Cells were loaded for 1 hour with ³⁶Cl⁻ and then diluted 1:20 into a medium containing unlabeled Cl⁻ and the indicated agents, and the efflux was followed. In both experimental series shown, 1 $\mu$M norepinephrine (NE) led to a significant increase in the rate of Cl⁻ efflux (here measured as the increase in Cl⁻ efflux after 1 minute, expressed relative to unstimulated cells). **(A)** 5 $\mu$M of the $\alpha_1$-adrenergic antagonist prazosin (PRA), which was without effect in itself, were fully able to inhibit norepinephrine-stimulated Cl⁻ efflux. (In parallel experiments with 40 $\mu$M of the $\beta$-adrenergic inhibitor propranolol no effect was seen [not shown].) **(B)** 100 $\mu$M of the phospholipase $A_2$ inhibitor mepacrine (MEP) were added. Values in **A** are from ref. 8.

Thus, although certain criteria which could be formulated for identification of the flux causing the membrane depolarization are seemingly fulfilled, it has not yet been shown that this Cl⁻ efflux constitutes a conductive pathway and that the cytosolic level of Cl⁻ is in excess of that predicted by a Nernst distribution (although our preliminary estimates indicate that the cytosolic Cl⁻ concentration is in excess of the 13 mM expected from a reported basal membrane potential of −60 mV).

The significance of the membrane depolarization is unknown. It may reflect a mechanism that alters cytosolic levels of ions. In the case of Cl⁻, it has been

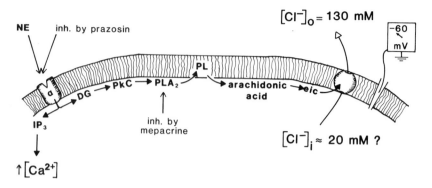

**FIGURE 2.** Proposed mechanism for the effect of norepinephrine on $Cl^-$ efflux. Working via $\alpha_1$-adrenergic receptors (FIG. 1A), the addition of norepinephrine leads to the production of $IP_3$ and ensuing mobilization of intracellular $Ca^{2+}$; this apparently has no effect on $Cl^-$ efflux. However, possibly via steps leading from diacylglycerol (DG) release, activation of protein kinase C (PkC) and through this of phospholipase $A_2$ ($PLA_2$) acting on phospholipids (PL) in the membrane, arachidonic acid is released, from which eicosanoids (eic) may be synthesized. These eicosanoids may (as suggested in other tissues) activate the $Cl^-$ efflux pathway.

suggested, for example, that cytosolic $Cl^-$ levels may affect the G-proteins involved in signal transduction.

## REFERENCES

1. NEDERGAARD, J. & O. LINDBERG 1982. Internat. Rev. Cytol. **74:** 187–286.
2. MOHELL, N., E. CONNOLLY & J. NEDERGAARD 1987. Am. J. Physiol. **253:** C301–C308.
3. GIRARDIER, L., J. SEYDOUX & T. CLAUSEN 1968. J. Gen. Physiol. **52:** 925–940.
4. GIRARDIER, L. & G. SCHNEIDER-PICARD 1983. J. Physiol. **335:** 629–641.
5. NÅNBERG, E., E. CONNOLLY & J. NEDERGAARD 1985. Biochim. Biophys. Acta **844:** 42–49.
6. CONNOLLY, E. & J. NEDERGAARD 1988. J. Biol. Chem. **263:** 10574–10582.
7. CONNOLLY, E., E. NÅNBERG & J. NEDERGAARD 1986. J. Biol. Chem. **261:** 14377–14385.
8. DASSO, L., E. CONNOLLY & J. NEDERGAARD 1989. Submitted for publication.
9. CONNOLLY, E., L. DASSO & J. NEDERGAARD 1989. *In* Thermoregulation: Research and Clinical Applications. P. Lomax & E. Schönbaum, eds.: 31–34. Karger, Basel.

# Effect of cAMP on Normal and Uremic Human Red Cell Chloride Conductance

R. D. LONDON, M. S. LIPKOWITZ,
AND R. G. ABRAMSON

*Mount Sinai School of Medicine*
*New York, New York 10029*

Recent studies indicate that cAMP regulates conductive pathways in diverse epithelia such as trachea, kidney, gallbladder, and shark rectal gland.[1-6] To determine whether this second messenger also modulates the ionic conductances of red blood cell (RBC) membranes, paired, washed normal human RBCs were incubated for 30 minutes at 37°C with 1 mM IBMX (3-isobutyl-l-methylxanthine, a phosphodiesterase inhibitor), without or with 1 mM 8-(4chlorophenylthio)-cAMP (8CPT-cAMP) or 10 $\mu$M forskolin (a direct activator of adenylate cyclase). Ghosts were then prepared without additives. Chloride and sodium conductances relative to potassium (GCl/GK and GNa/GK) were determined using the potential-sensitive fluorescent probe DiS-C3-(5).[7] In additional studies, the relative ionic conductances of RBC ghosts of normal and undialyzed uremic subjects were compared; in these studies, ghosts were prepared without prior incubation of the RBCs at 37°C. RBC cAMP levels were determined by a modification of the method of Rasmussen.[8]

In seven paired studies, GCl/GK was significantly higher in 8CPT-cAMP treated RBC ghosts than in ghosts prepared from untreated cells (0.38 ± 0.02 *vs* 0.22 ± 0.01). GNa/GK was not affected (0.61 ± 0.03 *vs* 0.53 ± 0.03). The effect of forskolin was virtually identical to that of 8CPT-cAMP: GCl/GK was significantly increased (0.33 ± 0.02 *vs* 0.17 ± 0.03), whereas GNa/GK was not affected (0.64 ± 0.05 *vs* 0.57 ± 0.06) (*n* = 7). The cAMP level (pm/ml RBCs) of forskolin-treated cells was significantly higher than that of untreated cells (56.6 ± 7.2 *vs* 26.8 ± 5.0), confirming that the effect of forskolin was via activation of adenylate cyclase and it was not nonspecific.

Because cAMP-mediated hormones (e.g., PTH and norepinephrine) are increased in uremia, it was postulated that cAMP levels and GCl/GK might be increased in RBC ghosts of uremic subjects. GCl/GK was significantly greater in RBC ghosts of 12 uremic subjects than in those of 15 normal subjects (0.55 ± 0.03 *vs* 0.35 ± 0.02); GNa/GK was not significantly different in the two populations (0.51 ± 0.04 *vs* 0.57 ± 0.05). The RBC cAMP level of eight uremic subjects was significantly greater than that of untreated RBCs of seven normal subjects (62.9 ± 8.9 *vs* 26.9 ± 5.0), but comparable to that of forskolin-treated normal RBCs.

These findings indicate that cAMP can selectively regulate relative chloride conductance in normal human RBCs. Furthermore, insofar as the RBC cAMP level is increased in uremia, these studies imply that the increase in GCl/GK in these cells may be due, in part, to increased levels of this messenger. Because chloride is at or above electrochemical equilibrium in diverse cells (e.g., muscle, nerve, and many epithelia), it is suggested that a generalized increase in cell cAMP concentration, and any associated increase in GCl/GK, could depolarize the membrane potential and thereby induce some of the diverse alterations in cell function that are seen in uremic subjects.

## REFERENCES

1. SMITH, P. L., M. J. WALSH, J. S. STOFF & R. A. FRIZZELL. 1982. J. Membr. Biol. **70:** 217–226.
2. FRIZZELL, R. A., G. RECHKEMMER & R. L. SHOEMAKER. 1986. Science **233:** 558–560.
3. GREGER, R., E. SCHLATTER & H. GOGELEIN. 1985. Pflüger's Arch. **403:** 446–448.
4. PETERSEN, K.-U. & L. REUSS. 1983. J. Gen. Physiol. **81:** 705–729.
5. SCHUSTER, V. L. 1987. J. Clin. Invest. **78:** 1621–1630.
6. LIPKOWITZ, M. S. & R. G. ABRAMSON. 1988. Kidney Internat. **33:** 421.
7. LIPKOWITZ, M. S. & R. G. ABRAMSON. 1987. Am. J. Physiol. **252:** F700–F711.
8. RASMUSSEN, H., W. LAKE & J. E. ALLEN. 1975. Biochim. Biophys. Acta. **411:** 63–73.

# Properties of an Anion Channel in Rat Liver Canalicular Membranes

M. SELLINGER,[a] S. A. WEINMAN, R. M. HENDERSON,
A. ZWEIFACH, J. L. BOYER,[b] AND J. GRAF

*Liver Center and Department of Cellular and
Molecular Physiology
Yale University, School of Medicine
New Haven, Connecticut 06510*

Previous measurements of $^{36}Cl$ uptake[1] from this laboratory have provided evidence for $Cl^-$ conductance in rat liver canalicular plasma membranes (cLPM). Because this membrane is not accessible to patch clamp pipettes, we have incorporated cLPM vesicles into a planar lipid bilayer system to further study this anion conductance.

## METHODS

cLPM were purified using rate zonal floatation and sucrose density gradient centrifugation.[2] Bilayers were prepared by the method of Mueller *et al.*[3] Fusion of vesicles with the bilayer was promoted by an osmotic gradient between the two compartments separated by the bilayer and stirring of the cis (vesicle containing) compartment after the addition of 1 mM $CaCl_2$.[4]

## RESULTS AND DISCUSSION

After incorporation of cLPM into the bilayer an anion channel was frequently observed. FIGURE 1 shows the ionic current through the channels observed at different holding potentials. A single channel conductance of 40 pS at 0 mV was determined in the presence of the 2:1 KCl gradient. A reversal potential greater than +15 mV corresponded closely to the $Cl^-$ gradient. The current voltage relationship showed Goldman rectification with a permeability ratio of $p_{Cl}/p_K$ greater than 13:1 (FIG. 2). The anion selectivity of the channel was investigated by the addition of different potassium salts to the trans-compartment. Only a small selectivity between several halides could be detected: $p_{Cl} \geq p_I > p_{Br} > p_F$. Channel kinetics were voltage dependent, with higher open probability at negative holding potentials.

In summary, an anion selective channel has been identified in cLPM with properties similar to those of intermediate, nonrectifying $Cl^-$ channels described in other epithelia.[5] The right-side out orientation of the vesicles suggests that the trans-compartment corresponds to the intracellular space.[6] With such an assump-

[a] Present address: Med. Klinik d. Universität Freiburg, Hugstetter Str. 55, D-7800 Freiburg, W. Germany.
[b] To whom correspondence should be addressed.

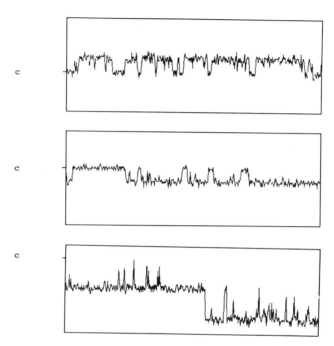

**FIGURE 1.** Ion currents through the channel at various holding potentials ($+40$, 0, and $-30$ mV, respectively) and at 2 : 1 KCl gradient. c indicates the closed state of the channel. Each box corresponds to 500 ms ($x$ axis) and 4 pA for the $y$-axis span.

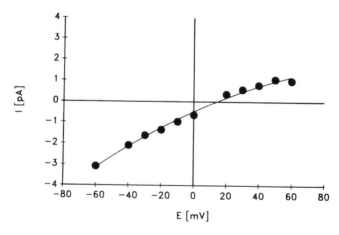

**FIGURE 2.** Current voltage relationship of a single anion channel at a 2 : 1 KCl gradient (cis:trans). Cl$^-$ current from cis to trans corresponds to negative current values. The line is a least squares fit according to the Goldman-Hodgkin-Katz equation giving a permeability ratio of $p_{Cl} : p_K$ greater than 13 : 1.

tion, depolarization at the canalicular membrane would increase channel activity. Therefore, this channel might compensate for the charge transfer occurring during bile acid secretion.[7,8]

## REFERENCES

1. MEIER, P. J., R. G. KNICKELBEIN, R. H. MOSELEY, J. W. DOBBINS & J. L. BOYER. 1985. J. Clin. Invest. **75:** 1256–1263.
2. MEIER, P. J., E. S. SZTUL, A. REUBEN & J. L. BOYER. 1984. J. Cell Biol. **98:** 991–1000.
3. MUELLER, P., D. RUDIN, H. T. TIEN & W. C. WESTCOTT. 1962. Circulation **26:** 1167–1171.
4. MILLER, C. & E. RACKER. 1976. J. Membr. Biol. **30:** 283–300.
5. GOEGELEIN, H. 1988. Biochim. Biophys. Acta **947:** 521–547.
6. LATORRE, R. 1986. The large calcium-activated potassium channel. *In* Ion Channel Reconstitution. Miller, C., ed.   : 431–467. Plenum Press, New York & London.
7. MEIER, P. J., A. S. MEIER-ABT, C. BARRETT & J. L. BOYER. 1984. J. Biol. Chem. **259:** 10614–10622.
8. WEINMAN, S. A., J. GRAF & J. L. BOYER. 1989. Am. J. Physiol. **256:** G826–G832.

# Hypotonic Medium-Activated Chloride Transport in Human Skin Fibroblasts

MICHELA RUGOLO

*Istituto Botanico*
*Dip. di Biologia*
*40126 Bologna, Italy*

In hypotonic media, most mammalian cells initially swell as nearly perfect osmometers, but subsequently regulate their volume (regulatory volume decrease, RVD). The shrinkage is produced by net loss of KCl combined with the exit of osmotic water. The general prevalence of this phenomenon suggests that it has physiologic importance; however, the underlying molecular mechanisms may be different in different cell types. In fact, in most of the red cells, the RVD seems to involve activation of a KCl cotransport,[1-4] except for Amphiuma red cells in

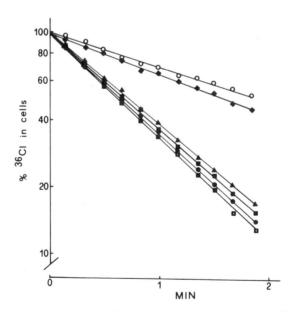

**FIGURE 1.** Effect of various inhibitors on $^{36}Cl^-$ efflux from human fibroblasts incubated in hypotonic medium. Fibroblasts were incubated in isotonic medium (300–310 mosmolal) containing 7 $\mu$Ci/ml of $^{36}Cl^-$ for 90 minutes in a waterbath at 37°C. Cells were washed and exposed to: (O) isotonic medium; (■) hypotonic medium (210 mosmolal); (▲) hypotonic medium containing Na gluconate instead of NaCl; (●) hypotonic medium plus 0.1 mM DIDS; (□) hypotonic medium plus 0.1 mM flurosemide; (♦) hypotonic medium plus 0.2 mM NPPB.

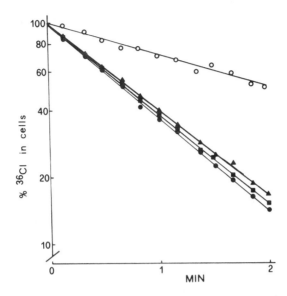

**FIGURE 2.** Effect of gramicidin D and quinine on $^{36}Cl^-$ efflux from human fibroblasts incubated in hypotonic medium. Experimental conditions as in FIGURE 1; (O) isotonic medium; (■) hypotonic medium; (▲) hypotonic medium plus 1 mM quinine; (●) hypotonic medium containing choline-Cl instead of NaCl, plus 1 $\mu$M gramicidin D.

which the KCl loss is achieved by a $K^+/H^+$ exchange, functionally coupled to $Cl^-/HCO_3^-$ exchange.[1,5] In human lymphocytes and Ehrlich ascites cells, activation of separate conductive $K^+$ and $Cl^-$ channels has been reported to occur during RVD.[6,7]

This report describes the effect of hypotonic medium on chloride transport in human skin fibroblasts. The semilogarithmic plot of FIGURE 1 shows that fibroblasts, preequilibrated with $^{36}Cl^-$, lose 50% of the isotope in 2 minutes and 0.6 minute in isotonic and hypotonic medium, respectively. The intracellular concentration of the isotope is reduced on cell swelling, and therefore the outward directed electrochemical gradient for $Cl^-$ is also reduced. It therefore suggests that the increased $Cl^-$ efflux reflects a volume-induced change in anion permeability. The substitution of $Cl^-$ in the hypotonic medium by the impermeant anion gluconate and the addition of furosemide did not significantly reduce the rate of $Cl^-$ efflux (FIG. 1). This finding argues against the involvement of $K^+/Cl^-$ cotransporter in RVD in fibroblasts. A simultaneous operativity of $K^+/H^+$ and $Cl^-/HCO_3^-$ exchanges seems unlikely, because $Cl^-$ efflux in hypotonic medium was unaffected by 4-4'-diisothiocyanatostilbene-2,2'-disulfonic acid (DIDS), as also shown in FIGURE 1. Further evidence for the involvement of a conductive pathway during RVD in human fibroblasts is provided in FIGURE 1, showing the inhibition observed in the presence of the $Cl^-$ channel blocker 5-nitro-2-(3-phenylpropylamine) benzoic acid (NPPB). FIGURE 2 shows that $^{36}Cl^-$ efflux in hypotonic medium is not affected by the addition of either the $K^+$ ionophore gramicidin D or the $K^+$ channel blocker quinine, indicating that the fluxes of $Cl^-$ and $K^+$ are not interdependent.

In conclusion, exposure of human fibroblasts to hypotonic medium stimulates a normally silent conductive pathway for $Cl^-$ transport, which operates independently of $K^+$ movements.

## REFERENCES

1. KREGENOW, F. M. 1981. Ann. Rev. Physiol. **43:** 493–505.
2. MC MANUS, T. J., M. HAAS, L. C. STARKE & C. Y. LYTLE. 1985. Ann. N. Y. Acad. Sci. **456:** 183–186.
3. DUNHAM, P. B. & J. C. ELLORY. 1981. J. Physiol. **318:** 511–530.
4. LAUF, P. K. 1985. J. Membr. Biol. **88:** 1–13.
5. CALA, P. M. 1985. Mol. Physiol. **8:** 199–214.
6. GRINSTEIN, S., A. ROTHSTEIN, B. SARKADI & E. W. GELFAND. 1984. Am. J. Physiol. **246:** C204–215.
7. HOFFMANN, E. K. 1986. Curr. Top. Membr. Transp. **30:** 125–180.

# The Na-K-Cl Cotransporter in the Kidney[a]

ROLF K. H. KINNE

*Max-Planck-Institut für Systemphysiologie*
*Rheinlanddamm 201*
*4600 Dortmund 1*
*Federal Republic of Germany*

Since its original description in the Ehrlich ascites tumor cells,[1] the concept of Na-K-Cl cotransport has revolutionized our understanding of transepithelial chloride transport and cell volume regulation. In this presentation some recent observations on Na-K-Cl cotransporters in the kidney and renal cells in culture will be reviewed. Three major items will thereby be touched upon: the occurrence of the transporter within the renal tubule and its potential physiologic role, the interaction of the transporter with its substrates and inhibitors, and finally the information currently available on its molecular properties.

## INTRARENAL DISTRIBUTION OF Na-K-Cl COTRANSPORTERS

Our current knowledge on the presence of Na-K-Cl cotransport in the kidney is summarized in FIGURE 1. The first tubular segments in which the existence of a Na-K-Cl cotransporter was demonstrated were the thick ascending limb of Henle's loop (TALH) in the rabbit and the corresponding diluting segment in Amphiuma. On the basis of results derived from studies of the intact epithelium, isolated TALH cells, and purified plasma membranes, a transport mechanism was identified in the luminal membrane of TALH cells that simultaneously translocates sodium, potassium, and chloride in a 1 : 1 : 2 stoichiometry.[2,3] In conjunction with a chloride channel and a KCl cotransporter in the contraluminal membrane this cotransporter mediates the active chloride reabsorption in this nephron segment.[3] The cotransporter displays in addition a high sensitivity towards "loop diuretics," and immobilization of the cotransporter by these compounds was identified as the mechanism of action of these potent diuretic drugs. Although the Na-K-Cl symport is apparently electrosilent,[4] its operation leads to the generation of a transepithelial electrical potential difference (lumen positive) that provides the driving force for the passive transport of magnesium, calcium, and potassium across the epithelium. The cotransporter also seems to be involved in transepithelial ammonium transport. Ammonium transport in the TALH is inhibited by furosemide, and ammonium can maintain active chloride transport in the complete absence of potassium.[5,6] The latter phenomenon suggests replacement of $NH_4^+$ for potassium at the potassium site of the cotransporter,[6] which also can be observed in isolated plasma membrane vesicles.[7] It is also noteworthy that in the intact animal, besides chloride, certain other anions such as nitrate and bromide are transported in the thick ascending limb of Henle's loop in a loop diuretic-sensitive

[a] Dedicated to William A. Brodsky, generous supporter and intellectual catalyst.

manner.[8] This relatively broad spectrum of the transport specificity is related to the specific properties of the transporter, which will be discussed in detail.

Finally, Na-K-Cl cotransport activity has also been observed in cell lines derived from rabbit or mouse thick ascending limb.[9-11] In transfected TALH cells isolated from rabbit outer medulla, expression of this transport system occurs at the luminal surface and can be maintained over an extended number of cell passages.[12]

In 1985 Sands et al.[13] discovered another renal site where sodium, potassium, and chloride transport is tightly coupled, the luminal membrane of the papillary surface epithelium. This epithelium is in contact with the pelvic space at its apical surface and with the interstitial space of the papillary (or inner medullary) tissue at its basolateral side. In these studies ouabain-induced cell swelling required the simultaneous presence of sodium, chloride, and potassium at the luminal surface, strongly suggesting a coupling of fluxes of these ions. In addition, bumetanide inhibited the swelling; interestingly already at $10^{-9}$ M the transport was completely inhibited. The papillary surface epithelium has been postulated to play an

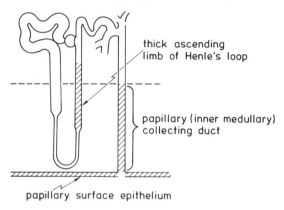

thick ascending limb of Henle's loop

papillary (inner medullary) collecting duct

papillary surface epithelium

**FIGURE 1.** Schematic representation of the renal epithelia in which Na-K-Cl cotransport has been described thus far.

important role in the urinary concentration mechanism; its transport properties and thus the role of the Na-K-Cl cotransporter in transepithelial transport remains to be determined.

Recently, Na-K-Cl cotransport was described in a third renal epithelium, the papillary collecting duct of the rat.[14] In collecting duct cells isolated from rat papilla, lactate production was found to be partly dependent on the presence of chloride, and this chloride-dependent lactate production was inhibited by bumetanide (FIG. 2). Furthermore, bumetanide showed no effect in the absence of sodium or the presence of ouabain. These findings suggested a coupling of sodium and chloride movement across the plasma membrane. Ion flux studies employing electron probe X-ray microanalysis corroborated this conclusion in demonstrating that chloride uptake was partly sodium dependent and bumetanide inhibitable (FIG. 3). It could be shown, in addition, that these chloride fluxes required the presence of potassium in the incubation medium. In addition, in the same experiments chloride and sodium seem to move in a 2 : 1 stoichiometry. Thus a coupled uptake of sodium, potassium, and chloride into the collecting duct cells had to be

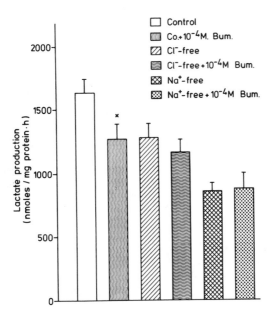

**FIGURE 2.** Effect of chloride, sodium, and bumetanide on lactate production in isolated rat inner medullary collecting duct cells. Data are taken from ref. 14.

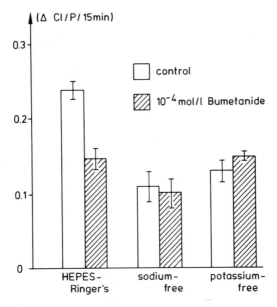

**FIGURE 3.** Chloride uptake into isolated rat inner medullary collecting duct cells. Uptake was determined by electron probe X-ray microanalysis. Data are taken from ref. 14.

postulated. Unfortunately in this cell preparation the location of this transport system within the plasma membrane could not be determined; similarly, the available data on ion transport in the papillary collecting duct do not allow this transporter to be assigned with certainty a role in transepithelial ion movements.

Na-K-Cl cotransport activity has also been demonstrated in two renal cell lines whose intrarenal origin is not defined. The proximal tubule-like LLC-PK₁ cell line possesses the cotransporter in the luminal cell membrane;[15,16] the distal tubule-like MDCK cell shows cotransport activity in its contraluminal cell membrane.[17] These two cell lines have contributed considerably to our understanding of the events taking place during the interaction of the cotransporter with the transported ions and inhibitors such as the loop diuretics.[18,19] Because mutants of the MDCK cell line are available which show a reduced transport activity and a similarly decreased piretanide binding capacity,[20] such cell lines might become important tools for the purification of the Na-K-Cl cotransporter.

## PROPERTIES OF THE RENAL Na-K-Cl COTRANSPORTERS

TABLE 1 lists the properties of the various cotransporters observed in the kidney and in the two renal cell lines, LLC-PK₁ and MDCK cells. The cotransporter in the latter cell line and in the luminal membrane of the thick ascending limb of Henle's loop has been studied most extensively. The affinity of the cotransporter to sodium is rather high in all systems investigated; apparent affinities between 0.5 and 9 mM have been observed.[2,3] There is positive cooperation between the sodium site, potassium site, and anion binding sites. Lithium is the only cation found thus far to interact with the sodium transport site too; it is also translocated by the carrier. There is also agreement that only one sodium site is present on the cotransporter.[22]

With regard to potassium, similarly only one site is postulated, again with a rather high affinity. The apparent $K_m$ derived from transport studies in the intact cell or in vesicles is between 0.3 and 9 mM. The specificity of the potassium site is broader than that of the sodium site. Besides rubidium, also $NH_4^+$ and in amphibia simple organic amines, such as choline, can bind to the transport system.[22] Ammonium competitively inhibits bumetanide-sensitive rubidium uptake into plasma membrane vesicles isolated from rabbit kidney outer medulla, and an $NH_4Cl$ gradient is capable of driving bumetanide-dependent sodium uptake in these vesicles.[7] Thus the Na-K-Cl cotransporter can also operate in a Na-NH₄-Cl mode, and secondary active transport of ammonium can be achieved. This uptake at the luminal side of TALH cells represents the first step in the chain of events comprising active $NH_4^+$ transport across this epithelium.[6] Furthermore, the affinity of the transporter for $NH_4^+$ lies in the concentration range found in the TALH lumen *in vivo,* supporting the role of the Na-K-Cl cotransporter in $NH_4^+$ reabsorption.

The interaction of the cotransporter with anions is more complex. Various studies have provided evidence that there are two binding sites at the transporter with different properties (TABLE 1). One site is highly specific; it accepts only bromide and chloride and shows also a high affinity for these anions. The apparent $K_m$ of this site is between 1 and 5 mM Cl. The second anion binding site is only operative when the first binding site is occupied by chloride or bromide. Then bromide, chloride, iodide, nitrate, and thiocyanate can interact with the second anion binding site.[23] At least for nitrate it has been demonstrated that this anion is also translocated by the cotransporter; thus NaCl-KNO₃ cotransport seems to be

**TABLE 1.** Properties of the Na-K-Cl Cotransporter in Various Renal Cells

| | Rabbit TALH[3,21,22,23] | Rat IMCD[14] | Rabbit PSE[13] | LLC-PK$_1$ Cells[15] | MDCK Cells[17] |
|---|---|---|---|---|---|
| Sodium binding site | | | | | |
| Affinity | 1.8–3.5 mM | ND | ND | 0.5 mM | 9 mM |
| Specificity | Na > Li ≫ NH$_4$ | ND | ND | ND | Na > Li |
| Potassium binding site | | | | | |
| Affinity | 0.3 mM | ND | ND | ND | 9 mM |
| Specificity | K = Rb > NH$_4$ > Cs | ND | ND | ND | K = Rb > NH$_4$ > Cs |
| Chloride binding site 1 | | | | | |
| Affinity | ~1.0 mM | ND | ND | ~5.1 mM | ND |
| Specificity | Br = Cl ≫ J = NO$_3$ = SCN | ND | ND | ND | Br = Cl |
| Chloride binding site 2 | | | | | |
| Affinity | ~15 mM | ND | ND | ~55.2 mM | ~48 mM |
| Specificity | Br > Cl > J = NO$_3$ = SCN | ND | ND | ND | ND |
| Interaction with loop diuretics | | | | | |
| Affinity (bumetanide) | ~$10^{-6}$ M | ~$10^{-8}$ M | ~$10^{-10}$ M | ~$10^{-6}$ M | ND |
| Specificity | bumetanide > piretanide > furosemide | | | | |

ABBREVIATIONS: IMCD = inner medullary collecting duct; ND = not done; PSE = papillary surface epithelium; TALH = thick ascending limb of Henle's loop.

possible. Accordingly, studies on the stoichiometry of the transporter yield a Hill coefficient of 2, when the chloride dependence of the cotransport is investigated in the presence of gluconate, but only of 1 when nitrate is present simultaneously in high concentrations.[15] Under physiologic conditions the chloride concentration in the lumen of the TALH, at the collecting duct and the papillary epithelium, is always high enough to saturate the high affinity-high specificity chloride site. Thus, as already stated, *in vivo* a relatively broad substrate specificity and low overall chloride affinity are observed.[3,8] Saturation of the first anion binding site is also required for the high affinity interaction of loop diuretics with the transport system (to be described).

With regard to the interaction of the Na-K-Cl cotransporter with loop diuretics, Schlatter *et al.*,[24] in an extensive study in the isolated perfused cortical thick ascending limb of rabbit kidney, determined the correlation between the structure

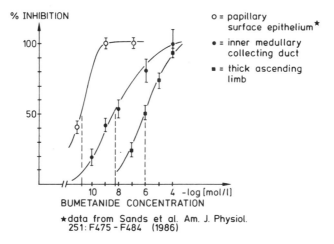

**FIGURE 4.** Bumetanide sensitivity of Na-K-Cl cotransport in various renal epithelia. Data are compiled from refs. 13, 14, and 42.

of the diuretics and their inhibitory potency on the active transepithelial chloride transport.

Their studies corroborated firstly that in this epithelium, as in other cells,[25,26] the sequence of potency of the best known diuretics—furosemide, bumetanide, and piretanide—is that bumetanide has the highest affinity to the Na-K-Cl cotransporter, followed by piretanide and furosemide. This sequence is typical for the Na-K-Cl cotransporter and distinguishes this system from other chloride transport systems that are loop diuretic-sensitive, such as Cl/OH exchanger, and KCl cotransport as well as from the transport systems that in the liver (and probably also the kidney) are involved in cellular uptake and secretion of these diuretics.[27]

As depicted in FIGURE 4, the absolute affinity of the Na-K-Cl cotransporter to the diuretics varies greatly, however. The system in the papillary surface epithelium shows the highest affinity, followed by the collecting duct, the thick ascending limb, and the renal cell cultures. From the studies of Schlatter *et al.*[24] it also could be delineated that all substitutes at the aromatic ring affect the potency; however, all potent diuretics have in position $C_1$ an acidic group or a small, strongly negatively charged group. The group at $C_1$ is probably the one that interacts with one of the chloride binding sites—most likely site 2—of the transporter.[28] Further interactions between the diuretics and the transporter are rather complex and currently difficult to define further. Studies using renal plasma membranes revealed binding sites for bumetanide and piretanide.[29-31] The interaction between some of the binding sites and the diuretics required sodium, potassium, and chloride.[29] The concentrations of sodium and potassium needed for maximal binding were similar to the affinities found in the transport studies.

Bumetanide binding exhibited a very peculiar behavior with regard to changes in chloride concentration.[29] Low concentrations of chloride stimulated binding, whereas higher concentrations inhibited it. This phenomenon is currently explained by the assumption that the high affinity-high specificity anion binding site has to be occupied by a chloride ion in order to allow interaction of the loop diuretic with the second anion binding site and the rest of the carrier. Thus a Na-K-Cl-diuretic complex is formed that is no longer capable of ion translocation.[28]

Determination of the number of binding sites and the turnover rate of the carrier has been difficult in the kidney, because no parallel studies on transport activity and number of binding sites are available as there are for the avian erythrocyte. In cell culture, turnover rates up to 500 $K^+$ ions/site per second have been determined.[20] However, the validity of these data is not clear hitherto, mainly because of the difficulties in correlating bumetanide binding sites with Na-K-Cl cotransport sites.[32]

## MOLECULAR BASIS OF Na-K-Cl COTRANSPORT

Attempts to identify the protein(s) that mediate the Na-K-Cl cotransport in the kidney or in other cells have employed mainly two strategies. One comprises the attempt to photolabel the transporter by using bumetanide or bumetanide derivatives. The other approach includes affinity chromatography on azidobumetanide or aminopiretanide columns. The results published so far from such investigations are compiled in TABLE 2. Haas and Forbush[33,34] consistently observed in dog kidney membranes labeling of a high affinity bumetanide binding site with an apparent molecular mass of 150 kDa and of a low affinity binding site with an apparent molecular mass of 50 kDa. The labeling of the former protein was strongly reduced in the absence of sodium, potassium, and chloride; the labeling of the low affinity binding site was affected by the removal of sodium and, to a small extent, by the lack of chloride.

A protein with a molecular mass in the 76-kDa range was identified by Feit *et al.*[35] as a putative transport protein in Ehrlich ascites cells based on its retention by an azidobumetanide affinity column. Under nondenaturing separation conditions the protein retained by the column showed a molecular mass of about 153 kDa, suggesting that the transporter has an oligomeric structure. Since Haas and Forbush determined the molecular mass of the labeled proteins under relatively mild denaturing conditions, the oligomers might have remained intact in their experiments and, therefore, in dog kidney and Ehrlich ascites cells the same proteins might have been identified.

**TABLE 2.** Tentative Molecular Mass of the Na-K-Cl Cotransporter

| Author | Method of Identification | Material | Molecular Mass | Comments |
|---|---|---|---|---|
| Haas and Forbush[34] | Photolabeling with [³H]-BSTBA high affinity potassium, sodium, and chloride dependence | Dog kidney membranes | ~150 kDa | Nonreducing SDS-PAGE |
| Haas and Forbush[33] | Photolabeling with [³H]-bumetanide, high affinity | Dog kidney membranes | ~150 kDa | |
| Feit et al.[35] | Affinity chromatography on bumetanide column | Ehrlich ascites cell plasma membranes | ~135 kDa | Cholate gel |
| Kinne et al. (unpublished observation) | Radiation inactivation | Rabbit kidney outer medulla microsomes | ~76 kDa ~83 kDa | Reducing SDS-PAGE |
| Deutscher et al. (unpublished observations) | Photolabeling with [³H]-bumetanide, chloride dependence | Shark rectal gland plasma membranes | ~42 kDa | Reducing SDS-PAGE |
| | Affinity chromatography on piretanide column, chloride-dependent labeling of isolated protein with [³H]-bumetanide | Shark rectal gland plasma membranes | ~42 kDa | Reducing SDS-PAGE |
| Haas and Forbush[34] | Photolabeling with [³H]-BSTBA low affinity, sodium dependence of labeling | Dog kidney membranes | ~50 kDa | Nonreducing SDS-PAGE |
| Jørgensen et al.[43] | Photolabeling with [³H]-bumetanide, competition with cold bumetanide | Rabbit kidney membranes | ~34 kDa | Reducing SDS-PAGE |

ABBREVIATIONS: BSTBA = 4-benzoyl-5-sulfamoyl-3-(3-thenyloxy)benzoic acid; SDS-PAGE = sodium dodecylsulfate polyacrylamide gel electrophoresis.

A molecular mass of 83 kDa has been attributed to the Na-K-Cl cotransporter in rabbit kidney outer medulla based on target size analysis employing radiation inactivation by accelerated electrons. In these experiments plasma membrane vesicles were isolated from rabbit kidney outer medulla which show chloride-dependent and bumetanide-inhibitable sodium transport.[23] These vesicles were rapidly frozen and at $-110°C$ were exposed to various doses of radiation. After rethawing, initial sodium uptake in the presence of a KCl gradient and a potassium-gluconate gradient was compared. As shown in FIGURE 5, a chloride-dependent sodium uptake decreased linearly with increasing radiation dose. From this correlation the dose required for inactivation of the transport system to 37% of its initial activity ($D_{37}$) was determined. In three experiments $D_{37}$ was reached at 18.3 Mrad (R. Kinne *et al.*, unpublished observations). From this value an apparent target size of 83 kDa was calculated.[36] Simultaneous measurements of sodium uptake in the absence of chloride and of the effect of irradiation on chloride permeability revealed no changes up to 10 Mrad. Therefore, the decrease in chloride-dependent transport was most likely due to inactivation of a transport system and was not related to unspecific changes in membrane permeability or changes in driving forces. The molecular mass obtained by irradiating intact membranes thus represents the minimum molecular weight required for chloride-dependent sodium translocation probably via the Na-K-Cl cotransporter and is close to the molecular mass determined for the transporter in Ehrlich ascites cells.

In TABLE 2 also, results using the shark rectal gland are included which were obtained in our Institute by Dr. Deutscher. In contraluminal membranes from this organ, which have a high density of transporters, low concentrations of [³H]bumetanide label a protein that under reducing conditions in SDS-PAGE has an apparent molecular mass of 42 kDa. Labeling in the intact membrane is reduced when chloride is replaced by gluconate and when the chloride concentration is increased above 200 mM. This dependence of labeling on the chloride concentration is similar to the one observed for binding of the diuretic to the transporter. A protein with a similar molecular mass is retained on an aminopiretanide affinity column and can be eluted by low concentrations of bumetanide. This protein, after isolation, can still be labeled with [³H]bumetanide and again exhibits first a stimulation and then a reduction in reactivity when the chloride concentration in the incubation medium is increased (J. Deutscher *et al.*, unpublished observa-

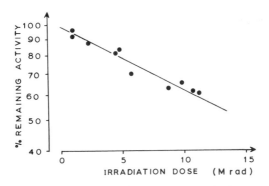

**FIGURE 5.** Radiation inactivation of rabbit renal outer medulla Na-K-Cl cotransporter. One typical experiment is shown.

tions). These properties of the protein strongly suggest that it is (part of) the Na-K-Cl cotransporter. The lower molecular weight in the rather primitive shark compared to much more evolved mammals might be due to several factors. One might be a lower degree of glycosylation, because in mammals certain lectins can inhibit the Na-K-Cl cotransporter.[37] It is also conceivable that the luminally localized cotransporter in mammals, which at least in some species also appears to be regulated by cAMP in higher vertebrates,[38,39] contains, in addition to the amino acid sequences required for ion translocation, amino acid sequences that code for the correct sorting and are targets of cAMP-dependent phosphorylation.

## CONCLUDING REMARKS

The results reported herein demonstrate that Na-K-Cl cotransport is not only confined to the cells of the thick ascending limb of Henle's loop but is also present in other renal cells and renal cell cultures. It occurs not only in luminal but also in contraluminal membranes and is involved in active transepithelial chloride transport and/or volume regulation of renal cells. The molecular nature of the transporter is far from being defined. Tools are now available, however, that should make it possible to characterize the protein(s) involved in ion cotransport. Knowledge about the composition and arrangement of the transporter in the membrane should then enable us to explain the wealth of information obtained in kinetic studies on coupling of the ion fluxes (or partial uncoupling[40,41]) and the strong cooperativity exerted by the different binding sites of the protein.

## REFERENCES

1. GECK, P., C. PIETRZYK, B. C. BURCKHARDT, B. PFEIFFER & E. HEINZ. 1980. Electrically silent cotransport of Na+, K+ and Cl− in Ehrlich cells. Biochim. Biophys. Acta **600:** 432–447.
2. KÖNIG, B., S. RICAPITO & R. KINNE. 1983. Chloride transport in the thick ascending limb of Henle's loop: Potassium dependence and stoichiometry of the NaCl cotransport system in plasma membrane vesicles. Pflügers Arch. **399:** 173–179.
3. GREGER, R. 1985. Ion transport mechanisms in thick ascending limb of Henle's loop of mammalian nephron. Physiol. Rev. **65:** 760–797.
4. HANNAFIN, J. A. & R. KINNE. 1985. Active chloride transport in rabbit thick ascending limb of Henle's loop and elasmobranch rectal gland: Chloride fluxes in isolated plasma membranes. J. Comp. Physiol. B **155:** 415–421.
5. GOOD, D. W., M. A. KNEPPER & M. B. BURG. 1984. Ammonia and bicarbonate transport by thick ascending limb of kidney. Am. J. Physiol. **247:** F35–F44.
6. GARVIN, J. L., M. B. BURG & M. A. KNEPPER. 1988. Active NH4+ absorption by the thick ascending limb. Am. J. Physiol. **255:** F57–F65.
7. KINNE, R., E. KINNE-SAFFRAN, H. SCHÜTZ & B. SCHOLERMANN. 1986. Ammonium transport in medullary thick ascending limb of rabbit kidney: Involvement of the Na+, K+, Cl−-cotransporter. J. Membr. Biol. **94:** 279–284.
8. KIIL, F., O. M. SEJERSTED & P. A. STEEN. 1980. Energetics and specificity of transcellular NaCl transport in the dog kidney. Int. J. Biochem. **12:** 245–250.
9. BURG, M., N. GREEN, S. SOHRABY, R. STEELE & J. HANDLER. 1982. Differentiated function in cultured epithelia derived from thick ascending limbs. Am. J. Physiol. **242:** C229–C233.
10. VALENTICH, J. D. & M. F. STOKOLS. 1986. An established cell line from mouse kidney medullary thick ascending limb. I. Cell culture techniques, morphology, and antigenic expression. Am. J. Physiol. **251:** C299–C311.

11. VALENTICH, J. D. & M. F. STOKOLS. 1986. An established cell line from mouse kidney medullary thick ascending limb. II. Transepithelial electrophysiology. Am. J. Physiol. **251:** C312–C322.
12. SCOTT, D. M., C. MACDONALD, H. BRZESKI & R. KINNE. 1986. Maintenance of expression of differentiated function of kidney cells following transformation by SV40 early region DNA. Exp. Cell Res. **166:** 391–398.
13. SANDS, J. M., M. A. KNEPPER & K. R. SPRING. 1986. Na-K-Cl cotransport in apical membrane of rabbit renal papillary surface epithelium. Am. J. Physiol. **251:** F475–F484.
14. GRUPP, C., I. PAVENSTADT-GRUPP, R. W. GRUNEWALD, J. B. STOKES III & R. K. H. KINNE. 1989. A Na-K-Cl cotransporter in isolated rat papillary collecting duct cells. Kidney Int. **36:** 201–209.
15. BROWN, C.D. A. & H. MURER. 1985. Characterization of a Na : K : 2Cl cotransport system in the apical membrane of a renal epithelial cell line (LCC-PK$_1$). J. Membr. Biol. **87:** 131–139.
16. AMSLER, K. & R. KINNE. 1986. Photoinactivation of sodium-potassium-chloride cotransport in LLC-PK$_1$/Cl 4 cells by bumetanide. Am. J. Physiol. **250:** C799–C806.
17. SAIER, M. H., JR., & D. A. BOYDEN. 1984. Mechanism, regulation and physiological significance of the loop diuretic-sensitive NaCl/KCl symport system in animal cells. Mol. Cell. Biochem. **59:** 11–32.
18. RINDLER, M. J., J. A. MCROBERTS & M. H. SAIER, JR. 1983. (Na$^+$,K$^+$)-cotransport in the Madin-Darby canine kidney cell line. Kinetic characterization of the interaction between Na$^+$ and K$^+$. J. Biol. Chem. **257:** 2254–2259.
19. MCROBERTS, J. A., S. ERLINGER, M. J. RINDLER & M. H. SAIER, JR. 1982. Furosemide-sensitive salt transport in the Madin-Darby canine kidney cell line. Evidence for the cotransport of Na$^+$, K$^+$ and Cl$^-$. J. Biol. Chem. **257:** 2260–2266.
20. GIESEN-CROUSE, E. M. & J.A. MCROBERTS. 1987. Coordinate expression of piretanide receptors and Na$^+$, K$^+$, Cl$^-$ cotransport activity in Madin-Darby canine kidney cell mutants. J. Biol. Chem. **262:** 17393–17397.
21. KINNE, R. K. H. 1988. Sodium cotransport systems in epithelial secretion. Comp. Biochem. Physiol. **90A:** 721–726.
22. KINNE, R., J. A. HANNAFIN & B. KONIG. 1985. Role of NaCl-KCl cotransport system in active chloride absorption and secretion. Ann. N.Y. Acad. Sci. **456:** 198–206.
23. KINNE, R., E. KINNE-SAFFRAN, B. SCHOLERMANN & H. SCHUTZ. 1986. The anion specificity of the sodium-potassium-chloride cotransporter in rabbit kidney outer medulla: Studies on medullary plasma membranes. Pflügers Arch. **407:** S168–S173.
24. SCHLATTER, E., R. GREGER & C. WEIDTKE. 1983. Effect of "high ceiling" diuretics on active salt transport in the cortical thick ascending limb of Henle's loop of rabbit kidney. Correlation of chemical structure and inhibitory potency. Pflügers Arch. **396:** 210–217.
25. O'GRADY, S. M., H. C. PALFREY & M. FIELD. 1987. Characteristics and functions of Na-K-Cl cotransport in epithelial tissues. Am. J. Physiol. **253:** C177–C192.
26. GECK, P. & E. HEINZ. 1986. The Na-K-2Cl cotransport system. J. Membr. Biol. **91:** 97–105.
27. PETZINGER, E., N. MULLER, W. FOLLMANN, J. DEUTSCHER & R. K. H. KINNE. 1989. Uptake of bumetanide into isolated rat hepatocytes and primary liver cell cultures. Am. J. Physiol. **256:** G78–G86.
28. HAAS, M. & T. J. MCMANUS. 1983. Bumetanide inhibits (Na + K + 2 Cl) co-transport at a chloride site. Am. J. Physiol. **245:** C235–C240.
29. FORBUSH, B., III, & H. C. PALFREY. 1983. [$^3$H]bumetanide binding to membranes isolated from dog kidney outer medulla. Relationship to the Na, K, Cl cotransport system. J. Biol. Chem. **258:** 11787–11792.
30. GIESEN-CROUSE, E. M., C. WELSCH, J. L. IMBS, M. SCHMIDT & J. SCHWARTZ. 1985. Characterization of a high affinity piretanide receptor on kidney membranes. Eur. J. Pharmacol. **114:** 23–31.
31. GIESEN-CROUSE, E. & J. L. IMBS. 1987. *In* Diuretics. II. Chemistry, Pharmacology, and Clinical Applications. J. B. Puschett & A. Greenberg, eds., : 182–187. Elsevier/North Holland Biomedical Press, Amsterdam.

32. GRIFFITHS, N. M. & N. L. SIMMONS. 1987. Attribution of [³H]bumetanide binding to the Na + K + Cl 'cotransporter' in rabbit renal cortical plasma membranes: A caveat. Quart. J. Exp. Physiol. **72:** 313–329.

33. HAAS, M. & B. FORBUSH, III. 1987. Na, K, Cl-cotransport system: Characterization by bumetanide binding and photolabelling. Kidney Int. **32:** S134–S140.

34. HAAS, M. & B. FORBUSH, III. 1987. Photolabeling of a 150-kDa (Na + K + Cl) cotransport protein from dog kidney with a bumetanide analogue. Am. J. Physiol. **253:** C243–C250.

35. FEIT, P. W., E. K. HOFFMANN, M. SCHIØDT, P. KRISTENSEN, F. JESSEN & P. B. DUNHAM. 1988. Purification of proteins of the Na/Cl cotransporter from membranes of Ehrlich ascites cells using a bumetanide-sepharose affinity column. J. Membr. Biol. **103:** 135–147.

36. BEAUREGARD, G. & M. POTIER. 1985. Temperature dependence of the radiation inactivation of proteins. Analyt. Biochem. **150:** 117–120.

37. HEIDENREICH, O., J. GREVEN & K. HEINTZE. 1982. Molecular actions of diuretics. Klin. Wochsenschr. **60:** 1258–1263.

38. HEBERT, S. C. & T. E. ANDREOLI. 1986. Control of NaCl transport in the thick ascending limb. Am. J. Physiol. **246:** F745–F766.

39. SILVA, P., B. KONIG, S. LEAR, J. EVELOFF & R. KINNE. 1987. Dibutyryl cyclic AMP inhibits transport dependent QO₂ in cells isolated from the rabbit medullary ascending limb. Pflügers Arch. **409:** 74–80.

40. ALVO, M., J. CALAMIA & J. EVELOFF. 1985. Lack of potassium effect on Na-Cl cotransport in the medullary thick ascending limb. Am. J. Physiol. **249:** F34–F39.

41. EVELOFF, J. L. & J. CALAMIA. 1986. Effect of osmolarity on cation fluxes in medullary thick ascending limb cells. Am. J. Physiol. **250:** F176–F180.

42. BAYERDORFFER, E. 1981. Gekoppelter NaCl-Transport im dicken aufsteigenden Teil der Henle'schen Schleife. Dissertation. Frankfurt am Main.

43. JØRGENSEN, P. L., J. PETERSEN & W. D. REES. Identification of a Na⁺, K⁺, Cl⁻-cotransporter protein of M_r 34,000 from kidney by photolabeling with [³H]bumetanide. The protein is associated with cytoskeleton components. Biochim. Biophys. Acta **775:** 105–110.

# Structure and Function of the Band 3 $Cl^-/HCO_3^-$ Transporter[a]

REINHART A. F. REITHMEIER, DEBRA M. LIEBERMAN,
JOSEPH R. CASEY, SANJAY W. PIMPLIKAR, PAMELA K.
WERNER, HILARIO SEE, AND CHARLES A. PIRRAGLIA

*MRC Group in Membrane Biology*
*Departments of Medicine and Biochemistry*
*University of Toronto*
*Toronto, Ontario*
*Canada M5S 1A8*

The Band 3 glycoprotein of the erythrocyte membrane is responsible for the electroneutral exchange of anions such as bicarbonate across the plasma membrane.[1,2] Band 3 makes up about 25% by weight of the protein in the erythrocyte membrane, and 10–20 mg of the protein can be purified from 1 unit (425 ml) of human blood using detergent solutions and standard chromatographic techniques.[3,4] With this amount of protein in hand it is possible to carry out a variety of protein-chemical studies on Band 3 in detergent solutions or after reconstitution into lipid vesicles. This paper briefly summarizes four aspects of the molecular characterization of Band 3 based on work carried out in my laboratory. The structure of Band 3 has been examined using (1) site-directed antibodies, (2) circular dichroism, (3) affinity chromatography, and (4) protein crystallization. This work has led to a better description of the structure of Band 3, a necessary prerequisite to an understanding of its mechanism of action.

## SITE-DIRECTED ANTIBODIES

The complete amino acid sequence of murine Band 3 has been deduced from the nucleotide sequence of cDNA transcribed from Band 3 mRNA isolated from the spleens of anemic mice.[5] Analyses of the murine sequence along with extensive studies of the homologous human protein have shown that Band 3 spans the membrane at least 8 and perhaps up to 14 times.[1,2,5] The amino terminus of Band 3 is exposed to the cytoplasm of the red cell, and a number of protease-sensitive sites and chemical labeling sites have been localized to one side of the membrane or the other (FIG. 1). We have used an immunologic approach to determine the orientation of the carboxyl-terminus of murine Band 3 with respect to the membrane.[6]

Polyclonal antibodies were raised in rabbits against a synthetic peptide conjugated to keyhole limpet hemocyanin which corresponds to the 12 amino acid carboxyl-terminal sequence of murine erythrocyte Band 3. The antibody reacted well with mouse or rat Band 3 but not with human or dog Band 3 on Western blots. The antibody also bound to murine ghost membranes applied directly to

[a] This research was supported by grants from the Medical Research Council of Canada and the National Institutes of Health.

**FIGURE 1.** Model for the folding of the membrane domain of murine Band 3 in the erythrocyte membrane. Homology with the known sequence of human Band 3 is indicated in boldface, whereas residues that are different are given in brackets. The folding arrangement is based on analysis of the sequence for prediction of hydropathy profiles and β-turn probabilities and the known protease-sensitive sites and chemical labeling sites. Abbreviations: Ac = acetylated amino terminus; Ch = chymotrypsin cleavage site; Tr = trypsin cleavage site; Pa = papain cleavage site; Pe = pepsin cleavage site; subscript i denotes in, whereas subscript o denotes out; *I = iodination site; me = reductive methylation site; *BS³ = BSSS cross-link site; Lys 581 and 558 are potential DIDS-reactive residues.

nitrocellulose but not to human ghosts, suggesting that mouse and human carboxyl Band 3 sequences are not identical. The antibody is specific for the carboxyl terminus of Band 3, because antibody binding is lost upon carboxypeptidase Y digestion of solubilized Band 3. The antibody also binds to carboxyl-terminal proteolytic fragments of Band 3. Chymotrypsin treatment of intact cells generates a 35,000-dalton fragment that extends from $Met_{577}$ in the mouse sequence to the carboxyl terminus of the protein. A 22,000-dalton fragment that can be generated by trypsin or chymotrypsin treatment of ghost membranes extends from $Ala_{762}$ to the carboxyl terminus (FIG. 1). Interestingly, vigorous chymotrypsin digestion of alkali-stripped membranes generates a 9,000-dalton cysteine-containing subfragment of the 35,000 chymotryptic fragment that does not bind the antibody

(FIG. 2). This proteolytic fragment, therefore, does not contain an intact carboxyl-terminal sequence and shows that chymotrypsin treatment of stripped ghosts can remove peptide material from the carboxyl-terminal region of Band 3. No such cleavage occurs with chymotrypsin treatment of unstripped ghost membranes. This finding suggests that extrinsic membrane proteins that are removed by the alkali stripping procedure protect the carboxyl-terminal region of Band 3 from digestion or that the alkaline stripping procedure exposes the carboxyl-terminal region of Band 3 to digestion.

The carboxyl terminus of Band 3 was localized to the cytoplasmic side of the erythrocyte membrane, because antibody binding as determined by immunofluorescence occurred in ghosts and permeabilized cells but not in intact red cells (FIG. 3). This finding was confirmed using competitive ELISA assays. Cells and resealed ghosts competed poorly for antibody binding to immobilized peptide-conjugate compared to ghost membranes or inside-out vesicles.

Localization of the carboxyl terminus to the cytoplasmic side of the erythrocyte membrane leads to the conclusion that Band 3 spans the erythrocyte membrane an even number of times. This limits the possible number of transmembrane segments to an even number between 8 and 14. It is clear that site-directed antibodies are a useful approach to determining directly the exposure of epitopes of membrane proteins to one side of a membrane or the other.

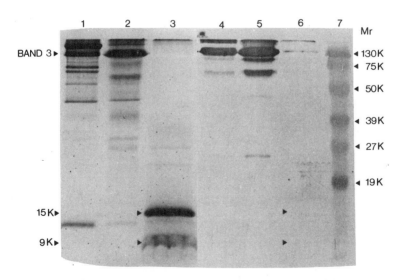

**FIGURE 2.** Lack of carboxyl-terminal–directed antibody binding to a 9,000-dalton chymotryptic fragment of mouse Band 3. *Lanes 1–3:* Coomassie blue stain of mouse ghost proteins. *Lane 1,* mouse ghost proteins; *lane 2,* mouse ghosts were stripped with 2 mM EDTA, pH 12; *lane 3,* alkali-stripped ghosts were digested with 1.5 mg/ml chymotrypsin for 1 hour at 37°C. *Lanes 4–6:* Corresponding immunoblot incubated with a 1/1,000 dilution of anticarboxyl-terminal peptide antiserum. The antibody-antigen complex was detected using goat anti-rabbit IgG conjugated to alkaline phosphatase and visualized with nitroblue tetrazolium. *Lane 7:* Prestained molecular weight standards.

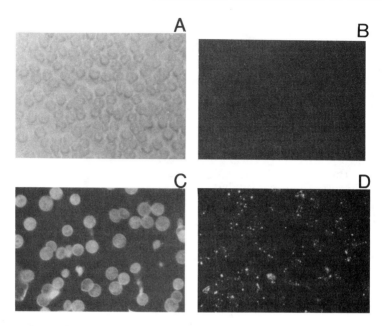

**FIGURE 3.** Immunofluorescence localization of the carboxyl terminus of Band 3 to the cytoplasmic side of the erythrocyte membrane. **(A/B)** Intact erythrocytes in 150 mM NaCl, 5 mM sodium phosphate, pH 7.4, were incubated with rabbit anticarboxyl-terminal peptide for 30 minutes at room temperature, and the complex was detected with fluorescein isothiocyanate conjugated to goat anti-rabbit IgG and viewed by light microscopy **(A)** and fluorescence microscopy **(B)** of the same cells. **(C)** Erythrocytes permeabilized with 100% ice-cold methanol, incubated as in **A/B** and visualized as in **B**. **(D)** Ghosts prepared from mouse erythrocytes were treated as in **A/B** and visualized as in **B**.

## CIRCULAR DICHROISM

Circular dichroism (CD) is a useful approach to determine the conformation of proteins. We have examined the conformation of Band 3 and its major proteolytic fragments using this technique. The Band 3 molecule in detergent solution (0.1% octaethylene glycol mono-n-dodecyl ether, $C_{12}E_8$) contains 46% $\alpha$-helix and 37% $\beta$-sheet, the remainder being random coil.[7] Interestingly, the membrane-associated domain of Band 3 is enriched in $\alpha$-helix (58%) relative to the helical content of the cytoplasmic domain (27%). Extensive digestion of ghost membranes with proteolytic enzymes and removal of extrinsic fragments left a membrane residue that was highly enriched (86–94%) in $\alpha$-helical peptide fragments. The portion of Band 3 buried in the membrane and protected from proteolytic degradation, therefore, has a very high helical content.

The studies described herein employed Band 3 in detergent solution. It is of considerable interest to examine the conformation of purified Band 3 in a lipid environment. In collaboration with Dr. Chan Jung (SUNY, Buffalo) we obtained CD spectra of human erythrocyte Band 3 reconstituted into phosphatidylcholine vesicles. FIGURE 4 shows CD spectra of reconstituted Band 3 and the effect of solubilization of the vesicles with $C_{12}E_8$. A fit of the data revealed that reconstituted Band 3 had a helical content of 35% and a $\beta$-sheet content of 31%. Addition of detergent increased the helical content slightly (38%). We will be examining the effects of substrates and inhibitors on the conformation of Band 3 in this system. The reincorporation of Band 3 into a lipid bilayer will also allow us to examine directly the effect of the lipid environment on the conformation of Band 3. This is of interest because the lipid composition of the erythrocyte membrane can influence anion transport.

## AFFINITY CHROMATOGRAPHY

Anion transport in erythrocytes is inhibited by stilbene disulfonates[1] due to the binding of these compounds to a single site in the Band 3 molecule that faces the cell exterior. We have synthesized an inhibitor affinity matrix by reacting 4-acetamido-4'-isothiocyanostilbene-2,2'-disulfonate (SITS) with Affi-Gel 102

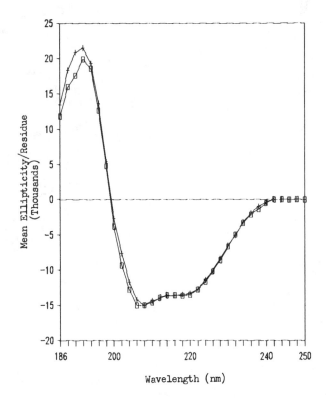

**FIGURE 4.** Far-ultraviolet circular dichroism spectra of purified Band 3 reconstituted into egg phosphatidylcholine vesicles after removal of $C_{12}E_8$ by absorption to SM2 Biobeads. o = reconstituted protein (0.12 mg/ml) in 1 mM sodium phosphate, pH 8.0; $x$ = same sample solubilized with 0.1% $C_{12}E_8$.

$$\Big\rangle - O - CH_2 - \overset{\overset{\displaystyle O}{\|}}{C} - NH - (CH_2)_2 - NH - \overset{\overset{\displaystyle S}{\|}}{C} - NH - \text{\phantom{}} - \overset{H \quad SO_3^-}{\underset{SO_3^- \quad H}{C = C}} - \text{\phantom{}} - NH - \overset{\overset{\displaystyle O}{\|}}{C} - CH_3$$

**FIGURE 5.** Structure of SITS-Affi-Gel 102 affinity matrix.

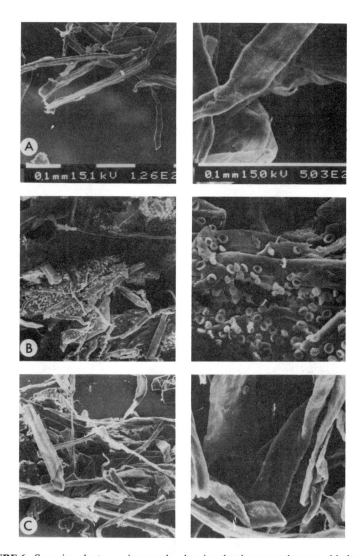

**FIGURE 6.** Scanning electron micrographs showing that intact erythrocytes bind to SITS-aminoethyl-cellulose. After removal of unbound cells samples were fixed with glutaraldehyde, dehydrated with ethanol, critical point dried, and examined under a scanning electron microscope. *Panel A,* cellulose fibers incubated with erythrocytes; *panel B,* SITS-cellulose incubated with erythrocytes; and *panel C,* SITS-cellulose incubated with DIDS-labeled erythrocytes.

(FIG. 5). Intact red blood cells bind to this matrix and to SITS-derivatized aminoethyl cellulose fibers (FIG. 6), confirming that the stilbene disulfonate binding site is accessible to immobilized ligand in intact erythrocytes. Pretreatment of cells with 4,4'-diisothiocyanostilbene-2,2'-disulfonate (DIDS) prevents immobilization of the erythrocytes on the matrix. This matrix is capable of binding solubilized Band 3, and the Band 3 can be eluted by free inhibitors.[8] The binding occurred in two stages.[8,9] Initially, Band 3 bound to the matrix with low affinity and the protein could be eluted with 2 mM 4-benzamido-4'-aminostilbene-2,2'-disulfonate (BADS). Band 3 bound to the matrix is converted in a temperature- and time-dependent manner to a tightly bound form that is no longer readily elutable by free inhibitor. This change may be due to a conformational change in Band 3 that results in tighter association of the Band 3 with the immobilized inhibitor.

**TABLE 1.** Effect of Ligand Density on Band 3 Binding to SITS-Affinity Resin

| Ratio of Ligand/Spacer[a] in Reaction Mix | Relative Amount[b] of Band 3 Bound | % Band 3 Eluted by BADS[c] | |
|---|---|---|---|
| | | 0°C | 37°C |
| 2 | 1.00 | 53% | 18% |
| 1 | 0.70 | 71% | 39% |
| 0.5 | 0.45 | 79% | 64% |
| 0.25 | 0.25 | 77% | 76% |
| 0.1 | 0.28 | 71% | 76% |
| 0 | 0.34 | ... | ... |

[a] Affi-Gel 102 contains 15 μM spacer arm per milliliter of resin. SITS was added at the listed molar ratios and allowed to react at 37°C for 1 hour. Free spacer arms were blocked with a twofold excess of 3-sulfophenylisothiocyanate. Binding was performed using 50 μl resin and 1 ml of Band 3 at 1 mg/ml as described previously.[8]

[b] The relative amount of Band 3 bound to the matrix was determined by boiling the washed resin with an equal volume of Laemmli sample buffer and resolving the eluted protein on sodium dodecyl sulfate gels, staining with Coomassie blue, and scanning the Band 3 peak.

[c] After binding the Band 3 to the resin at 0°C, the resin was washed with buffer and incubated on ice or at 37°C for 15 minutes. Band 3 was eluted at 0°C with 2 volumes of 2 mM BADS in 5 mM sodium phosphate, and tightly bound protein was eluted by boiling in 2 volumes Laemmli sample buffer.

Band 3 has a primarily dimeric structure in the membrane and in nonionic detergent solutions. It is possible that the weak binding is due to binding of Band 3 to the matrix via one subunit, whereas the tight binding is due to both subunits of a dimer binding to the resin. We found that Band 3 dimers with one site filled with DIDS can be bound to the matrix via the unoccupied subunit but that Band 3 could not be converted to the tightly bound form.[9] To examine this further a series of matrices were prepared that had a decreasing ligand concentration. As the ligand density decreased, it was found that less Band 3 was bound and that the lower the ligand density the more readily Band 3 was eluted by the free inhibitor. Significantly, the ability of Band 3 to be converted to the high affinity form decreased as the ligand density decreased (TABLE 1). We propose that Band 3 binds initially through a single subunit to the matrix. This subunit can undergo a temperature-dependent conformational change which allows the second subunit to attach to the matrix if the ligand density is high enough. Band 3 with DIDS labeled in one subunit can bind to the matrix in the low affinity form. It cannot be converted to the high affinity form either because the one site is blocked with DIDS or because the occupation of one site with DIDS prevents the conformational change in the partner subunit.

We have used this affinity matrix to identify and purify stilbene disulfonate binding proteins from kidney.[10] A 130,000-dalton protein in the brush border membrane of canine kidney cortex was found to bind specifically to this matrix and to be readily eluted by 2 mM BADS. Binding of the protein is inhibited by the free inhibitor BADS (1 mM), by reaction of the protein with 50 $\mu$M DIDS and

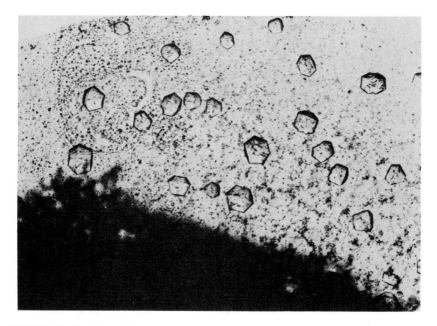

**FIGURE 7.** Crystals of human erythrocyte Band 3. Crystals were grown at 4°C by the hanging drop method in $C_{12}E_8$ using ammonium sulfate as the precipitant. The longest dimension of the crystals is 0.1 mm. The dark area is due to precipitated protein.

interestingly by 100 mM sodium bicarbonate. This protein can also be covalently labeled with [$^3$H]-H$_2$-DIDS, and the labeling is prevented by 1 mM free BADS. Further characterization of this protein is underway.[11] It is a glycoprotein containing high mannose and complex N-linked sugar chains, and it can be deglycosylated to a 110,000-dalton protein. The 130,000-dalton protein is contained in the brush border membrane fraction and not in the basolateral membrane fraction. It is also present in a membrane fraction prepared from the LLC-PK$_1$ kidney epithelial cell line. Amino-terminal sequence analysis and enzymatic studies have shown that this stilbene disulfonate-binding protein is aminopeptidase N, rather than an anion transporter.

## CRYSTALLIZATION

To understand the mechanism of anion transport at the molecular level a high resolution structure must be obtained. We have been involved with Wayne F. Anderson (Vanderbilt University) in attempts to produce crystals of human erythrocyte Band 3 that are suitable for X-ray diffraction studies. Some preliminary results of crystallization trials have produced microcrystals of Band 3. Band 3 was purified in the presence of $C_{12}E_8$ and enzymatically deglycosylated. FIGURE 7 is a photograph of crystals obtained by the hanging drop method using ammonium sulfate as the precipitant. This result is encouraging, and we are now employing other detergents, small amphiphiles, other precipitants, and buffer conditions to optimize crystal growth.

## REFERENCES

1. PASSOW, H. 1986. Molecular aspects of Band 3 protein-mediated anion transport across the red blood cell membrane. Rev. Physiol. Biochem. Pharmacol. **103:** 61–223.
2. JAY, D. & L. C. CANTLEY. 1986. Structural aspects of the red cell anion exchange protein. Ann. Rev. Biochem. **55:** 511–538.
3. LUKACOVICS, M. F., M. B. FEINSTEIN, R. I. SHA'AFI & S. PERRIE. 1981. Purification of stabilized Band 3 protein of the human erythrocyte membrane and its reconstitution into liposomes. Biochemistry **20:** 3145–3151.
4. LIEBERMAN, D. M. & R. A. F. REITHMEIER. 1983. Characterization of the stilbenedisulfonate binding site of the Band 3 polypeptide of human erythrocyte membranes. Biochemistry **22:** 4028–4033.
5. KOPITO, R. R. & H. F. LODISH. 1985. Primary structure and transmembrane orientation of the murine anion exchange protein. Nature **316:** 234–238.
6. LIEBERMAN, D. M. & R. A. F. REITHMEIER. 1988. Localization of the carboxyl terminus of Band 3 to the cytoplasmic side of the erythrocyte membrane using antibodies raised against a synthetic peptide. J. Biol. Chem. **263:** 10022–10028.
7. OIKAWA, K., D. M. LIEBERMAN, & R. A. F. REITHMEIER. 1985. Conformation and stability of the anion transport protein of human erythrocyte membranes. Biochemistry **24:** 2843–2848.
8. PIMPLIKAR, S. W. & R. A. F. REITHMEIER. 1986. Affinity chromatography of Band 3, the anion transport protein of erythrocyte membranes. J. Biol. Chem. **261:** 9770–9778.
9. PIMPLIKAR, S. W. & R. A. F. REITHMEIER. 1988. Studies on the interaction of matrix-bound inhibitor with Band 3, the anion transport protein of human erythrocyte membranes. Biochim. Biophys. Acta **942:** 253–261.
10. PIMPLIKAR, S. W. & R. A. F. REITHMEIER. 1988. Identification, purification and characterization of a stilbenedisulfonate binding glycoprotein from canine kidney brush border membranes. J. Biol. Chem. **236:** 4485–4493.
11. SEE, H., J. H. M. CHARUK & R. A. F. REITHMEIER. 1988. Identification and purification of the major concanavalin A-binding protein of renal brush border membranes. American Society of Nephrology, 21st Annual Meeting Abstracts, p. 16A, Abstract 139.

# Characteristics of the Binding Site for Extracellular Substrate Anions in Human Red Blood Cell Band 3[a]

MICHAEL L. JENNINGS

*Department of Physiology and Biophysics*
*University of Texas Medical Branch*
*Galveston, Texas 77550*

The transport of $HCO_3^-$, $Cl^-$, and other inorganic anions across the red blood cell membrane is catalyzed by the major transmembrane protein known as band 3 or capnophorin. (See reference 1 for a comprehensive review of the literature through 1986.) The sequences of mouse[2] and human[3] band 3 are known; the 95-kDa polypeptide consists of a 43-kDa water-soluble N-terminal cytoplasmic domain and a 52-kDa hydrophobic membrane domain. The former is not necessary for anion transport;[1,4] its function is to serve as an attachment site for the membrane skeleton.[5] Although three-dimensional structural information is still lacking, it is possible to make some limited statements about the relation between structure and the transport function of the protein. The goal of this paper is to summarize our current understanding of the binding site for extracellular substrate anions.

It is well established that each band 3 polypeptide has a single binding site for the stilbenedisulfonate class of anion transport inhibitors.[1] The stilbenedisulfonates act as competitive inhibitors of anion exchange; moreover, they are inhibitory only from the extracellular solution, and they bind with highest affinity only to the outward-facing conformation of the protein.[1] It is believed, therefore, that the binding site for extracellular substrate anions is contained in the stilbenedisulfonate site. The stilbenedisulfonate site is not considered to be identical with the substrate site; transported anions are generally much smaller and are expected to occupy only a portion of the stilbenedisulfonate site.

The anion exchange catalyzed by band 3 consists mainly of a tightly coupled, electrically silent one-for-one exchange. Although the kinetics of the exchange are complex, especially at high substrate concentrations,[6,7] there is widespread (although not universal; see ref. 8) agreement that the exchange takes place by way of a "ping-pong" mechanism, in which the exchanging anions take turns crossing the membrane.[9–15] An intracellular anion binds to an inward-facing configuration of the protein, is transported outward by way of a conformational change in the protein, is released, and then an extracellular anion is bound and transported inward. For $Cl^-$, at least at low temperature, binding and release are much more rapid than is translocation.[15] For more slowly transported anions such as $SO_4^=$, translocation is presumably rate limiting at all temperatures.

---

[a] This work was supported by NIH Grant R01-GM2861.

## AFFINITY OF THE TRANSPORT SITE FOR CHLORIDE

Early experiments addressing the issue of substrate affinity in band 3 by Gunn et al.[16] showed that the $Cl^-$ self-exchange flux, with the $Cl^-$ concentration approximately the same on each side of the membrane, is half-maximal at about 30 mM. Subsequent measurements of $Cl^-$ self-exchange in red cells and resealed ghosts in several laboratories have produced similar values for the half-maximal $Cl^-$ concentration under symmetric conditions (e.g., ref. 17). The $K_{1/2}$ measured under these conditions represents a weighted average of the dissociation constants for binding to the inward-facing and outward-facing states.[9,18,19] Direct measurements with nuclear magnetic resonance detect a binding process with a dissociation constant 60–90 mM.[13] This value also represents a weighted average of inward- and outward-facing dissociation constants; the weighting factors depend on the asymmetry of the translocation process.[1,9,14,18,19] In the absence of other influences, the dissociation constant determined from magnetic resonance should equal the $K_{1/2}$ for $Cl^-$-$Cl^-$ exchange under symmetric conditions. The smaller $K_{1/2}$ for the flux measurement may reflect the effect of self-inhibition.[1,6] In any case, neither the NMR measurement nor the transport measurement gives an estimate of the true dissociation constant for binding of $Cl^-$ to either the inward-facing or the outward-facing state.

Gunn and Fröhlich[10] measured the $Cl^-$-$Cl^-$ exchange flux at fixed intracellular and varying extracellular $Cl^-$ concentration; the $K_{1/2}$ for extracellular $Cl^-$ is about 3 mM. This number, however, does not represent the dissociation constant for binding to outward-facing sites, because, as the extracellular $Cl^-$ concentration is raised, the proportion of outward-facing states decreases. Thus, saturation of the flux by extracellular $Cl^-$ reflects not only increased occupancy of outward-facing sites but also a decrease in the total number of such states as the extracellular $Cl^-$ concentration increases. The $K_{1/2}$ for external $Cl^-$ in this kind of experiment is always lower than the true dissociation constant for extracellular sites. It now appears[19] that the low $K_{1/2}$ for external $Cl^-$ at fixed internal $Cl^-$ is a consequence not of a high external affinity, but rather of an asymmetry in the translocation rate constants (inward rate constant higher than outward rate constant).

## ADVANTAGES OF STUDYING TRANSPORT UNDER VERY ASYMMETRIC CONDITIONS

The main point of the foregoing discussion is to emphasize that real substrate affinities are very difficult to determine from transport measurements or from NMR experiments, because the transport protein can exist in either inward-facing or outward-facing states. Therefore, even if translocation is rate limiting and if there is no interference from "spectator" ions, a series of transport experiments does not generally provide a reliable estimate of substrate binding affinity. There is one exception to this rather discouraging rule: situations in which either influx or efflux (but not both) is much slower than any other step in the catalytic cycle. In a $Cl^-$-$Cl^-$ exchange experiment, neither influx nor efflux by itself limits the overall turnover rate. However, during $Cl^-$ exchange for a much more slowly transported substrate, the turnover of the catalytic cycle is limited by the slow anion. For example, when $Cl^-$-loaded cells are suspended in a $Cl^-$-free medium that contains no penetrating anions except $SO_4^=$, the outward $Cl^-$ gradient will recruit essentially all the transporters into the outward-facing conformation, independent of the extracellular $SO_4^=$ concentration.[12] Under these conditions, the $K_{1/2}$ for extracellular $SO_4^=$ is a good measure of the true dissociation constant for binding (if other effects, such as spectator competition and surface potential, are taken into account).

Because of the simplifying effect of the outward $Cl^-$ gradient, we have used $Cl^-$-$SO_4^=$ exchange measurements to examine the characteristics of the outward-facing substrate anion binding site in band 3. The goal of the work is to determine which amino acid residues are important for substrate binding. We also have examined the question of whether the substrate binding site lies in a cleft that extends some distance across the membrane.

## $SO_4^=$ INFLUX

The net exchange of $Cl^-$ for $SO_4^=$ across the red cell membrane was first measured 60 years ago by Mond and Gerz.[20] More recently, work in our laboratory[12,21,22] and by Milanick and Gunn[23,24] has established some of the characteristics of the band 3-mediated exchange of $Cl^-$ for $SO_4^=$. The transport is almost entirely stilbenedisulfonate sensitive and, at neutral and acid extracellular pH, consists of an exchange of one $Cl^-$ ion for one $SO_4^=$ plus one $H^+$ ion,[21] as predicted in the original titratable carrier model of Gunn.[25] The cotransport of $H^+$ and $SO_4^=$ is a consequence of the fact that the form of the protein that transports $SO_4^=$ has an extra $H^+$ bound, most likely to a glutamate residue.[26] Milanick and Gunn,[23] in a study of the effect of extracellular $SO_4^=$ and pH on $Cl^-$-$Cl^-$ exchange at 0°C, found that $H^+$ and $SO_4^=$ can bind to the outward-facing state of the protein in either order, and that the binding of one ion increases the affinity of the other by a factor of about 10. In a subsequent study, the same authors[24] measured $SO_4^=$ influx into $Cl^-$-loaded cells as a function of extracellular pH, and found that the half-maximal $SO_4^=$ concentration was about 4–5 mM at all pH values less than 9. This result is somewhat surprising in light of the earlier work indicating that proton binding increases $SO_4^=$ affinity, although the latter work was not intended to be a detailed study of the effect of pH on $SO_4^=$ affinity.

## GLUTAMATE AS SPECTATOR ANION

The influx experiments of Milanick and Gunn[24] were carried out in a medium containing gluconate, which has been shown by Knauf and Mann[7] to be a weak inhibitor of red cell anion exchange ($K_I$ about 50 mM). Another commonly used substitute anion, citrate, is not a strong inhibitor of transport at neutral pH; however, it inhibits significantly at acid pH.[23,27] To try to find a substitute anion that interacts less with the transporter, we chose glutamate because it has two negative and one positive charge. The hope was that the positive charge prevents it from binding strongly to the anion binding site of band 3. Initial studies were conducted at an extracellular pH of 5, in order to protonate the titratable group involved in $H^+$-$SO_4^=$ cotransport. FIGURE 1 shows the time course of $SO_4^=$ influx at extracellular pH 5 into cells initially containing 100 mM $Cl^-$. The time course is linear for at least 30 seconds. Double reciprocal plots of the flux versus $SO_4^=$ concentration are linear over the concentration range used; the flux saturates with a $K_{1/2}$ of about 0.8 mM. This $K_{1/2}$, obtained in a glutamate medium, is considerably lower than that measured in a gluconate or sucrose-citrate medium (ref. 24; Jennings, data not shown). We feel that the higher $K_{1/2}$ observed in the gluconate or citrate media is a consequence of competition from these anions.

## TITRATABLE CARBOXYL GROUP NEAR THE EXTRACELLULAR SUBSTRATE SITE

The $K_{1/2}$ and $V_{max}$ for extracellular $SO_4^=$ were determined at extracellular pH 5.0, 6.0, and 7.0 (FIG. 2). The effects of pH on the maximum influx are identical to those found by Milanick and Gunn;[24] the pH dependence of the $V_{max}$ almost certainly reflects the protonation of a carboxyl group. We found that the $K_{1/2}$ for extracellular $SO_4^=$ is a decreasing function of pH, consistent with the idea that

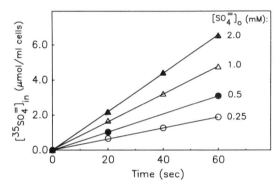

**FIGURE 1.** Time course of $^{35}SO_4$ influx at different extracellular $SO_4^=$ concentrations. Human red cells were washed three times in unbuffered 150 mM KCl, then once in 140 mM Na-glutamate, 28 mM glutamic acid, pH 5.0, to remove extracellular $Cl^-$. Because of the low extracellular pH, there is very little loss of intracellular $Cl^-$ during this wash. Immediately after the final wash, 100-$\mu l$ cells were suspended in 5 ml of the Na-glutamate/glutamic acid medium. The indicated concentration of [$^{35}S$] $Na_2SO_4$ (1 mCi/mmol) was then added at zero time. The time course of influx of tracer was measured at 20°C, as described previously.[12] The indicated $SO_4^=$ concentrations represent the total sulfate present, including the $\sim 10\%$ that exists as the ion pair $NaSO_4^-$.[23]

protonation of the carboxyl group associated with $H^+$-$SO_4^=$ cotransport increases the affinity of the transport site for extracellular $SO_4^=$ several-fold. (As already mentioned herein, an effect of extracellular pH in this range on the $K_{1/2}$ of $SO_4^=$ influx was not detected by Milanick and Gunn,[24] possibly because gluconate is a stronger competitor at low pH.)

## ROLE OF AMINO GROUPS

Amino groups are known to be associated with the transport pathway through band 3, but the function of these groups in the transport process is not clear. There is a reactive lysine residue denoted Lys $a$, which is believed to be either

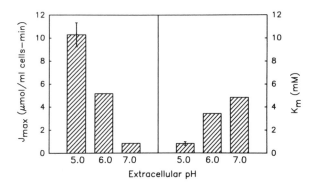

**FIGURE 2.** Extracellular pH dependence of $SO_4^=$ influx into Cl-loaded cells. Cells were prepared and the influx of $^{35}SO_4^=$ was determined at 20°C, as described for Figure 1. The maximal flux ($J_{max}$) and apparent dissociation constant ($K_m$) for extracellular $SO_4^=$ were determined from double reciprocal plots, which were linear over the concentration range used (0.25–8 mM $SO_4^=$). Because influx is rate limiting, the extracellular $SO_4^=$ concentration ($K_{1/2}$) that gives half-maximal flux is equal to the $K_m$. The flux media (excluding the $Na_2SO_4$) were the following: 140 mM Na-glutamate, 28 mM glutamic acid, pH 5.0; 125 mM Na-glutamate, 25 mM glutamic acid, 25 mM Bistris, pH 6.0; and 125 mM Na-glutamate, 25 mM glutamic acid, 25 mM Bistris, titrated with NaOH to pH 7.0. In all cases the intracellular pH was 7.2–7.3. The data at extracellular pH 5 represent the average of four separate experiments, each of which included four different extracellular $SO_4^=$ concentrations. The data at pH 6 and pH 7 are from single experiments with four different $SO_4^=$ concentrations.

Lys 539 or 542 (numbers from the human sequence in ref. 3). Lys *a* is readily dinitrophenylated, and the $pK_a$ of this lysine residue is apparently quite low.[1] The maximum rate of $Cl^-$ exchange is not affected by extracellular pH in the range in which this lysine residue is titratable,[28] indicating that the charge on this lysine does not detectably affect the maximum rate of $Cl^-$-$Cl^-$ exchange.

In addition to Lys *a,* there is at least one more lysine residue at the stilbenedisulfonate site.[29] This residue, Lys *b,* is in the C-terminal 150 residues[30] and is possibly identical to the lysine (Lys 851) that Kawano and coworkers[31] recently showed to be modified with pyridoxal phosphate in intact cells. Although the reaction of this lysine with PLP is correlated with inhibition of transport, the role of this residue in transport is unknown.

One problem with interpreting the effects of chemical modification of lysine residues on the function of the protein is that most of the modifications completely (or very nearly completely) inhibit transport. Thus, the functional characteristics of the modified protein cannot be studied. The modified residue cannot be inferred to be essential, because the chemical modification often introduces steric bulk and/or negative charge. We have applied methods for modifying lysine residues under conditions in which the transport is not completely inhibited. One such method uses the membrane-impermeant active ester cross-linker bis(sulfosuccinimidyl)suberate, or BSSS, which was developed by Staros.[32] BSSS reacts with lysine residues in intact cells to create an intramolecular cross-link between the same two papain fragments as are cross-linked by $H_2DIDS$.[33] Reaction of BSSS completely prevents $H_2DIDS$ covalent binding, as would be expected if the BSSS and $H_2DIDS$ react with the same amino groups. BSSS, at higher concentrations, also forms an intermolecular cross-link,[32] which does not detectably alter anion transport.[34]

Unlike $H_2DIDS$, BSSS does not introduce any new negative charge into the protein; instead, the amino group is converted into an uncharged amide attached to an aliphatic hydrocarbon. We found that in cells in which essentially all copies of band 3 are modified with BSSS, anion transport is not completely inhibited. However, the alkaline limb of the extracellular pH dependence of the maximal rate of $Cl^--Br^-$ exchange is shifted by about 5 units.[33] These results were interpreted as evidence that, although the amino groups themselves are not absolutely essential for anion translocation, one or both of them provide positive charge that strongly influences the transport rate. When the flux in the modified cells is carried out at extracellular pH 8, the transport is strongly inhibited because of the missing positive charge. In the same BSSS-pretreated cells, $Cl^--Br^-$ exchange is not detectably inhibited when the transport is measured at extracellular pH 6. It appears, then, that lowering the extracellular pH activates transport by replacing some of the positive charge that was removed by the BSSS treatment. Even though the maximal rate of $Cl^--Br^-$ exchange is essentially normal in the modified cells at extracellular pH 6, the $K_{1/2}$ for extracellular Br is several-fold higher than normal,[33] suggesting that the affinity for binding of extracellular substrate has been altered by the BSSS.

To investigate the role of the BSSS-reactive amino groups in extracellular substrate anion binding under conditions in which influx is clearly rate limiting, we measured the effect of the modification on net $Cl^--SO_4^=$ exchange. FIGURE 3 shows the extracellular pH dependence of the $SO_4^=$ influx into $Cl^-$-loaded cells (intracellular pH 7.2). In the modified cells, the extracellular pH dependence of the influx is even more striking than that in control cells, indicating that in order to

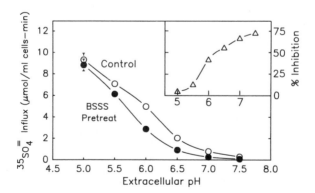

**FIGURE 3.** Extracellular pH dependence of $SO_4^=$ influx into Cl-loaded cells (intracellular pH 7.2). The cells were pretreated with or without 5 mM BSSS at pH 7.4, as described previously.[33] The extracellular media all contained 80 mM $Na_2SO_4$ and were buffered at the indicated pH with combinations of citrate, MOPS, bistris, and tris. At a $SO_4^=$ concentration of 80 mM, there is insignificant interference by the buffers. Each point represents the mean of four influx determinations (two each on two preparations of cells); the standard deviation was smaller than the size of the symbol except at pH 5. The inset shows the percentage of inhibition of the flux at each pH. The pH refers to that of the flux medium; BSSS pretreatment in all cases was at pH 7.4.

activate the influx in the modified cells, a second titration must take place. The need for the second titration is illustrated by plotting the flux in the modified cells relative to the control flux (FIG. 3). At neutral pH, the flux is inhibited by about 75%, but as the pH is lowered, the percentage of inhibition is much lower. At extracellular pH 5, the influx in the modified cells (at 80 mM extracellular $SO_4^=$) is essentially the same as in unmodified cells.

The results in FIGURE 3 are entirely as expected from our previous work with monovalent anion exchange in BSSS modified cells.[33] Even though two outward-facing amino groups are modified covalently in a way that removes their charge, the maximum influx at extracellular pH 5 is essentially normal. Therefore, the amino groups themselves are not absolutely necessary for the translocation event. However, as was found with monovalent anion exchange, the transport at neutral pH is inhibited by the modification, indicating that the positive charge contributed by one or both the lysines does have a role in transport, and that replacing some or all of this charge by lowering the pH restores the influx to the control level.

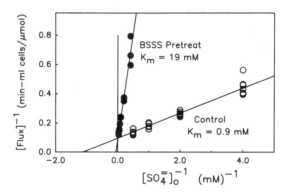

**FIGURE 4.** Double reciprocal plots of $SO_4^=$ influx into Cl-containing cells at a function of extracellular $SO_4^=$ concentration in 140 mM Na-glutamate, 28 mM glutamic acid, pH 5.0, 20°C. Influxes were carried out as in FIGURE 1, except that higher $SO_4^=$ concentrations were used for the BSSS-treated cells.

The main point of FIGURE 3 is that we have found conditions (extracellular pH 5, 80 mM $SO_4^=$) in which the $SO_4^=$ influx into Cl⁻ containing cells is essentially normal, despite the removal of two positive charges from the stilbenedisulfonate site. Under these conditions, then, it is possible to determine the effect of the amino group modification on the extracellular $SO_4^=$ affinity of the transport site. FIGURE 4 is a double reciprocal plot of $SO_4^=$ influx at extracellular pH 5 in control and BSSS-treated cells. The modification clearly has a large effect (over 20-fold) on the $K_{1/2}$ under conditions in which the maximum flux is at least as high as that in the control cells. We conclude that the positive charge on the reactive amino groups does influence extracellular substrate affinity, although we cannot rule out the possibility that a steric effect of the aliphatic chain also is introduced between the modified amino groups. Previous studies by Wieth et al.[28] and by Falke and Chan[14] had led to the conclusion that the essential positive charge at the stilbene-disulfonate site is provided by one or more arginine residues. The present work indicates that the charge on an amino group also influences substrate binding.

## EFFECT OF MEMBRANE POTENTIAL ON SUBSTRATE
## BINDING AND TRANSLOCATION

There are several indications that the transport pathway through band 3 is largely hydrophilic but interrupted by a thin permeability barrier. The translocation event takes place when the barrier moves relative to the anion. The movement of barrier relative to anion does not require the anion itself to move; instead, a gate could open in front of a bound anion and another gate could close behind the anion. Alternatively, the anion, complexed with protein-bound positive charge, could jump across a rate-limiting barrier. The alternative to the thin barrier model is the "molecular zipper" model, in which the transport pathway is mainly hydrophobic but contains an array of paired charges.[35] The transport of an ion, in this model, takes place as a succession of exchanges of a transported anion with a protein-bound negative charge. The evidence in favor of a thin barrier model includes distance measurements between the stilbenedisulfonate site and the cytoplasmic side of the membrane;[36] carboxyl group chemical modification studies indicating that a single carboxyl group span the permeability barrier;[26] nuclear magnetic resonance studies showing that some inhibitors of transport seem to act as "channel blockers" that prevent access between the anion binding site and the extracellular medium;[37] and evidence that anion conductance through band 3 involves the tunneling of the anion through the permeability barrier.[38,39]

If the outward-facing transport site does in fact lie in a hydrophilic cleft that extends some distance from the geometric outer surface of the bilayer, then it would be expected that the anion experiences some of the transmembrane potential as it passes from the bulk medium through the cleft toward the binding site. Accordingly, the local anion concentration should be affected by the membrane potential. In intact red cells the membrane potential can be altered with gramicidin or valinomycin, with no detectable effect on either monovalent anion exchange or the exchange of $Cl^-$ for $SO_4^=$.[19,24,38,40] Grygorczyk et al.[41] recently showed that membrane voltage does affect the $^{36}Cl$ efflux from mouse band 3 expressed in Xenopus oocytes. The effects of voltage indicate that about $-0.1$ elementary charge moves during the rate-limiting translocation event. The effects of the membrane potential on band 3-mediated $Cl^-$ tracer fluxes in the Xenopus expression system contrast with the lack of observable effects in the native red cell.[19,24,40]

To determine the effect of the membrane potential on extracellular substrate binding, conditions must be found in which anion influx is rate limiting and in which the membrane potential can be clamped with ionophore. These two conditions are somewhat difficult to satisfy. For example, it is well known that gramicidin induces a cation conductance that is much larger than the anion conductance of the membrane.[38] However, at an extracellular pH of 5, at which the $SO_4^=$ influx is optimal, we have found a very significant $H^+$ conductance, which severely compromises the range of membrane potentials available. The $H^+$ conductance can be demonstrated by measuring gramicidin-mediated $^{86}Rb^+$ efflux from $SO_4^=$-loaded cells into pH 5 media containing N-methyl glucamine as the only cation (Jennings, data not shown). The rate of $^{86}Rb^+$ efflux vastly exceeds the rate of net $SO_4^=$ efflux, suggesting that there is a significant $H^+$ influx. It is not yet clear whether this influx is catalyzed by gramicidin, but there is certainly precedent for a gramicidin-mediated $H^+$ conductance.[42]

After trying various conditions of extracellular pH and temperature with both valinomycin and gramicidin, we found that at extracellular pH 6, the $H^+$ permeability mediated by valinomycin is sufficiently low that the $K^+$ conductance exceeds that of all other ions. To help ensure that the ionophore really does have a major influence on membrane potential, the $^{35}SO_4^=$ influx experiments were conducted on cells loaded with nonradioactive $SO_4^=$. Sulfate and atmospheric $HCO_3^-$ were the only intracellular permeant anions. The $SO_4^=$ conductance is considerably smaller than that of $Cl^-$,[43] and the replacement of the intracellular $Cl^-$ with $SO_4^=$ makes the $K^+$ conductance in the presence of valinomycin much higher than that of the anions present.

The goal of these experiments is to measure the effect of membrane potential on $^{35}SO_4^=$ influx under conditions in which influx is the rate-limiting step in the catalytic cycle. If the intracellular $Cl^-$ has been replaced by $SO_4^=$, how can we be sure that influx is still rate limiting? This question has two answers, one theoretical and one experimental. The conditions were chosen so that the intracellular pH is 1.3 units higher than the extracellular pH. The pH gradient implies that there is an outward $HCO_3^-$ gradient of about 20-fold. Even though the intracellular $HCO_3^-$ concentration is much less than 1 mM, $HCO_3^-$ is transported so much more rapidly than $SO_4^=$ that the outward $HCO_3^-$ gradient should recruit most of the transporters into an outward-facing state. In keeping with this theoretical argument, we find experimentally that the $K_{1/2}$ and $V_{max}$ for $SO_4^=$ influx from a $Cl^-$-free medium are not significantly different between $SO_4^=$-loaded and $Cl^-$-loaded cells, *as long as the intracellular pH is much higher than the extracellular pH.* Therefore, even though the main intracellular anion is $SO_4^=$, the conditions of the influx experiments come close to satisfying the ideal that the influx branch of the catalytic cycle is rate limiting.

FIGURE 5 shows the initial influx of $^{35}SO_4^=$ as a function of extracellular $SO_4^=$ concentration in a Na-glutamate medium, pH 6.0, in the presence and absence of 2 $\mu$M valinomycin. The valinomycin does not change the maximum influx; if anything, the maximum influx is actually increased by valinomycin, although the effect is quite small. The major effect of valinomycin is on the half-maximal concentration of $SO_4^=$, which increases from 3.8 mM in the absence of valinomycin to 6.6 mM in its presence. If the membrane potential in the presence and absence of ionophore were known, it would be possible to calculate the fraction of the transmembrane potential difference that acts on the $SO_4^=$ anion as it moves from the bulk solution to the outward-facing substrate binding site. The following two paragraphs describe our attempts to estimate the membrane potential.

To estimate the membrane potential in the Na-glutamate medium, pH 6, in the presence of valinomycin, the efflux of $^{86}Rb$ was measured and compared with the efflux from the same cells suspended in a high-$K^+$ medium. In cells treated identically to those used for the $SO_4^=$ influx experiments (but preloaded with $^{86}Rb$), the valinomycin-mediated $^{86}Rb$ efflux into a K-glutamate medium, pH 6, was 5–8 times more rapid than that into a Na-glutamate medium, pH 6 (two experiments, not shown). If the $^{86}Rb$ flux obeys the constant field equation, the membrane potential is calculated to be about $-70$ mV in the Na-glutamate medium. Evidence exists that such estimates of potential can be misleading, because the loaded form of valinomycin is transported more rapidly than the empty form.[44] However, even if the Rb permeability coefficient is overestimated by a factor of two in the high-K medium, the membrane potential in the Na-glutamate medium is still calculated to be about $-50$ mV.

In the absence of ionophore, the membrane potential is harder to estimate. We measured the $^{86}Rb$ efflux under the conditions of the $SO_4^=$ influx experiments ($SO_4^=$-loaded cells, intracellular pH 7.2, extracellular pH 6, Na-glutamate medium) and found it to be quite slow; the rate constant is about 0.025 per hour. This rate constant corresponds to a permeability coefficient of about $3 \times 10^{-10}$ cm/s.

The $SO_4^=$ conductive permeability coefficient, estimated from the increase in net $SO_4^=$ efflux induced by valinomycin under the same conditions, is only about twice as high as the Rb permeability coefficient. Therefore, the outward $SO_4^=$ gradient should make the membrane potential slightly positive in the absence of ionophore. If the potential is $+10$ mV in the absence of valinomycin and $-50$ mV in its presence, then a change of 60 mV causes a shift in the $K_m$ for external $SO_4^=$ by a factor of about 1.7. If $SO_4^=$ is assumed to enter the transport pathway unpaired with any cation, and if glutamate is assumed not to compete significantly for the transport site, then it may be calculated that $SO_4^=$ moves through about 11% of the transmembrane electric field in going from the external bulk solution to the outward-facing transport site. If glutamate does compete with $SO_4^=$, 11% is an under-

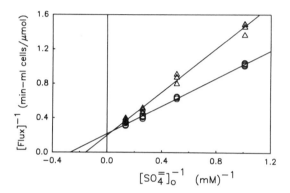

**FIGURE 5.** Effect of valinomycin-mediated change in membrane potential on $SO_4^=$ influx. Cells were loaded with $SO_4^=$ by washing three times at a 20:1 dilution in 110 mM $K_2SO_4$, with a 5-minute incubation at 37°C before each centrifugation to allow the $SO_4^=$ to enter the cells. After a final wash in $SO_4^=$-free Na-glutamate, pH 5, 0.2 ml of a 50% suspension of cells was resuspended in 5 ml of 125 mM Na-glutamate, 25 mM glutamic acid, 25 mM Bistris, pH 6.0. Valinomycin (2 $\mu$M final; 0.2% ethanol final) was then added, and the influx of $SO_4$ was initiated by adding 0.5–4 mM [$^{35}$S] $Na_2SO_4$. The $K_{1/2}$ for the control fluxes in 0.2% ethanol (*circles*) is about 1.7 times lower than that for the fluxes in the presence of valinomycin (*triangles*). Separate controls showed that the net $SO_4^=$ efflux induced by the valinomycin does not contribute significantly to the total extracellular $SO_4^=$ concentration in the medium.

estimate, because the negative potential should help drive out the monovalent competitor (although not as strongly as it drives out the divalent substrate).

The foregoing calculations of membrane potential are subject to numerous uncertainties, and the amount of membrane potential drop between external medium and extracellular binding site is similarly uncertain. We are reasonably sure, however, that the effect of membrane potential on the $K_{1/2}$ for extracellular $SO_4^=$ is real; it was observed in three experiments in addition to the one shown in FIGURE 5.

These studies provide further support for the idea that the extracellular anion binding site in band 3 lies some distance from the outer surface of the membrane. If the ion feels 11% of the transmembrane potential on the way to the outward-facing binding site, the site itself would be expected to lie much more than 11% of the way across the membrane, because the resistance of the access channel is low compared with that of the rate-limiting barrier. In the absence of any independent estimates of the geometry of the mouth of the channel, it is impossible to draw further inferences about the real distance between the outside solution and the binding site.

It is of interest that there is no detectable effect of membrane potential on the maximum $SO_4^=$ influx at low extracellular pH. This finding is in agreement with those of Milanick and Gunn[24] and indicates that the rate-limiting step in $SO_4^=$ influx involves no net charge movement in the electric field across the membrane. Therefore, if the $SO_4^=$ actually is moving during the rate-limiting step, then two positive charges must move with it. Alternatively, the $SO_4^=$ could be stationary during the rate-limiting step; if so, whatever moves in the protein to give the $SO_4^=$ access to the intracellular solution must not carry net charge perpendicular to the membrane.

## CONCLUSIONS

We have provided the following new information about the extracellular binding site for $SO_4^=$ in human red cell band 3.

1. At low pH the site has higher affinity than was previously believed: the dissociation constant is about 0.8 mM. The higher affinity is revealed by using glutamate rather than citrate or gluconate as a spectator anion.

2. The affinity of $SO_4^=$, as measured in net anion exchange experiments, is increased by lowering the pH; the simplest interpretation of this effect is that it is an electrostatic consequence of the neutralization of the carboxyl group associated with $H^+$-$SO_4^=$ cotransport.

3. Removal of the charge on two amino groups at the stilbenedisulfonate site by chemical modification with BSSS does not have a major effect on the maximum rate of $SO_4^=$ influx at low extracellular pH. However, the lysine modification lowers the affinity for extracellular $SO_4^=$ binding by a factor of at least 20.

4. Variations in the membrane potential do not detectably affect the maximum rate of $SO_4^=$ influx, but the affinity of extracellular $SO_4^=$ for the outward-facing transport site is sensitive to potential. It is estimated that $SO_4^=$, in moving from the bulk solution to the binding site for extracellular anions, experiences about 11% of the transmembrane potential change. This finding supports the notion that there is an access channel between the outward-facing substrate binding site and the bulk solution.

## ACKNOWLEDGMENTS

The experiments in FIGURE 4 were performed by Mini Locke; the remaining experiments were performed with the assistance of Richard Schulz. I thank Drs. Otto Fröhlich and Malcolm Brodwick for helpful discussions.

## REFERENCES

1. PASSOW, H. 1986. Rev. Physiol. Biochem. Pharmacol. **103:** 61–203.
2. KOPITO, R. R. & H. F. LODISH. 1985. Nature **316:** 234–238.

3. TANNER, M. J. A., P. G. MARTIN & S. HIGH. 1988. Biochem. J. **256:** 703–712.
4. LEPKE, S. & H. PASSOW. 1976. Biochim. Biophys. Acta **455:** 353–370.
5. BENNETT, V. & P. J. STENBUCK. 1980. J. Biol. Chem. **255:** 6424–6432.
6. DALMARK, M. 1976. J. Gen. Physiol. **67:** 223–234.
7. KNAUF, P. A. & N. A. MANN. 1986. Am. J. Physiol. **251** (Cell Physiol. **20**): C1–C9.
8. SALHANY, J. M. & P. B. RAUENBUEHLER. 1983. J. Biol. Chem. **258:** 245–249.
9. FROHLICH, O. & R. B. GUNN. 1986. Biochim. Biophys. Acta **864:** 169–194.
10. GUNN, R. B. & O. FROHLICH. 1979. J. Gen. Physiol. **74:** 351–374.
11. FURUYA, W., T. TARSHIS, F.-Y. LAW & P. A. KNAUF. 1984. J. Gen. Physiol. **83:** 657–681.
12. JENNINGS, M. L. 1982. J. Gen. Physiol. **79:** 169–185.
13. FALKE, J. J., R. J. PACE & S. I. CHAN. 1984. J. Biol. Chem. **259:** 6481–6491.
14. FALKE, J. J. & S. I. CHAN. 1985. J. Biol. Chem. **260:** 9537–9544.
15. FALKE, J. J., K. J. KANES & S. I. CHAN. 1985. J. Biol. Chem. **260:** 9545–9551.
16. GUNN, R. B., M. DALMARK, D. C. TOSTESON & J. O. WEITH. 1973. J. Gen. Physiol. **61:** 185–206.
17. HAUTMANN, M. & K. F. SCHNELL. 1985. Pflügers Arch. **405:** 193–201.
18. FROHLICH, O. 1982. J. Membr. Biol. **65:** 111–123.
19. KNAUF, P. A. 1989. *In* The Red Cell Membrane: A Model for Solute Transport. B. U. Raess & G. Tunnicliff, eds. Humana Press, in press.
20. MOND, R. & H. GERZ. 1929. Pflügers Arch. **221:** 623–632.
21. JENNINGS, M. L. 1976. J. Membr. Biol. **28:** 187–205.
22. JENNINGS, M. L. 1980. *In* Membrane Transport in Erythrocytes. Lassen, U. V., H. H. Ussing & J. O. Wieth, eds.:450–463. Munksgaard, Copenhagen.
23. MILANICK, M. A. & R. B. GUNN. 1982. J. Gen. Physiol. **79:** 87–113.
24. MILANICK, M. A. & R. B. GUNN. 1984. Am. J. Physiol. **247** (Cell Physiol. **16**): C247–259.
25. GUNN, R. B. 1972. *In* Oxygen Affinity of Hemoglobin & Red Cell Acid-Base Status. M. Rorth & P. Astrup, eds.:823–827. Munksgaard, Copenhagen.
26. JENNINGS, M. L. & S. AL-RHAIYEL. 1988. J. Gen. Physiol. **92:** 161–178.
27. KAUFMANN, E., G. EBERL & K. F. SCHNELL. 1986. J. Membr. Biol. **91:** 129–146.
28. WIETH, J. O., O. S. ANDERSEN, J. BRAHM, P. J. BJERRUM & C. L. BORDERS, JR. 1982. Phil. Trans. R. Soc. Lond. B **299:** 383–399.
29. JENNINGS, M. L. & H. PASSOW. 1979. Biochim. Biophys. Acta **554:** 498–519.
30. JENNINGS, M. L., M. P. ANDERSON & R. MONAGHAN. 1986. J. Biol. Chem. **261:** 9002–9010.
31. KAWANO, Y., K. OKUBO, F. TOKANAGA, T. MIYATA, S. IWANAGA & N. HAMASAKI, N. 1988. J. Biol. Chem. **263:** 8232–8238.
32. STAROS, J. V. 1982. Biochemistry **23:** 3950–3955.
33. JENNINGS, M. L., R. MONAGHAN, J. S. NICKNISH & S. M. DOUGLAS. 1985. J. Gen. Physiol. **86:** 653–669.
34. JENNINGS, M. L. & J. S. NICKNISH. 1985. J. Biol. Chem. **260:** 5472–5479.
35. BROCK, C., M. J. A. TANNER & C. KEMPF. 1983. Biochem. J. **213:** 577–586.
36. RAO, A., P. MARTIN, R. A. F. REITHMEIER & L. C. CANTLEY. 1979. Biochemistry **18:** 4505–4516.
37. FALKE, J. J. & S. I. CHAN. 1986. Biochemistry **25:** 7895–7898.
38. FROHLICH, O. 1984. J. Gen. Physiol. **84:** 877–893.
39. KNAUF, P. A. F.-Y. LAW & P. J. MARCHANT. 1983. J. Gen. Physiol. **81:** 95–126.
40. WIETH, J. O., J. BRAHM & J. FUNDER. 1980. Ann. N.Y. Acad. Sci. **341:** 394–418.
41. GRYGORCZYK, R., W. SCHWARTZ & H. PASSOW. 1987. J. Membr. Biol. **99:** 127–136.
42. DECKER, E. R. & D. G. LEVITT. 1988. Biophys. J. **53:** 25–32.
43. KNAUF, P. A., G. F. FUHRMANN, S. ROTHSTEIN & A. ROTHSTEIN. 1977. J. Gen. Physiol. **69:** 363–386.
44. BENNEKOU, P. & P. CHRISTOPHERSEN. 1986. J. Membr. Biol. **93:** 221–227.

# Heterogeneity of Anion Exchangers Mediating Chloride Transport in the Proximal Tubule[a]

PETER S. ARONSON AND SHIU-MING KUO

*Section of Nephrology*
*Departments of Medicine,*
*and Cellular and Molecular Physiology*
*Yale University School of Medicine*
*New Haven, Connecticut 06510*

A significant fraction of proximal tubule $Cl^-$ transport occurs via an active, transcellular process.[1,2] Intracellular $Cl^-$ activity in the proximal tubule cell is above the level at which $Cl^-$ would be in electrochemical equilibrium across the luminal membrane. Taken together, these findings indicate that one or more processes capable of mediating uphill $Cl^-$ uptake across the luminal membrane of the proximal tubule cell must be present.

## CHLORIDE-FORMATE EXCHANGER

Using microvillus membrane vesicles isolated from the rabbit renal cortex as an experimental system, we attempted to identify those processes that might account for uphill $Cl^-$ uptake across the luminal membrane of the proximal tubule cell. We could not demonstrate the presence of $Na^+$-$Cl^-$ cotransport or $Na^+$-$K^+$-$2Cl^-$ cotransport as possible mechanisms of uphill $Cl^-$ transport in these membrane vesicles.[3]

In previous studies we had similarly found an absence of $Na^+$-$Cl^-$ cotransport or $Na^+$-$K^+$-$2Cl^-$ cotransport in microvillus membrane vesicles isolated from rabbit ileum, but had been able to demonstrate the presence of $Na^+$-$H^+$ and $Cl^-$-$HCO_3^-$ exchangers.[4,5] The parallel operation of these exchangers gave rise to $Na^+$-coupled $Cl^-$ transport that was $CO_2$ dependent and acetazolamide sensitive.[5]

In contrast, although $Cl^-$-$HCO_3^-$ exchange could readily be demonstrated in renal basolateral membrane vesicles,[6,7] the activity of this transport process in renal microvillus membrane vesicles was equivocal.[7–9] However, imposing an outward formate gradient stimulated the initial rate of $Cl^-$ uptake much more appreciably, and it induced a transient uphill accumulation of $Cl^-$ above its eventual level of equilibrium uptake, indicating the presence of a $Cl^-$-formate exchange process.[8] Further confirming the presence of $Cl^-$-formate exchange was the finding that imposing an outward $Cl^-$ gradient induced the transient uphill accumulation of formate. $Cl^-$ self-exchange, exchange of $Cl^-$ with other halides such as $Br^-$ and $I^-$, and exchange of $Cl^-$ with $NO_3^-$ could be demonstrated, but outward gradients of acetate, propionate, butyrate, $HCO_3^-$, lactate, PAH, succinate, and $SO_4^=$ did not cause comparable stimulation of $Cl^-$ uptake, reflecting the narrow anion specificity of the $Cl^-$-formate exchanger.[8,9]

[a] This work was supported by U.S.P.H.S. grants DK-17433 and DK-33793.

Reabsorption of significant quantities of $Cl^-$ by a process involving exchange for formate requires a mechanism for recycling formate back across the luminal membrane into the proximal tubule cell. In fact, we found that imposing an inside-alkaline pH gradient caused the uptake of formate against its concentration gradient,[8] suggesting transport of formate via nonionic diffusion of formic acid. Whereas uptake of formate in exchange for $Cl^-$ was sensitive to inhibition by high concentrations of disulfonic stilbenes, uptake of formate driven by an inside-alkaline pH gradient was not stilbene sensitive, as predicted for passive nonionic diffusion of formic acid.

As shown in FIGURE 1, uptake of formic acid into the cell by nonionic diffusion followed by exchange of intracellular formate for intraluminal $Cl^-$ results in the net entry of $H^+$ and $Cl^-$ across the luminal membrane. In essence, formate serves as a necessary coupling factor to allow $H^+$-coupled $Cl^-$ absorption. In support of this notion, we found that when $^{82}Br$ was used as a tracer for $Cl^-$, its uphill accumulation in renal microvillus vesicles was stimulated by an inside-alkaline pH gradient in the presence of a physiologic concentration of formate, but not by the same pH gradient in the absence of formate.[8] To the extent that active $H^+$ secretion across the luminal membrane occurs by $Na^+$-$H^+$ exchange, as is largely

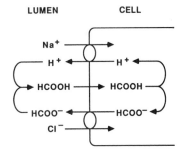

**FIGURE 1.** Schematic model of active $Cl^-$ uptake across the luminal membrane via the $Cl^-$-formate exchanger.

the case in the proximal tubule, the net process is electroneutral $Na^+$-coupled $Cl^-$ transport (FIG. 1). This process is driven by the inward $Na^+$ gradient that results from the primary active extrusion of $Na^+$ across the basolateral membrane via the $Na^+,K^+$-ATPase.

Next, we found that addition of physiologic concentrations of formate to the bath and lumen of isolated, perfused proximal tubules from the rabbit markedly stimulated the rate of NaCl absorption.[10] The stimulation of NaCl absorption by formate was not associated with any change in the transepithelial electrical potential difference, consistent with a process of electroneutral NaCl absorption. Similar concentrations of acetate did not stimulate NaCl absorption, consistent with the anion specificity of the $Cl^-$-formate exchanger as just discussed. The formate-stimulated component of NaCl absorption was inhibited by the addition of DIDS to the lumen or ouabain to the peritubular bath, as predicted for a model in which stilbene-sensitive $Cl^-$-formate exchange is indirectly coupled to the luminal membrane $Na^+$ gradient that is generated by ouabain-sensitive $Na^+$ extrusion across the basolateral membrane. These findings strongly supported the concept that $Cl^-$-formate exchange is a major mechanism for transcellular $Cl^-$ reabsorption in the intact proximal tubule under physiologic conditions.

## CHLORIDE-OXALATE AND SULFATE (OXALATE)-CARBONATE EXCHANGERS

Imposing an outward oxalate gradient significantly enhanced the rate of $Cl^-$ transport into rabbit renal microvillus membrane vesicles, suggesting the presence of $Cl^-$-oxalate exchange.[9] The presence of $Cl^-$-oxalate exchange was confirmed by the finding that imposing an outward $Cl^-$ gradient markedly stimulated the rate of oxalate uptake and caused the transient uphill accumulation of oxalate above equilibrium. Interestingly, an outward gradient of formate also stimulated uphill oxalate uptake, indicating the presence of formate-oxalate exchange.[9] The simplest hypothesis to explain these findings is that the processes of $Cl^-$-formate exchange, $Cl^-$-oxalate exchange, and formate-oxalate exchange represent different modes of a single anion exchanger. Indeed, we found that the dose response curve for inhibition of $Cl^-$-oxalate exchange by each of several different substrates and inhibitors was identical to that for inhibition of formate-oxalate exchange, suggesting that these two processes are modes of a single transport system.[9] However, $Cl^-$-formate exchange was less sensitive to inhibition by DIDS and was more sensitive to furosemide than were the processes of $Cl^-$-oxalate exchange and formate-oxalate exchange. Moreover, concentrations of oxalate that saturated the processes of $Cl^-$-oxalate exchange and formate-oxalate exchange caused only slight inhibition of $Cl^-$-formate exchange. Studies utilizing $K^+$ gradients in the presence of the $K^+$ ionophore valinomycin to vary the transmembrane electrical potential difference indicated that $Cl^-$-formate exchange was electroneutral, whereas $Cl^-$-oxalate exchange and formate-oxalate exchange were electrogenic in a manner consistent with exchange of one monovalent anion for divalent oxalate.

We therefore concluded that at least two different anion exchangers are involved in mediating $Cl^-$, formate and oxalate transport across the luminal membrane, as summarized in TABLE 1. According to this view, one transporter, the $Cl^-$-formate exchanger, accepts $Cl^-$ and formate as substrates, has little or no affinity for oxalate, is more sensitive to inhibition by furosemide than by DIDS, and is electroneutral. A second transporter, the $Cl^-$ (formate)-oxalate exchanger, also accepts $Cl^-$ and formate as substrates, but has a high affinity for oxalate, is highly sensitive to inhibition by DIDS, and is electrogenic.

For $Cl^-$-oxalate exchange to play an important role in mediating active transcellular $Cl^-$ reabsorption in the proximal tubule, a mechanism for recycling oxalate from lumen to cell is necessary. We therefore tested for other possible mechanisms of oxalate transport in rabbit renal microvillus membrane vesicles.[11] We found that oxalate was a poor substrate for $Na^+$ cotransport in these membrane vesicles. $SO_4^=$-$CO_3^=$ exchange had previously been demonstrated in microvillus and basolateral membrane vesicles isolated from rat and bovine kidney cortex.[12-14] We demonstrated that oxalate was a readily transported, high affinity substrate for the $SO_4^=$-$CO_3^=$ exchanger present in rabbit renal microvillus and basolateral membrane vesicles.[11,15] At least two modes of oxalate transport could take place via this transporter: oxalate-$CO_3^=$ exchange and oxalate-$SO_3^=$ exchange. As summarized in TABLE 1, the $SO_4^=$(oxalate)-$CO_3^=$ exchanger has properties that clearly distinguish it from the $Cl^-$-formate and $Cl^-$-oxalate exchangers. For example, the $SO_4^=$(oxalate)-$CO_3^=$ exchanger does not mediate $Cl^-$ transport,[11,15] in obvious contrast to the $Cl^-$-formate and $Cl^-$-oxalate exchangers. In its ability to mediate transport of $SO_4^=$, the $SO_4^=$(oxalate)-$CO_3^=$ exchanger differs from the $Cl^-$-formate and $Cl^-$-oxalate exchangers, for which $SO_4^=$ is a poor substrate.[9]

**TABLE 1.** Contrasting Properties of Three Luminal Membrane Anion Exchangers

| | $Cl^-$-Formate Exchanger | $Cl^-$-Oxalate Exchanger | $SO_4^=$(Oxalate)-$CO_3^=$ Exchanger |
|---|---|---|---|
| Substrates | $Cl^-$, formate<br>Little or no affinity for oxalate | $Cl^-$, formate<br>High affinity for oxalate<br>Little or no affinity for $SO_4^=$ | $SO_4^=$, oxalate, $CO_3^=$<br>High affinity for $SO_4^=$ and oxalate<br>Little or no affinity for $Cl^-$ |
| Inhibitors | Furosemide > DIDS<br>Low affinity for DIDS | DIDS >> furosemide<br>High affinity for DIDS | High affinity for DIDS |
| Exchange Modes | $Cl^-$-$Cl^-$, $Cl^-$-formate<br>but not $Cl^-$-oxalate | $Cl^-$-oxalate, formate-oxalate<br>but not $Cl^-$-$Cl^-$, $Cl^-$-formate, $Cl^-$-$SO_4^=$ | $SO_4^=$-$CO_3^=$, oxalate-$CO_3^=$, $SO_4^=$-oxalate<br>but not any $Cl^-$ exchange modes |
| Electrogenicity | Electroneutral<br>$1A^-$:1 $A^-$ | Electrogenic<br>1 $A^-$:1 $Ox^=$ | Electroneutral<br>1 $A^=$:1 $A^=$ |

The $SO_4^=$(oxalate)-$CO_3^=$ exchanger can in principle mediate the recycling of oxalate from lumen to cell necessary to support $Cl^-$ absorption by $Cl^-$-oxalate exchange, as illustrated in FIGURE 2. The active secretion of $H^+$ leads to the generation of an outward gradient of $CO_3^=$ across the luminal membrane (left panel of FIG. 2). Uphill transport of oxalate into the cell across the luminal membrane can therefore take place by oxalate-$CO_3^=$ exchange via the $SO_4^=$(oxalate)-$CO_3^=$ exchanger. The resulting outward gradient of oxalate may then drive $Cl^-$ absorption via the $Cl^-$-oxalate exchanger. To the extent that active $H^+$ secretion occurs by $Na^+$-$H^+$ exchange, the net process is $Na^+$-coupled $Cl^-$ absorption with a 2:1 stoichiometry. Alternatively (right panel of FIG. 2), the outward gradient of $SO_4^=$ across the luminal membrane resulting from the operation of the $Na^+$-$SO_4^=$ cotransporter[16,17] can drive uphill oxalate uptake via the oxalate-sulfate exchange mode of the $SO_4^=$(oxalate)-$CO_3^=$ exchanger. Again, the resulting outward gradient of oxalate may then drive $Cl^-$ absorption via the $Cl^-$ oxalate exchanger, so that the net process is $Na^+$-coupled $Cl^-$ absorption with a 2:1 stoichiometry. Whether

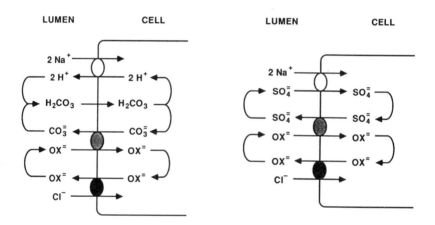

**FIGURE 2.** Schematic models of active $Cl^-$ uptake across the luminal membrane via the parallel operation of the $Cl^-$-oxalate exchanger (*black oval*) and the $SO_4^=$(oxalate)-$CO_3^=$ exchanger (*shaded oval*).

appreciable $Cl^-$ absorption in the intact proximal tubule under physiologic conditions actually takes place via the luminal membrane $Cl^-$-oxalate exchanger remains to be tested.

## SUMMARY AND CONCLUSION

Three distinct anion exchangers are described that directly or indirectly mediate $Cl^-$ transport across the luminal membrane of the proximal tubule cell. Studies on the intact proximal tubule indicate that the $Cl^-$-formate exchanger is a major mechanism for $Cl^-$ transport under physiologic conditions. As just discussed, the physiologic importance of the $Cl^-$-oxalate and $SO_4^=$(oxalate)-$CO_3^=$ exchangers in mediating $Cl^-$ transport across the luminal membrane of the proximal tubule cell is currently unknown.

These three anion exchangers are part of a larger group of at least eight distinct anion transporters in the proximal tubule that share with erythrocyte Band 3 the properties of stilbene sensitivity and/or the ability to mediate anion exchange.[18] It is tempting to speculate that these proximal tubule anion transporters are members of a family of proteins structurally related to the prototypic anion exchanger, erythrocyte Band 3. If this is true, comparing the structures of these anion transporters with each other and with Band 3 should provide important insight into the molecular basis for differences in substrate and inhibitor specificity within this family of transport proteins.

## REFERENCES

1. GIEBISCH G. & P. S. ARONSON. 1986. The proximal nephron. *In* Physiology of Membrane Disorders, 2nd edition. T. E. Andreoli, J. F. Hoffman, D. D. Fanestil & S. G. Schultz, eds.: 669–700. Plenum Medical Book Co. New York, NY.
2. SCHILD L., G. GIEBISCH, L. KARNISKI & P. S. ARONSON. 1986. Chloride transport in the mammalian proximal tubule. Pflügers Arch. **407** (Suppl. 2): S156–S159.
3. SEIFTER, J. L., R. KNICKELBEIN & P. S. ARONSON. 1984. Absence of Cl-OH exchange and Na-Cl cotransport in rabbit renal microvillus membrane vesicles. Am. J. Physiol. **247**: F753–F759.
4. KNICKELBEIN, R., P. S. ARONSON, W. ATHERTON & J. W. DOBBINS. 1983. Sodium and chloride transport across rabbit ileal brush border. I. Evidence for Na-H exchange. Am. J. Physiol. **245**: G504–G510.
5. KNICKELBEIN, R., P. S. ARONSON, C. M. SCHRON, J. SEIFTER & J. W. DOBBINS. 1985. Na and Cl transport across rabbit ileal brush border. II. Evidence for Cl:HCO3 exchange and mechanism of coupling. Am. J. Physiol. **249**: G236–G245.
6. GRASSL, S. M., L. P. KARNISKI & P. S. ARONSON. 1985. Cl-HCO3 exchange in rabbit renal cortical basolateral membrane vesicles (abstr.) Kidney Intern. **27**: 282.
7. GRASSL, S. M., P. D. HOLOHAN & C. R. ROSS. 1987. Cl⁻-HCO3̄ exchange in rat renal basolateral membrane vesicles. Biochim. Biophys. Acta **905**: 475–484.
8. KARNISKI, L. P. & P. S. ARONSON. 1985. Chloride/formate exchange with formic acid recycling: A mechanism of active chloride transport across epithelial membranes. Proc. Natl. Acad. Sci. USA **82**: 6362–6365.
9. KARNISKI, L. P. & P. S. ARONSON. 1987. Anion exchange pathways for Cl⁻ transport in rabbit renal microvillus membranes. Am. J. Physiol. **253**: F513–F521.
10. SCHILD, L., G. GIEBISCH, L. P. KARNISKI & P. S. ARONSON. 1987. Effect of formate on volume reabsorption in the rabbit proximal tubule. J. Clin. Invest. **79**: 32–38.
11. KUO, S. -M. & P. S. ARONSON. Pathways for oxalate transport in rabbit renal microvillus membrane vesicles. In preparation.
12. PRITCHARD, J. B. & J. L. RENFRO. 1983. Renal sulfate transport at the basolateral membrane is mediated by anion exchange. Proc. Natl. Acad. Sci. USA **80**: 2603–2607.
13. PRITCHARD, J. B. 1987. Sulfate-bicarbonate exchange in brush-border membranes from rat renal cortex. Am. J. Physiol. **252**: F346–F356.
14. TALOR, Z., R. M. GOLD, W.-C. YANG & J. A. L. ARRUDA. 1987. Anion exchanger is present in both luminal and basolateral membranes. Eur. J. Biochem. **164**: 695–702.
15. KUO, S.-M & P. S. ARONSON. 1988. Oxalate transport via the sulfate-HCO3 exchanger in rabbit renal basolateral membrane vesicles. J. Biol. Chem. **263**: 9710–9717.
16. LÜCKE, H., G. STANGE & H. MURER. 1979. Sulfate-ion/sodium-ion co-transport by brush-border membrane vesicles from rat kidney cortex. Biochem. J. **182**: 223–229.
17. SCHNEIDER, E. G., J. C. DURHAM & B. SACKTOR. 1984. Sodium-dependent transport of inorganic sulfate by rabbit renal brush-border membrane vesicles. J. Biol. Chem. **259**: 14591–14599.
18. ARONSON, P. S. 1989. The renal proximal tubule: A model for diversity of anion exchangers and stilbene-sensitive anion transporters. Ann. Rev. Physiol. **51**: 419–441.

# Two Gene Products Encoding Putative Anion Exchangers of the Kidney

S. L. ALPER, F. C. BROSIUS III, A. M. GARCIA,
S. GLUCK, D. BROWN, AND H. F. LODISH

*MIT and Whitehead Institute*
*Cambridge, Massachusetts 02142;*
*Beth Israel Hospital*
*Massachusetts General Hospital and Harvard Medical School*
*Boston, Massachusetts 02115 and 02129;*
*and*
*Washington University and Jewish Hospital*
*St. Louis, Missouri 63110*

We have characterized the products of two structurally and functionally related genes encoding putative anion exchangers of the rodent kidney. The band 3 gene[1] is transcribed in mouse[2] and rat kidney in two or three forms, respectively, which differ in their 5' ends.[3] The major kidney transcript in each species lacks exons 1–3. In mouse, this transcript extends ~100 nucleotides 5'-ward of exon 4, but it does not include sequence from intron 3. The transcript encodes a kidney band 3 (KB3) polypeptide that is truncated at its amino terminus when compared to erythroid band 3. The likely initiation site of KB3 is Met 80 of the erythroid sequence. When this 850 amino acid polypeptide is expressed in *Xenopus* oocytes from *in vitro* transcribed RNA, it mediates $^{36}$Cl uptake (TABLE 1) completely inhibited by 1 $\mu$M DIDS.[3] Antibodies directed against several domains of KB3, and used in indirect immunofluorescence studies, localize KB3 to the basolateral plasma membrane of a subset of collecting duct cells. All KB3-positive cells express vacuolar H$^+$ ATPase at the apical plasma membrane and thus represent Type A, acid-secreting intercalated cells.[4]

Band 3-related protein (B3RP) is a 1237 amino acid polypeptide 67% identical in the amino acid sequence to KB3 of its putative membrane-associated domain, and 33% identical to KB3 in the overlapping portion of its putative cytoplasmic domain.[5] Specific anti-B3RP antipeptide antibodies recognize glycoproteins of $M_r$ 130 and 155 kd encoded by B3RP cDNA in COS cells (FIG. 1). At least some of the $M_r$ 130-kd B3RP appears to be a precursor of the 155-kd form. In the presence of tunicamycin, both forms are reduced to a putative apoprotein of $M_r \sim 125$ kd. B3RP mRNA, unlike band 3 mRNA, is expressed in a wide range of epithelia. When B3RP is expressed in *Xenopus* oocytes from *in vitro* transcribed RNA, it mediates incremental $^{36}$Cl uptake by the oocytes (TABLE 1). Studies are underway

TABLE 1. KB3 and B3RP Increase $^{36}$Cl Uptake into Microinjected *Xenopus* Oocytes[a]

|  | H$_2$O Injected | RNA Injected | Fold Stimulated | n |
|---|---|---|---|---|
| KB3 | 1.29 ± 0.04 | 8.03 ± 0.53 | 6.2 | 10 |
| B3RP | 0.8 ± 0.09 | 2.31 ± 0.35 | 2.9 | 9 |

[a] Typical experiment, with Cl$^-$ uptake (nmol Cl$^-$/oocyte · h) expressed as mean ± SEM.

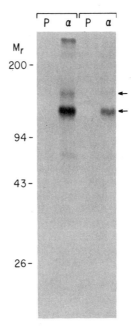

**FIGURE 1.** Fluorograph of SDS PAGE-fractionated immunoprecipitates of detergent-solubilized $^{35}S$-methionine-labeled COS cells expressing B3RP. Lanes 1 and 3 are preimmune sera. Lane 2 is serum directed against residues 426–440 of B3RP. Lane 4 is serum directed against residues 625–639 of B3RP.

to localize B3RP in the kidney using indirect immunofluorescence with the aforementioned antibodies and others.

## REFERENCES

1. KOPITO, R. R., M. A. ANDERSSON & H. F. LODISH. 1987. J. Biol. Chem. **262:** 8035–8040.
2. KOPITO, R. R., M. A. ANDERSSON & H. F. LODISH. 1987. Proc. Natl. Acad. Sci. USA **84:** 7149–7153.
3. BROSIUS, F. C. III, S. L. ALPER, A. M. GARCIA & H. F. LODISH. 1989. J. Biol. Chem. **264:** 7784–7787.
4. ALPER, S. L., J. NATALE, S. GLUCK, H. F. LODISH & D. BROWN. 1989. Proc. Natl. Acad. Sci. USA **86:** 5429–5433.
5. ALPER, S. L., R. R. KOPITO, S. M. LIBRESCO & H. F. LODISH. 1988. J. Biol. Chem. **263:** 17092–17099.

# Chloride Uptake into High Resistance MDCK Cells

KURT AMSLER AND B. AVERY INCE

*Department of Physiology and Biophysics*
*Mt. Sinai School of Medicine*
*New York, New York 10029*

Movement of chloride across several epithelia is enhanced by cAMP elevation.[1] The cellular mechanisms mediating this process in various tissues are still under investigation. Confluent populations of the renal epithelial cell line, Strain I MDCK cells, exhibit stimulatable transepithelial chloride secretion. We have obtained a clonal line derived from MDCK cells and have examined the chloride uptake pathways across the apical membrane of confluent populations under control conditions and upon elevation of cAMP.

The uptake rate for chloride across the apical membrane of confluent populations of 5D11 cells was approximately 2 nmol chloride per milligram of protein per minute and was linear for at least 2 minutes (FIG. 1). Uptake was accelerated two- to fourfold by a 2-minute pretreatment with 10 $\mu$M forskolin plus 1 mM 1-methyl-3-isobutyl xanthine (FM; a procedure that should elevate cAMP). Since, under these conditions, uptake was linear for at least 1 minute, all further experiments measured chloride uptake after 1 minute.

We examined the abilities of several anion transport inhibitors to affect uptake in control and FM-treated populations. A putative chloride channel blocker, N-

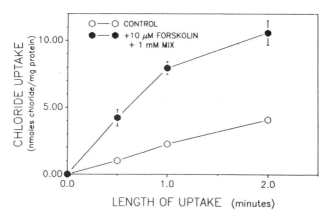

**FIGURE 1.** Uptake of chloride across the apical membrane of basal and 10 $\mu$M forskolin plus 1 mM 1-methyl-3-isobutyl xanthine-treated confluent 5D11 cell populations. Cells were seeded onto 35-mm tissue culture dishes and maintained at 37°C for 10 days in alphaMEM supplemented with 10% fetal calf serum. Uptake was performed at 37°C in a Cl-free balanced salt solution to which was added $^{36}$Cl (final concentration 1 mM and 0.5 $\mu$Ci/ml). Populations were treated with FM for 2 minutes prior to initiation of uptake. Uptake was terminated by aspiration of uptake solution and rapid rinsing in ice-cold saline solution. Cell protein and retained radioactivity were solubilized by addition of 0.2% sodium dodecyl sulfate. Radioactivity was measured by liquid scintillation counting, and protein was determined by a fluorometric assay.

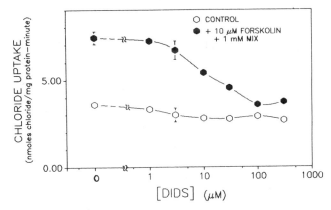

**FIGURE 2.** Effect of varying DIDS concentrations on chloride uptake rate into basal and 10 $\mu$M forskolin plus 1 mM 1-methyl-3-isobutyl xanthine-treated confluent 5D11 cell populations. DIDS and FM were added 2 minutes prior to initiation of uptake. Uptake was measured over a 1-minute period.

phenylanthranilic acid (PAA), inhibited uptake into control populations with a $K_{I,0.5}$ of approximately 20 $\mu$M. Uptake into FM-treated populations exhibited a partial inhibition by PAA with an identical $K_{I,0.5}$. The PAA-insensitive component of chloride uptake was equal to the increment in uptake rate produced by FM treatment. DIDS, an inhibitor of anion antiporters and other anion transporters, inhibited the stimulated portion of uptake in FM-treated populations with a $K_{I,0.5}$ of approximately 8 $\mu$M, but had no effect in control populations (FIG. 2). Bumetanide had no effect on uptake under either condition.

Both basal and FM-stimulated uptake rates were identical when measured using either Cl-free HBSS or a Na-, K-, and Cl-free sucrose solution. The anion specificities of uptake into control and FM-treated cell populations were identical ($I \geq Br > Cl > HCO_3 \geq SO_4$).

These results demonstrate that chloride can be taken up across the apical membrane of confluent 5D11 cells. Uptake is enhanced several-fold when cells are treated with compounds that elevate cAMP. Basal and FM-stimulated uptake components appear to be mediated by distinct pathways on the basis of differential inhibitor sensitivities, but both components exhibit identical anion selectivities. There is little contribution of cation-chloride symport to the observed uptake rate under either condition, but both chloride-anion antiport and conductive chloride movements remain possible pathways mediating either or both uptake components.

## ACKNOWLEDGMENT

The 5D11 cells used in this study were a kind gift from Dr. G. Ojakian, Downstate Medical School, Brooklyn, New York.

## REFERENCES

1. GREGER, R., E. SCHLATTER, F. WANG & J. N. FORREST, Jr. 1984. Pflüeger's Arch. **420:** 376.
2. BROWN, C. D. A. & N. L. SIMMONS. 1981. Biochim. Biophys. Acta **649:** 427.

# Reconstitution of the Rabbit Parotid Basolateral Membrane Na/K/Cl-Dependent Bumetanide Binding Site

A. CORCELLI[a,b] AND R. JAMES TURNER[b,c]

[a]Istituto Di Fisiologia Generale
Università Di Bari
Bari, Italy

[b]Clinical Investigations and Patient Care Branch
National Institute of Dental Research
National Institutes of Health
Bethesda, Maryland 20892

Na/K/Cl cotransporters have been found in a number of secretory and absorptive epithelia where this transporter is thought to play a major role in transepithelial chloride fluxes related to salt and water movements. We recently demonstrated the presence of such a cotransport system in a basolateral membrane vesicle (BLMV) preparation from the rabbit parotid.[1] In addition, we identified a high affinity ($K_d \approx 3$ $\mu$M) binding site for the loop diuretic bumetanide in this preparation and provided evidence that this site is identical to the bumetanide inhibitory site on the cotransporter.[2] Bumetanide binding to this site shows a hyperbolic dependence on [Na] and [K], and a biphasic dependence on [Cl] similar to that observed for high affinity bumetanide binding sites in several other tissues.[2] The purpose of our on-going studies, of which this is an initial report, is to further characterize the parotid Na/K/Cl cotransporter by extracting it from its native membrane environment and reconstituting it into artificial liposomes. In the present experiments we have employed [³H]-bumetanide binding to follow the cotransporter (or more precisely its bumetanide binding moiety) through a series of extraction, solubilization, and reconstitution steps.

## METHODS

A modification of the method of Wolosin[3] was used for reconstitution. BLMVs (1 mg protein/milliliter using the Bio-Rad protein assay kit with bovine IgG as the standard) from the rabbit parotid[1] were suspended in Buffer A (10 mM HEPES titrated to pH 7.4 with Tris and containing 100 mM mannitol and 1 mM EDTA) plus 1 mM Triton X-100, left on ice for 30 minutes, and then centrifuged at 100,000 g for 15 minutes. The resulting supernate (5 ml) was diluted fivefold with Buffer A containing sufficient Triton X-100, NaCl, and KCl to yield final concentrations of 5, 1, and 4 mM, respectively. This mixture was transferred to a flask in which egg

[c] Address for correspondence: Dr. R. J. Turner, Bldg. 10, Rm. 1A06, National Institutes of Health, Bethesda, MD 20892.

phosphatidylcholine and liver phosphatidylinositol (Avanti Polar Lipids, Pelham, Alabama) had been evaporated from chloroform under argon as a thin film (final concentrations 0.14% and 0.01%, respectively; final lipid to protein ratio $\approx$ 10:1 w/w). This clear solution was left on ice for 1 hour and then shaken overnight at 4°C with 3.75 g of Biobeads SM-2 (Bio-Rad, Richmond, California). The resulting turbid suspension was separated from the Biobeads by aspiration into a glass pipette and centrifuged at 150,000 g for 60 minutes.

[³H]-bumetanide binding (1 $\mu$M) was determined by a rapid filtration method as previously described.[2] Binding was measured in Buffer A in the presence of either 100 mM Na gluconate, 95 mM K gluconate and 5 mM KCl, or 100 mM NaCl plus 100 mM KCl. Because high affinity bumetanide binding is dramatically inhibited by high chloride concentrations (ref. 2, $K_{1/2} \approx$ 20 mM), specific binding was taken to be the difference between these two determinations.

## RESULTS AND DISCUSSION

FIGURE 1 shows the results of a typical reconstitution experiment. The total specific binding of 1 $\mu$M bumetanide in the starting BLMV, the supernate from the

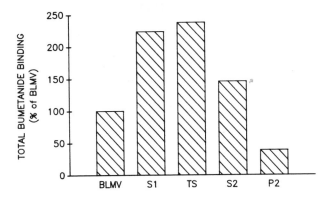

**FIGURE 1.** Total specific binding of 1 $\mu$M [³H]-bumetanide in 5 mg of rabbit parotid basolateral membrane vesicles (BLMV, 1 mg/ml) and in various fractions obtained during the extraction and reconstitution of the high affinity bumetanide binding site from this preparation. (See text.) Specific binding was determined in a 20-$\mu$l aliquot of each fraction.[2] Total specific binding in each fraction has been normalized to that of the BLMV.

1 mM Triton extraction (S1), the turbid suspension obtained after Triton removal with Biobeads (TS), and the supernate (S2) and pellet (P2) obtained after centrifuging this suspension at 150,000 g for 60 minutes are illustrated. In additional studies we demonstrated that the increase in specific binding observed in the 1 mM Triton extract, S1 (FIG. 1), is due to a fivefold decrease in the $K_d$ of high affinity bumetanide binding relative to the starting BLMV and that approximately 90% of the high affinity bumetanide binding sites are present in this extract (R. J. Turner and J. N. George, unpublished results). No specific binding was detectable after this extract was diluted in the presence of 5 mM Triton X-100 plus exogenous lipids (see Methods, data not shown); however, specific binding was com-

pletely recovered in the turbid suspension obtained after Triton removal with Biobeads (TS). Approximately 20% of this specific binding sedimented after centrifugation at 150,000 g for 60 minutes (FIG. 1), and an additional 20% could be sedimented by extending the centrifugation time to 90 minutes (data not shown). Similar recoveries have been reported by others[4] using related reconstitution methods. When the same reconstitution procedure was carried out in the absence of the BLMV extract, no specific bumetanide binding was observed in S1, TS, S2, or P2 (not shown).

We have not yet been able to demonstrate the presence of Na/K/Cl cotransport in the proteoliposomes obtained with the procedure just described. Whether this is due to the insensitivity of our flux measurements or to an actual dissociation of bumetanide binding activity from Na/K/Cl cotransport is currently under investigation.

## ACKNOWLEDGMENT

We thank Dr. Marie Therese Paternoster for many helpful discussions.

## REFERENCES

1. TURNER, R. J., J. N. GEORGE & B. J. BAUM 1986. J. Membr. Biol. **94:** 143–152.
2. TURNER, R. J. & J. N. GEORGE 1988. J. Membr. Biol. **102:** 71–77.
3. WOLOSIN, J. M. 1980. Biochem. J. **189:** 35–44.
4. KOHNE, W., C. W. M. HAEST & B. DEUTICKE 1981. Biochim. Biophys. Acta **664:** 108–120.

# Chinese Hamster Ovary Cell Mutants Deficient in an Anion Exchanger Functionally Similar to Erythroid Band 3[a]

ADA ELGAVISH,[b] JEFFREY D. ESKO,[c] RAYMOND
FRIZZELL,[d] DALE BENOS,[d] AND ERIC SORSCHER[d]

*Departments of Pharmacology,[b] Biochemistry,[c] and
Physiology and Biophysics[d]
University of Alabama School of Medicine
Birmingham, Alabama 35294*

Physiologic studies have demonstrated functional similarities between erythroid and nonerythroid anion exchange systems. As in the erythrocyte, electroneutral carrier-mediated anion transport systems with similar substrate specificities, pH dependence, and sensitivity to derivatives of stilbene disulfonic acid have been demonstrated in a variety of nonerythroid cells.[1–14] However, some differences in kinetic behavior and variations in the extent of pH or $Na^+$ dependence suggest structural variation of the anion exchange membrane protein, possibly resulting in an adaptation to the function of the carrier in the particular cell system.

As a prerequisite to studies of structure-function relationships of the various domains of the anion exchanger in a nonerythroid cell, we isolated 11 Chinese hamster ovary cell (CHO) mutants (*pgs* C) deficient in an anion exchange activity.[5,6] These mutants define a unique complementation group, but have been categorized in different classes by kinetic criteria.[5] Our studies in the wild-type CHO cells[5,6] indicate the presence of an anion exchanger manifesting several similarities to the erythroid band 3 exchanger: (1) Extracellular $Cl^-$ is a competitive inhibitor of $SO_4^{2-}$ influx and stimulates $SO_4^{2-}$ efflux, suggesting that the mechanism of transport is $SO_4^{2-}/Cl^-$ exchange. (2) Anion exchange inhibitor DIDS (4,4'-diisothiocyano-stilbene-2,2'-disulfonic acid) inhibits $SO_4^{2-}$ influx in a dose-dependent manner. Half maximal inhibition is achieved at 0.06 $\mu$M DIDS. (3) Low extracellular pH markedly stimulates $SO_4^{2-}$ influx. A sixfold decrease in the apparent Km is observed at $pH_{out}$ 5.5 as compared to $pH_{out}$ 7.5. Studies over a broad range of extracellular $SO_4^{2-}$ concentrations indicate the presence of three components of this transport activity in CHO cells: two high-affinity low-capacity systems, one in the range 0.5 $\mu$M $< [SO_4^{2-}]_{out} < 50$ $\mu$M and one in the range 50 $\mu$M $< [SO_4^{2-}]_{out} < 150$ $\mu$M, and a low-affinity high-capacity system (at $[SO_4^{2-}]_{out} > 150$ $\mu$M). These properties have not previously been reported for the erythroid band 3 transporter. The availability of mutants deficient in these activities has enabled us to carry out studies which suggest that the high-affinity systems are functionally independent of the low-affinity system but that all systems are dependent on the same anion exchange protein. Studies in a mutant that lacks all the components of the transport activity indicated that the anion exchanger may be instrumental in the regulation of the intracellular pH in CHO cells.

*a* This work was supported by a grant from the Cystic Fibrosis Foundation (to A.E.) and by NIH grant DK 38518. Ada Elgavish is the recipient of a Career Investigator Award from the American Lung Association/American Lung Association of Alabama.

To characterize this nonerythroid anion exchange membrane protein, we used an affinity-purified antibody produced against a DIDS-binding site-containing peptide that had been synthesized according to the amino acid sequence discovered by Kopito and Lodish[8] for erythroid band 3. The antibody inhibited $SO_4^{2-}$ influx but not $Na^+$ influx into the CHO cells (TABLE 1).

Plasma membranes were isolated from wild-type CHO cells and separated by SDS-gel electrophoresis. The Coomassie blue staining pattern of proteins (FIG. 1) was similar to that of other isolated plasma membranes.[2] Plasma membranes isolated from wild-type as well as mutant (605) CHO cells were separated by SDS-

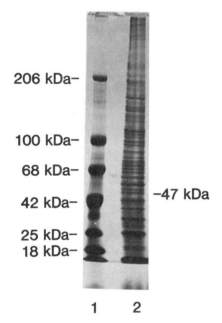

**FIGURE 1.** Coomassie blue staining profile of the membrane proteins isolated from wild-type Chinese hamster ovary (CHO) cells. Plasma membranes from cultured wild-type CHO cells were isolated using the method described by Kartner et al.,[17] in the presence of a cocktail of protease inhibitors containing 0.1 mM phenylmethyl-sulfonyl fluoride (PMSF), 1 μg/ml leupeptin, and 2 μg/ml pepstatin. The membrane proteins were separated by electrophoresis on an SDS-polyacrylamide gradient gel (5–12.5%) and stained with Coomassie blue. The staining pattern is shown for standard proteins (*lane 1*) and membrane proteins isolated from wild-type CHO cells (*lane 2*).

gel electrophoresis under reducing or nonreducing conditions. The gels were transferred to nitrocellulose and probed with the affinity-purified antibody produced against the DIDS-binding site-containing peptide. The antibody interacted with one 47-kDa membrane protein on Western blots of the wild-type CHO membrane proteins, under reducing or nonreducing conditions, in the presence of protease inhibitors (FIG. 2). The antibody interacted with an identical plasma membrane protein isolated from a mutant that lacks anion exchange activity, that is, mutant 605 (FIG. 2).

**TABLE 1.** Effect on Membrane Transport of an Affinity-Purified Antibody Produced against a DIDS-Binding Site-Containing Polypeptide[a]

| | DIDS-Sensitive Sulfate Influx (pmol/mg protein) | Sodium Influx (pmol/mg protein) |
|---|---|---|
| (A) Without antibody | 417.9 ± 21 | 1,070 ± 400 |
| (B) With antibody | 181.6 ± 11 | 950 ± 300 |

[a] Transport was measured essentially as previously described.[5,6] For sulfate influx measurements, the growth medium was removed and the cells were washed three times in a medium containing 150 mM sodium gluconate, 1 mM calcium gluconate, 0.1 mM sodium sulfate, 10 mM Tris HEPES, pH 7.5. The cells were then incubated for 2 minutes in the same medium with [$^{35}$S]-Na$_2$SO$_4$ and with (B) or without (A) 156 nM antibody, in the presence or absence of 0.1 mM DIDS (4,4'-diisothiocyanostilbene-2,2'-disulfonic acid). Finally, cells were extracted with 1 ml of 0.1 N NaOH for 30 minutes at room temperature, and the amount of radioactivity in 0.5 ml sample of extract was measured by liquid scintillation spectrometry. Results given are the DIDS-sensitive component of the sulfate influx, that is, total influx in the absence of the inhibitor minus the influx in its presence. This component represents influx via the anion exchanger.[5] For sodium influx measurements, the growth medium was removed and the cells were washed three times in a medium containing 1 mM NaCl, 150 mM N-methyl glucamine chloride, 1 mM calcium gluconate, 10 mM Tris HEPES, pH 7.5. The cells were then incubated for 2 minutes in the same medium, in the presence (B) or absence (A) of 156 nM antibody. $^{22}$NaCl was then added and the incubation continued for 30 seconds. Cells were then extracted and the amount of radioactivity was measured as described above. Each value is the mean ± standard deviation of triplicate samples.

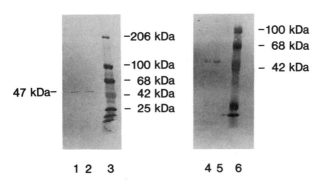

**A. Reducing conditions    B. Non-reducing conditions**

**FIGURE 2.** Western blot analysis of wild-type and mutant (605) Chinese hamster ovary cell plasma membranes with an antibody produced against a DIDS-binding site-containing peptide. Plasma membranes and prestained standard proteins from Diversified Biotech were subjected to electrophoresis under reducing (A) or nonreducing (B) conditions. The gel was transferred to nitrocellulose and probed with an affinity-purified antibody produced against a DIDS-binding site-containing peptide synthesized according to the amino acid sequence discovered by Kopito and Lodish[8] for the erythroid band 3 exchanger. The sequence of the DIDS-binding site-containing peptide was: Thr-Phe-Ser-Lys-Leu-Ile-Lys-Ile-Phe-Gln-Asp-Tyr. Goat anti-rabbit alkaline phosphatase conjugated second antibody and reagents BCIP (5-bromo-4-chloro-3-indolyl phosphate p-toluidine salt) and NBT (p-nitro blue tetrazolium chloride) from BIO-RAD were used to detect the antigens bound to the nitrocellulose. The staining pattern on Western blots of proteins subjected to electrophoresis under reducing (A) or nonreducing (B) conditions is shown in each case for prestained standard proteins (*lanes 3 and 6*) and membrane proteins isolated from wild-type (*lanes 2 and 5*) and mutant (*lanes 1 and 4*) CHO cells.

In conclusion, our results suggest that the 47-kDa protein may be the anion exchange protein in the CHO cell. However, we cannot exclude the possibility that the gene product defined by the CHO anion exchange-deficient mutants forms complexes with itself or other proteins that manifest three kinetic modes. The 47-kDa membrane protein appears to be expressed in the mutant that lacks anion exchange activity (mutant 605) (FIG. 2), suggesting that the defect is not in the domain recognized by the antibody or is not substantial enough to render the membrane protein unrecognizable by the antibody. A polypeptide of similar molecular weight has been suggested to be part of the anion transport system in hepatocytes.[15] Moreover, immunoautoradiographic analysis[16] revealed that antibodies to the band 3 erythroid anion exchanger (95 kDa) interact with three polypeptides of 38, 48, and 60 kDa in cultured human lung cells. Thus, band-3-like proteins in nonerythroid cells may be variants of different molecular weights adapted to each cell type and its function.

## REFERENCES

1. ELGAVISH, A., J. B. SMITH, D. J. PILLION & E. MEEZAN. 1985. J. Cell. Physiol. **125:** 243–250.
2. ELGAVISH, A., D. R. DIBONA, P. NORTON & E. MEEZAN. 1987. Am. J. Physiol. **253:** C416–C425.
3. ELGAVISH, A. & E. MEEZAN. 1988. Biochem. Biophys. Res. Commun. **152:** 99–106.
4. ELGAVISH, A. & E. MEEZAN. 1989. Am. J. Phys. **256:** C486–C496.
5. ELGAVISH, A., J. D. ESKO & A. KNURR. 1988. J. Biol. Chem. **263:** 18607–18613.
6. ESKO, J. D., A. ELGAVISH, W. H. TAYLOR & J. L. WEINKE. 1986. J. Biol. Chem. **261:** 15725–15733.
7. HOFFMAN, E. K. 1982. Phil. Trans. R. Soc. Lond. **299:** 519–535.
8. KOPITO, R. R. & H. F. LODISH. 1985. Nature **316:** 234–238.
9. KNAUF, P. A. 1979. Curr. Top. Membr. Transp. **12:** 251–363.
10. KNICKELBEIN, R. G., P. S. ARONSON & J. W. DOBBINS. 1985. J. Membr. Biol. **88:** 199–204.
11. OLSNES, S., J. LUDT, T. I. TONNESSEN & K. SANDVIG. 1987. J. Cell. Physiol. **132:** 192–202.
12. PASSOW, H. 1986. Rev. Physiol. Biochem. Pharmacol. **103:** 62–186.
13. SCHNEIDER, E. G., J. C. DURHAM & B. SACKTOR. 1984. J. Biol. Chem. **259:** 14591–14599.
14. VON DIPPE, P. & D. LEVY. 1982. J. Biol. Chem. **257:** 4381–4385.
15. CHENG, S. & D. LEVY. 1980. J. Biol. Chem. **255:** 2637–2640.
16. KAY, M. M. B., C. M. TRACEY, J. R. GOODMAN, J. C. CONE & P. S. BASSEL. 1983. Proc. Natl. Acad. Sci. USA **80:** 6882–6886.
17. KARTNER, N., N. ALAN, M. SWIFT, M. BUCHWALD & J. R. RIORDAN. 1977. J. Membr. Biol. **36:** 191–211.

# Propionate/Bicarbonate Exchange in Human Small Intestinal Brush Border Membrane Vesicles[a]

J. M. HARIG,[b,c] K. H. SOERGEL,[d] AND K. RAMASWAMY[d]

*Gastroenterology Sections*
*University of Chicago[b]*
*Chicago, Illinois 60637*
*and*
*Medical College of Wisconsin[d]*
*Milwaukee, Wisconsin 53226*

Short chain fatty acids (SCFA) (mainly acetic, propionic, and n-butyric acids) are produced by anaerobic bacterial fermentation of carbohydrate and protein in the gastrointestinal tracts of most higher animals. In man, these SCFA are readily absorbed during *in vivo* perfusions of jejunum,[1] ileum,[2] and colon.[3] Absorption of these SCFA is accompanied by a rise in gastrointestinal luminal pH and [$HCO_3^-$]. It has generally been assumed that SCFA cross the intestinal brush border membrane via nonionic diffusion of the protonated acid. However, the existence of an anion antiporter that exchanges luminal SCFA anions for $HCO_3^-$ is entirely consistent with data obtained from human perfusion studies (because at luminal pH 7–8 nearly all SCFA exist as anions).

We used brush border membrane vesicles (BBMV) to study mechanisms of SCFA transport in human small intestine. These BBMV were obtained from organ donor small intestine using modified $CaCl_2$ precipitation and vesiculation methods previously established by us as useful for human intestinal transport studies.[4] These studies were performed by measuring uptake of [1-$^{14}$C] propionic acid (Dupont-New England Nuclear) with a rapid filtration technique into human jejunal and ileal BBMV. Fourteen different organ donor intestines were used for this study.

TABLE 1 summarizes SCFA transport mechanisms in human small intestinal BBMV. The results of our studies demonstrate a $SCFA^-/HCO_3^-$ anion exchange mechanism in human small intestine. The major characteristics of the $SCFA^-/HCO_3^-$ antiporter are as follows: (1) Propionate uptake is markedly stimulated by intravesicular $HCO_3^-$, resulting in a large (up to 30-fold) and fast "overshoot" phenomena (peak uptake at about 6 seconds). (2) Propionate uptake into $HCO_3^-$-loaded BBMV is a saturable process. (3) [$^{14}$C] propionate uptake undergoes trans-stimulation by unlabeled intravesicular propionate. (4) $HCO_3^-$-stimulated propionate uptake is inhibited by the known anion exchange inhibitors SITS and DIDS. (5) Propionate uptake is inhibited by other SCFA in a somewhat specific manner (TABLE 2). (6) Diffusional uptake of propionate occurs in gluconate-loaded vesi-

[a] This work was supported by P.H.S. Grants AM-07267, AM-33349, and DK-26678 and the Veterans Administration.

[c] Address for correspondence: James M. Harig, M.D., Clinical Nutrition Research Unit, Gastroenterology Section, University of Chicago Hospitals, Box 223, 5841 S. Maryland Ave., Chicago, IL 60637.

**TABLE 1.** Characteristics of Short-Chain Fatty Acid Transport in Human Small Intestinal Brush Border Membrane Vesicles

1. *Two Mechanisms*:
   $SCFA^-/HCO_3^-$ antiporter – major pathway
   Diffusion (presumed nonionic) – minor pathway
2. *SCFA$^-$/HCO$_3^-$ Exchanger*:
   Saturable: apparent $Km = 14$ mM, $V_{max} = 125$ nmol $\times$ mg protein$^{-1} \times 3$ s$^{-1}$ (ileum)
   Magnitude of transport: Jejunum > Ileum
   $Na^+$ independent
   Inhibited by stilbenes – SITS, DIDS
   High degree of specificity for SCFA and other structurally related organics (TABLE 2)
   Transstimulated by $HCO_3^-$ and propionate but not $Cl^-$ or gluconate

**TABLE 2.** Organic Anion Effects on [1-$^{14}$C] Propionate into Human Ileal Brush Border Membrane Vesicles (Uptake of 250 $\mu$M [1-$^{14}$C] propionate was measured at 3 seconds)

| Organic Anion (10 mM) | % Inhibition | Organic Anion (10 mM) | % Inhibition |
|---|---|---|---|
| Formate | 0 | Caproate | 0 |
| Acetate | 40 | Succinate | 0 |
| Propionate | 40 | L-alanine | 0 |
| n-Butyrate | 32 | Lactate | 0 |
| Isobutyrate | 24 | Oxalate | 0 |
| n-Valerate | 28 | Propionamide | 0 |
| Isovalerate | 29 | Propylamine | 0 |
| Benzoate | 40 | Propiolic acid | 0 |
| Gamma-aminobutyrate | 33 | B-hydroxybutyrate | 0 |
| | | Tricarbyllic acid | 0 |

cles; however, this diffusional uptake represents only about 25% of $HCO_3^-$-stimulated propionate transport even at high (75 mM) concentrations. (7) Replacement of intravesicular $HCO_3^-$ with large amounts of buffer markedly attenuates propionate uptake. (8) The activity of this anion exchanger is much greater than that of $Cl^-/HCO_3^-$ exchanger described in human ileal BBMV.[5]

Future studies will be directed at studying this exchanger in human colon and colon-cancer-derived cell lines.

**REFERENCES**

1. SCHMITT, M. G., K. H. SOERGEL & C. M. WOOD. 1976. Gastroenterology **70:** 211–215.
2. SCHMITT, M. G., K. H. SOERGEL, C. M. WOOD & J. J. STEFF. 1978. Am. J. Dig. Dis. **22:** 340–347.
3. RUPPIN, H., S. BAR-MEIR, K. H. SOERGEL, C. M. WOOD & M. G. SCHMITT. 1980. Gastroenterology **78:** 1500–1507.
4. RAJENDRAN, V. M., S. A. ANSARI, J. M. HARIG, M. B. ADAMS, A. H. KHAN & K. RAMASWAMY. 1985. Gastroenterology **89:** 1298–1304.
5. RAMASWAMY, K., M. CHUNG & J. A. BARRY. 1988. Gastroenterology **94:** 366.

# Identification of an Overexpressed Putative Anion Transport Protein in *Drosophila* Kc Cells Resistant to Disulfonic Stilbenes

MARILYN M. SANDERS, KATHLEEN JOHN-ALDER, AND
ANN C. SHERWOOD

*Department of Pharmacology*
*UMDNJ-Robert Wood Johnson Medical School*
*Piscataway, New Jersey 08854*

Disulfonic stilbene derivatives including 4,4'-diisothiocyanostilbene-2,2'-disulfonate (DIDS) and anthranilic acid derivatives like flufenamic acid (FFA) inhibit a major anion uptake system in *Drosophila* Kc cells.[1] Furthermore, these drugs induce all the characteristics of the heat shock response,[2] including inhibition of DNA synthesis and arrest of cell growth.

As one approach to investigating the suggested relationship between growth control, anion transport, and the heat shock response, we selected variants of Kc cells resistant to the growth-inhibiting properties of DIDS. The variants were selected over a period of months using gradually increasing concentrations of drug in the medium. This step selection protocol is known to be effective in producing variants that overexpress polypeptides functioning to overcome the growth-inhibiting effects of the drug.[3]

The variants show several-fold resistance to growth inhibition by DIDS, and they have increased steady-state chloride content. In addition, the DIDS-resistant cells contain 3 more than 10-fold overexpressed polypeptides with molecular weights of 46, 62, and 123 kDa. All three overexpressed polypeptides copurify with the detergent-insoluble cytoskeleton, and the 46-kDa polypeptide has been identified as a member of the intermediate filament family of proteins.[4] A monoclonal antibody to the 46-kDa protein, called Ah6,[4] was used to confirm our identification of the 46-kDa overexpressed protein. This antibody unexpectedly showed strong reaction with the overexpressed 62- and 123-kDa polypeptides as well, indicating that all three polypeptides share the Ah6-recognized epitope.

To determine whether any exterior membrane proteins were overproduced in the DIDS-resistant variants, intact cells were labeled on the outside with sulfo-NHS-biotin.[5] FIGURE 1 (left panel) shows that the 123-kDa polypeptide reacted with this reagent, showing it is an exterior membrane protein. To determine whether it could be an anion transporter, cells were reacted first with sulfo-NHS-acetate in the presence of FFA to substitute all reactive sites on the cell exterior that were not protected by FFA. The sulfo-NHS-acetate and FFA were washed out and the cells were then reacted with sulfo-NHS-biotin. FIGURE 1 (right panel) shows more than 20-fold protection of the 123-kDa protein by FFA, both in normal cells and in the DIDS-resistant variants. These experiments identify the 123-kDa polypeptide as FFA binding and thus a putative anion transport protein, and they suggest that anion transport function may be linked to growth regulation in *Drosophila*.

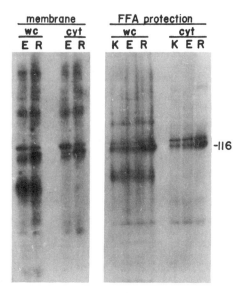

**FIGURE 1.** Overexpressed membrane and FFA-binding proteins in DIDS-resistant variants. Lanes labeled membrane: Intact cells were labeled with sulfo-NHS-biotin, and proteins from whole cells (wc) and purified cytoskeletons (cyt) were displayed on SDS gels. Biotinylated membrane proteins were detected with [$^{125}$I]-avidin. An autoradiogram is shown. Lanes labeled FFA protection: Intact cells were labeled first with sulfo-NHS-acetate in the presence of 100 $\mu$M FFA and then with sulfo-NHS-biotin as described in the text. Whole cell (wc) and purified cytoskeleton (cyt) proteins were displayed on SDS gels and biotinylated proteins were detected with [$^{125}$I]-avidin. E = parent clone from which variants were selected; R = DIDS-resistant variant clone; K = Kc line parent. Position of migration of a 116-kDa molecular weight marker is indicated on the right.

## REFERENCES

1. SHERWOOD, A. C., K. JOHN-ALDER & M. M. SANDERS. 1988. J. Cell. Physiol. **136:** 500–506.
2. SHERWOOD, A. C., K. JOHN-ALDER & M. M SANDERS. 1989. *In* Stress-Induced Proteins. UCLA symposia on Molecular and Cellular Biology, New Series. M. Pardue, J. Feramisco & S. Lindquist, eds. **96:** 117–128. Alan R. Liss Inc., New York.
3. SCHIMKE, R. T. 1984. Cell **37:** 705–713.
4. FALKNER, F. G., H. SAUMWEBER & H. BIESSMANN. 1981. J. Cell Biol. **91:** 175–183.
5. INGALLS, H. M., C. M. GOODLOE-HOLLAND, & E. J. LUNA. 1986. Proc. Natl. Acad. Sci. USA **83:** 4779–4783.

# Modulation of Anion Exchange by Cardiac Glycosides and Cytochalasin E in Separated Rabbit Kidney Medullary Collecting Duct (MCD) Cells[a]

A. JANOSHAZI AND A. K. SOLOMON

*Biophysical Laboratory*
*Harvard Medical School*
*Boston, Massachusetts 02115*

In view of the evidence[1,2] that the anion exchange protein on the basolateral face of MCD cells is immunochemically similar to red cell band 3, Janoshazi *et al.*[3] characterized the physicochemical properties of the MCD protein and found them to be closely similar to red cell band 3. These data were obtained from the fluorescence enhancement at equilibrium after the specific stilbene anion exchange inhibitor DBDS (4,4'-dibenzamido-2,2'-stilbene disulfonate) is bound to its site on the MCD cell anion exchange protein. We now have measured the DBDS binding kinetics by the stopped-flow method and related them to the kinetics of $Cl^-$ exchange in MCD cells by showing that the $K_I$ (0.5 ± 0.1 $\mu$M) for $H_2$-DIDS (4,4'-diisothiocyano-2,2'-dihydrostilbene disulfonate) inhibition of DBDS fluorescence enhancement kinetics is very close to the $K_I$ of 0.9 ± 0.1 $\mu$M for $H_2$-DIDS inhibition of $Cl^-$ exchange in MCD cells, measured by the $Cl^-$-sensitive fluorescent dye SPQ (6-methoxy-N-(3-sulfopropyl)quinolinium).[4] This close relation provides strong support for the conclusion that ligands that modulate DBDS binding kinetics also modulate anion exchange kinetics.

Because Drenckhahn *et al.*[1] had reported that actin, isoforms of ankyrin, and spectrin are colocalized with the anion exchange protein on the basolateral surface of MCD cells, we used cytochalasin E (CE, which binds to a specific site on the spectrin/actin/band 4.1 complex in the red cell) as a probe for interactions between the cytoskeleton and the anion exchange protein in MCD cells. CE significantly modulates DBDS binding kinetics in MCD cells, with a $K_I$ of 0.08 ± 0.005 $\mu$M, consistent with red cell cytoskeleton binding affinities. CE also slows $Cl^-/HCO_3^-$ exchange in MCD cells by a factor of about 2 (FIG. 1). These experiments indicate that there is a mechanism in MCD cells by which intracellular events can modulate membrane anion exchange.

The cardiac glycoside cation transport inhibitor ouabain also modulates DBDS binding kinetics in MCD cells with a $K_I$ of 0.003 ± 0.001 $\mu$M. Ten $\mu$M ouabain increases the time constant of $Cl^-$ exchange from a control value of $\tau_{Cl^-}$ of 0.3 ± 0.02 second to $\tau_{Cl^-}$ of 0.56 ± 0.06 second (30 mM $HCO_3^-$). Digitoxigenin, which binds to MCD cells much more rapidly than does ouabain (FIG. 2), has a much lower affinity, with a $K_I$ of 0.09 ± 0.004 $\mu$M. These experiments show that the cation transport protein and the anion exchange protein in MCD cells are linked

[a] This work was supported in part by the American Heart Association, both National and Massachusetts Affiliate, and the American Cancer Society, Massachusetts Division, Inc.

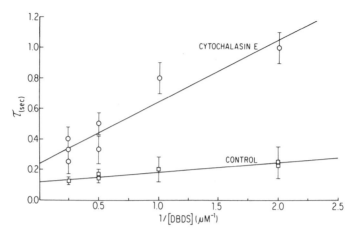

**FIGURE 1.** Cytochalasin E (CE) effect on Cl$^-$ flux of MCD cells. Fresh SPQ-loaded MCD cells were incubated for 30 minutes with 2 $\mu$M CE, and Cl$^-$/HCO$_3^-$ exchange with 30 mM NaHCO$_3$ was measured in the stopped-flow apparatus at 20–23°C. Data fitted by least squares to a single exponential with time constant, $\tau_{Cl^-}$ of 0.20 $\pm$ 0.04 second; + CE, $\tau_{Cl^-}$ of 0.50 $\pm$ 0.08 second.

by a mechanism through which conformational information can flow and, together with the CE results, are consistent with a regulatory system by which events within the cell can control the traffic of anions and cations across the basolateral membrane.

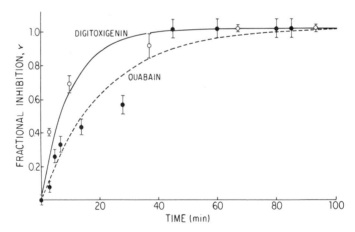

**FIGURE 2.** Time dependence of digitoxigenin and ouabain effects on DBDS binding kinetics. Previously frozen MCD cells were mixed with 0.1 $\mu$M digitoxigenin or 1.0 $\mu$M ouabain at 20–23°C and the time course of fluorescence enhancement was measured in the stopped-flow apparatus. The fraction of the maximum inhibition (70% for digitoxigenin; 50% for ouabain) was fitted to a single exponential by nonlinear least squares to give $\tau_{digitoxigenin}$ of 8 $\pm$ 2 minutes; $\tau_{ouabain}$ of 21.3 $\pm$ 0.9 minutes.

## REFERENCES

1. DRENCKHAHN, D., K. SCHLUTER, D. P. ALLEN & V. BENNETT. 1985. Nature **230:** 1287.
2. SCHUSTER, V. L., S. M. BONSIB & M. L. JENNINGS. 1986. Am J. Physiol. **251:** C347.
3. JANOSHAZI, A., D. M., OJCIUS, B. KOHE, J. L. SEIFTER & A. K. SOLOMON. 1988. J. Membr. Biol. **103:** 188.
4. ILLSLEY, N. P. & A. VERKMAN. 1987. Biochemistry **26:** 1215.

# Outward Potassium/Chloride Cotransport from Mammalian Cortical Neurons[a]

SCOTT M. THOMPSON

*Department of Neurology*
*College of Physicians and Surgeons*
*Columbia University*
*New York, New York 10032*

In the mammalian central nervous system, inhibitory synaptic transmission is mediated by neurotransmitters, such as gamma-aminobutyric acid (GABA), which activate a conductance that is selectively permeable to $Cl^-$ ions. The resting membrane potential ($V_M$) of cortical neurons is typically 5–15 mV less negative than the $Cl^-$ equilibrium potential ($E_{Cl^-}$), suggesting that there is an active, outwardly directed $Cl^-$ transport system in these cells that maintains an intracellular $[Cl^-]$ lower than expected for a passive distribution. In these experiments, I have attempted to characterize this neuronal $Cl^-$ transport mechanism.

Hippocampal neurons were grown in culture using the cultured slice procedure. Because mammalian neurons are so small, $E_{Cl^-}$ must be indirectly estimated from the equilibrium potential of $Cl^-$ currents activated by GABA, either iontophoretically applied to the cell or synaptically released from endogenous inhibitory neurons. GABA increases $Cl^-$ permeability and causes a current to flow whose magnitude and polarity are determined by the $Cl^-$ driving force ($V_M - E_{Cl^-}$). Neurons were impaled with micropipettes for measurement of $V_M$, and single-electrode voltage-clamp techniques were used to measure $Cl^-$ current. We may assume that the GABA equilibrium potential equals $E_{Cl^-}$. (For justification see ref. 3, where these data are described more completely.)

Furosemide ($5 \times 10^{-4}$ M) was found to reduce $Cl^-$ driving force $69 \pm 20\%$ ($n = 7$) by shifting $E_{Cl^-}$ towards the resting $V_M$ (FIG. 1A). In contrast, SITS has no effect on $E_{Cl^-}$ in these neurons.[1,2] Furthermore, previous experiments showed that active $Cl^-$ efflux is electroneutral.[2] These data suggested that a cation/chloride cotransport process was responsible for outward $Cl^-$ transport. The effect of varying the extracellular concentrations of $Na^+$ and $K^+$ on maintenance of the $Cl^-$ gradient was therefore examined. Reducing the extracellular $Na^+$ concentration from 140 to 70 mM had no effect on $E_{Cl^-}$ despite significantly reducing the energy barrier for outward $Na^+$ transport, suggesting the $Na^+$ ions are not cotransported with $Cl^-$. Reducing the extracellular $K^+$ concentration ($[K^+]_0$) from control (5.8 mM) to 1 mM, however, caused a significant increase in $Cl^-$ driving force due to a 15-mV hyperpolarizing shift in $E_{Cl^-}$ (FIG. 1B). Because the extracellular $[Cl^-]$ was constant, the intracellular $[Cl^-]$ must have decreased in low $[K^+]_0$. When furosemide was applied while simultaneously lowering $[K^+]_0$, the shift in $E_{Cl^-}$ was much smaller than that when $[K^+]_0$ is lowered in the absence of furosemide (FIG. 2). These data suggest that active transport is responsible for lowering the intracellular $[Cl^-]$ when $[K^+]_0$ is reduced, and that $K^+$ is involved in the cotransport reaction.

Taken together, these data imply that active $Cl^-$ extrusion from mammalian

---

[a] This work was supported by NATO (1986) and NIH Postdoctoral fellowships.

120

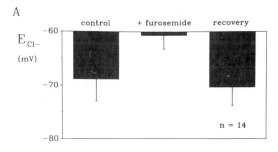

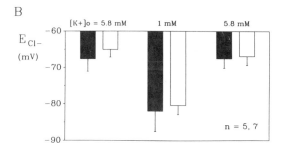

**FIGURE 1.** Effects of furosemide ($5 \times 10^{-4}$ M) **(A)** and reductions in $[K^+]_0$ **(B)** on the mean chloride equilibrium potential ($E_{Cl^-}$) ($\pm$SD) as determined by iontophoresis of GABA (*open bars*) or by synaptic release of endogenous GABA (*filled bars*). Membrane potential was held constant at $-60$ mV in these experiments; the driving force for $Cl^-$ is therefore given by the amplitude of the bars. Furosemide decreased, and low $[K^+]_0$ increased, $Cl^-$ driving force by altering $E_{Cl^-}$.

cortical neurons is mediated by outwardly directed potassium/chloride cotransport. Furthermore, this transport system is stimulated by a reduction in $[K^+]_0$. Inhibition of outward $Cl^-$ transport by elevated $[K^+]_0$ may be important in the genesis and spread of epileptic discharge.[2,3]

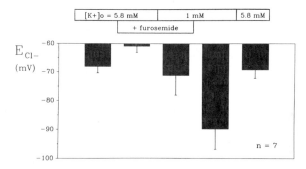

**FIGURE 2.** Effect of furosemide on the shift in the mean $E_{Cl^-}$ that results when $[K^+]_0$ is reduced is shown as in FIGURE 1. Application of furosemide decreases $Cl^-$ driving force and greatly reduces the shift in $E_{Cl^-}$ when $[K^+]_0$ is reduced from 5.8 to 1 mM. Washout of furosemide from the low $[K^+]_0$ saline results in a much larger hyperpolarizing shift in $E_{Cl^-}$. $E_{Cl^-}$ recovers on return to control saline solution.

## ACKNOWLEDGMENT

It is a pleasure to thank Drs. B. Gähwiler, R. Deisz, and D. Prince for their essential contributions to this work.

## REFERENCES

1. MISGELD, U., R. A. DEISZ, H. U. DODT & H. D. LUX. 1986. Science (Wash., DC) **232:** 1413–1415.
2. THOMPSON, S. M., R. A. DEISZ & D. A. PRINCE. 1988. Neurosci. Lett. **89:** 49–54.
3. THOMPSON, S. M. & B. H. GÄHWILER. 1989. J. Neurophysiol. **61:** 512–523.

# Effect of Disulfonic Stilbene Anion-Channel Blockers on Contractility of Vascular Smooth Muscle

AIMIN ZHANG, BELLA T. ALTURA, AND
BURTON M. ALTURA[a]

*Department of Physiology*
*State University of New York*
*Health Science Center*
*Brooklyn, New York 11203*

Although the role of cations in contractility of vascular smooth muscle has been studied extensively, very little information exists with respect to the potential role of anions. Our previous studies indicate that lowering of extracellular chloride by replacement with less permeant anions (isethionate and acetate) can decrease contractility of vascular smooth muscle, suggesting that $Cl^-$ anion movement across the vascular smooth muscle membrane significantly modulates its contractility.[1] Although it is uncertain whether chloride anions are passively or actively transported across the membrane of vascular smooth muscle, an exchanging anion carrier in rabbit aorta has been demonstrated.[2] In the present study, we used disulfonic stilbene anion-channel blockers (4-acetamido-4'-isothiocyano-2,2'-stilbene disulfonic aid [SITS] and 4,4'-diisothiocyano-2, 2'-stilbene disulfonate [DIDS]), which are inhibitors of choice in a variety of well-documented anion transport systems, to study the influence of anions on the responses of isolated rat aortas and portal veins to depolarization ($K^+$) and norepinephrine.

Rat aortas and portal veins were prepared and treated as described before.[3] After equilibration in normal Krebs-Ringer bicarbonate,[3] tissues were exposed to KCL or norepinephrine. Cumulative dose-response curves to these agonists were then obtained before and during incubation with either SITS or DIDS (400–600 $\mu M$) for 45 minutes in the absence of light. Both anion-channel blockers decreased the contractile tensions induced by $K^+$ and norepinephrine; resting tension and spontaneous mechanical activity, however, were not affected. In addition, these anion-channel blockers shifted the agonist-induced cumulative dose-response curves in rat aortas and portal veins rightward in a parallel manner. The inhibition by SITS and DIDS was reversible and varied with the tissue and stimuli. DIDS was clearly more potent in portal veins, and SITS was more potent in rat aortas. These results suggest either that differences between the anion exchangers may exist in arterial and venous smooth muscles or that the mechanism of interaction of anion-channel blockers is more complicated than currently believed.

Consistent with $Cl^-$ substitution experiments in which contraction was potentiated by more permeant anions and attenuated by less permeant anions,[4] our results suggest that $Cl^-$ movement and membrane anion permeability are required for tension. The responses of rat aortas and portal veins to norepinephrine were found more sensitive to disulfonic stilbene anion-channel blockers than were the

---

[a] Address for correspondence: Dr. B. M. Altura, Box 31, SUNY Health Science Center, 450 Clarkson Avenue, Brooklyn, NY 11203.

123

responses to KCL, further indicating an importance of $Cl^-$ for normal contractility; $Cl^-$ efflux has been thought to contribute to norepinephrine-induced depolarization in vascular smooth muscle. It is possible, however, that elevation of extracellular chloride induced by KCl stimulation may act to compete with anion-channel blockers in vascular smooth muscle.

The physiologic role and precise relation between the anion transport system and the cation transport system in vascular smooth muscle is not clear at the present time. In guinea pig atria, it has been demonstrated that uptake of $^{36}$chloride and $^{45}$calcium was inhibited by SITS,[5] whereas studies[4] in vascular smooth muscle demonstrate that replacement of $Cl^-$ with $NO_3^-$ potentiates $K^+$-induced contraction while enhancing influx of $Ca^{2+}$. These studies suggest that both $Cl^-$ and $Ca^{2+}$ may be intimately involved in tension development, or that the simultaneous movement of both an anion and a cation is required for excitation-contraction coupling. One possibility is that anion transport provides a force to facilitate $Ca^{2+}$ movement against osmotic and electrical resistance. A block of anion transport across the vascular smooth muscle membrane and a subsequent increase in membrane resistance to $Ca^{2+}$ could account for the decrease of contractility. Recently, a $Ca^{2+}$-dependent $Cl^-$ efflux mechanism has been described in vascular smooth muscle.[3] Because an exchanging anion carrier seems to be involved in regulation of intracellular pH, and an electroneutral $Ca^{2+}$-$H^+$ antiporter has been suggested, it is possible that disulfonic stilbene anion-channel blockers act on vascular smooth muscle by influencing an intracellular $Ca^{2+}$ pool.

## REFERENCES

1. ZHANG, A., B. T. ALTURA & B. M. ALTURA. 1988. Interaction of removal of extracellular $Mg^{2+}$ and $Cl^-$ ion on tone and contractility of vascular smooth muscle. FASEB J. **2:** A 757.
2. ALTURA, B. T. & B. M. ALTURA. 1975. Pentobarbital and contraction of vascular smooth muscle. Am. J. Physiol. **229:** 1635–1641.
3. GERSTHEIMER, F. P., M. MUHLEISEN, D. NEHRING & V. A. W. KREYE. 1987. A chloride-bicarbonate exchanging anion carrier in vascular smooth muscle of the rabbit. Pflüger's Arch. **409:** 60–66.
4. KAMM, K. E. & R. CASTEELS. 1979. Activation of arterial smooth muscle in the presence of nitrate and other anions. Pflüger's Arch. **381:** 63–69.
5. MINOCHERHOMJEE A-E-V-.M. D. ROUFOGALIS & J. H. MCNEILL. 1985. Effect of disulfonic stilbene anion-channel blockers on the guinea-pig myocardium. Can. J. Physiol. Pharmacol. **63:** 912–917.

# Influence of Chloride and Other Anions on Nervous Control of Vascular Smooth Muscle by Neurotransmitters

AIMIN ZHANG, BELLA T. ALTURA, AND
BURTON M. ALTURA[a]

*Department of Physiology*
*State University of New York*
*Health Science Center*
*Brooklyn, New York 11203*

Neurotransmitters released from autonomic nerve terminals play important roles in the regulation of vascular smooth muscle *in vivo* and *in vitro*. Studies on neurotransmitter secretion in the CNS and at neuromuscular junctions have suggested that chloride ion transport across synaptic cell membranes may participate in neurotransmission.[1,2] Studies were undertaken with rat aortas and portal veins to determine if changes in the anionic environment could exert any influence on neurotransmitters released from autonomic nerve terminals in vascular smooth muscle.

When rat aortas and portal veins equilibrated in normal Krebs-Ringer bicarbonate (NKRB)[3] were incubated in a modified Krebs-Ringer bicarbonate (SCN-KRB) in which the chloride was replaced by an isoosmolar amount of sodium thiocyanate, the resting tension of these tissues rapidly increased to a contractile state, and the spontaneous mechanical activity of the portal vein disappeared followed by spasm (FIG. 1). When the rat aorta and portal vein were returned to NKRB, large contractions were recorded, consisting of two components (fast and slow phases), which were followed by slow relaxation (10–20 minutes in duration) (FIG. 1). These contractile responses could be inhibited by alpha-adrenergic antagonists (e.g., $10^{-5}$M phentolamine) and denervation by 6-hydroxydopamine (6-OHDA). $\beta$-adrenergic, serotoninergic, cholinergic, and histaminergic antagonists had no effect. Cyclooxygenase inhibitors and local anesthetics also failed to affect these SCN-induced contractions. Because phentolamine in high concentrations might attenuate mechanical, neurogenic responses by a nonspecific effect on the membrane properties of smooth muscle,[4] EGTA, calcium removal, and verapamil were tested. EGTA (5 mM) and $Ca^{2+}$-free solution abolished both the increases in resting tension and the secondary contraction.

The most likely explanation for the contraction induced by replacement of $SCN^-$ by $Cl^-$ is neurotransmitter release from the autonomic nerve terminals, such as norepinephrine. The amplitudes of these SCN-KRB–induced contractions could be modified by other anions. The order of potency of these anions followed the lytropic series, where acetate>isethionate>$Cl^-$>$Br^-$>$NO_3$>$I^-$ (FIG. 2). Collectively, these observations suggest that norepinephrine release from sympathetic nerve endings (SNE) seems to have resulted from nerve terminal depolarization brought about by efflux of intracellular SCN rather than an

---

[a] Address for correspondence: Dr. B. M. Altura, Box 31, SUNY Health Science Center, 450 Clarkson Avenue, Brooklyn, NY 11203.

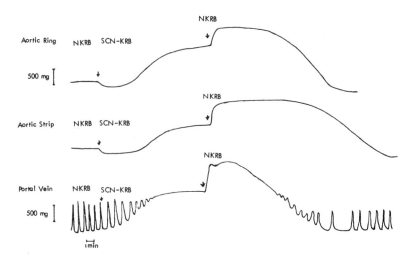

**FIGURE 1.** Influence of extracellular anions on tension development in rat aortas and portal veins. The tissues were first incubated in normal Krebs-Ringer bicarbonate (NKRB), then modified Krebs-Ringer bicarbonate (SCN-KRB) followed by NKRB. Subsequent changes of different anion-containing media are indicated by *arrows*. *Bars* at the left represent tension in milligrams.

effect on inactivation, or reversal, of neuronal reuptake, or other actions of the replacement anions.

The neuronal intracellular concentration of an anion depends on either its passive distribution across the membrane at electrochemical equilibrium or an anion transport system. Reduction of its extracellular concentration should cause its efflux until equilibrium is reestablished, and this efflux should be accompanied

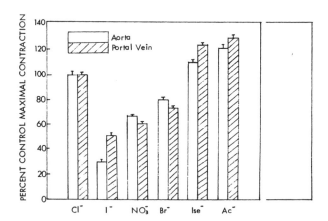

**FIGURE 2.** Effect of different anions on anion-related contractile responses of rat aortas and portal veins. The tissues were first incubated in NKRB, then SCN-KRB followed by NKRB (control) or anion-substituted media.

ɔy depolarization. The rate of efflux and the amount of depolarization would depend on the ratio of extracellular and intracellular log concentration of the anions and membrane permeability. Although membrane-anion permeability of SNE is unknown, it has been demonstrated that $SCN^-$ and $I^-$ are much more permeable than are other anions in cortical neurones and synaptosomes.[5,6]

Although anion-elicited neurotransmitter release is not likely to occur physiologically, this technique could prove valuable in unmasking a role for chloride ions in modulation of excitation-secretion coupling mechanisms in vascular smooth muscle, and as a model suitable for studying peripheral nervous control of vascular smooth muscle by neurotransmitters.

## REFERENCES

1. NILES, W. D. & D. O. SMITH. 1982. Anion blockers inhibit impulse-evoked quantal transmitter release at neuromuscular junction. J. Physiol. **329:** 203–210.
2. TURNER, J. D., J. A. BOAKES, J. A. HARDY & M. A. VIRMANI. 1987. Efflux of putative transmitters from superfused rat brain slice induced by low chloride ion concentrations. J. Neurochem. **48:** 1060–1068.
3. ALTURA, B. M. & B. T. ALTURA. 1974. Magnesium and contraction of arterial smooth muscle. Microvasc. Res. **7:** 145–155.
4. HOSHI, K. & S. FUJNO. 1980. Effect of alpha blockers on blood pressure and on the $Ca^{2+}$-contracture of cat aortic strips. Jap. J. Pharmacol. **30:** 427–435.
5. KANNE, B. I. 1978. Active transport of $\alpha$-aminobutyric acid by membrane vesicles isolated from rat brain. Biochemistry **17:** 1207–1211.
6. KELLY, J. S., K. KRNJEVIC, M. E. MORRIS & G. K. W. YIM. 1969. Anion permeability of cortical neurones. Exp. Brain Res. **7:** 11–31.

# Characteristics of $Na^+/H^+$ and $Cl^-/HCO_3^-$ Antiport Systems in Human Ileal Brush-Border Membrane Vesicles

K. RAMASWAMY,[a] J. M. HARIG, J. G. KLEINMAN, AND
M. S. HARRIS

*Gastroenterology Division*
*V. A. Medical Center and*
*The Medical College of Wisconsin*
*Milwaukee, Wisconsin 53295*

Perfusion studies of human ileum *in vivo* suggest the presence of a pathway for NaCl absorption involving the simultaneous exchange of $Na^+$ for $H^+$ and $Cl^-$ for $HCO_3^-$.[1] Studies using brush-border membrane vesicles (BBMV) from rat and rabbit intestine demonstrated convincing evidence for the presence of $Na^+/H^+$ and $Cl^-/HCO_3^-$ antiporters.[2-5] However, the presence of these antiporters has not been elucidated in the human ileum at the membrane level. We have established that intestinal tissue from organ donors provides an excellent source for BBMV to study nutrient absorption at the membrane level.[6,7] The aim of the present study was to use BBMV from human ileum to determine the existence of $Na^+/H^+$ and $Cl^-/HCO_3^-$ exchange systems in the apical membrane.

BBMV were isolated by the divalent cation-precipitation method and vesiculated as described previously.[6,7] Uptake of $^{22}Na$ and $^{36}Cl$ into BBMV was measured by the rapid filtration method. $H^+$ transport was estimated from the variations of pH gradients between intra- and extravesicular environment as determined from fluorescence quenching of acridine orange.

## $Na^+/H^+$ EXCHANGE

Concentrative uptake of $^{22}Na$ (fourfold overshoot above equilibrium) was observed under conditions of an outward proton gradient ($pH_{in}$ 5.5; $pH_{out}$ 7.5) (TABLE 1). Voltage clamping ($K_{in}^+ = K_{out}^+$ + valinomycin) reduced the uptake of $^{22}Na$ by 40–50%, indicating the presence of $Na^+$ conductance. Dissipation of the acridine orange fluorescence quench in ileal vesicles with a preformed pH gradient ($pH_{in}$ 5.5; $pH_{out}$ 7.5) was accelerated by either external $Na^+$ or voltage clamping in the absence of $Na^+$. The effects of $Na^+$ and voltage clamping were additive under the foregoing conditions. In the absence of a pH gradient, acridine orange quenching was induced by intravesicular $Na^+$ as well as an interior negative membrane potential. In voltage-clamped BBMV, pH-driven $Na^+$ uptake was inhibited by amiloride ($K_i = 140$ $\mu M$). The apparent $K_m$ and $V_{max}$ values for $Na^+$ uptake were

[a] Address for correspondence: K. Ramaswamy, Ph.D., VA Medical Center, Gastroenterology Section/111C, 5000 W. National Avenue, Milwaukee, WI 53295.

**TABLE 1.** Effect of pH Gradients on $^{22}$Na Uptake by Human Ileal BBMV[a]

|  | $^{22}$Na Uptake (pmol/mg protein) |
|---|---|
| $7.5_{in}/7.5_{out}$ | $234 \pm 10$ |
| $5.5_{in}/7.5_{out}$ | $2,302 \pm 88$ |
| $7.5_{in}/5.5_{out}$ | $147 \pm 10$ |
| $5.5_{in}/5.5_{out}$ | $166 \pm 10$ |

[a] BBMV were preloaded with 100 mM tetramethylammonium gluconate, 100 or 115 mM mannitol, and 50 mM Tris-HEPES buffer (pH 7.5) or 50 mM Tris-Mes buffer (pH 5.5); $^{22}$NaCl, 1 mM; time of incubation, 6 seconds.

27 mM and 940 nmol/mg protein per minute, respectively. Li$^+$ and NH$_4^+$, but not Cs$^+$, K$^+$, Rb$^+$, or choline$^+$, inhibited pH-gradient–driven $^{22}$Na uptake. These results demonstrate in human ileal BBMV the presence of a Na$^+$/H$^+$ exchanger and conductive transport pathways for Na$^+$ and H$^+$.

## Cl$^-$/HCO$_3^-$ EXCHANGE

An inside alkaline pH gradient (pH $8_{in}$, pH $6.5_{out}$) stimulated Cl$^-$ uptake into BBMV. When an outwardly directed bicarbonate gradient was imposed in addition to the inside alkaline pH gradient, a twofold overshoot of uptake above equilibrium was observed, suggesting a Cl$^-$/HCO$_3^-$ exchange process (TABLE 2). A valinomycin-induced K$^+$ diffusion potential (inside positive) enhanced Cl$^-$ uptake rate, consistent with a conductive pathway for Cl$^-$ transport. DIDS inhibited pH and HCO$_3^-$ gradient stimulated Cl$^-$ uptake rate (72% inhibition at 10 mM). K$_m$ and V$_{max}$ values for Cl$^-$ uptake were 13.3 mM and 130 nmol/mg protein per minute, respectively. Thiocyanate, nitrate, iodide, and bromide inhibited Cl$^-$ uptake rate between 65–75%, whereas fluoride, lactate, para-aminohippurate, phosphate, and sulfate did not exert appreciable inhibition. These results demonstrate the presence of Cl$^-$/HCO$_3^-$ exchange and a conductive pathway for Cl$^-$.

Our results provide conclusive evidence for the presence of both Na$^+$/H$^+$ and Cl$^-$/HCO$_3^-$ antiporters in human ileal brush-border membrane. Future research will focus on determining the role of these antiporters in NaCl absorption and will study possible direct Na$^+$ and Cl$^-$ coupling.

**TABLE 2.** Effect of pH and Bicarbonate Gradients on $^{36}$Cl Uptake[a]

|  | $^{36}$Cl Uptake (pmol/mg protein) |
|---|---|
| $8.0_{in}/8.0_{out}$ | $463 \pm 20$ |
| $8.0_{in}/6.5_{out}$ | $1,356 \pm 100$ |
| HCO$_3$,$8.0_{in}/6.5_{out}$ | $4,611 \pm 208$ |

[a] BBMV were preloaded with 150 mM KHCO$_3$ or K gluconate, 100 mM Tris-HEPES buffer (pH 8.0); tetramethylammonium $^{36}$chloride, 5 mM. Incubation media contained 150 mM K gluconate and 100 mM Tris-HEPES (8.0) or 150 mM K gluconate, 50 mM mannitol, and 100 mM Tris-Mes (6.5) or 3 mM KHCO$_3$, 147 mM K gluconate, 50 mM mannitol, and 100 mM Tris-Mes (6.5); time of incubation, 6 seconds.

## REFERENCES

1. TURNBERG, L. A., F. A. BIEBERDORF, S. G. MORASOWSKI, & J. S. FORDTRAN. 1970. J. Clin. Invest. **49:** 557–567.
2. LIEDKE, C. & U. HOPFER. 1982. Am. J. Physiol. **242:** G263–271.
3. LIEDKE, C. & U. HOPFER. 1982. Am. J. Physiol. **242:** G272–280.
4. KNICKLEBEIN, R., P. S. ARONSON, W. ATHERTON & J. W. DOBBINS. 1983. Am. J. Physiol. **245:** G504–G510.
5. KNICKLEBEIN, R., P. ARONSON, C. SCHRON, J. SEIFTER & J. DOBBINS. 1985. Am. J. Physiol. **249:** G236–G245.
6. RAJENDRAN, V. M., J. M. HARIG, M. B. ADAMS & K. RAMASWAMY. 1987. Am. J. Physiol. **252:** G33–G39.
7. KLEINMAN, J. G., J. M. HARIG, J. A. BARRY & K. RAMASWAMY. 1988. Am. J. Physiol. **255:** G206–G211.

# Na-Dependent HCO$_3$$^-$ Transport and Cl$^-$/HCO$_3$$^-$ Exchange in Ciliary Epithelium

J. MARIO WOLOSIN,[a,b] JOSEPH A. BONANNO,[c] AND
TERRY E. MACHEN[c]

*Departments of Ophthalmology and Physiology and Biophysics*
*Mount Sinai School of Medicine[a]*
*New York, New York 10029*
*and*
*Department of Physiology-Anatomy*
*University of California[c]*
*Berkeley, California 94720*

Active secretion of fluid by the epithelium covering the ciliary body (cb) contributes to the formation of aqueous humor. This epithelium consists of two layers of cells closely apposed and metabolically communicated to each other through their apical membranes. The inner, pigmented epithelial (pe) layer faces basally the vascularized cb stroma; the outer nonpigmented epithelial (npe) layer faces basally the posterior humoral space and is believed to be the main entity responsible for the secretory activity in which bicarbonate metabolism plays a central role.[1,2] In the rabbit, *in vivo*, the [HCO$_3$$^-$) proximal to the cb exceeds that found both in the plasma and at distal humoral locations.

The elucidation of molecular events underlying this activity has been hindered by both the complex epithelial arrangement and the intricate anatomy of the ciliary structures. Therefore, the development of simplified models, amenable to quantitative study, is being actively sought. We have found that when cb dissected from pigmented, Dutch Belted rabbits are incubated with the esterified form of 2′,7′ biscarboxyethyl 5(6)-carboxyfluorescein, a pH probe, entrapped fluorescence develops only in npe. Thus, using microscope-based fluorophotometry it was possible to monitor npe cell pH (pH$_i$), while a whole strip of cb was perfused by solutions of varying composition (FIG. 1). The initial studies reported here were aimed at identifying mechanisms of HCO$_3$$^-$ transport.

In CO$_2$ (5%)/HCO$_3$$^-$-rich Ringer's solution (BRR, pH 7.45), the pH$_i$ is 7.10 ± 0.11 (± SD, $n$ = 11). Replacement of Cl$^-$ by gluconate (FIG. 2, Exp. II) elicited rapidly (~1 minute) a 0.6-unit pH$_i$ increase (cell [HCO$_3$$^-$] raising from 11 to 44 mM), which was followed by a gradual decrease to a new stable pH$_i$ about 0.2 units higher than that in Cl$^-$ medium. At this piont or after a 2-hour incubation in Cl$^-$-free solution, replacement of 71 mM Na by K$^+$ increased pHi by 0.55 ± 0.1 ($n$ = 4) units, raising cell [HCO$_3$$^-$] to about 70 mM. Replacement of 71 mM Na$^+$ by N-methyl-d-glucamine (NMDG), in contrast, decreased pH$_i$ by 0.05 unit. The rapid effect of high [K$^+$] on pH$_i$ was abolished in Na$^+$-free NMDG medium. The

[b] Address for correspondence: J. Mario Wolosin, Department of Ophthalmology, Box 1183, Mount Sinai School of Medicine of New York, One Gustave Levy Place, New York, NY 10029.

[K$^+$] increase maneuver also increased pH$_i$ in Cl$^-$ medium (FIG. 2, Exp. I, +0.35 ± 0.07, $n$ = 4). This increase was diminished to 0.17 ± 0.03 ($n$ = 4) in CO$_2$/HCO$_3^-$-free Ringer's solution (BFR, HCO$_3^-$ replaced by HEPES, 5-N-2-hydroxyethyl-piperazine-N'-2-ethanesulfonic acid). The addition of 0.5 mM dihydro 4,4'-diisothiocyano-stilbene-2,2'-disulfonate (H$_2$-DIDS) elicited a 0.15 pH$_i$ increase and inhibited (80–100%) the increases elicited by both high [K$^+$] and removal of Cl$^-$ (FIG. 2, Exp. III).

Inferences on the nature of the HCO$_3^-$ transport mechanism were also derived from studies of pH$_i$ recovery after cell acidification by NH$_4^+$ prepulses. In BFR, pH$_i$ recovery was fully Na$^+$ dependent, largely inhibited by amiloride and insensitive to H$_2$-DIDS. In BRR pH$_i$ recovery was also fully dependent on Na$^+$ but proceeded fourfold faster, was insensitive to amiloride, and was slowed 52% by H$_2$-DIDS.

Taken together these results indicate that npe pH$_i$ is influenced by a Cl$^-$/HCO$_3^-$ exchanger and a Na$^+$-dependent Cl$^-$-independent HCO$_3^-$ transport that either is cotransported with K$^+$ or, given the effect of K$^+$ in cell potential, carries a net negative change. Because pH is higher in BFR than in BRR media, it can be concluded that HCO$_3^-$ is actively secreted from the npe.

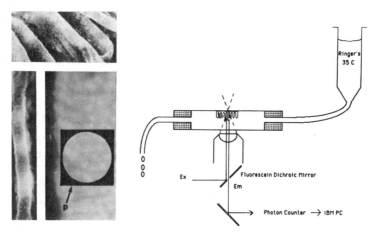

**FIGURE 1.** Experimental methods. (*Left*) Micrographs of dissected ciliary body strips loaded with BCECF and photographed under epifluorescent illumination using 4× (*upper photo*), 20× (*bottom left photo*), and 100× (*bottom right photo*) objectives. (*Right*) Schematic description of experimental arrangement used to determine the pH$_i$ of nonpigmented cells in whole ciliary body strips. The stromal side of BCECF-loaded strips was affixed to the upper lid coverslip of a perfusion chamber by means of cyanoacrylate glue. Thus, the ciliary ridges faced the objective and were exposed to the 36°C perfusate. Alternating pulses (~ 0.2 seconds in duration) of 439- or 495-nm monochromatic light were used for sample excitation. Emission at wavelengths longer than 525 nm was directed to a computerized photon counting system able to establish the ratio between the intensities derived from the 495 and 439 excitations (R 495/439). A pinhole, described schematically in the left side micrograph, was introduced in the light path to limit the number of illuminated cells (10–20). Ratios were converted to actual pH$_i$s by calibration against the ratios measured after equalizing cellular to extracellular pH. This was achieved by the use of a high K$^+$ (105 mM) solution containing the H$^+$/K$^+$ exchanger nigericin.

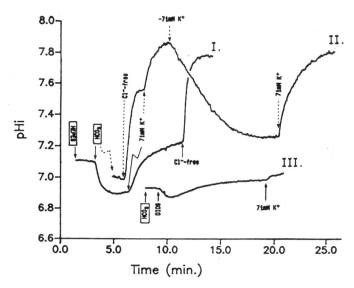

**FIGURE 2.** Effect of Cl removal and elevated K on the pH$_i$ of nonpigmented cells. Experiment I (*thin-line arrows*) was started in BFR (HEPES) which was then replaced by BRR (HCO$_3$) at *second arrow*. At the *third arrow* 71 mM Na were replaced by K. At the *fourth arrow* all Cl was replaced by gluconate while maintaining the high [K]. Experiment II (*broken-line arrows*) was started in BRR. At the *second arrow* all Cl was replaced by gluconate. At the *third arrow* 71 mM of the Na gluconate were replaced by K gluconate. At the *fourth* the Na gluconate (−71 mM K) was reintroduced, and 10 minutes later, at the *fifth arrow*, the replacement of Na by K salt was made again. Experiment III was started in BRR (HCO$_3$). The perfusate was then complemented with 0.5 mM H$_2$-DIDS (DIDS), and after 10 minutes, while maintaining the H$_2$-DIDS concentration, 71 mM Na were replaced by K.

### REFERENCES

1. COLE, D. F. 1977. Secretion of aqueous humor. Exp. Eye Res. 25 Suppl. 161–176.
2. KINSEY, V. E. 1971. Ion movement in ciliary processes. *In* Membranes and Ion Transport, Vol 3. Bittar, E. E., ed.: 161–175. Wiley Interscience, New York.

# Ion Permeability and Pump Regulation[a]

P. LORENTZON,[b] S. J. HERSEY,[c] B. WALLMARK,[b] AND
G. SACHS[d]

*[b]Hassle AB Göteborg*
*Göteborg, Sweden*

*[c]Emory University*
*Atlanta, Georgia*

*[d]Cure Wadsworth VA Hospital and UCLA*
*Los Angeles, California*

Ion pumps can be classified in various ways. For analysis of pumping regulation by passive ion permeabilities, the simplest classification of pumps consists of uniport or countertransport (antiport) categories.

A uniport pump transfers charge in one direction across a membrane; therefore, in the absence of a compensating conductance, it generates a potential rather than an ion gradient. When a conductance is present, an ion gradient is produced with a compensatory decrease in potential. The usual equation,

$$\Delta\bar{\mu}_x = \Delta\psi + RT/nF \ln [x]_o/[x]_i,$$

shows that for a pump the electrochemical gradient is a constant and is a sum of the potential and ion gradient terms. The relative contributions of the osmotic gradient and potential terms for a given membrane will reflect the size of the conductance of the membrane as well as the turnover rate of the pump. For an electroneutral pump or an electrogenic pump in a membrane of high conductance (for other than the transported ion), the potential term becomes minimal and only an ion gradient is generated.

With a countertransport pump, two situations can be envisaged. An electroneutral pump requires a supply of the countertransported ions at either face of the membrane, Na and K for the Na pump and H and K for the gastric proton pump, for example. In addition, if the stoichiometries of the transported ions are unequal, the pump will be electrogenic; with the uniport pump, a conductance must be present in the membrane to compensate for charge transfer by the enzyme.

Electrogenic pumps, whether uniport or antiport, can be regulated by the supply of transported ions to either face of the enzyme ($H^+\cdot Na^+\cdot K^+$ or $Ca^{2+}$), by modulation of ionic conductance of the membrane, and by direct regulation of the pump itself, as by phosphorylation of a regulating subunit such as phospholamban[1] or by membrane potential. Regulation of an electroneutral, countertransport pump is independent of membrane conductance, but it does depend on the supply of transported ions. Furthermore, if the individual steps of the pump are voltage sensitive, even though equal numbers of ions are exchanged, voltage may affect the rate of the pump. For example, in electroneutral H,K ATPase, it is thought that transport of H and therefore K (to satisfy electroneutrality) involves opposite charge movement through the pump, and if either one of these steps of the

[a] This work was supported by VA SMI and NIH DDK.

enzyme is rate limiting, voltage may effect ATPase turnover and hence acid secretion by the gastric mucosa. There are therefore several means by which an ion pump can be regulated.

Studies of regulation of transport ATPases are most commonly performed after enzyme isolation. This raises the inevitable question about possible artifacts during isolation or loss of possible regulatory processes. If it is possible to determine factors that regulate ion pumps in intact cells or even permeabilized cells, then data obtained in isolated membrane particles can be confirmed.

Recent work on the vacuolar proton pump present in parafollicular cells indicates that acidification in granules of this cell type apparently depends on activation of a Cl conductance. This was shown on isolated membrane vesicles and in permeabilized cells.[2] Therefore, in this cell type the granule membrane pump is limited by an activatable parallel ion pathway, a Cl conductance. The gastric H,K ATPase is a model of an ion pump also regulated by parallel ion permeation pathways that is amenable to study at various levels of cellular organization.[3]

In terms of cell content, the gastric proton pump is the major ATPase that uses mitochondrially generated ATP. It is an electroneutral exchange ATPase that

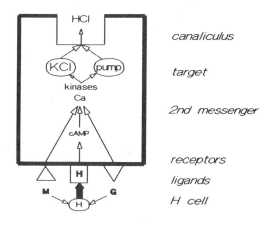

**FIGURE 1.** Cartoon of the mechanism of stimulation of the parietal cell showing the insertion/activation of KCl and H,K ATPase into the secretory membrane of the parietal cell.

exchanges cytosolic $H_3O^+$ for luminal $K^+$. Various ions can act as surrogates for $K^+$, particularly $NH_4^+$. $Na^+$ acts as a surrogate for $H_3O^+$.[4] Vesicle studies have shown that when isolated from the resting parietal cell, ATPase activity and $H^+$ transport are restricted by a low permeability to KCl,[5] but when isolated from stimulated tissue, KCl restriction appears to be relaxed.[6] In addition to a change in ion permeability that apparently must take place, the parietal cell undergoes a considerable morphologic transition between rest and stimulation. Here, the ATPase that is present in smooth surfaced cytoplasmic vesicles appears to translocate to microvilli on the cell surface (more precisely the infolding of the membrane termed the secretory canaliculus). This is illustrated conceptually in FIGURE 1. It is not known if the change in KCl permeability depends on this association with the plasma membrane or if the change precedes association. Neither is it known if the only mechanism of activation of the enzyme *in situ* is a change in the KCl permeability of the membrane, or if other factors are also involved.[7] Evidence from vesicle studies is consistent with this ATPase being a neutral pump, whereas much classical evidence from frog gastric mucosa has been interpreted on the basis of electrogenic H secretion.[8] This short review explores the use of $NH_4^+$ as a probe to activate the H,K ATPase and the electrogenicity of the enzyme.

Three models of enzyme function have been employed for these studies. The first, dealing with the action of $NH_4^+$ on the isolated enzyme, will describe data obtained with isolated resting ATPase containing vesicles that show the properties of $NH_4^+$ as a surrogate for $K^+$ as well as providing evidence for voltage sensitivity of the K limb of the H,K ATPase. The second will use $^{14}CO_2$ production from labeled glucose to determine whether the addition of $NH_4^+$ can mimic stimulation by dbcAMP in intact rabbit parietal cells, and finally permeabilized cells will be used to confirm that it is a change in ion permeability that determines turnover of the H,K ATPase derived from vesicle and intact cell data.

## ACTION OF $NH_4^+$ ON GASTRIC ATPase

In resuspended, lyophilized membranes, the gastric ATPase responds to changes of $K^+$ in a biphasic manner, first with stimulation and then with inhibi-

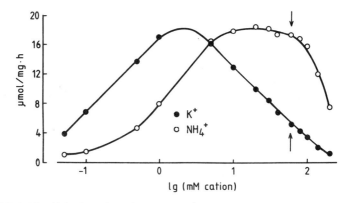

**FIGURE 2.** The biphasic action of $K^+$ and $NH_4^+$ on activity of the gastric ATPase in ion-permeable vesicles.

tion. A similar response is obtained with changing $NH_4^+$ concentration.[9] FIGURE 2 shows that the biphasic response is shifted to the right when ammonium ion is compared to potassium ion. However, virtually the same maximal activity is obtained with both ions. These data show that $NH_4^+$ is an effective surrogate for $K^+$, but has lower affinity with respect to both stimulation and inhibition. Inhibition can be explained as due to the cations binding to a dephospho- form of the enzyme. At 60 mM cation, where K-stimulated activity has almost been abolished, $NH_4^+$ is still fully activating. However, $K^+$ is still a better stimulant of dephosphorylation (FIG. 3).

An abbreviated cycle of the enzyme can be portrayed as follows (FIG. 4). The enzyme in the presence of protons and Mg ATP phosphorylates to form first an $E_1$-P (ion binding sites cytosolic) and then an $E_2$-P conformer (ion binding sites luminal). This latter form turns over very slowly in the absence of luminal K.

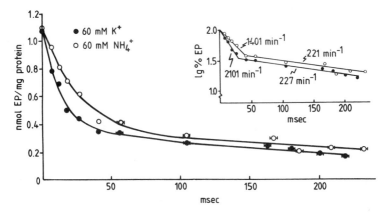

**FIGURE 3.** Stimulation of dephosphorylation of preformed phosphoenzyme by the addition of either $K^+$ or $NH_4^+$ at equal concentrations of 60 mM. This should be compared to FIGURE 2, where at this concentration, potassium inhibits and ammonium stimulates ATP turnover.

However, in the presence of luminal $K^+$ there is rapid dephosphorylation to first an $E_2 \cdot K$ and then an $E_1 \cdot K$ form of the enzyme. ATP reacts with these forms of the enzyme to catalyze the displacement of K and the reinitiation of phosphorylation. Because 60 mM $K^+$ is still better than $NH_4^+$ in stimulating the rate-limiting step of dephosphorylation, but effectively inhibits ATPase activity, this effect must be due to binding of K to a dephospho- form of the enzyme. The rate of exchange of K across the enzyme in the absence of ligands is equal to the overall turnover of the enzyme; therefore, it is likely that the rate-limiting step is the change from $E_2 \cdot K$ to $E_1 \cdot K$. Because it requires more $NH_4^+$ to inhibit the ATPase, the binding of this ion to the dephospho- forms of the enzyme is of lower affinity. The data presented herein may be equivalent to the evidence for kinetic occlusion of the K site obtained for the NaK ATPase 15 years ago.[10]

The ATPase in unlyophilized vesicles is not activated by $K^+$ added to the medium, because in these intact, inside-out membrane vesicles, there is insufficient $K^+$ permeability to provide activation at the luminal site. This is illustrated in FIGURE 5 where in the absence of ionophores, virtually no $K^+$ activation is seen over a wide range of K concentrations. In contrast, in leaky vesicles the charac-

**FIGURE 4.** An abbreviated catalytic cycle of H,K ATPase, illustrating the two major conformations of the enzyme alternating as the enzyme is phosphorylated and dephosphorylated.

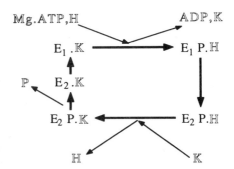

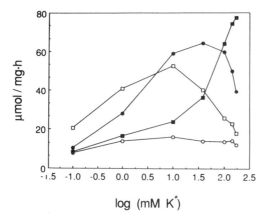

**FIGURE 5.** Effect of valinomycin and valinomycin with tetrachlorsalicylanilide on the ATPase activity of ion-tight gastric vesicles. The presence of valinomycin prevents the $K^+$ inhibition of turnover. TCS restores the inhibition by collapsing the potential gradient generated in the presence of valinomycin. $\bigcirc\!-\!\bigcirc$ = $K^+$ alone, $\blacksquare\!-\!\blacksquare$ = $K^+$ + valinomycin, $\bullet\!-\!\bullet$ = $K^+$ + valinomycin + TCS, and $\square\!-\!\square$ = leaky membranes.

teristic biphasic effect of $K^+$ is seen. With $NH_4^+$, no ionophore is necessary to demonstrate activation of the ATPase.

In the presence of valinomycin, an electrogenic $K^+$ ionophore, there is progressive activation of the ATPase and virtually no inhibition, even at 175 mM. A difference between lyophilized vesicles and intact vesicles in the presence of valinomycin is the generation of a potential directed positively inward in the ionophore-treated membranes. If this potential is collapsed by the further addition of an electrogenic protonophore, TCS, inhibition by $K^+$ is restored (FIG. 5). One explanation for this finding is that the rate-limiting step, the change from the K liganded form of $E_2$ to $E_1$, is charge carrying and therefore sensitive to voltage. The interior positivity of the vesicles therefore stimulates the conversion of $E_2 \cdot K$ to $E_1 \cdot K$ and inhibits the formation of $E_2 \cdot K$, thus decreasing inhibition by high $K^+$ (FIG. 6). Accordingly the enzyme is probably charge carrying in the K transport direction, and because it is overall an electroneutral enzyme, it must be charge carrying in the H pumping direction as well. Indeed, adsorption of ATPase vesicles onto bilayers has shown that a capacitative potential develops with the addition of ATP, but no potential is present when K is added.[11] This behavior is expected of a net electroneutral process where, however, the ion-translocating steps carry equal charges in opposite directions.

The concept of an electroneutral pump with voltage-sensitive, ion-motive steps may partly resolve the conflict between data in intact frog mucosa showing voltage sensitivity of acid secretion in the presence of high K solutions in the

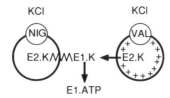

**FIGURE 6.** Diagram explaining the effect of potential on the gastric H,K ATPase. (*Left*) The $E_2 \cdot K$ form of the pump converting to the $E_1$ form is slow in the absence of potential, and (*right*) the presence of valinomycin generates a potential, increasing the rate of loss of the $E_2 \cdot K$ form.

secretory solution and data obtained in mammalian vesicles demonstrating electroneutrality. If in the frog mucosa the rate of enzyme turnover in the H transport direction were rate limiting, then application of current to provide increasingly negative secretory side potentials would stimulate acid secretion. With increasing voltage, the $E_2 \cdot K$ form of the enzyme would start to predominate, and reduction in acid secretion would be observed. Data such as these have been obtained in intact frog mucosa.[8]

## ACTION OF $NH_4^+$ ON RESPIRATION IN INTACT GLANDS

The effect of ammonium on the ATPase is of interest not only because of the ability to define the steps of the enzyme reaction that are inhibited by K, but also because the permeability of $NH_3$ allows bypass of permeability restrictions on the enzyme. FIGURE 7 explains the interaction of $NH_4^+$ with the ATPase in ion-tight vesicles and presumably in the intact cell. The free permeability of $NH_3$ allows entry into the vesicle without the need for an ion pathway. Inside the vesicle, $NH_3$

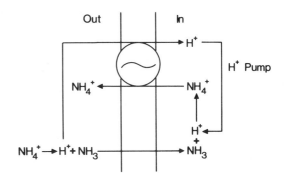

FIGURE 7. Diagram showing the interaction of the $K^+$ surrogate, $NH_4^+$, with the ATPase.

picks up a proton, forms $NH_4^+$, and then is extruded by the ATPase. This results in activation of the enzyme in the absence of activation of ion pathways, but also abolishes any proton gradient that the pump can generate at saturating $NH_4^+$ concentrations.

The major aerobic metabolism of the parietal cell and indeed of gastric glands is due to the energy requirements of acid secretion, specifically the turnover of the H,K ATPase. Measurement of either $O_2$ consumption or labeled $CO_2$ production from glucose provides an alternative method for measuring stimulation of acid secretion as compared to uptake of a weak base such as aminopyrine or benzylamine. In the presence of $NH_4^+$, there is no acidification by gastric glands, but the production of labeled $CO_2$ should represent, in part, turnover of the ATPase.

If the addition of $NH_4^+$ stimulates respiration in unstimulated gastric glands to about the same level that a more usual secretagogue stimulates respiration, and if it can be shown that this stimulation is sensitive to inhibitors of ATPase, then it is evidence for the concept that KCl permeability of the ATPase membrane is rate limiting for acid secretion. Activation of KCl permeability then represents the major mechanism for stimulation of the enzyme.

Stimulated glands consume about 60 nM glucose per milligram of dry weight per hour. This corresponds to an ATP production of 2.2 $\mu$mol/mg per hour (FIG. 8). Of that consumption 75% is blocked by inhibitors of the H,K ATPase, such as SCH 28080 and omeprazole. Hence about 1.6 $\mu$mol ATP is dedicated to gastric ATPase in stimulated parietal cells. The addition of 30 mM $NH_4^+$ to stimulated glands neither inhibits nor stimulates glucose utilization. The pH gradient is abolished. This can be observed in gastric glands where acid secretion is monitored on an image analyzer by the uptake of acridine orange, and the metachromatic shift is plotted in pseudocolor. FIGURE 9 shows the region of acidification and a rapid loss of acid space with the addition of the cation. A similar conclusion, namely, that acidification is abolished, can be reached by observing the effect of omeprazole as compared to that of SCH 28080 after the addition of ammonium. Omeprazole requires an acid space for its action on the H,K ATPase,[12] SCH 28080, although accumulated in acid spaces, does not depend on acid-catalyzed conversion for its activity as a K competitive antagonist.[13] As also shown in FIGURE 8, the action of

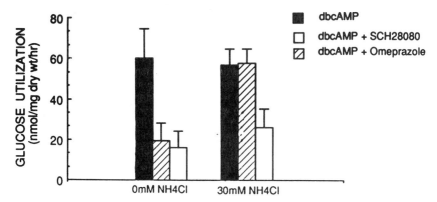

**FIGURE 8.** $^{14}C\text{-}CO_2$ production from glucose by intact, stimulated rabbit gastric glands. The addition of the inhibitors of the H,K ATPase, SCH 28080 and omeprazole, shows that about 75% of the production is due to turnover of the H,K ATPase. The addition of $NH_4^+$ does not further stimulate $CO_2$ production, but it abolishes the action of the acid space-dependent inhibitor omeprazole, leaving intact the inhibition due to SCH 28080.

omeprazole is abolished in the presence of $NH_4^+$, whereas that of SCH 28080 is only slightly affected.

The addition of $NH_4^+$ to unstimulated glands enhances glucose utilization to close to the level obtained with the addition of dbcAMP, and then no further stimulation occurs on the addition of dbcAMP (FIG. 10). $NH_4^+$ stimulates respiration in this preparation as effectively as does the secretagoge, dbcAMP. It can therefore be concluded that using $NH_4^+$ as a probe for ion pathways regulating the H,K ATPase shows that in gastric glands activation of KCl permeability is both a necessary and a sufficient condition for activation of the H,K ATPase.

It is possible to compare the predicted ATP production to ATP utilization by the ATPase, as discussed in the next section. From the foregoing, the ATP production for an ATP/glucose ratio of 36 that is sensitive to either omeprazole or SCH 28080 is 1.6 $\mu$mol/mg per hour. ATP utilization by the H,K ATPase measured under ion permeability unrestricted conditions (TABLE 1) corresponds to 1.4

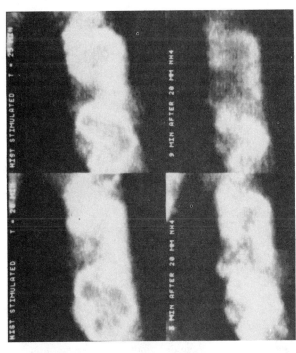

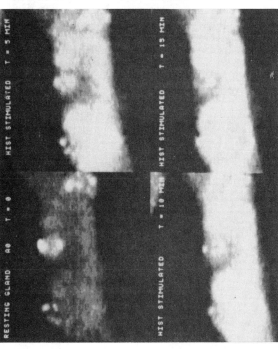

**FIGURE 9.** A pair of image analysis pictures using the red shift of acridine orange (blue in pseudocolor, light in black and white), showing first, the stimulation of acid secretion by histamine in rabbit gastric glands (*left*), and then the collapse of the pH gradient due to the addition of $NH_4^+$ (*right*). (Imaging was performed by D. Blissard.)

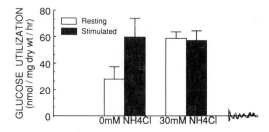

**FIGURE 10.** Effect of the addition of $NH_4^+$ to unstimulated and stimulated rabbit gastric glands, showing that the permeable $NH_3$/$NH_4^+$ couple stimulates respiration of this preparation to the level induced by the potent secretagog dbCAMP.

$\mu$mol ATP used/mg per hour. The correlation of these two numbers suggests that in the gland preparation, ATPase activity is limited largely by ion permeabilities, as will be discussed further.

## ENZYME ACTIVITY IN PERMEABILIZED GLANDS

When gastric glands or parietal cells are permeabilized with digitonin, they retain their ability to generate a proton gradient with the addition of ATP.[14,15] Two components are evident in the response to ATP.[15] The first component reflects a compartment that acidifies without the addition of valinomycin. This compartment, therefore, must allow access of $K^+$ to the luminal face of the H,K ATPase. It is likely that this compartment represents a component of enzyme that is associated with stimulated HCl secretion and therefore has a KCl permeability pathway.[15] A second phase of acidification is induced by the addition of valinomycin (FIG. 11). This component of acidification reflects a fraction of the H,K ATPase that is restricted by $K^+$ permeability and therefore reflects the resting or unstimulated component of enzyme. These data on proton transport by permeabilized glands are mirrored by direct measurements of H,K ATPase activity in this preparation.

Selective determination of H,K ATPase activity in permeabilized gastric glands is made either as K-stimulated ATPase in the presence of ouabain and Na-free media or through the use of SCH 28080 or omeprazole as selective inhibitors.[16] Based on proton transport measurements, H,K ATPase activity should have multiple components. One component should depend on the presence of KCl alone, representing that fraction of the ATPase that is associated with stimulation of KCl permeability. This fraction of enzyme activity could contain, as a subfraction, ATPase associated with membranes that have been damaged and

**TABLE 1.** ATPase Activity in Permeable Gastric Glands[a]

|  | ATPase Activity (nmol/min-mg protein) | |
|  | Resting | Stimulated |
| --- | --- | --- |
| Basal (K free) | 17 | 18 |
| 90 mM KCl | 27 | 40 |
| KCl + nigericin | 68 | 67 |
| 90 mM $NH_4Cl$ | 59 | 52 |

[a] Glands were pretreated with cimetidine (resting) or histamine (stimulated) before permeabilizing and enzyme assay.

therefore have lost KCl restriction. This subfraction would be insensitive to ome-prazole, because no acidification, and thus no inhibitor activation, would occur in the damaged membranes. A large number of experiments have indicated that about 90% of K-stimulated ATPase is inhibited by omeprazole and thus capable of forming proton gradients. A second major component of H,K ATPase activity should be stimulated by KCl only in the presence of ionophores, such as valino-mycin or nigericin, or by $NH_4^+$ in the absence of ionophores. This component would represent the fraction of enzyme in a resting or KCl impermeant compart-ment. Data reflecting the distribution of H,K ATPase activity in unstimulated and prestimulated permeabilized glands are presented in TABLE 1.

It can be seen that, indeed, there are two major fractions of K-stimulated ATPase activity. One fraction is found in the presence of KCl alone, and a second fraction in the presence of KCl plus an ionophore (nigericin). The addition of 90 mM $NH_4Cl$ stimulates the enzyme more than 90 mM KCl and nearly to the level obtained with nigericin plus KCl. These data are consistent with the concept that H,K ATPase activity in permeabilized glands is restricted by K permeability. The data in TABLE 1 show also that prestimulation results in an increase in the fraction of activity seen with KCl alone, but no increase in maximal activity as found with

**FIGURE 11.** Acidification by permea-ble rabbit gastric glands measured as the quench of acridine orange fluorescence. Glands incubated in KCl medium show acidification in response to ATP, and a second component of acidification when valinomycin is added to bypass the lim-ited KCl permeability.

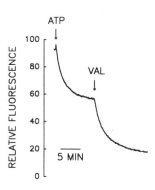

nigericin. Moreover, the activity seen with $NH_4Cl$ does not change with prestimu-lation. These results then are consistent with the concept that the mechanism of activation of acid secretion is mainly an activation of a KCl pathway and relaxa-tion of K restriction.

This paper has presented data using a variety of models, reflecting various degrees of organization of the acid secretory complex of the parietal cell. The data using $NH_4^+$ and ionophores as probes of the functional state of the complex show that the major restriction on turnover of the gastric proton pump is an associated cation permeability. Regulation of this ATPase, therefore, is mediated mainly by changing the properties of a parallel pathway. Additionally, by comparing $K^+$ and $NH_4^+$ activation of the enzyme, kinetic occlusion of the $E_2 \cdot K$ form of this enzyme is deduced, and the deocclusion is shown to be voltage sensitive. This suggests that although the pump is electroneutral, the ion transport steps are equally elec-trogenic. Because the rate-limiting step is voltage sensitive, it is suggested that the pump activity could be modulated by changes in membrane potential arising from changes in ion conductances. With the H,K ATPase, this effect would be small as compared to the role of membrane potential in modulating a fully electrogenic uniport mechanism.

## REFERENCES

1. FLEISCHER, S. & M. INUI. 1988. Progr. Clin. Biol. Res. **273:** 435–450.
2. BARASCH, J., M. R. GERSHON, E. NUNEZ, H. TAMIR & Q. AL-AQAWTI. 1988. J. Cell. Biol. **107:** 2137–2147.
3. SACHS, G., H. H. CHANG, E. RABON, R. SCHACKMAN, M. LEWIN & G. SACCOMANI. 1976. J. Biol. Chem. **251:** 7890–7896.
4. POLVANI, C., G. SACHS & R. BLOSTEIN. 1989. Biophys. J. **55:** 337a.
5. SCHACKMAN, R., A. SCHWARTZ, G. SACCOMANI & G. SACHS. 1977. J. Membr. Biol. **32:** 361–381.
6. WOLOSIN, J. M. & J. G. FORTE. 1985. J. Membr. Biol. **83:** 261–272.
7. NANDI, J. M. A. ZHOU & T. K. RAY. 1987. Biochemistry **26:** 4264–4272.
8. REHM, W. S. 1989. This conference. N.Y. Acad. Sci.
9. LORENTZON, P., G. SACHS & B. WALLMARK. 1988. J. Biol. Chem. **263:** 10705–10710.
10. POST, R. L., C. HEGYVARY & S. KUME. 1972. J. Biol. Chem. **247:** 6530–6540.
11. FENDLER, K., HVD. HIJDEN, G. NAGEL, J. J. H. M. DE PONT & E. BAMBERG. 1988. Prog. Clin. Biol. Res. **268A:** 501–510.
12. SACHS, G., E. CARLSSON, P. LINDBERG & B. WALLMARK. 1988. Ann. Rev. Pharmacol. Toxicol. **28:** 269–284.
13. WALLMARK B. BRIVING, C., J. FRYKLUND, K. MUNSON, R. JACKSON, J. MENDLEIN, E. RABON & G. SACHS. 1987. J. Biol. Chem. **262:** 2077–2084.
14. PEREZ, A., D. BLISSARD, G. SACHS & S. J. HERSEY. 1989. Am. J. Physiol. **256:** 9299–9305.
15. HERSEY, S. J., L. I. STEINER, S. MATHERAVIDATHU & G. SACHS. 1988. Am. J. Physiol. **254:** G856–863.

# Pumps and Pathways for Gastric HCl Secretion

JOHN G. FORTE, DAVID K. HANZEL, TETSURO
URUSHIDANI, AND J. MARIO WOLOSIN

*Department of Molecular and Cell Biology*
*University of California*
*Berkeley, California 94720*

The cell biology of HCl secretion has come into focus within the last 10–15 years. Some of the major characters have been identified, and structural-functional dynamics of the secretory cycle have been described. We know that the H,K-ATPase is the primary pump,[1,2] delivering $H^+$ in exchange for $K^+$ by an ATP-driven catalytic cycle.[3–6] Early ultrastructural observation suggested that there were major differences in the morphologic configuration of resting and stimulated gastric parietal cells.[7,8] Systematic studies on the resting/stimulated transition led to the hypothesis that cytoplasmic tubulovesicles of the resting parietal cell fuse with the apical plasma membrane when HCl secretion is stimulated.[8] This so-called membrane recycling hypothesis was supported by biochemical[9–11] and immunocytochemical[12] experiments that sought to localize the H,K- pump within membrane compartments of the parietal cell. In homogenates from resting, or nonsecreting, gastric mucosa the H,K-ATPase is highly localized to the microsomal membrane vesicles that are derived from the cytoplasmic tubulovesicles of the parietal cell.[9,11] In maximally stimulated gastric mucosa, H,K-ATPase disappears from the microsomes and is "transferred" to a fraction of larger, denser membrane vesicles, which we originally called stimulation-associated (s.a.) vesicles and which we now know to be derived from the expanded apical and canalicular plasma membrane of the stimulated parietal cell.[9,11]

The observations just cited provide an enlightened view of the membrane recycling hypothesis of HCl secretion whereby pumps are recruited from a quiescent storage compartment into the apical plasmalemma where they engage in active secretion. However, several enormous gaps exist in understanding how these membrane transformations and secretory activation come about. These include (1) the sequence of biochemical events between secretagogue-receptor interaction and parietal cell activation, (2) the means by which the massive parietal cell membrane transformations are regulated and morphologically organized, (3) the transport properties of the various membrane compartments, that is, tubulovesicles, resting and stimulated apical plasma membrane, and recycled vesicles, (4) the pathways for recycling membranes and segregating proteins when secretion is inhibited, and (5) the translation of the biochemical and transport data from isolated vesicles into the physiologic context of intact parietal cells. We will present the recent progress that has been made in addressing some of these questions in parietal cell biology.

## STIMULATION-DEPENDENT REDISTRIBUTION OF MEMBRANES

When we first observed that maximal histamine stimulation of the stomach was associated with a change in distribution of H,K-ATPase activity from one type of membrane to another, we proposed that this redistribution represented a biochemical index of the ultrastructural events related to parietal cell stimulation.[9,10] This proposal has now been substantiated by our studies of membrane redistribution under conditions of defined physiologic secretory activity. These experiments were designed to examine the process of secretion-dependent membrane redistribution using the *in situ* rabbit stomach to directly measure the $H^+$ secretory rate and as the source of mucosal tissue for subsequent membrane separations and biochemical tests. A fistula was inserted into the stomach of the anesthetized rabbit, and gastric juice was collected and titrated. Animals were treated with (1) no secretagogues (resting), (2) a maximal dose of histamine (stimulated), or (3) a potent histamine-$H_2$ antagonist, SKF 93479.[13] At designated times and conditions, animals were killed by anesthetic overdose, the stomachs were removed, and the gastric mucosa was homogenized and fractionated as previously described.[14]

The distribution of our marker enzyme for H,K-ATPase among the crude cell fractions harvested by differential centrifugation is shown in FIGURE 1 for the various states of rest, histamine stimulation, and $H_2$-receptor inhibition. It is clear that a secretion-related change occurred in the distribution of pump enzyme between the crude microsomal fraction (P3) and the low-speed fraction containing apical membranes (P1). For each stomach we calculated a P1/P3 ratio, that is, the

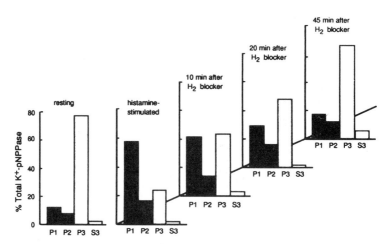

**FIGURE 1.** Distribution of total H,K-ATPase activity among crude cell fractions isolated from rabbit gastric mucosal homogenates taken at rest (nonsecreting), at maximal histamine-stimulated acid secretion, and at various times after $H_2$-receptor inhibition. The potent $H_2$-receptor antagonist SFK 93497[13] was used. Actual rates of $H^+$ secretion were measured via a gastric fistula for each stomach prior to tissue fractionation. Ouabain-insensitive $K^+$-stimulated pNPPase was used as the marker of H,K-ATPase activity. Centrifugal forces for harvesting cell fractions were: P1, 4,000 × g × 10 minutes; P2, 14,500 g × 10 minutes; P3, 50,000 × g × 90 minutes; S3, remaining supernatant. At rest most of the $K^+$-pNPPase was in the P3 fraction (crude microsomes), whereas for maximally stimulated stomach most of the activity was distributed to the low-speed P1 fraction. The activity redistributed back to P3 as $H^+$ secretion was progressively inhibited.

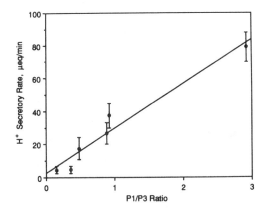

**FIGURE 2.** Correlation between the HC1 secretory rate and the membrane redistribution index of H,K-ATPase (P1/P3 ratio) for rabbit gastric mucosa. Experimental procedures were similar to those described in FIGURE 1. Anesthetized rabbits were prepared with gastric fistulas, and HCl secretion was measured. $H^+$ secretory rates were altered by stimulation with histamine (20 $\mu$mol subcutaneously) or inhibition by the histamine $H_2$-receptor antagonist SKF 93473 (12.5 $\mu$mol intravenously). At designated times the HCl secretory rates were recorded, and the stomach was immediately taken for homogenization and membrane fractionation as described in FIGURE 1. To quantify the extent of redistribution of H,K-ATPase among the membrane fractions, we introduced the P1/P3 ratio, that is, (total K-pNPPase activity in the low-speed P1 pellet)/(total K-pNPPase activity in the microsomal P1 pellet). There is good correlation between the HCl secretory activity and the P1/P3 ratio; the solid line, determined by the method of least squares, has an R = 0.96.

ratio of total H,K-ATPase activity in these two particulate fractions. For resting, or nonsecretory, stomachs the P1/P3 ratio was very low, that is, the majority of total H,K-ATPase was in the P3 microsomal fraction. The higher the rate of acid secretion, the higher the P1/P3 ratio, that is, more H,K-ATPase moved from the microsomal fraction to the apical membrane-containing fraction. FIGURE 2 shows the measured rates of $H^+$ secretion, along with the P1/P3 ratio for the pump enzyme, at various stages of HCl secretion and inhibition. These results unequivocally establish the validity of the P1/P3 ratio as a biochemical index of HCl secretion, and they are wholly consistent with the membrane recycling model for secretion.[8]

Of course, P1 and P3 are very crude fractions, separated only by differential centrifugation. Further density gradient purification conveniently provides membrane vesicle fractions that are much more highly purified and homogeneous. SDS-PAGE is shown in FIGURE 3 for purified tubulovesicles that are derived from the crude P3 fraction by sucrose density gradient centrifugation. The dominant tubulovesicular protein clearly is the H,K-ATPase. (The faintly staining glycoprotein at 60–80 kDa that has been identified as a $\beta$-subunit is also seen.) Apical membrane vesicles can be purified from the P1 fraction by separation on a Ficoll density gradient.[14] As shown in FIGURE 3, these apical membranes are also rich in H,K-ATPase, but at least two other major bands are apparent, an 80-kDa protein and the 45-kDa actin. The significance of these latter two bands is emphasized below.

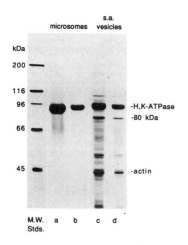

**FIGURE 3.** SDS-PAGE of purified stimulated-associated (s.a.) vesicles (apical membranes) and purified microsomes (tubulovesicles). The s.a. vesicles were purified from the low-speed P1 fraction from stimulated stomach using a Ficoll density gradient (see ref. 14). Tubulovesicles were purified from the microsomal fraction of resting stomach on a sucrose density gradient (see ref. 6). *Lanes a and b* represent 50 and 25 μg, respectively, of microsomes; *lanes c and d* are 50 and 25 μg of s.a. vesicles. The only bands visible in the purified microsomes are the H,K-ATPase at 94 kDa and the poorly stained 60–80-kDa glycoprotein which is probably the β-unit for the ATPase. These bands are also seen in s.a. vesicles, but there are also prominent bands at 80 and 45 kDa (actin).

## K AND CL TRANSPORT PROPERTIES OF PARIETAL CELL MEMBRANES

Early work from our own laboratory and that of George Sachs clearly showed that purified gastric microsomal vesicles, which we now know to be derived from tubulovesicles, showed characteristics of low permeability to $K^+$ and $Cl^-$. Thus, relatively high concentrations of KCl and/or treatment with $K^+$ ionophores (e.g., valinomycin) were required to provide sufficient intravesicular $K^+$ as substrate for the ATP-driven $H^+/K^+$ exchange of the pump. Although data gathered on these purified microsomal preparations were of interest, they did not provide an immediate answer to the question of how the H,K-ATPase operates to produce gastric secretion *in situ*.

When the secretion-dependent redistribution of H,K-ATPase was first identified, we also demonstrated a concomitant increase in the intrinsic permeability of the H,K-ATPase-containing membranes to KCl.[9,10] This latter observation is a very important element in the scenario of how HCl secretion is turned on.

Initial studies led us to suggest that enhanced KCl permeability of stimulated apical membrane vesicles was due to an electroneutral KCl symport.[15] However, more systematic studies indicate that the observed interdependence between $K^+$ and $Cl^-$ fluxes is due to electrical coupling, and that these two ions move through independent conductive pathways.[16,17] One of the approaches used in these studies was to monitor ion permeabilities that were electrically coupled to $H^+$ efflux from vesicles under conditions of exclusive $H^+/K^+$ counterflow or conditions of $H^+$-$Cl^-$ co-efflux.[16] The design of the experiments is straightforward: (1) A proton gradient is formed using ATP and the intrinsic vesicular proton pump with the desired ionic conditions; (2) the pump is then turned off by eliminating ATP (glucose/hexokinase) or inhibitors; and (3) the rate of $H^+$ efflux is monitored under the established set of internal and external ionic conditions.

The experiment shown in FIGURE 4 provides evidence for a $K^+$ conductance pathway in apical membrane vesicles that is absent in tubulovesicles. The experiments were carried out in the absence of a readily permeating anion (i.e., isethionate replaced $Cl^-$. After the pump was turned off and a protonophore (TCS) was added to increase $H^+$ conductance, rapid $H^+$ efflux from apical membrane vesicles was promoted by external $K^+$ (in contrast to other cations like $Na^+$ or $Li^+$), and the rate of $H^+$ efflux was directly proportional to external $K^+$ concentration. These data are most consistent with an endogenous $K^+$ conductance pathway in stimulated apical membrane vesicles with the more rapid $H^+$ efflux occurring via $H^+/K^+$ electrically coupled exchange. In contrast, $H^+$ efflux from tubulovesicles was relatively little influenced by external $K^+$ (FIG. 4B), thus verifying low $K^+$ conductivity for these membranes.

For a measure of $Cl^-$ permeability, pump-generated accumulation of HCl was first accomplished in a highly concentrated suspension of vesicles. Then the pump was turned off, the vesicles were diluted into the desired medium, and $H^+$ efflux was followed. The data of FIGURE 5 show the relative $H^+$ efflux at various concentrations of external $Cl^-$ (relative $H^+$ efflux at 150 mM $Cl^- = $ unity). The individual data points show that $H^+$ efflux was very sensitive to external $[Cl^-]$ in the apical membrane vesicles, whereas $H^+$ efflux from tubulovesicles was relatively independent of external $[Cl^-]$. The efflux of $H^+$ can be described by the Goldman flux

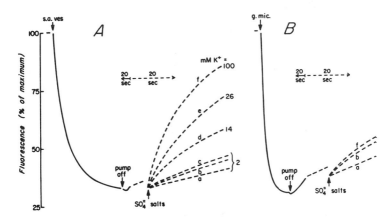

**FIGURE 4.** Evaluation of cation conductance in H,K-ATPase-rich vesicles from resting or stimulated parietal cells using electrically coupled $H^+$ efflux. The general method involves (1) the generation of an $H^+$ gradient in a $Cl^-$-free system using the vesicular ATP-driven H,K- pump as measured by the acridine orange fluorescence quenching method, and (2) then following the gradient relaxation (proton leak rate) after the pump is turned off under conditions where $H^+$ conductance is not limiting (i.e., when the protonophore TCS was added). The vesicles were preincubated in 200 mM potassium isethionate in order to preload intravesicular $K^+$ as a substrate for the H,K-pump. At the indicated points, vesicles were diluted 100-fold into a 200-mM sucrose medium containing 2 μM TCS, 4 mM Tris-TES buffer (pH 7.2), 0.4 mM Mg ATP, 0.04 mM EDTA, 2 μM acridine orange, and 10 mM glucose. Vesicular proton uptake then proceeded to a maximum $H^+$ gradient as shown by fluorescence quenching. At this point the pump was turned off (hexokinase added to consume all ATP), and the rate of $H^+$ efflux was followed as various external cations were added. At the times indicated by the *dashed lines* the speed of the chart recorder was increased to provide a better visual record of the $H^+$ efflux rate. **(A)** Recordings for stimulated apical membrane vesicles (s.a. ves.). Extravesicular conditions for $H^+$ efflux were (a) 2 mM $K^+$, balance sucrose; (b) 2 mM $K^+$, 100 mM $Li^+$; (c) 2 mM $K^+$, 100 mM $Na^+$; (d) 14 mM $K^+$; (e) 26 mM $K^+$; and (f) 100 mM $K^+$. The rate of $H^+$ efflux was clearly increased as a function of extravesicular $K^+$ concentration. **(B)** Recordings for resting tubulovesicles (g. mic.). Incubation and $H^+$ uptake conditions were identical to those described in **A**. Extravesicular conditions for $H^+$ efflux were (a) 2 mM $K^+$, balance sucrose; (b) 2 mM $K^+$, 100 mM $Li^+$; and (c) 100 mM $K^+$. See reference 16 for further experimental details.

equation and is a function of the membrane potential as indicated in the inset in
FIGURE 5. If $Cl^-$ conductance were high such that $P_{Cl} > P_H$, then the membrane
potential, and $H^+$ efflux in turn, would be sensitive to the $Cl^-$ gradient in a
predictable way. The dashed line in FIGURE 5 shows values for relative $H^+$ efflux
that were calculated using the $Cl^-$ equilibrium potential as the membrane potential. The correspondence of these data clearly demonstrates a high $Cl^-$ conductance for isolated apical membrane vesicles. It is important to stress the differences between the two functionally distinct populations of H,K-ATPase-rich
membranes. Tubulovesicles from resting cells are virtually devoid of rapid $K^+$ (cf.
FIG. 4) or $Cl^-$ (cf. FIG. 5) flux pathways that are evident in stimulated apical
membrane vesicles.

The observations on membrane properties just cited can be used to construct a
model for the activation of HCl secretion that expands upon the morphologically
founded membrane recycling hypothesis. As depicted in FIGURE 6, the parietal
cell at rest has an abundant supply of H,K-ATPase, but these pumps are con-

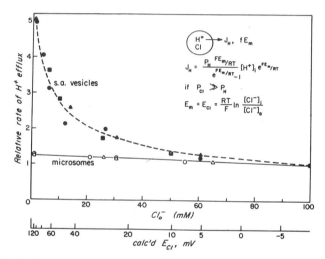

**FIGURE 5.** Evaluation of $Cl^-$ conductance in parietal cell-stimulated apical membrane
vesicles (s.a. vesicles) and resting tubulovesicles (microsomes). The general design of these
experiments as in FIGURE 4 was to form an $H^+$ gradient using the H/K-pump, then to turn off
the pump and monitor $H^+$ efflux after the vesicles were diluted into media of varying $Cl^-$
concentration. For $H^+$ uptake conditions concentrated vesicles were incubated in 20 mM
KCl, 240 mM sucrose, 40 mM Tris-TES buffer (pH 7.2), 4 mM Mg, 8 mM ATP, 0.8 mM
EDTA, 20 mM glucose, and 36 $\mu$M acridine orange. After 3 minutes hexokinase was added
to stop the pump, and the vesicles were diluted 20-fold into various $Cl^-$ concentrations for
measurement of $H^+$ efflux. The measured rates of $H^+$ efflux at any given $Cl^-$ concentration
are shown as relative rates normalized to the rate of $H^+$ efflux at 100 mM $Cl^-$ (set to unity).
For s.a. vesicles $H^+$ efflux was clearly increased as external $[Cl^-]$ was increased. Efflux of
$H^+$ from microsomes was relatively insensitive to external $[Cl^-]$. *Inset:* $H^+$ efflux ($J_H$) is
described by the Goldman equation and is a function of the membrane potential $E_m$. In very
high vesicular $Cl^-$ permeability, such that $P_{Cl} > P_H$, then $E_m$ would be dominated by the $Cl^-$
concentration gradient. R, T, and F have their usual meaning. The *dotted line* is the relative
$H^+$ efflux calculated from the Goldman equation assuming that $E_m$ is determined by the $Cl^-$
gradient ($E_{Cl}$). To calculate the $E_{Cl}$ shown on the lower abscissa, intravesicular $[Cl^-]$ was
assumed to be 75 mM throughout. For further experimental details see reference 16.

A. Morphological Representation

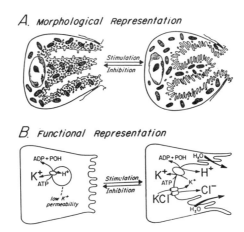

B. Functional Representation

**FIGURE 6.** Schematic representations of parietal cell activation as the cells are transformed from rest to active HCl secretion. **(A)** *Morphologic representation:* Membrane transformation between cytoplasmic tubulovesicles and apical plasma membrane bordering the secretory canaliculi from the basis of the membrane recycling hypothesis of HCl secretion (see refs. 8 and 11). **(B)** *Functional representation:* In the resting cell, tubulovesicles contain H,K-ATPase, but because of low membrane permeability to $K^+$ (and $Cl^-$), there is very little $H^+$ accumulation and virtually no ATP turnover. Cell activation brings about a fusion of tubulovesicles with the apical plasma membrane transferring the H,K-ATPase to that membrane domain. In addition, participation of conductive pathways (possibly activated?) for $K^+$ and $Cl^-$ movement provides the means for KCl movement into the secretory canaliculi. The $H^+/K^+$ exchange pump recycles $K^+$ back into the cytoplasm with the net effect of HCl transfer and ATP turnover. Water flux into the canaliculus is osmotically driven by net solute flux.

tained within a cytoplasmic membrane domain, the tubulovesicles, which have intrinsic low permeability to $K^+$ and $Cl^-$. Consequently, there is no net HCl secretion, and ATP turnover is low. Stimulation by secretagogues leads to a recruitment of tubulovesicles into the apical plasma membrane domain, greatly expanding the canalicular membrane and providing a rich supply of H,K-ATPase. There is an important simultaneous appearance of pathways for $K^+$ and $Cl^-$ movement across the apical plasma membrane. The operation of KCl flux, from cell to lumen, in concert with the $H^+/K^+$ exchange powered by the pump, effects net HCl secretion.

It must be pointed out that several details concerning this picture of HCl secretion remain obscure and/or controversial. The relocation of the pump enzyme is well established, but the nature and disposition of the $K^+$ and $Cl^-$ transport proteins are uncertain. For example, it is not certain if KCl transport pathways must be activated or if they are always functionally present and the secretagogue-dependent membrane transformation simply places the KCl transport systems in parallel with the H,K-ATPase. There are also questions concerning the nature of the pathways for $K^+$ and $Cl^-$ transport in the activated cell. We have presented evidence that $K^+$ and $Cl^-$ pathways are conductive in stimulated apical membrane vesicles; similar conclusions have been reached by Gunther *et al.*[18] in their studies of apical membrane vesicles. However, a number of investigations have reported *either* $K^+$ or $Cl^-$ conductances at the apical membrane, not both, and they suggest that electroneutral KCl cotransporter may account for the high KCl fluxes that are present in the stimulated apical membrane of parietal cells.[19–22]

## STIMULUS-SECRETION COUPLING

The general area of stimulus-secretion coupling, that is, how external signals are translated into a physiologic secretory response, has grown enormously in the last 5 years. For the parietal cell there is a strong body of evidence to show that histamine-receptor activation occurs via pathways involving cAMP.[22–25] However, the specific details of cell activation and modulation of secretory activity are most likely complex, involving multiple nodes of control. For example, in addition to cAMP, recent experiments have demonstrated alterations in intracellular levels of $Ca^{2+}$ and inositol tris-phosphate ($IP_3$) associated with cholinergic, gastrinergic, and even histaminergic receptor activation.[26–29] It is certainly of importance to identify the activating intermediates such as cAMP, $IP_3$, diacylglycerol, $Ca^{2+}$, and the like, but a complete picture of parietal cell activation is necessary to elucidate the interplay among constituitive cytoplasmic and membrane elements that lead to HCl secretion.

Several laboratories have initiated a search for effector proteins in parietal cell activation.[30–32] Model systems have consisted of isolated parietal cells, or whole gastric glands, into which $^{32}P$ is incorporated and the changes in protein phosphorylation associated with stimulation are followed. Results presented thus far have produced a highly varied picture, with a number of phosphoproteins being cited as possible candidates. Most of these studies were not especially selective in what was taken to assay, and it has been difficult to determine where many of the reported phosphoproteins reside within the cell, but this avenue of research is just beginning. The approach taken in our laboratory, which will be described herein, was to identify stimulation-related phosphoproteins within specified cell fractions that had defined functional activities, such as those already described herein. For our experimental protocol, rabbit isolated gastric glands were preincubated with $^{32}Pi$, washed, and then subjected to experimental test. Control or cimetidine-treated glands were compared with those to which secretagogues were added. After designated times, the glands were homogenized and separated into cell fractions.

When gastric glands were treated with secretagogues acting via the cAMP pathway (e.g., histamine, IBMX, dibutyryl cAMP, and forskolin), there was an increase in $^{32}P$ incorporation into two proteins, 80 kDA and 120 kDa, in the apical membrane fraction,[32–34] as exemplified in FIGURE 7. The 120-kDa protein was also found in the cytosol, and its transfer to the apical plasma membrane was seen to occur concomitant with stimulation of HCl secretion. We were especially interested in the 80-kDa phosphoprotein, because it copurified with, and was restricted to, the fraction harvested as apical membrane vesicles. In fact, the apical membrane fraction, whether from resting or stimulated glands, is always rich in the 80-kDa protein (80K), even though the amount of $^{32}P$ incorporation into 80K and the amount of H,K-ATPase associated with s.a. vesicles are both correlated with stimulation. Conversely, microsomes are virtually devoid of 80K. (For example, see FIGURE 3.) This suggests that 80K is a permanent resident of the apical plasma membrane, and that it is excluded from the endocytic recycling of H,K-ATPase into the tubulovesicles when stimulated parietal cells return to the resting state.

Isoelectric focusing, or 2-D electrophoresis, of apical membrane vesicular proteins reveals that 80K is focused at about pH 6.7, with a series of spots streaking from the main spot toward the anode.[32,34] We have now succeeded in separating 80K into a series of six distinct isoforms whose isoelectric points vary by 0.03–0.04 pH units (FIG. 8). Peptide mapping has established that the individually focused bands are indeed isomers of the same protein.[34] Invariably the [32]P radioactivity was always associated with the three most acidic isoforms. Moreover, we have ascertained that serine residues are the site of phosphorylation,[34] consistent with stimulation via cAMP/protein kinase A pathway. We are still in the process of working out the interplay among the 80-kDa isoforms and the enhanced phosphorylation associated with stimulation of HCl secretion.

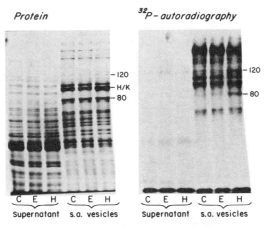

**FIGURE 7.** Analysis of proteins and [32]P-labeled phosphoproteins in cell fractions from resting and stimulated rabbit gastric glands. Isolated gastric glands were prelabeled with [32]P and divided into aliquots for treatment to produce control, nonsecreting glands (C, $10^{-4}$ M cimetidine) or glands maximally stimulated by the histamine/cyclic AMP pathway (H, $10^{-4}$ M histamine + $10^{-5}$ M isobutylmethylxanthine). In the experiment shown here an aliquot of glands was also treated with $10^{-7}$ M epidermal growth factor (E). After treatment for 40 minutes, the glands were homogenized and separated into cell fractions; here we show the s.a. vesicle fraction, rich in apical membranes, and the cytosolic (supernatant) fraction. Invariably, the s.a. vesicle fraction from glands stimulated via the histamine/cyclic AMP pathway showed increased [32]P-labeling of an 80-kDa protein and a 120-kDa protein. The 120-kDa protein can also be identified in the supernatant fraction, but stimulation did not effect changes in labeling in the supernatant.

Recently, we have developed monoclonal antibodies against the purified 80-kDa protein (anti-80K).[35] When sections of gastric mucosa were probed with anti-80K and visualized by immunoperoxidase labeling, intense staining of parietal cells was observed; chief cells, mucous neck cells, and surface cells were virtually devoid of stain (FIG. 9). The anti-80K was localized to a network of interconnecting structures emanating from the apical pole and wending throughout the parietal cell, exclusive of the nucleus. This specific pattern of staining is more clearly visible when isolated glands are stained with a fluorescent-labeled second antibody to anti-80K, such as that shown in FIGURE 10, where the morphology is most

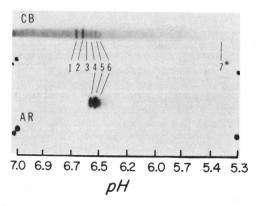

**FIGURE 8.** Isoelectric focusing of purified 80K and its autoradiograph. 80K was purified from $^{32}$P-labeled apical membranes by preparative and one-dimensional SDS-PAGE, electroeluted, and separated on an IEF gel with a narrow pH range. Six protein bands (C.B.) were identified in the neutral region of the gel (1–6). Radioactivity (A.R.) was found only in bands 4, 5, and 6. Peptide mapping revealed that all six bands were isomers of the 80K protein; a distant seventh band was not an isomer. (From ref. 34.)

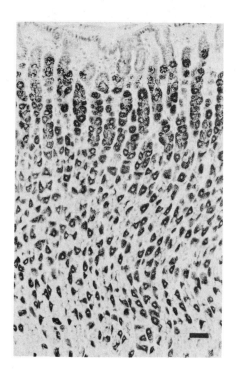

**FIGURE 9.** Distribution of 80K phosphoprotein in cells of rabbit gastric mucosa. Thin sections (7 $\mu$m) of gastric mucosa, fixed by freeze-substitution, were probed with anti-80K followed by a second antibody linked with peroxidase. Peroxidase staining occurs exclusively in parietal cells. Surface cells, mucous neck cells, and chief cells are devoid of the stain. Scale bar = 20 $\mu$m. (From ref. 35.)

consistent with a localization to the canalicular network that forms the apical surface of parietal cells.

We have also identified a characteristic pattern of staining for F-actin in gastric glands.[35,36] This is of particular interest because of suggested involvement of the cytoskeleton in the membrane transformations associated with stimulation of the parietal cell.[37] Because of the consistent similarities in the staining patterns of F-actin and 80K, we performed a double labeling experiment so that the two fluorescent probes could be visualized in the same gland using different filters. As demonstrated in FIGURE 11B, F-actin is localized as a fine layer outlining the

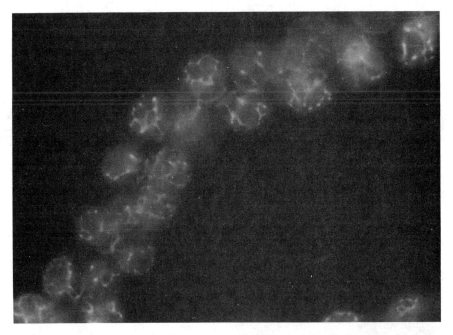

**FIGURE 10.** Distribution of 80K phosphoprotein within parietal cells. Rabbit isolated gastric glands were fixed with paraformaldehyde and then permeabilized with Triton X-100. The fixed glands were probed with a monoclonal anti-80K followed by a second antibody conjugated with FITC. The distribution of fluorescence was localized to a network of interconnecting filaments within the parietal cell. The general appearance of the 80K staining resembles the system of apical secretory canaliculi that extend from the luminal surface throughout the parietal cell.

entire gastric lumen, including the most luminal aspects of parietal, chief, and neck cells. In addition, parietal cells have a more extensive pattern of F-actin staining in the form of a fibrillar network that is distributed in the region of the secretory canaliculi. It is clear that this latter distribution of F-actin is identical to the distribution of 80K in the parietal cell (FIG. 11A).

We have pointed out herein that actin and 80K were major components of apical membranes isolated from parietal cells (FIG. 3). Moreover, their morphologic colocalization in parietal cells suggested a possible interaction between 80K and actin microfilaments. Recent experiments have demonstrated that 80K retains a colocalization with actin networks in detergent-resistant cytoskeletons from gastric glands. This "actin-binding" property is the first hint of a function for 80K; the effect of phosphorylation on actin-binding properties remains to be uncovered.

Important characteristics of 80K include restriction to the apical membrane of parietal cells; colocalization with actin networks; and specific phosphorylation with the stimulation of gastric acid secretion and translocation of H,K-ATPase into the apical membrane. These features suggest that 80K may function to facilitate the incorporation of the H,K-ATPase into the apical membrane and that this process may be regulated by phosphorylation. An attractive possibility is that 80K may serve as a link, or anchoring protein, between the apical membrane and underlying microfilaments.

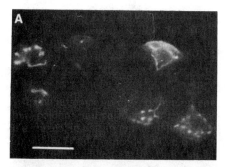

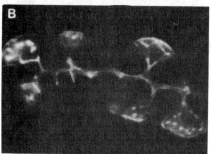

**FIGURE 11.** Simultaneous localization of 80K **(A)** and F-actin **(B)** in the same gastric gland. Fixed glands were probed for 80K with mouse anti-80K, followed by rabbit anti-mouse-IgG labeled with Texas-Red. The same glands were probed with FITC-phalloidin to localize F-actin. The two images were produced with filters appropriate for each fluorophore. F-actin can be found in networks lining the canaliculi of parietal cells as well as along the entire glandular lumen. 80K is found exclusively within the parietal cell canaliculi and thus represents a subset of F-actin distribution. Scale bar = 20 $\mu$M. (From ref. 35.)

## SUMMARY

Data reviewed herein show that the HCl-secreting parietal cell is an exaggerated example of dynamic membrane transformation. Recruitment and recycling of membrane provide the means for the massive redistribution of the gastric proton pump, the H,K-ATPase, from one membrane domain (cytoplasmic tubulovesicles) to another (apical plasma membrane) as a function of parietal cell activation and inactivation. Functional activation of HCl secretion requires not only the redistribution of pump protein, but also the participation of pathways for the rapid flux of $K^+$ and $Cl^-$ across the apical membrane. In apical plasma membrane vesicles from stimulated cells these pathways appear to be conductive and can operate independently. Thus, our model for the parietal cell proposes that $K^+$ and $Cl^-$ flux from cell to lumen, operating in parallel and in concert with ATP-driven $H^+/K^+$ exchange, provides the concentration and osmotic forces required for net HCl secretion. Whether and how the $K^+$ and $Cl^-$ pathways are activated by stimulation and/or how they get to the apical membrane domain remain important questions.

With respect to mechanisms of parietal cell activation, secretagogue-coupled elevation of cAMP and activation of protein kinase A form the basis of a well-established second messenger pathway. Several laboratories have identified various proteins that are phosphorylated concomitant with parietal cell stimulation, representing numerous candidates for effectors in stimulus-secretion coupling. Here, we emphasized the possible involvement of an 80-kDa protein whose phosphorylation was correlated with the cAMP pathway of HCl secretion. Immunocytolocalization of the 80-kDa phosphoprotein to the apical membrane and associated actin microfilaments prompted our suggestion that this protein might serve as a linkage between plasma membrane and cytoskeleton. Search for a possible role for the 80-kDa phosphoprotein in apical surface organization, stability, and turnover should represent an important thrust of research. Further understanding of the mechanism of cell activation will require a more complete elaboration of the functional role of many activation-related proteins.

## REFERENCES

1. GANSER, A. L. & J. G. FORTE. 1973. Biochem. Biophys. Acta **307:** 169–180.
2. FORTE, J. G., A. L. GANSER, R. C. BEESLEY & T. M. FORTE. 1975. Gastroenterology **69:** 175–189.
3. SACHS, G., H. H. CHANG, E. RABON, R. SCHACKMANN, M. LEWIN & G. SACCOMANI. 1976. J. Biol. Chem. **251:** 7690–7698.
4. SCHACKMANN, R. A., A. SCHWARTZ, G. SACCOMANI & G. SACHS. 1977. J. Membr. Biol. **32:** 361–381.
5. LEE, H. C., H. BREITBART, M. BERMAN & J. G. FORTE. 1979. Biochem. Biophys. Acta **553:** 107–131.
6. LEE, H. C. & J. G. FORTE. 1978. Biochem. Biophys. Acta **508:** 339–356.
7. HELANDER, H. F. & B. HIRSCHOWITZ. 1972. J. Cell Biol. **63:** 951–961.
8. FORTE, T. M., T. E. MACHEN & J. G. FORTE. 1977. Gastroenterology **73:** 941–955.
9. WOLOSIN, J. M. & J. G. FORTE. 1981. J. Biol. Chem. **256:** 3149–3152.
10. WOLOSIN, J. M. & J. G. FORTE. 1981. FEBS Lett. **125:** 208–212.
11. FORTE, J. G., J. A. BLACK, T. M. FORTE, T. E. MACHEN & J. M. WOLOSIN. 1981. Am. J. Physiol. **241:** G349–G358.
12. SMOLKA, A., H. F. HELANDER & G. SACHS. 1984. Am. J. Physiol. **245:** G389–396.
13. BLAKEMORE, R. C., T. H. BROWN, G. J. DURAN, C. R. GANELLIN, M. E. PARSONS, A. C. RASMUSSEN & D. A. RAWLINGS. 1981. Br. J. Pharmacol. **74:** 200P.
14. HIRST, B. H. & J. G. FORTE. 1985. Biochem. J. **231:** 641–649.
15. WOLOSIN, J. M. & J. G. FORTE. 1983. J. Membr. Biol. **71:** 195–207.
16. WOLOSIN, J. M. & J. G. FORTE. 1984. Am. J. Physiol. **246:** C537–C545.
17. WOLOSIN, J. M. & J. G. FORTE. 1985. J. Membr. Biol. **83:** 261–272.
18. GUNTHER, R. D., S. BASSELIAN & E. RABON. 1987. J. Biol. Chem. **262:** 13966–13972.
19. DEMAREST, J. R., D. D. F. LOO & G. SACHS. 1988. Biophys. J. **53:** 58a.
20. SHOEMAKER, R. L., P. J. VELDKAMP & G. SACCOMANI. 1988. Biophys. J. **53:** 525a.
21. HERSEY, S. J., L. STEINER, S. MATERAVIDATHU & G. SACHS. 1988. Am. J. Physiol. **254:** G856–G863.
22. MALINOWSKA, D. H. & J. CUPPOLETTI. 1988. FASEB J. **2:** A718, Abstract #2447.
23. HARRIS, J. & D. ALONSO. 1965. Fed. Proc. **24:** 1368–1376.

24. SOLL, A. H. 1979. Am. J. Physiol. **237:** E444–E450.
25. CHEW, C. S., S. F. HERSEY, G. SACHS & T. BERGLINDH. 1980. Am. J. Physiol. **238:** G312–G320.
26. NEGULESCU, P. A. & T. E. MACHEN. 1988. Am. J. Physiol. **254:** 130–140.
27. CHEW, C. S. 1986. Am. J. Physiol. **250:** G814–G823.
28. CHIBA, T. S., K. FISHER, J. PARK, E. B. SEQUIN, B. W. AGRANOFF & T. YAMADA. 1988. Am. J. Physiol. **255:** G99–G105.
29. PFEIFFER, A., H. ROCHLITZ, A. HERZ & G. PAUMGARTNER. 1988. Am. J. Physiol. **254:** G622–G629.
30. CHEW, C.S. & M. R. BROWN. 1987. Am. J. Physiol. **253:** G823–G829.
31. MODLIN, I. M., M. ODDSDOTTIR, T. E. ADRIAN, M. J. ZDON, K. A. ZUCKER & J. R. GOLDENRING. 1987. J. Surg. Res. **42:** 348–353.
32. URUSHIDANI, T., D. K. HANZEL & J. G. FORTE. 1987. Biochem. Biiophys. Acta **930:** 209–219.
33. CUPPOLETTI, J., G. SACHS & D. H. MALINOWSKA. 1986. Fed. Proc. **45:** 1676.
34. URUSHIDANI, T., D. K. HANZEL & J. G. FORTE. 1989. Am. J. Physiol. **256:** G1070–G1081.
35. HANZEL, D. K., T. URUSHIDANI & J. G. FORTE. 1989. Am. J. Physiol. **256:** G1082–G1089.
36. WOLOSIN, J. M., C. OKAMOTO, T. M. FORTE & J. G. FORTE. 1983. Biochem. Biophys. Acta **761:** 171–182.
37. BLACK, J. A., T. M. FORTE & J. G. FORTE. 1982. Gastroenterology **83:** 595–604.

# Regulation of CA II and H+,K+-ATPase Gene Expression in Canine Gastric Parietal Cells

VIRGINIA W. CAMPBELL AND TADATAKA YAMADA

*Department of Internal Medicine*
*The University of Michigan Medical Center*
*Ann Arbor, Michigan 48109–0682*

The gastric parietal cell is a highly specialized and differentiated cell dedicated to the rapid production of voluminous amounts of HCl in response to the binding of acid secretagogues at specific receptors at the cell surface.[1] Stimulation of this cell by the association of acetylcholine (or an analogue such as carbachol), histamine, or gastrin to its particular receptor initiates a series of intracellular activating events, including the induction of gene expression of proteins involved in morphologic transformation, acid production and secretion, and ionic transport processes.[1,2] Acid secretagogues apparently lead to activation of the gastric proton pump, H+,K+-ATPase, resulting in its insertion into the secretory membrane of the cell[2,3] and to the activation of $Na^+/H^+$ exchange at the basolateral surface[4] as the cell becomes transformed from a nonsecreting to an acid-secreting cell. Associated with the activation of these proton transporters is a large increase in cytosolic pH, which may be handled within the cell via the action of carbonic anhydrase, combining $OH^-$ with $CO_2$ to form $HCO_3^-$, followed by the subsequent activation of the basolateral $Cl^-/HCO_3^-$ exchanger.[4,5] Recently, it was shown that both basolateral $Na^+/H^+$ and $Cl^-/HCO_3^-$ exchange activities are necessary for acid secretion across the apical membrane of these cells.[4]

In a series of studies with freshly isolated canine gastric parietal cells we examined the expression of genes activated by acid secretagogues, using cloned cDNA probes for the proton pump itself, the gastric H+,K+-ATPase,[6] and for carbonic anhydrase II (CA II).[7,8] As a probe for a control gene, the expression of which did not vary during the course of our experiments, we used a human cDNA encoding a ubiquitin extension protein, ubiquitin carboxyl-terminal precursor (UBCP).[9,10] Our results indicate that acid secretagogue-specific receptor activation in parietal cells triggers coordinate gene expression of both H+,K+-ATPase and CA II and that the induction of CA II gene expression is independent of hydroxyl ion generation resulting from H+,K+-ATPase activation.[11]

## STEADY-STATE mRNA STUDIES OF H+,K+-ATPase AND CA II GENE EXPRESSION

In Northern and dot blotting experiments[10–13] we recently investigated the action of each of the three principal acid secretagogues, carbachol, histamine, and gastrin, on the pattern of expression of genes in gastric parietal cells. Each stimulant rapidly induced the expression of CA II mRNA (FIG. 1).[10] Carbachol (0.1 mM) increased steady-state levels of CA II mRNA to a peak after 20 minutes of incubation, after which there was a decline toward basal mRNA levels within 1

**A.**                                           **B.**

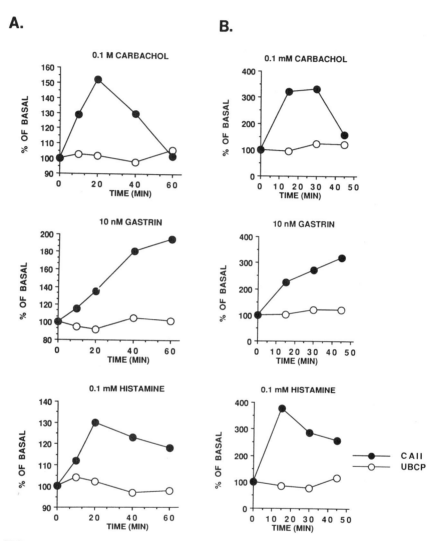

**FIGURE 1.** Relative changes in **(A)** total mRNA concentrations and **(B)** gene transcription rates of CA II and UBCP with time of incubation in carbachol, gastrin, or histamine. The mRNA data shown, expressed as a percentage of basal, are from cells obtained from a single dog and are similar to results obtained from three other cell preparations. mRNA levels were analyzed by densitometric scanning of dot blots. In other experiments we established the high degree of correlation between hybridizible mRNA quantified by dot blots and by Northern transfers. The relative rate of CA II and UBCP gene transcription for each stimulant, expressed as a percentage of basal, represents the mean value of two separate RNA isolations. Each set of nuclei was assayed in duplicate; thus, each point graphed represents the mean value from four assays. (From ref. 10.)

hour. Similarly, the addition of 0.1 mM histamine to isolated gastric parietal cells induced a modest but reproducible increase over steady-state basal levels in CA II mRNA which peaked around 20 minutes after the addition of secretagogue with a subsequent decline toward basal. In related experiments, when isolated parietal cells were stimulated with 10 nM gastrin (G17), steady-state levels of CA II mRNA exhibited sustained increases that reached levels about twofold over basal within 1 hour. Similar changes in $H^+,K^+$-ATPase mRNA levels were observed with the various acid secretagogues.

## NUCLEAR TRANSCRIPTION STUDIES OF CA II GENE EXPRESSION

To examine whether differences in the observed steady-state levels of CA II mRNA after acid secretagogue addition to parietal cells resulted from differences in transcriptional rates or posttranscriptional events, we performed nuclear runoff experiments.[10,14,15] Data from studies examining the relative transcription levels in nuclei isolated at various times after gastric parietal cells were incubated in each stimulant are shown in FIGURES 1 and 2. CA II gene transcription appeared to reach an early peak between 15 and 30 minutes after the addition of either carbachol or histamine. The increase was transient and decreased rapidly within 45 minutes. Transcription of the CA II gene induced by gastrin increased progressively over the 45-minute period. Moreover, the addition of carbachol, gastrin, or histamine to isolated parietal cells had no effect on either the transcription of UBCP mRNA or its steady-state level.

## STUDIES WITH OMEPRAZOLE, A SPECIFIC $H^+,K^+$-ATPase INHIBITOR

Activation of parietal cell receptors results in the accumulation of $H^+$ into the secretory canaliculi by an $H^+,K^+$-ATPase-dependent process. This accumulation is paralleled by a large increase in $HCO_3^-$, presumably resulting from the action of CA II on the $OH^-$ that accompanies $H^+$ generation. The $HCO_3^-$ produced is excreted at the basolateral surface of the cell through a $Cl^-/HCO_3^-$ exchanger[5,16] that simultaneously provides the $Cl^-$ needed to accompany $H^+$ for acid secretion. Having shown that acid secretagogues induce the genes encoding these two enzymes in coordinate fashion, we undertook studies to examine whether the CA II gene is directly induced by secretagogue action or is secondarily induced as a result of the action of the $H^+,K^+$-ATPase during the process of acid secretion.[12] For these studies we used omeprazole, an agent known to inhibit $H^+,K^+$-ATPase specifically and irreversibly.[17]

When cells from the same preparation of canine gastric parietal cells were incubated with 0.1 mM omeprazole, levels of both CA II and $H^+,K^+$-ATPase mRNAs increased nearly twofold over basal. The time course for changes in steady-state levels of CA II mRNA induced by 0.1 mM carbachol showed no difference whether or not the cells had been pretreated with omeprazole. In both instances, CA II mRNA levels reached peaks within 20 minutes of stimulation with carbachol. The increase in $H^+,K^+$-ATPase mRNA with carbachol stimulation was nearly identical, with or without pretreatment with omeprazole.

Our results led us to conclude that the induction of CA II gene expression with carbachol stimulation is independent of the generation of $H^+$ (or $OH^-$) ions by the action of $H^+,K^+$-ATPase, because the kinetics of induction of CA II mRNA from

**0.1 M CARBACHOL**

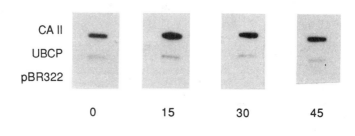

CA II
UBCP
pBR322

0          15          30          45

**10 nM GASTRIN**

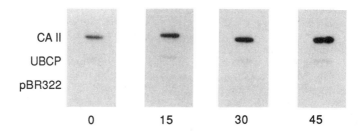

CA II
UBCP
pBR322

0          15          30          45

**0.1 M HISTAMINE**

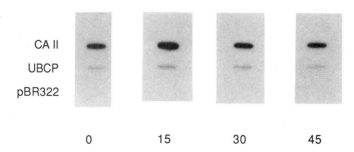

CA II
UBCP
pBR322

0          15          30          45

**FIGURE 2.** Runoff transcription analysis of isolated canine gastric parietal treated with various acid secretagogues. Isolated cells were treated with a single dose of carbachol, gastrin (G17), or histamine for the time indicated. Nuclei were isolated and stored at $-70°C$ until runoff reactions were performed. Approximately equal amounts of purified $^{32}P$-labeled RNA transcripts, measured by the number of counts per minute per milliliter of runoff products added to each filter for nuclei that were being directly compared, were hybridized to individual nitrocellulose filter strips containing the indicated DNAs. Quantitation of the hybridization signals was determined by autoradiography followed by densitometry. (From ref. 10.)

the time of carbachol addition are the same whether or not the $H^+,K^+$-ATPase has been inhibited. Basal levels of $H^+,K^+$-ATPase and CA II gene expression appear to be dependent on factors different from those in operation in acid secretagogue-stimulated cells. The level of both CA II and $H^+,K^+$-ATPase mRNA increased almost twofold over basal when parietal cells were incubated for 20 minutes in omeprazole, indicating that the intracellular concentration of the proton pump itself may be important in regulating basal expression of both enzymes.

## SUMMARY

Acid secretagogue-specific receptor activation in parietal cells triggers rapid and coordinate gene expression of gastric $H^+,K^+$-ATPase and of CA II. The rapid rise in steady-state levels of CA II mRNA is due to new transcription of the CA II gene in stimulated cells. Although the presumed function of CA II in activated parietal cells is to catalyze the generation of $HCO_3^-$ from $OH^-$, regulation of CA II gene expression appears to be independent of the generation of $H^+$ (and $OH^-$) through the action of $H^+,K^+$-ATPase.

## REFERENCES

1. SOLL, A. H. & T. BERGLINDH. 1987. Physiology of isolated gastric glands and parietal cells: Receptors and effectors regulating function. In Physiology of the Gastrointestinal Tract, 2nd Edition, Vol. 1. Johnson, L. R., ed. : 883–909. Raven Press, New York.

2. FORTE, J. G. & J. M. WOLOSIN. 1987. HCl secretion by the gastric oxyntic cell. In Physiology of the Gastrointestinal Tract, 2nd Edition, Vol. 1. Johnson, L. R., ed. : 853–863. Raven Press, New York.

3. HERSEY, S. J. 1979. Intracellular pH measurements in gastric mucosa. Am. J. Physiol. 237: E82–E89.

4. MUALLEM, S., D. BLISSARD, E. E. CRAGOE, JR. & G. SACHS. 1988. Activation of the $Na^+/H^+$ and $Cl^-/HCO_3^-$ exchange by stimulation of acid secretion in the parietal cell. J. Biol. Chem. 263: 14703–14711.

5. MUALLEM, S., C. BURNHAM, D. BLISSARD, T. BERGLINDH & G. SACHS. 1985. Electrolyte transport across the basolateral membrane of the parietal cells. J. Biol. Chem. 260: 6641–6653.

6. SHULL, G. E. & J. B. LINGREL. 1986. Molecular cloning of the rat stomach $(H^+ + K^+)$-ATPase. J. Biol. Chem. 261: 16788–16791.

7. CURTIS, P. J., E. WITHERS, D. DEMUTH, R. WATT, P. J. VENTA & R. E. TASHIAN. 1983. The nucleotide sequence and derived amino acid sequence of cDNA coding for mouse carbonic anhydrase II. Gene 25: 325–332.

8. CURTIS, P. J. 1983. Cloning of mouse carbonic anhydrase mRNA and its induction in mouse erythroleukemic cells. J. Biol. Chem. 258: 4459–4463.

9. LUND, P. K., B. M. MOATS-STAATS, J. G. SIMMONS, E. HOYT, A. J. D'ERCOLES, F. MARTIN & J. J. VAN WYK. 1986. Nucleotide sequence analysis of a cDNA encoding human ubiquitin reveals that ubiquitin is synthesized as a precursor. J. Biol. Chem. 260: 7609–7613.

10. CAMPBELL, V. W., J. DEL VALLE, M. HAWN, J. PARK & T. YAMADA. 1989. Carbonic anhydrase II gene expression in isolated canine gastric parietal cells. Am. J. Physiol. 256: G631–G636.

11. CAMPBELL, V. W. & T. YAMADA. 1987. Time-dependent alterations in tissue specific gene expression: Studies with isolated canine gastric parietal cells. Gastroenterology 92: 1337.

12. CAMPBELL, V. W. & T. YAMADA. 1988. Regulation of $H^+,K^+$-ATPase gene expression in canine gastric parietal cells by omeprazole. Gastroenterology **94:** A57.
13. THOMAS, P. S. 1980. Hybridization of denatured RNA and small DNA fragments transferred to nitrocellulose. Proc. Natl. Acad. Sci. USA **77:** 5201–5205.
14. LINIAL, M., N. GUNDERSON & M. GROUDINE. 1985. Transcription of c-myc in bursal lymphoma cells requires continuous protein synthesis. Science **230:** 1126–1132.
15. SPINDLER, S. R., S. H. MELLON & J. D. BAXTER. 1982. Growth hormone gene transcription is regulated by thyroid and glucocorticoid hormones in cultured rat pituitary tumor cells. J. Biol. Chem. **257:** 11627–11622.
16. PARADISO, A. M., R. Y. TSIEN, J. P. DEMAREST & T. E. MACHEN. 1987. $Na^+/H^+$ and $Cl^-/HCO_3^-$ exchange in rabbit oxyntic cells using fluorescence microscopy. Am. J. Physiol. **253:** C30–C36.
17. IM, W. B., D. P. BLAKEMAN & J. P. DAVIS. 1985. Irreversible inactivation of rat gastric $(H^+-K^+)$ATPase in vivo by omeprazole. BBRC **126:** 78–82.

# Transmembrane Segments of the P-Type Cation-Transporting ATPases

## A Comparative Study[a]

ROBERT K. NAKAMOTO, RAJINI RAO,
AND CAROLYN W. SLAYMAN

*Departments of Human Genetics and Cellular and Molecular
Physiology
Yale School of Medicine
New Haven, Connecticut 06510*

Since the Na,K-ATPase was discovered by Skou[1] in 1957, a vast literature has accumulated on the distribution, structure, reaction mechanism, and physiologic function of the $E_1E_2$- or P-type[2] family of ATPases. These enzymes serve as primary cation transporters in cells ranging from bacteria to higher plants and animals. They have in common a large catalytic subunit (ca. 100 kDa) which is an integral membrane protein, deeply embedded in the lipid bilayer and requiring detergents for solubilization. In addition, they share a reaction cycle that alternates between at least two major conformational states ($E_1$ and $E_2$), with transient phosphorylation of a conserved aspartyl residue located near the middle of the polypeptide chain:

$$E_1 \xrightarrow{\text{ATP}} E_1P$$
$$E_2 \xleftarrow{\quad P_i \quad} E_2P$$

During the past few years, much has been learned about the primary structure of the P-type ATPases, mainly through gene cloning and sequencing. Consistent with their common mechanism, the enzymes possess 10–16 regions of conserved amino acid sequence, concentrated for the most part in hydrophilic domains that are believed to be exposed at the cytoplasmic surface of the membrane.[3,4] Indeed, strong chemical evidence exists that several of the most conspicuous regions participate directly in nucleotide binding and phosphorylation (e.g., ref. 5).

Less is known about the membrane-embedded portions of the P-type ATPases. As will be detailed, the enzymes share a very similar profile of alternating hydrophilic and hydrophobic regions, and it is likely that the hydrophobic stretches serve to anchor the protein in the bilayer; they may also assemble into a channel-like structure through which cations pass as they are transported across the membrane. If so, then at least some of the determinants of cation specificity and/or stoichiometry may be found in the hydrophobic regions. Much work lies ahead to gather the direct structural and functional information that will be needed to evaluate this possibility. In the meantime, it has seemed useful to carry out a

[a] This work was supported by NIH research grant GM15761, NIH postdoctoral fellowship GM11146 to R.K.N., and a James Hudson Brown Postdoctoral Fellowship from the Yale School of Medicine to R.R.

comparative sequence analysis of P-type ATPases that pump different cations (H, K, Na, and Ca), asking whether any striking similarities or differences can be found in the hydrophobic regions that may correlate with transport properties. Although sequence comparisons have been carried out previously by other authors (see, for example, refs. 3 and 4), we believe this is the first time that the whole family of ATPases has been looked at with a common analytic method in an attempt to define the length, boundaries, and homologies of the membrane-spanning segments.

## CHOICE OF ATPases FOR THIS STUDY

Among the primary sequences that have been published for the various P-type ATPases, 10 have been selected for the present study (TABLE 1). Evidence regarding their cation specificity can be summarized as follows.

**TABLE 1.** Properties of P-Type ATPases[a]

| Enzyme | Molecular Weight[b] (kDa) | Transport Stoichiometry (per ATP hydrolyzed) |
|---|---|---|
| E. coli K-ATPase | 72 | nd |
| Neurospora H-ATPase | 100 | 1H |
| Arabidopsis H-ATPase | 104 | nd |
| Leishmania H-ATPase | 107 | nd |
| Gastric H,K-ATPase | 114 | 1H:1K or 2H:2K |
| Na,K-ATPase | 110 | 3Na:2K |
| Sarcoplasmic reticulum Ca-ATPase | 110 | 2Ca:2H |
| Plasma membrane Ca-ATPase | 129–132 | 1Ca:1-2H |

[a] References are provided in the text.

[b] Molecular weights calculated from deduced amino acid sequences; values for E. coli K-ATPase and mammalian plasma-membrane Na,K-ATPase are for the catalytic subunit only.

Escherichia coli *K-ATPase.* When E. coli is starved for potassium, a high-affinity K transport system is derepressed consisting of three subunits: KdpA, B, and C, with calculated molecular weights of 59, 72, and 20 kDa.[6] All three are hydrophobic proteins, located in the bacterial inner membrane,[7] and together they possess K-dependent ATPase activity.[8] Although the precise interactions among the subunits are not yet clear, KdpB is phosphorylated by ATP[9] and exhibits clear-cut sequence homology to the other P-type ATPases;[6] thus, it appears to be the major catalytic polypeptide. Mutational evidence suggests that KdpA may be an auxiliary K-binding protein.[8] The Kdp system is extremely specific for potassium and does not accept even Rb as a substrate.[10] Its stoichiometry has not been measured experimentally, nor has it been determined whether the inward K transport is coupled to the flux of any other ion. From an energetic point of view, however, Epstein et al.[8] calculated that electrogenic potassium transport could have a stoichiometry as high as 2K/ATP, whereas electroneutral transport (e.g., K/H exchange or K-anion symport) would require a stoichiometry of 1K/ATP.

*Plasma Membrane H-ATPases of* Neurospora crassa, Saccharomyces cerevisiae, *and* Schizosaccharomyces pombe. This is a closely related trio of proton-transporting ATPases, with calculated molecular weights of 99.5–99.8 kDa and an overall sequence homology of 66%.[11–14] All three enzymes form a $\beta$-aspartyl phosphate reaction intermediate,[15,16] and all are sensitive to vanadate.[17–19] Following reconstitution into liposomes, the fungal plasma-membrane ATPases function convincingly as electrogenic H pumps.[20–23] Information on stoichiometry is most detailed for the Neurospora enzyme, where electrophysiologic analysis of intact cells clearly indicates a value of 1H/ATP;[24] likewise, a kinetic analysis of transport by the same enzyme in membrane vesicles has given values between 0.82 and 1.23.[25] Recently, a second ATPase gene (PMA2) has been cloned and sequenced from *S. cerevisiae* by Schlesser *et al.*[26] The new gene encodes a 102-kDa protein that may also function in proton transport, because it shares 90% homology with the original PMA1 ATPase; the protein has not yet been isolated biochemically, however, and it is not essential for normal vegetative growth.

*Plasma Membrane H-ATPase of* Arabidopsis thaliana. The plasmalemma of higher plant cells has for many years been known to contain an ATP-dependent, electrogenic proton pump, which has been studied both in intact cells and in isolated membrane vesicles.[27,28] More recently, the corresponding vanadate-sensitive ATPase has been purified from red beet, corn, and oat roots and shown to be a 100-kDa integral membrane protein with an aspartyl phosphate reaction intermediate.[29–33] Thus, the plant ATPase is quite clearly a member of the P-type group. The fact that enzyme activity is modestly stimulated by K, Rb, and $NH_4$ has led to suggestions that influx of these ions may be coupled to the efflux of H. Alkali cations are not required for ATPase activity or for proton pumping, however, and most investigators now believe that they merely exert a nonspecific salt effect on enzyme turnover.[28]

During the last year, Harper *et al.*[34] have cloned and sequenced a gene that appears to encode the plasma-membrane H-ATPase of the cruciform plant *Arabidopsis thaliana*. The 104-kDa Arabidopsis polypeptide contains short amino acid sequences known to be conserved throughout the family of P-type ATPases, and it shows widespread regions of homology (36% amino acid identity in all) to the plasma-membrane H-ATPase of Neurospora. Six random tryptic fragments derived from the oat plasma-membrane H-ATPase can easily be located within the Arabidopsis polypeptide, with only a few amino acid substitutions that presumably reflect the species difference. Interestingly, Harper *et al.*[34] have also cloned and partially sequenced a second Arabidopsis gene encoding a closely related product; whether the second polypeptide is an isoform of the plasma-membrane H-ATPase or a transporter for a different cation (e.g., Ca) remains to be determined.

*Plasma Membrane H-ATPase of* Leishmania donovani. *L. donovani* is a parasitic protozoan with a complex life cycle, alternating between a flagellated promastigote that lives in the intestine of the insect vector and a nonflagellated amastigote that is sequestered intracellularly within the lysosomes of mammalian macrophages. Thus, the cells must cope with an extremely acidic environment, especially in the amastigote state. Several years ago, Zilberstein and Dwyer[35] reported that the uptake of glucose and proline by *L. donovani* occurs by H-linked cotransport, driven by the proton gradient across the surface membrane. It has therefore seemed likely on indirect grounds that the primary pump of the Leishmania membrane is a proton-transporting ATPase, although the enzyme remains to be isolated and characterized biochemically. In 1987, Meade *et al.*[36] described the isolation of a pair of tandemly repeated ATPase genes from *L. donovani*. One of the genes was sequenced and found to encode a 107-kDa polypeptide with pronounced homology to other members of the P-type class; by

Northern blotting, it was shown to be expressed about equally in the promastigote and amastigote forms of the organism. For purposes of the comparisons to be presented, this sequence is referred to provisionally as the Leishmania H-ATPase.

*Gastric H,K-ATPase.* Multicellular animals have evolved at least four kinds of P-type ATPases to meet the physiologic needs of various cells and tissues. In gastric mucosa, for example, the apical and canalicular surfaces of the parietal cell contain an H,K-ATPase that is responsible for proton secretion into the stomach lumen. Biochemical characterization has established that this enzyme, like other members of the P-type group, reacts by way of an aspartyl phosphate intermediate and is sensitive to vanadate.[37] Sequence information is now available for the rat gastric ATPase, which is 114 kDa in size and has a surprising 62% homology to the Na,K-ATPase of sheep kidney.[38]

The transport cycle of the gastric enzyme is not yet completely clear. Membrane vesicle preparations carry out an electroneutral exchange of H for K, but whether the stoichiometry is $1:1$[39] or $2:2$[40] is currently under debate. There is general agreement that a conductive pathway for Cl must be present for acid secretion to occur,[41] and it has been proposed that the Cl conductance may somehow be functionally associated with the ATPase.[42] In this regard, Shull and Lingrel[38] have pointed out a small open reading frame located just upstream of the rat gastric ATPase gene; if expressed *in vivo,* this reading frame would encode a 4700-Da peptide that is rich in positively charged residues and might conceivably function as part of the Cl pathway.

*Plasma Membrane Na,K-ATPase.* In nearly all animal cells, the plasma membrane contains an ouabain- and vanadate-sensitive ATPase that pumps Na out of the cell in exchange for K. The resulting inward Na gradient drives a wide variety of physiologic processes, and the high intracellular K concentration serves to activate many enzymes. Since its discovery 30 years ago, the Na,K-ATPase has been well characterized biochemically (reviewed in ref. 43). It has a 110-kDa catalytic polypeptide (called $\alpha$) that contains all of the known ligand-binding sites and reacts with ATP to form the aspartyl phosphate intermediate; in addition, there is a closely associated 55-kDa subunit ($\beta$) that may be involved in biogenesis. Under ordinary physiologic conditions, the transport cycle of the Na,K-ATPase involves the outward movement of three Na ions coupled to the inward movement of two K ions and is thus electrogenic.[44] By altering the assay conditions, however, the system can be made to display other modes of transport such as H/K exchange,[45] Na/H exchange,[45] and even Na-anion cotransport.[46] The stoichiometries in the latter cases are not yet clear.

Primary sequences are available for both $\alpha$ and $\beta$ subunits from a growing list of species, including Torpedo,[47] chick,[48,49] and several mammals.[50-55] In mammals, evidence began to emerge as early as 1979[56] for multiple forms of the Na,K-ATPase that can be distinguished by their tissue distribution, sensitivity to cardiac glycosides, and immunologic properties. Consistent with this notion, three $\alpha$-subunit genes have now been isolated, both from rat[57] and from man;[58] the corresponding proteins are 85% homologous with one another, suggesting that their functional differences will turn out to be relatively subtle. For purposes of the present study, the sheep kidney $\alpha$-subunit[50] has arbitrarily been selected to represent the Na,K-ATPase group.

*Ca-ATPase of Sarcoplasmic Reticulum.* Mammalian muscle contains a well-characterized ATPase that pumps calcium from the cytoplasm into the vesicles of the sarcoplasmic reticulum. This enzyme has been purified, reconstituted, and shown to react by way of a $\beta$-aspartyl phosphate intermediate (reviewed in ref.

59). Its most likely transport stoichiometry involves an electrogenic exchange of 2Ca with 2H for each ATP hydrolyzed.[59] MacLennan and co-workers[60,61] have cloned two rabbit genes, one encoding the Ca-ATPase of fast twitch skeletal muscle and the other encoding the ATPase of slow twitch and cardiac muscle; the latter form is also expressed in the microsomes of nonmuscle tissues including kidney, brain, and liver.[62,63] Both proteins are about 110 kDa in size, and they are 84% homologous to one another. The slow-twitch/cardiac sequence[60] has been used in the study to be described.

*Plasma Membrane Ca-ATPase.* In addition to the specialized Ca-ATPase of sarcoplasmic reticulum, the plasma membrane of most mammalian cells contains a calmodulin-stimulated ATPase that transports Ca out of the cell, helping to control the intracellular Ca concentration. This enzyme has a large catalytic subunit, forms a phosphorylated intermediate, and is sensitive to vanadate (reviewed in ref. 64). Although debate about the details of the transport cycle continues, the prevailing view favors a stoichiometry of 1Ca : 1–2H : ATP.[65,66] Recently, Shull and Greeb[67] have sequenced two genes from rat brain that appear to encode isoforms of the plasma membrane Ca-ATPase on the basis of size (129 and 132 kDa), amino acid sequence, and putative calmodulin binding sites; the corresponding proteins, which have been named PMCA1 and 2, are 82% homologous to one another. The sequence of PMCA1 has been used in the analyses to be described.

## METHOD OF ANALYSIS

During the past few years, the notion has emerged that the transbilayer segments of integral membrane proteins are mainly $\alpha$-helices. From a theoretic point of view, free energy calculations predict that a hydrophobic $\alpha$-helix should insert into the bilayer much more readily than does a random coil or $\beta$-sheet;[68] moreover, direct structural investigations of bacteriorhodopsin[69,70] and the photosynthetic reaction center[71] have yielded data completely consistent with this prediction. Accordingly, algorithms[68,72] have been developed to locate transmembrane segments of integral membrane proteins on the basis of their amino acid sequence and have been widely used to construct structural models of membrane transporters. All of the algorithms work by scanning a protein sequence from the N-terminus to the C-terminus, picking out segments that are long enough ($\geq 20$ amino acids) and hydrophobic enough to be transmembrane helices.

The work to be described herein has employed the algorithm of Goldman, Engelman, and Steitz (GES), which was well documented in a recent review.[68] We have used a test window of 20 amino acids, based on the number of residues needed for an $\alpha$-helix to span a 30-Å lipid bilayer, and have set 20 kcal/mol as a threshold value for the free energy of transfer of a peptide to water, above which value the peptide is likely to be in the membrane. As pointed out by others, neither of these assumptions is completely reliable. The thickness of the membrane can vary by more than a factor of 2, depending on its lipid composition,[73] and the minimum number of residues needed to cross the bilayer will vary accordingly. Furthermore, if a helix is bent or tilted with respect to the plane of the bilayer, a longer segment may fit within the membrane,[68] while in a packed array of helices, the inner ones (which contact other helices rather than the lipids of the bilayer) could easily be shorter than the "minimum" length.[74] Nevertheless, when the aforementioned values for test window and threshold are used, the GES algorithm gives predictions for the 11 helices of the photosynthetic reaction center that are in remarkable agreement with the experimentally determined crystal structure.[68] This result lends plausibility to models based on the careful use of hydropathy analysis.

## RESULTS

FIGURE 1 illustrates the hydropathy profiles obtained for eight P-type ATPases with the use of the GES algorithm; the results for the plasma membrane H-ATPases of *S. cerevisiae* and *S. pombe* are very similar to those for the Neurospora plasma membrane H-ATPase, and have not been included in the figure. In every case, the profile includes a hydrophilic N-terminus followed in turn by four fairly well-defined hydrophobic regions, a large hydrophilic domain occupying the central one-third of the protein, and a C-terminal domain containing two to six hydrophobic peaks. Because of differences in the length of the various polypeptides, which range from 682 amino acid residues (72 kDa) in the case of the *E. coli* KdpB subunit to 1,198 residues (132 kDa) for the larger isoform of the plasma-membrane Ca-ATPase, it is not surprising that gaps must be introduced to achieve optimal alignment of the hydropathy profiles. Specifically, the hydrophilic N-terminus varies in length from 32 residues in the *E. coli* K-ATPase to 115 residues in the fungal plasma-membrane H-ATPases; the central hydrophilic domain, from 295 residues in *E. coli* to 483–516 residues in the mammalian enzymes; and the C-terminus, from 0 residues in *E. coli* to 107 residues in the Arabidopsis H-ATPase. The functional significance of these differences is not yet clear.

It is relatively easy to predict the lengths and boundaries of the membrane-spanning segments at the N-terminal end of the various polypeptides, because the hydropathy profiles are quite clear in this region. The results are illustrated in FIGURES 2–5. In each case, underlining denotes a sequence for which the calculated free energy of transfer to water exceeds the threshold value of 20 kcal/mol; acidic residues are enclosed in circles and basic residues in squares; and vertical dotted lines indicate the current best guess for the ends of the transmembrane segments, based on homologies among the ensemble of polypeptides as well as on the calculated GES hydrophobicity values.

*Segment I* ranges in length from 20 amino acids in the muscle Ca-ATPase to about 27 amino acids in *E. coli,* where the C-terminal boundary is poorly defined. Its most interesting feature is a glutamyl residue which, in the Neurospora enzyme, is labeled by N,N-dicyclohexylcarbodiimide (DCCD), a well-known inhibitor of proton translocation in the $F_0F_1$ ATPases of mitochondria, chloroplasts, and bacteria.[75] This residue is conserved in the plasma membrane H-ATPases of *S. cerevisiae* and *S. pombe* and also in the H-ATPase of Arabidopsis; and a glutamate is found three residues away (one turn of the helix) in the H-ATPase of Leishmania. Although it is tempting to speculate that glutamate-129 of Neurospora may constitute part of the pathway by which protons cross the membrane, two lines of evidence indicate that this notion may be overly simplistic. First, contrary to popular belief, DCCD is not a diagnostic inhibitor of proton transport, because it also blocks the mammalian Na,K- and Ca-ATPases.[76–78] Second, Portillo and Serrano[79] recently used site-directed mutagenesis to replace glutamate-129 of the yeast ATPase with glutamine or leucine, and saw no obvious reduction of enzyme activity or proton transport when ATPase production was directed by a plasmid-borne copy of the mutated gene. If it can be confirmed by direct DNA sequencing that the mutation has not reverted during the growth of the yeast, this

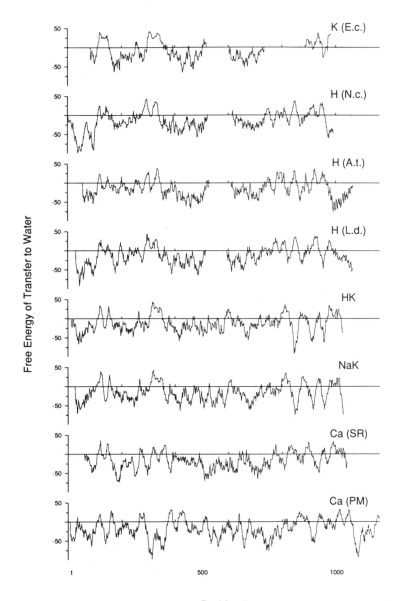

Residue #

**FIGURE 1.** Hydropathy plots for eight P-type ATPases generated by the algorithm of Engelman, Steitz, and Goldman,[68] as described in the text. Symbols denote the *E. coli* K-ATPase (subunit B),[6] plasma membrane H-ATPase of *Neurospora crassa*,[12] plasma membrane H-ATPase of *Arabidopsis thaliana*,[34] H-ATPase of *Leishmania donovani*,[36] H,K-ATPase of rat gastric mucosa,[38] Na,K-ATPase of sheep kidney,[50] Ca-ATPase of rabbit slow-twitch muscle,[60] and plasma membrane Ca-ATPase of rat brain (isoform 2).[67]

**SEGMENT I**

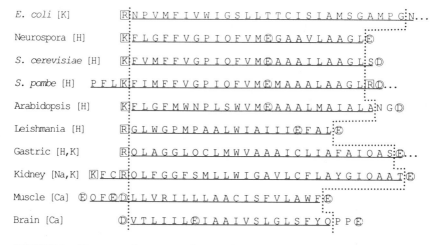

**FIGURE 2.** Alignment of sequences for membrane-spanning segment I. The *underlining* denotes a peptide with a free energy of transfer to water 20 kcal/mol, suggesting that it is located in the membrane. *Vertical dotted lines* indicate the predicted boundaries of the membrane-spanning segments. Acidic residues are enclosed in *circles*, and basic residues in *squares*.

observation will constitute strong evidence against an obligatory role for glutamate-129 in transport.

*Segment II* is relatively short, comprising 20–21 amino acids, and is flanked on both sides by acidic groups. Near the N-terminal end of segment II, a negatively

**SEGMENT II**

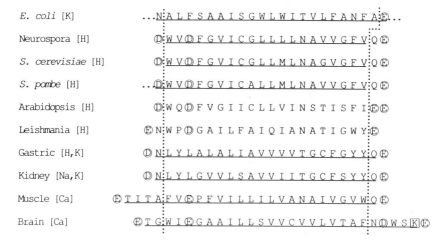

**FIGURE 3.** Alignment of sequences of membrane-spanning segment II.

**SEGMENT III**

```
E. coli [K]        R R K T P N D I A L T I L L I A L T I V F L L A T A T L W P F S ...
Neurospora [H]   S G H F T D V L N G I G T I L L I L V I F T L L I V W V S S F Y R S N
S. cerevisiae [H] Q G H F T D V L N G I G I I L L V L V I A T L L L V W T A C F Y R T N
S. pombe [H]     T G H F T D V L N G I G T I L L V L V L L T L F C I Y T A A F Y R S V R
Arabidopsis [H]  V G H F Q K V L T S I G N F C I C S I A I G I A I D I V V M Y P I Q H R
Leishmania [H]     R R V M F S L C A I S ... F M L C M C C F I Y L L A R ... F Y D T F ...
Gastric [H,K]      D H F V D I I A G L . A I L F G A T F F V V A M C I G Y T F L ...
Kidney [Na,K]      D I D H F I H I I T G V . A V F L G V S F F I L S L I L D Y T W L ...
Muscle [Ca]        D D F G D Q L S K V . I S L I C I A V W I I N I G H F N D
Brain [Ca]       K L A V Q I G K A G L L M S A I . T V I I L V L Y F V I D T F W V Q K R
```

**FIGURE 4.** Alignment of sequences of membrane-spanning segment III.

charged residue (aspartate in the Neurospora, *S. cerevisiae*, *S. pombe*, Arabidopsis, and Leishmania enzymes; glutamate in the mammalian Ca-ATPases) appears in seven of the sequences examined.

*Segment III* is longer (25–28 amino acids) and contains a motif (TILLI) that is present in the *E. coli* K-ATPase and the fungal H-ATPases; a related motif (TVIIL) is found in the same position in the mammalian plasma membrane Ca-ATPase.

*Segment IV* is the best conserved of the transmembrane helices, and Shull *et al.*[50] have suggested that it might form part of an energy transduction pathway. This segment is quite long (26–30 amino acids), but possesses either one or two proline residues near the middle that could introduce kinks in the helix. The motif "PVGLP" of the fungal enzymes becomes "PIAMP" in Arabidopsis, and it picks

**SEGMENT IV**

```
E. coli [K]          ... T V L V A V L V C L I P T T I G G L L S A S A V A G M S R
Neurospora [H]   N P I V Q I L D F T L A I T I I G V P V G L P A V V T T T M A V G A A Y L A K K
S. cerevisiae [H] G I V R I L R Y T L G I T I I G V P V G L P A V V T T T M A V G A A Y L A K K
S. pombe [H]     R L A R L L D Y T L A I T I I G V P V G L P A V V T T T M A V G A A Y L A D K
Arabidopsis [H]    D G I D N L L V L L I G G I P I A M P T V L S V T M A I G S H R L S Q Q
Leishmania [H]   ... R H A L Q F A V V V L V V S I P I A L D I V V T T T L A V G S K H L S K
Gastric [H,K]    ... R A M V F F M A I V V A Y V P D G L L A T V T V C L S L T A K R
Kidney [Na,K]    ... D A V I F L I G I I V A N V P D G L L A T V T V C L T L T A K R
Muscle [Ca]        F K I A V A L A V A A I P D G L P A V I T T C L A L G T R R
Brain [Ca]       Q Y F V K F F I I G V T V L V V A V P D G L P L A V T I S L A Y S V K K
```

**FIGURE 5.** Alignment of sequences of membrane-spanning segment IV.

SEGMENT V

| | |
|---|---|
| Neurospora [H] | R S L N I E L V V F I A I F A D V A T L A I A Y D N |
| *S. cerevisiae* [H] | D N S L D I D L I V F I A I F A D V A T L A I A Y D N |
| *S. pombe* [H] | R N Q L L N L E L V V F I A I F A D V A T L A I A Y D N |
| Arabidopsis [H] | ...D F S A F M V L I I A I L N D G T I M T I S K D |
| Leishmania [H] | ...Q F F H L P V L M F M L I T L L N D G C L M T I G Y D |
| Gastric [H,K] | E P L A A Y S Y F Q I G A I Q S F A G F A D Y F T A M A Q E |
| Kidney [Na,K] | R L I S M A Y G Q I G M I Q A L G G F F T Y F V I M A E |
| Muscle [Ca] | ...A I G C Y V G A A T V G A A A W W F I A A D G G P R V S F Y Q |
| Brain [Ca] | K N I L G H A F Y Q L V V V F T L L F A G E K F F I D |

**FIGURE 6.** Alignment of sequences of membrane-spanning segment V.

up a negatively charged residue in Leishmania (PIALE) and the mammalian ATPases (PEGLL, PEGLP).

It is much more difficult to settle upon the number of membrane-spanning segments in the C-terminal portion of the P-type ATPases, and estimates have ranged as high as seven.[26] Part of the difficulty arises from the presence, at the end of the large central hydrophilic domain, of a region that falls below the 20-kcal threshold in most of the enzymes but above it in a few; if this region is in fact located in the membrane, it is long enough to contain two bilayer-spanning segments. In the present study, we have chosen to focus instead on the more hydrophobic regions nearer the C-terminal end, where a comparison of the ensemble of sequences reveals four putative membrane-spanning segments in the eukaryotic ATPases and two in the *E. coli* ATPase (FIGS. 6–9).

SEGMENT VI

| | |
|---|---|
| Neurospora [H] | K W N L P K L W G M S V L L G V V L A V G T W I T V T T M Y A Q G E N |
| *S. cerevisiae* [H] | K W N L P R L W G M S I I L G I V L A I G S W I T L T T M F L P K G G I I Q |
| *S. pombe* [H] | K W N L P R L W G L S T V I G I V L A I G T W I T N T T M I A Q G Q |
| Arabidopsis [H] | K L K E I F A T G I V L G G Y Q A A I M S V I F F W A A H K |
| Leishmania [H] | K W N L P V V F V S A S I L A A V A C G S S L M L L W I G L E G Y S S Q |
| Gastric [H,K] | Y Q Q Y T C Y T V F F I S I E M C Q I A D V L I R K |
| Kidney [Na,K] | R K I V E F T C H T A F F V S I V V V Q W A D L V I C K |
| Muscle [Ca] | D C A I F E S P Y P M T M A L S V L V T I E M |

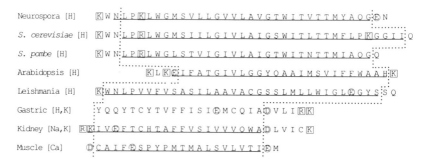

**FIGURE 7.** Alignment of sequences of membrane-spanning segment VI.

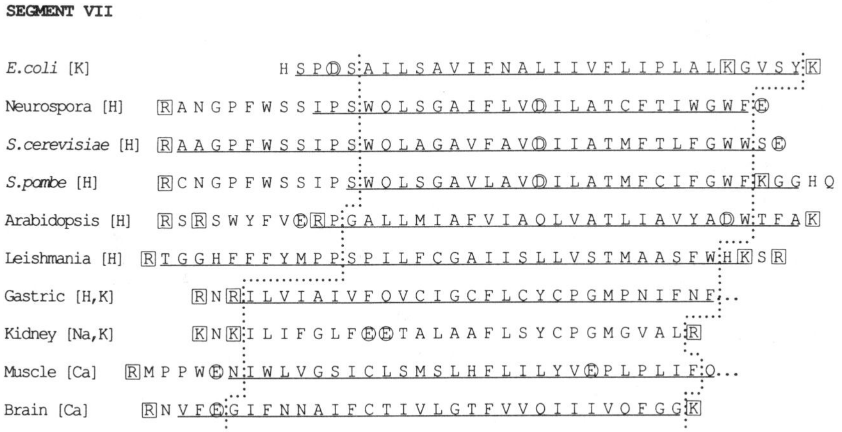

**SEGMENT VII**

```
E.coli [K]                         H S P Ⓓ S⋮A I L S A V I F N A L I I V F L I P L A L Ⓚ G V S Y Ⓚ
Neurospora [H]      Ⓡ A N G P F W S S I P S⋮W Q L S G A I F L V Ⓓ I L A T C F T I W G W F Ⓔ
S.cerevisiae [H]    Ⓡ A A G P F W S S I P S⋮W Q L A G A V F A V Ⓓ I I A T M F T L F G W W S Ⓔ
S.pombe [H]         Ⓡ C N G P F W S S I P S⋮W Q L S G A V L A V Ⓓ I L A T M F C I F G W F Ⓚ G G H Q
Arabidopsis [H]     Ⓡ S Ⓡ S W Y F V Ⓔ Ⓡ P⋮G A L L M I A F V I A Q L V A T L I A V Y A Ⓓ W⋮T F A Ⓚ
Leishmania [H]      Ⓡ T G G H F F F Y M P P⋮S P I L F C G A I I S L L V S T M A A S F W⋮H Ⓚ S Ⓡ
Gastric [H,K]        Ⓡ N Ⓡ I L V I A I V F Q V C I G C F L C Y C P G M P N I F N F⋮. .
Kidney [Na,K]        Ⓚ N Ⓚ I L I F G L F Ⓒ Ⓒ T A L A A F L S Y C P G M G V A L Ⓡ
Muscle [Ca]         Ⓡ M P P W Ⓒ N I W L V G S I C L S M S L H F L I L Y V Ⓟ P L P L I F Q . . .
Brain [Ca]          Ⓡ N V F Ⓓ G I F N N A I F C T I V L G T F V V Q I I I V Q F G G Ⓚ
```

**FIGURE 8.** Alignment of sequences of membrane-spanning segment VII.

*Segment V* ranges from 20–26 residues in length. It barely misses the free-energy threshold in the fungal enzymes (with values of 19.32–19.90 kcal/mol) and in the gastric H,K-ATPase (18.24 kcal/mol), and it seems to be absent altogether in *E. coli*. A possibly significant feature of segment V is the aspartyl residue that is conserved in the fungal, Arabidopsis, Leishmania, and gastric proton-transporting enzymes and also in the Ca-ATPase of sarcoplasmic reticulum; this residue is not found in the plasma membrane Na,K- or Ca-ATPases.

*Segments VI* (20–33 amino acids) and *VII* (23–27 amino acids) are difficult to align with confidence, although weak patterns of homology can be discerned. *Segment VIII* (19–30 amino acids) is somewhat easier to deal with, mainly because of the conserved tryptophan residue that was originally pointed out by Shull

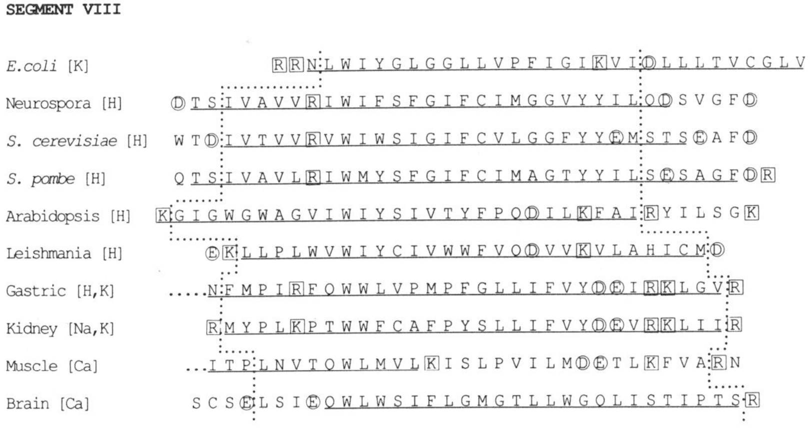

**SEGMENT VIII**

```
E.coli [K]                    Ⓡ Ⓡ N⋮L W I Y G L G G L L V P F I G I Ⓚ V I⋮Ⓓ L L L T V C G L V
Neurospora [H]      Ⓓ T S⋮I V A V V Ⓡ I W I F S F G I F C I M G G V Y Y I L⋮Ⓓ Ⓓ S V G F Ⓓ
S. cerevisiae [H]   W T Ⓓ⋮I V T V V Ⓡ V W I W S I G I F C V L G G F Y Y Ⓔ M⋮S T S Ⓔ A F Ⓓ
S. pombe [H]        Q T S⋮I V A V L Ⓡ I W M Y S F G I F C I M A G T Y Y I L⋮S Ⓔ S A G F Ⓓ Ⓡ
Arabidopsis [H]     Ⓚ G I G W G W A G V I W I Y S I V T Y F P Q Ⓓ I L Ⓚ F A I⋮Ⓡ Y I L S G Ⓚ
Leishmania [H]        Ⓒ Ⓚ L L P L W V V I Y C I V W W F V Q Ⓓ V V Ⓚ V L A H I C M⋮Ⓓ
Gastric [H,K]       . . . . . N⋮F M P I Ⓡ F Q W W L V P M P F G L L I F V Y Ⓓ Ⓔ I Ⓡ Ⓚ L G V⋮Ⓡ
Kidney [Na,K]       Ⓡ⋮M Y P L Ⓚ P T W W F C A F P Y S L L I F V Y Ⓓ Ⓔ V Ⓡ Ⓚ L I I⋮Ⓡ
Muscle [Ca]         . . . I T P⋮L N V T Q W L M V L Ⓚ I S L P V I L M Ⓓ Ⓔ T L Ⓚ F V A Ⓡ N
Brain [Ca]          S C S Ⓔ L S I Ⓔ Q W L W S I F L G M G T L L W G Q L I S T I P T S⋮Ⓡ
```

**FIGURE 9.** Alignment of sequences of membrane-spanning segment VIII.

and Greeb;[67] it is also of interest because of the potential for ion pairs in the Arabidopsis and Leishmania ATPases (D/K) and even more conspicuously in the H,K- and Na,K-ATPases (DE/RK).

## SUMMARY

The transmembrane segments predicted for the Neurospora H-ATPase are laid out diagrammatically in FIGURE 10. Although the eight segments have arbitrarily been compressed into rectangles of the same size, they range in length from 20 residues (II) to 30 residues (IV and VI), so the corresponding helices must vary in length as well. Notable features of the model include the charged residues located just outside the plane of the membrane, with a clear excess of negative charges (5−, 1+) at the extracellular surface and a slight excess of positive charges (4+, 3−) at the cytoplasmic surface. There are also a conspicuous number of bulky residues (tryptophan, phenylalanine, and tyrosine) just inside the plane of the membrane. Within the bilayer, most of the helices are noticeably amphipathic, consistent with the expectation that at least some of them stack together to form a channel-like structure with a hydrophobic surface and a hydrophilic core.

The charged residues predicted to lie within the membrane are listed in TABLE 2, which is a summary of data from eight of the P-type ATPases; the S. cerevisiae and S. pombe enzymes have not been included because they are nearly identical in this respect to the Neurospora enzyme. Interestingly, all of the ATPases have more membrane-embedded negative charges (5 to 8) than positive ones (0 to 4), a pattern that may be connected with their role as cation transporters. Certainly, other unrelated transport proteins have a rather different pattern of positive and negative charges: for example, the mammalian glucose transporter (1+, 2−),[80] Na-glucose transporter (3+, 3−),[81] and the E. coli lac permease (11+, 7−).[82]

The actual positioning of the negative charges in the P-type ATPases does not make it easy to single out the functionally important ones, however. The glutamyl residue in segment I is present in the fungal, plant, and Leishmania H-ATPases

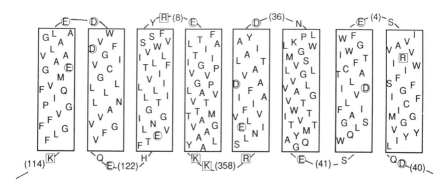

**FIGURE 10.** Model for the membrane domain of the Neurospora plasma membrane H-ATPase, based on data from FIGURES 2–9. The N- and C-termini of the ATPase are known to be exposed at the cytoplasmic surface of the membrane (S. M. Mandala and C. W. Slayman, J. Biol. Chem., in press).

TABLE 2. Charged Residues in the Membrane Segments of the P-Type ATPases

| | Membrane Segment | | | | | | | Charged Residues | |
|---|---|---|---|---|---|---|---|---|---|
| | I | II | III | IV | V | VI | VII | VIII | + | − |
| *Neurospora* [H] | E | D | — | E | D,E | K | D | R | 2 | 6 |
| *Arabidopsis* [H] | E | D | — | E,K | D | — | D | D,K | 2 | 6 |
| *Leishmani* [H] | E | D | E | R | D | E | — | D,K | 2 | 6 |
| Gastric [H,K] | — | — | E | D | D | E | — | D,E,R,R,K | 3 | 6 |
| Kidney [Na,K] | — | — | E | E | — | E | E,E | D,E,R,K,K | 3 | 7 |
| Muscle [Ca] | — | E | E | E,K | D,R | E | E | D,E,K,K | 4 | 8 |
| Brain [Ca] | E | E | E | D | — | — | — | E | 0 | 5 |

but not in the gastric H,K-ATPase. The same is true for the aspartate in segment II, except that it also appears in the muscle and brain Ca-ATPases. A glutamate is found at one end of segment III in the *E. coli* and fungal enzymes and at the other end in Arabidopsis; in segment IV, another glutamate appears in a well-conserved region in the Leishmania and mammalian enzymes but not in the bacterial, fungal, or plant ones. At the C-terminal end of the molecule, as already mentioned, the aspartate in segment V may be of interest, because it is conserved in the fungal, plant, Leishmania, and gastric proton-transporting ATPases, but it is also found in the muscle Ca pump. The aspartate of segment VII is found only in the fungal enzymes; however, a glutamate appears in a different position of the muscle Ca-ATPase sequence and two glutamates appear in the gastric sequence. Finally, conspicuous ion pairs are present in segment VIII of the Arabidopsis, Leishmania, and three of the mammalian sequences, but not in the rest, whereas a lone glutamate is found in segment VIII of *S. cerevisiae* but not in the other two fungi. Thus, it is practically impossible to discern any correlation between the positioning of charged residues and the nature of the cation being pumped, although at least part of the answer may lie in the flexible cation specificity of some P-type ATPases. In any event, it seems clear that a detailed model for the mechanism of cation transport will have to await direct structural information coupled with systematic dissection of the transmembrane domain by site-directed mutagensis.

## REFERENCES

1. SKOU, J. C. 1957. Biochim. Biophys. Acta **23:** 394–401.
2. PEDERSEN, P. L. & E. CARAFOLI. 1987. Trends Biochem. Sci. **12:** 146–150.
3. GREEN, N. M., W. R. TAYLOR & D. H. MACLENNAN. 1988. *In* Ion Pumps. W. J. Stein, ed. : 15–24. Alan R. Liss, Inc., New York.
4. SERRANO, R. 1988. Biochim. Biophys. Acta **947:** 1–28.
5. WALDERHAUG, M. O., R. L. POST, G. SACCOMANI, R. T. LEONARD & D. P. BRISKIN. 1985. J. Biol. Chem. **260:** 3852–3859.
6. HESSE, J. E., L. WIECZOREK, K. ALTENDORF, A. S. REICIN, E. DORUS & W. EPSTEIN. 1984. Proc. Natl. Acad. Sci. **81:** 4746–4750.
7. LAIMINS, L. A., D. B. RHOADS, K. ALTENDORF & W. EPSTEIN. 1978. Proc. Natl. Acad. Sci. **75:** 3216–3219.
8. EPSTEIN, W., V. WHITELAW & J. HESSE. 1978. J. Biol. Chem. **253:** 6666–6668.
9. EPSTEIN, W., L. LAIMINS & J. HESSE. 1979. *In* 11th Int. Cong. Biochemistry, p. 449.
10. RHOADS, D. B., A. WOO & W. EPSTEIN. 1977. Biochim. Biophys. Acta **469:** 45–51.
11. SERRANO, R., M. C. KIELLAND-BRANDT & G. R. FINK. 1986. Nature **319:** 689–693.

12. HAGER, K. M., S. M. MANDALA, J. W. DAVENPORT, D. W. SPEICHER, E. J. BENZ & C. W. SLAYMAN. 1986. Proc. Natl. Acad. Sci. **83:** 7693–7697.
13. ADDISON, R. 1986. J. Biol. Chem. **261:** 14896–14901.
14. GHISLAIN, M., A. SCHLESSER & A. GOFFEAU. 1987. J. Biol. Chem. **262:** 17549–17555.
15. DAME, J. B. & G. A. SCARBOROUGH. 1981. J. Biol. Chem. **256:** 10724–10730.
16. AMORY, A. & A. GOFFEAU. 1982. J. Biol. Chem. **257:** 4723–4730.
17. BOWMAN, B. J. & C. W. SLAYMAN. 1979. J. Biol. Chem. **254:** 2928–2934.
18. DUFOUR, J. P., M. BOUTRY & A. GOFFEAU. 1980. J. Biol. Chem. **255:** 5735–5741.
19. FOURY, F., A. AMORY & A. GOFFEAU. 1981. Eur. J. Biochem. **119:** 395–400.
20. VILLALOBO, A., M. BOUTRY & A. GOFFEAU. 1981. J. Biol. Chem. **256:** 12081–12087.
21. SERRANO, R. 1983. Arch. Biochem. Biophys. **227:** 1–8.
22. PERLIN, D. S., K. KASAMO, R. J. BROOKER & C. W. SLAYMAN. 1984. J. Biol. Chem. **259:** 7884–7892.
23. GOORMAGHTIGH, E., C. CHADWICK & G. A. SCARBOROUGH. 1986. J. Biol. Chem. **261:** 7466–7471.
24. GRADMANN, D., U. P. HANSEN, W. S. LONG, C. L. SLAYMAN & J. WARNCKE. 1978. J. Membrane Biol. **39:** 333–367.
25. PERLIN, D. S., M. J. D. SAN FRANCISCO, C. W. SLAYMAN & B. P. ROSEN. 1986. Arch. Biochem. Biophys. **248:** 53–61.
26. SCHLESSER, A., S. ULASZEWSKI, M. GHISLAIN & A. GOFFEAU. 1988. J. Biol. Chem. **263:** 19480–19487.
27. SPANSWICK, R. 1981. Ann. Rev. Plant Physiol. **32:** 267–289.
28. SZE, H. 1985. Ann. Rev. Plant Physiol. **36:** 175–208.
29. BRISKIN, D. P. & R. J. POOLE. 1983. Plant Physiol. **71:** 507–512.
30. BRISKIN, D. P. & R. J. POOLE. 1983. Plant Physiol. **72:** 1133–1135.
31. SCALLA, R., A. AMORY, J. RIGAUD & A. GOFFEAU. 1983. Eur. J. Biochem. **132:** 525–530.
32. VARA, F. & R. SERRANO. 1983. J. Biol. Chem. **258:** 5334–5336.
33. ANTHON, G. E. & R. M. SPANSWICK. 1986. Plant Physiol. **81:** 1080–1085.
34. HARPER, J. F., T. K. SUROWY & M. R. SUSSMAN. 1989. Proc. Natl. Acad. Sci., **86:** 1234–1238.
35. ZILBERSTEIN, D. & D. M. DWYER. 1985. Proc. Natl. Acad. Sci. 82:1716–1720.
36. MEADE, J. C., J. SHAW, S. LEMASTER, G. GALLAGHER & J. R. STRINGER. 1987. Mol. Cell. Biol. **7:** 3937–3946.
37. FALLER, L. D., A. SMOLKO & G. SACHS. 1985. In The Enzymes of Biological Membranes. A. Martonosi, ed. : 431–448. Plenum, New York.
38. SHULL, G. E. & J. B. LINGREL. 1986. J. Biol. Chem. **261:** 16788–16791.
39. REENSTRA, W. W. & J. G. FORTE. 1981. J. Membrane Biol. **61:** 55–60.
40. RABON, E. C., T. L. MCFALL & G. SACHS. 1982. J. Biol. Chem. **257:** 6296–6299.
41. CUPPOLETTI, J. & G. SACHS. 1984. J. Biol. Chem. **261:** 16788–16791.
42. TAKEGUCHI, N. & Y. YAMAKAZI. 1986. J. Biol. Chem. **261:** 2560–2566.
43. SKOU, J. C., J. G. NORBY, A. B. MAUNSBACH & M. ESMANN. 1988. The Na⁺,K⁺ Pump. Alan R. Liss, Inc., New York.
44. POST, R. L. & P. C. JOLLY. 1957. Biochim. Biophys. Acta **25:** 118–128.
45. POLVANI, C. & R. BLOSTEIN. 1988. J. Biol. Chem. **263:** 16757–16763.
46. DISSING, S. & J. F. HOFFMAN. 1982. Biophys. J. **41:** 188a.
47. KAWAKAMI, K., S. NOGUCHI, M. NODA, H. TAKAHASHI, T. OHTA, M. KAWAMURA, H. NOJIMA, K. NAGANO, T. HIROSE, S. INAYAMA, H. HAYASHIDA, T. MIYATA & S. NUMA. 1985. Nature **316:** 733–736.
48. TAKEYASU, K., M. M. TAMKUN, N. R. SIEGEL & D. M. FAMBROUGH. 1987. J. Biol. Chem. **262:** 10733–10740.
49. TAKEYASU, K., M. M. TAMKUN, K. J. RENAUD & D. M. FAMBROUGH. 1988. J. Biol. Chem. **263:** 4347–4354.
50. SHULL, G. E., A. SCHWARTZ & J. B. LINGREL. 1985. Nature **316:** 691–695.
51. SHULL, G. E., L. K. LANE & J. B. LINGREL. 1986. Nature **321:** 429–431.
52. MERCER, R. W., J. W. SCHNEIDER, A. SAVITZ, J. EMANUEL, E. J. BENZ & R. LEVENSON. 1986. Mol. Cell. Biol. **6:** 3884–3890.

53. OVCHINNOKOV, Y. A., N. N. MODYANOV, N. E. BROUDE, K. E. PETROKHIN, A. V. GRISHIN, N. M. ARZAMAZOVA, N. A. ALDANOVA, G. S. MONASTYRSKAYA & E. D. SVERDLOV. 1986. FEBS Lett. **201:** 237–245.
54. KAWAKAMI, K., T. OHTA, H. NOJIMA & K. NAGANO. 1986. J. Biochem. **100:** 389–397.
55. KAWAKAMI, K., H. NOJIMA, T. OHTA & K. NAGANO. 1986. Nuc. Acids Res. **14:** 2833–2844.
56. SWEADNER, K. J. 1979. J. Biol. Chem. **254:** 6060–6067.
57. SHULL, G. E., J. GREEB & J. B. LINGREL. 1986. Biochemistry **25:** 8125–8132.
58. SHULL, M. M. & J. B. LINGREL. 1987. Proc. Natl. Acad. Sci. **84:** 4039–4043.
59. INESI, G. 1985. *In* The Enzymes of Biological Membranes. A. Martoniso, ed. : 157–191. Plenum, New York.
60. MacLENNAN, D. H., C. J. BRANDL, B. KORCZAK & N. M. GREEN. 1985. Nature **316:** 696–700.
61. BRANDL, C. J., N. M. GREEN, B. KORCZAK & D. H. MacLENNAN. 1986. Cell **44:** 597–607.
62. LYTTON, J. & D. H. MacLENNAN. 1988. J. Biol. Chem. **263:** 15024–15031.
63. GUNTESKI-HAMBLIN, A. M., J. GREEB & G. E. SHULL. 1988. J. Biol. Chem. **263:** 15032–15040.
64. CARAFOLI, E. 1985. *In* The Enzymes of Biological Membranes. A. Martonosi, ed. : 235–248. Plenum, New York.
65. NIGGLI, V., E. SIGEL & E. CARAFOLI. 1982. J. Biol. Chem. **257:** 2350–2356.
66. SMALLWOOD, J. I., D. M. WAISMAN, D. LAFRENIERE & H. RASMUSSEN. 1983. J. Biol. Chem. **258:** 11092–11097.
67. SHULL, G. E. & J. GREEB. 1988. J. Biol. Chem. **263:** 8646–8657.
68. ENGELMAN, D. M., T. A. STEITZ & A. GOLDMAN. 1986. Ann. Rev. Biophys. Biophys. Chem. **15:** 321–353.
69. HENDERSON, R. & P. N. T. UNWIN. 1975. Nature **257:** 28–32.
70. LEIFER, R. & D. HENDERSON. 1983. J. Mol. Biol. **163:** 451–463.
71. DIESENHOFER, J., O. EPP, K. MIKI, R. HUBER & H. MICHEL. 1985. Nature **318:** 618–624.
72. KYTE, J. & R. F. DOOLITTLE. 1982. J. Mol. Biol. **157:** 105–132.
73. LEWIS, B. A. & D. M. ENGELMAN. 1983. J. Mol. Biol. **166:** 203.
74. LODISH, H. F. 1988. Trends Biochem. Sci. **13:** 332–334.
75. SUSSMAN, M. R., J. E. STRICKLER, K. M. HAGER & C. W. SLAYMAN. 1987. J. Biol. Chem. **262:** 4569–4573.
76. PEDEMONTE, C. H. & J. H. KAPLAN. 1986. J. Biol. Chem. **261:** 3632–3639.
77. PEDEMONTE, C. H. & J. H. KAPLAN. 1986. J. Biol. Chem. **261:** 16660–16665.
78. PICK, U. & E. RACKER. 1979. Biochemistry **18:** 108–113.
79. PORTILLO, F. & R. SERRANO. 1988. EMBO J. **7:** 1793–1798.
80. MUECKLER, M., C. CARUSO, S. A. BALDWIN, M. PANICO, I. BLENCH, H. R. MORRIS, W. J. ALLARD, G. E. LIENHARD & H. F. LODISH. 1985. Science **229:** 941–945.
81. HEDIGER, M. A., M. J. COADY, T. S. IKEDA & E. M. WRIGHT. 1987. Nature **330:** 379–381.
82. FOSTER, D. L., M. BOUBLIK & H. R. KABACK. 1983. J. Biol. Chem. **258:** 31–34.

# Vacuolar and Cell Membrane H⁺-ATPases of Plant Cells[a]

ROGER M. SPANSWICK

*Section of Plant Biology*
*Division of Biological Sciences*
*Cornell University*
*Ithaca, New York 14853*

Although the idea of $H^+$ transport has arisen from time to time in the literature on ion transport in plants and indeed influenced the original formulation of Mitchell's chemiosmotic hypothesis,[1] only within the last 10 years has convincing evidence become available to support the existence of $H^+$ pumping ATPases in plant cell membranes. The reasons for this have to do with both the experimental difficulties involved in studying $H^+$ transport in intact systems and the nature of the prevalent hypotheses for ion transport.

The experimental problems arise because it is not possible to use tracers to measure the fluxes of $H^+$ due to the exchange of $H^+$ with $H_2O$ and the high fluxes of water across the cell membrane. Furthermore, even the measurement of net $H^+$ fluxes is compromised by the net fluxes of $CO_2$ associated with respiration and, in the case of green tissues, photosynthesis. Added to this are the possible complications presented by the high cation exchange capacity of plant cell walls and back fluxes of $H^+$ via cotransport systems.

In the early 1960s, the major hypotheses concerning ion transport in plants were heavily influenced by the counterparts for animal systems. Thus, the resting electrical properties of plant cell membranes were postulated to reflect the passive diffusion of ions down gradients established by neutral ion pumps for the major ions[2] and did not involve $H^+$ at all. Similarly, by analogy to the work on $Na^+,K^+$-ATPase from animal systems,[3] stimulation of ATPase activity by $K^+$ was taken to indicate the involvement of this enzyme in $K^+$ transport.[4] The possible role of a membrane ATPase in $K^+$ transport is still a matter of debate.[5]

The realization that $H^+$ transport might play a central role in plant membrane transport arose from the convergence of ideas from three different areas. First, attempts to explain the resting electrical properties in terms of passive ion diffusion had led to evidence for the existence of an electrogenic ion pump that could produce potentials as large as $-300$ mV in some cells,[6] a value far in excess of any possible diffusion potential. Because experimental evidence appeared to exclude the involvement of the major ions,[7] it was postulated by default that $H^+$ was the ion transported out of the cell electrogenically. Second, the presumed problems faced by plant cells in regulating cytoplasmic pH under conditions in which both the exterior and the vacuole were acid and the cytoplasm electrically negative led to the hypothesis that active $H^+$ transport was necessary.[8] Third, extension of the chemiosmotic hypothesis to include secondary active transport systems driven by $\Delta\bar{\mu}H^+$[9] provided a working hypothesis that has been productive in linking the transport of $H^+$ to secondary transport of small organic molecules and the major ions via cotransport systems or porters.[10,11] This provided the context for a fresh

[a] This work was aided by National Science Foundation Grant number DMB-8716363.

approach to the study of plant membrane ATPases as electrogenic $H^+$ pumps. Although I will concentrate on studies using membranes isolated from tissues of higher plants, I should point out that the properties of the enzymes from fungi are very similar and that the work on fungi has proceeded in parallel, and often ahead of, the work on higher plants. (See chapters by C. L. and C. W. Slayman in this volume.)

## THE TONOPLAST ATPase

In retrospect, it is evident that several groups independently conceived the idea of applying to plant membrane vesicles the methods pioneered by workers in the field of bioenergetics to measure pH and electrical gradients across the membranes of vesicles isolated from bacteria and organelles. The application of these methods to cell membrane vesicles requires (1) the production of sealed vesicles and (2) orientation of the vesicles so that the catalytic site for ATP hydrolysis is on the exterior surface of the vesicle. If these conditions are fulfilled, considerations of the direction of transport in the intact cell necessary to maintain cytoplasmic pH near neutrality suggest that ATP-dependent $H^+$ transport would result in acidification of the vesicle interior whether the vesicle derived from the plasma membrane or the vacuolar membrane (tonoplast). Detection of the resulting pH gradient should be possible by measuring the fluorescence of dyes such as 9-aminoacridine, quinacrine, or acridine orange which, because they are weak bases, accumulate in acid compartments and have the property that this accumulation quenches the fluorescence.[12] Thus, not only was a technique available to measure the net transport of $H^+$, but also the use of isolated membrane vesicles had the potential to eliminate the problems posed by $CO_2$ exchange in intact cells and by the cation exchange capacity of the cell wall.

The first clue that membrane ATPase activity might indeed be associated with $H^+$ transport in isolated membrane vesicles was the stimulation of ATPase uptake observed by Sze[13] using a microsomal preparation from tobacco callus. Observations were also made of ATP-dependent uptake of $SCN^-$,[14,15] which suggested the presence of an electrogenic ion pump. The first observation of ATP-stimulated $H^+$ transport was from Hager's laboratory.[16] As with the other groups, they suggested that the transport was associated with the plasma membrane. Publication of our own work[17,18] was delayed sufficiently for us to realize that the unusual properties of the transport and its associated ATPase activity in a microsomal preparation from corn roots were similar to the ATPase activity associated with intact isolated vacuoles.[19,20] Specifically, it was observed that both the transport and ATPase activity were stimulated by anions, particularly $Cl^-$. $NO_3^-$ was an exception because it inhibited activity. Furthermore, $H^+$ transport was stimulated almost equally by choline chloride and KCl, whereas the plasma membrane ATPase, as identified by surface labeling of protoplasts and recovery of membranes on sucrose density gradients,[21,22] was indeed stimulated by $K^+$ but was not inhibited by $NO_3^-$.[23] The tonoplast origin of the vesicles was verified by the observation that they equilibrated on a sucrose density gradient at a peak density lower than that for the plasma membrane,[18] as subsequently confirmed by the position on a gradient of vesicles prepared from intact vacuoles isolated from red beets.[24]

The preparation that has been characterized most extensively is that from red beet.[24–27] In addition to the properties just mentioned, it has been demonstrated that the transport is not inhibited by vanadate[24] and that the $H^+/ATP$ stoichiometry is 2.0.[25] The substrate for the enzyme appears to be $MgATP$,[26] and the pH optimum is 7.0.[25] ATP hydrolysis is inhibited competitively by ADP but only weakly by free ATP, whereas AMP and $P_i$ have only small effects.[26] The properties of the enzyme from this and other tissues have been reviewed by Sze.[28]

The electrogenic nature of the transport observed using uptake of $SCN^-$ [14,15] has been confirmed by using oxonol dyes[29,30] and, more elegantly, by observing the effect of ATP on the clamp current in vacuoles from barley leaves measured using a patch electrode in the "whole-cell" (i.e., whole-vacuole) mode and with symmetric KCl solutions.[31,32]

The stimulation of $H^+$ transport by $Cl^-$ is complicated in that the kinetics of a preparation from corn roots indicate an effect proportional to $Cl^-$ concentration in addition to an effect that saturates with a $K_m$ for KCl between 4 and 5 mM.[29] Martinoia et al.[33] observed similar kinetics for the influx of $Cl^-$ into isolated barley vacuoles. In corn, there is a close correspondence between the reduction in the membrane potential, measured with oxonol VI, and the stimulation of the initial rate of $H^+$ influx as a function of $Cl^-$ concentration.[29] This suggests that the ATPase may be limited by the magnitude of the membrane potential. However, the stimulation of ATPase activity by $Cl^-$ is also evident in the deoxycholate-solubilized enzyme and in the reconstituted enzyme (although the enzyme was reconstituted without purification).[34] Therefore, it also appears to be necessary to postulate a direct effect of the anion on the enzyme. Indeed, in the presence of gramicidin, which abolishes both the potential and the pH gradients, the $K_m$ for KCl of the ATPase in native membrane vesicles is 4.38 mM.[29] Thus, it apparently matches the saturable component of $H^+$ transport. However, if the saturable effect of KCl stimulation of $H^+$ transport represents only a direct stimulation of the ATPase, it is difficult to account for the fact that the membrane potential becomes less positive rather than more positive as $H^+$ transport increases.[29]

At this stage, it does not seem possible to present an entirely satisfactory explanation of the relations between KCl concentration and the rate of $H^+$ transport. However, several pieces of experimental evidence exist that provide a basis for further work in this area.

First, the saturable component of $Cl^-$-stimulated $H^+$ transport in corn root tonoplast vesicles is inhibited by 4,4'-diisothiocyano-2,2'-stilbenedisulfonic acid (DIDS) and 4-acetamido-4'-isothiocyano-2,2'-stilbenedisulfonic acid (SITS),[29] which are inhibitors of the red blood cell anion transporter and anion channel blockers. Interpretation of the saturable component of ATPase stimulation solely in terms of a reduction in the membrane potential by passive influx of $Cl^-$ through a channel that can be blocked by DIDS or SITS is complicated by the fact that both SITS[29] and DIDS[35] appear to inhibit ATPase activity directly; however, there is disagreement as to whether they do[35] or do not[29] inhibit the KCl-stimulated component of the ATPase activity. Recent patch-clamp studies on isolated vacuoles indicate that the predominant slowly voltage-activated channels, which are cation channels, although with low selectivity, are not open at the potentials prevailing in the systems used to study $H^+$ transport.[31,32,36,37] However, Hedrich and Neher[38] have observed rapidly activated channels at positive membrane potentials under conditions corresponding to low $Ca^{2+}$ concentrations. The selectivity for these channels was also low ($P_K/P_{Cl}$ about 6). The effects of SITS or DIDS on these channels apparently are unknown.

Second, Schumaker and Sze[39] have suggested that the tonoplast also contains an anion/$H^+$ cotransport system that is inhibited by DIDS. If this system were an antiport, it would not aid us in explaining the saturable component of $H^+$ uptake by DIDS, because its inhibition would tend to increase rather than decrease the net influx of $H^+$. However, an anion/$H^+$ antiport would be strongly electrogenic and might account for the large effect of $Cl^-$ on the membrane potential. The tonoplast also contains an $H^+$-translocating pyrophosphatase[40–42] which has a lower activity than does the ATPase but is oriented in the same direction. It is separable from the ATPase by column chromatography.[43,44] In this case, the activity of the enzyme is unaffected by the anion. However, there is still a stimulation of $H^+$ transport by anions ($Cl^-$, $Br^-$, and $NO_3^-$) that also reduce the membrane potential.[45] Further experiments are required to elucidate the possible contributions of anion stimulation of the ATPase, ion channels, and anion cotransport systems to the kinetics of $H^+$ transport.

Structurally, the tonoplast $H^+$-ATPase has multiple subunits. In regard to both structure and transport properties, it appears to be very similar to the ATPases from lysosomes, endoplasmic reticulum, coated vesicles, and fungal vacuoles.[46]

## PLASMA MEMBRANE ATPase

Following the discovery of the tonoplast ATPase, there remained the task of demonstrating that the vanadate-sensitive ATPase associated with the plasma

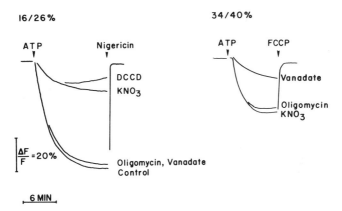

**FIGURE 1.** ATP-dependent quenching of quinacrine fluorescence with low density (tonoplast) or high density (plasma membrane) vesicles from red beets. Low density (16/26% sucrose step gradient interface) membrane vesicles were assayed in the presence of KCl, and high density (34/40% interface) vesicles were assayed in the presence of 100 mM $KNO_3$. Seventy micrograms of membrane protein were added for each assay. Inhibitors were added at the following concentration: oligomycin (5 $\mu$g/ml), vanadate (50 $\mu$M), $KNO_3$ (50 mM), N,N′-dicyclohexylcarbodiimide (DCCD) (50 $\mu$M), Nigericin (0.5 $\mu$M), and carbonyl cyanide trifluorophenylhydrazone (FCCP) (1 $\mu$M). Oligomycin was included to illustrate the small amount of contamination by mitochondria. (From ref. 24, with permission.)

membrane was involved directly in $H^+$ transport. As with the tonoplast enzyme, this was achieved almost simultaneously by several groups.[28] In earlier work, we had been unsuccessful in demonstrating $H^+$ transport in plasma membranes purified from corn roots.[47] However, we have since been able to separate vanadate-sensitive $H^+$-transport activity from $NO_3^-$-sensitive activity in red beets[24] (FIG. 1), soybean roots,[48] zucchini fruits,[49] and finally corn roots.[50] The enzyme has been solubilized and reconstituted[51,52] recently with prior purification.[53]

The plasma membrane ATPase from higher plants is very similar to that from fungi. It has only a single subunit with a molecular weight of about 90 kD,[53] it forms a phosphorylated intermediate,[54] and it is inhibited by vanadate. Cross-linking studies performed *in vitro* indicate that the enzyme may be able to form dimers and trimers. Although this does not mean that trimers form *in vivo*, it is consistent with the target size of 228,000 estimated from radiation inactivation analysis.[55]

Investigation of the kinetics of the purified enzyme[56] demonstrated that vanadate is an uncompetitive inhibitor ($K_i = 3 \ \mu M$). The extent of inhibition is dependent on both pH and potassium concentration. An $E_1$-$E_2$ type model for $H^+$ transport can account for the observed kinetics without invoking the transport of $K^+$ by the enzyme.

It should now be possible to begin the task of relating the kinetic properties of the enzyme to the electrophysiologic properties of the plasma membrane. Because the existence of electrically conducting plasmodesmata[57] makes it difficult to investigate the electrophysiologic properties of the plasmalemma in intact higher plant tissues, patch clamping of intact protoplasts will be necessary for this work.

## ROLE OF $H^+$-ATPases IN A CHEMIOSMOTIC SCHEME FOR TRANSPORT IN PLANT CELLS

The realization that $H^+$ pumps might play a central role in plant membrane transport made it possible to formulate a chemiosmotic scheme[10,11,58] that has provided a working hypothesis for nearly two decades. As can be seen from a current version of this scheme (FIG. 2), $H^+$ fulfills the role played by $Na^+$ in animal systems, particularly in providing the driving force for secondary active transport systems.

The rapid progress made in identifying and characterizing the primary $H^+$ transport systems has provided a sound basis for this hypothesis. Initial progress in obtaining evidence for $H^+$ cotransport systems is also gaining momentum. The first evidence related not to ions but to sugars and amino acids, which were shown to produce depolarizations of the membrane potential in plant cells from many different tissues.[10,59] However, the demonstration of an associated $H^+$ flux was often more difficult. An example is the depolarization produced by sucrose in the cells of developing soybean cotyledons.[60] A 62-kD sucrose binding protein that may be the transport molecule for sucrose in this tissue has recently been identified.[61]

Progress with identifying cotransport systems for ions has been slower, partly because of the ambiguity of interpreting electrical effects when the transported entities are themselves charged. However, strong evidence has been obtained for a $H^+/Cl^-$ cotransport system at the plasmalemma of *Chara corallina* with a stoichiometry of $2H^+/1Cl^-$.[62] In higher plants, evidence for cotransport systems for anions is often contradictory,[58] because in many tissues the transport of phosphate[63] or nitrate[64] is associated with a slow hyperpolarization rather than a depolarization. We have preliminary evidence that, in the case of nitrate in corn roots, the hyperpolarization may be preceded by a rapid, transient depolarization (McClure, Kochian, and Spanswick, unpublished). The hyperpolarization might well result from an effect on the plasma membrane ATPase of the cotransported $H^+$, because the pH optimum of the ATPase[50] (6.5) is lower than that of the cytoplasmic pH. Hence an influx of $H^+$ would result in an increase in the activity of the electrogenic $H^+$ pump.

Again, experiments with isolated membrane vesicles may be more useful in identifying $H^+$ cotransport systems for the major ions than experiments with intact cells. For example, good evidence exists for a proton-coupled $Ca^{2+}$ transport system at the tonoplast.[65,66]

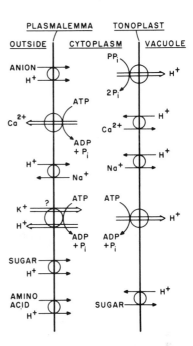

**FIGURE 2.** A scheme for the transport of ions and small organic molecules across the plasma membrane and tonoplast of plant cells based on an extension of the chemiosmotic hypothesis.[9] (Modified from ref. 56.)

## REFERENCES

1. MITCHELL, P. 1961. Coupling of phosphorylation to electron and hydrogen transfer by a chemi-osmotic type of mechanism. Nature **191:** 144–148.
2. DAINTY, J. 1962. Ion transport and electrical potentials in plant cells. Ann. Rev. Plant Physiol. **13:** 379–402.
3. SKOU, J. C. 1965. Enzymatic basis for active transport of $Na^+$ and $K^+$ across cell membrane. Physiol. Rev. **45:** 596–617.
4. FISHER, J. & T. K. HODGES. 1969. Monovalent ion stimulated adenosine triphosphatase from oat roots. Plant Physiol. **44:** 385–395.
5. BRISKIN, D. P. 1986. Plasma membrane $H^+$-transporting ATPase: Role in potassium ion transport? Physiol. Plant. **68:** 159–163.
6. SPANSWICK, R. M. 1981. Electrogenic ion pumps. Ann. Rev. Plant Physiol. **32:** 267–289.
7. SPANSWICK, R. M. 1974. Evidence for an electrogenic pump in *Nitella translucens*. II. Control of the light-stimulated component of the membrane potential. Biochim. Biophys. Acta **332:** 387–398.
8. SMITH, F. A. & J. A. RAVEN. 1979. Intracellular pH and its regulation. Ann. Rev. Plant Physiol. **30:** 289–311.
9. MITCHELL, P. 1970. Membranes of cells and organelles: Morphology, transport and metabolism. Symp. Soc. Exp. Biol. **20:** 121–166.

10. POOLE, R. J. 1978. Energy coupling for membrane transport. Ann. Rev. Plant Physiol. **29:** 437–460.
11. SLAYMAN, C. L. 1974. Proton pumping and generalized energetics of transport: A review. *In* Membrane Transport in Plants. U. Zimmermann & J. Dainty, eds. : 107–119. Springer-Verlag, New York.
12. SCHULDINER, S., H. ROTTENBERG & M. AVRON. 1972. Determination of $\Delta$pH in chloroplasts. 2. Fluorescent amines as a probe for the determination of $\Delta$pH in chloroplasts. Eur. J. Biochem. **25:** 64–70.
13. SZE, H. 1980. Nigericin-stimulated ATPase activity in microsomal vesicles of tobacco callus. Proc. Natl. Acad. Sci. USA **77:** 5904–5908.
14. SZE, H. & K. A. CHURCHILL. 1981. $Mg^{2+}$/KCL-ATPase of plant plasma membranes is an electrogenic pump. Proc. Natl. Acad. Sci. USA **78:** 5578–5582.
15. RASI-CALDOGNO, F., M. I. DE MICHELIS & M. C. PUGLIARELLO. 1981. Evidence for an electrogenic ATPase in microsomal vesicles from pea internodes. Biochim. Biophys. Acta **642:** 37–45.
16. HAGER, A., R. FRENZEL & D. LAIBLE. 1980. ATP-dependent proton transport into vesicles of microsomal membranes of *Zea mays* coleoptiles. Zeit. Naturforsch. **39:** 783–793.
17. DUPONT, F. M., A. B. BENNETT & R. M. SPANSWICK. 1982. Proton transport in microsomal vesicles from corn roots. *In* Plasmalemma and Tonoplast: Their Function in the Plant Cell. D. Marmé, E. Marrè and R. Hertel, eds. : 409–415. Elsevier Biomedical Press, Amsterdam, The Netherlands.
18. DUPONT, F. M., A. B. BENNETT & R. M. SPANSWICK. 1982. Localization of a proton-translocating ATPase on sucrose gradients. Plant Physiol. **70:** 1115–1119.
19. WALKER, R. R. & R. A. LEIGH. 1981. Characterization of a salt-stimulated ATPase activity associated with vacuoles isolated from storage roots of red beet (*Beta vulgaris* L.). Planta **153:** 140–149.
20. ADMON, A. B., B. JACOBY & E. E. GOLDSMITH. 1981. Some characteristics of the Mg-ATPase of isolated red beet vacuoles. Plant Sci. Lett. **22:** 89–96.
21. GALBRAITH, D. W. & D. H. NORTHCOTE. 1977. The isolation of plasma membrane from protoplasts of soybean suspension cultures. J. Cell Sci. **24:** 295–310.
22. PERLIN, D. S. & R. M. SPANSWICK. 1980. Labeling and isolation of plasma membranes from corn leaf protoplasts. Plant Physiol. **65:** 1053–1057.
23. PERLIN, D. S. & R. M. SPANSWICK. 1981. Characterization of ATPase activity associated with corn leaf plasma membranes. Plant Physiol. **68:** 521–526.
24. BENNETT, A. B., S. D. O'NEILL & R. M. SPANSWICK. 1984. $H^+$-ATPase activity from storage tissue of *Beta vulgaris*. I. Identification and characterization of an anion-sensitive $H^+$-ATPase. Plant Physiol. **74:** 538–544.
25. BENNETT, A. B. & R. M. SPANSWICK. 1984. $H^+$-ATPase activity from storage tissue of *Beta vulgaris*. II. $H^+$/ATPase. Plant Physiol. **74:** 545–548.
26. BENNETT, A. B., S. D. O'NEILL, M. EILMANN & R. M. SPANSWICK. 1985. $H^+$-ATPase activity from storage tissue of *Beta vulgaris*. III. Modulation of ATPase activity by reaction substrates and products. Plant Physiol. **78:** 495–499.
27. POOLE, R. J., D. P. BRISKIN, Z. KRATKY & R. M. JOHNSTONE. 1984. Density gradient localization of plasma membrane and tonoplast from storage tissue of growing and dormant red beet. Characterization of proton-transport and ATPase in tonoplast vesicles. Plant Physiol. **74:** 549–556.
28. SZE, H. 1985. $H^+$-translocating ATPases: Advances using membrane vesicles. Ann. Rev. Plant Physiol. **36:** 175–208.
29. BENNETT, A. B. & R. M. SPANSWICK. 1983. Optical measurements of $\Delta$pH and $\Delta\psi$ in corn root membrane vesicles: Kinetic analysis of $Cl^-$ effects on a proton-translocating ATPase. J. Membr. Biol. **71:** 95–107.
30. LEW, R. R. & R. M. SPANSWICK. 1985. Characterization of anion effects on the nitrate-sensitive ATP-dependent proton pumping activity of soybean (*Glycine max* L.) seedling root microsomes. Plant Physiol. **77:** 352–357.
31. HEDRICH, R., U. I. FLUGGE & J. M. FERNANDEZ. 1986. Patch-clamp studies of ion transport in isolated plant vacuoles. FEBS Lett. **204:** 228–232.

32. COYAUD, L., A. KURKDJIAN, R. KADO & R. HEDRICH. 1987. Ion channels and ATP-driven pumps involved in ion transport across the tonoplast of sugarbeet vacuoles. Biochim. Biophys. Acta **902:** 263–268.

33. MARTINOIA, E., M. J. SCHRAMM, G. KAISER, W. M. KAISER & U. HEBER. 1986. Transport of anions in isolated barley vacuoles. I. Permeability to anions and evidence for a Cl⁻-uptake system. Plant Physiol. **80:** 895–901.

34. BENNETT, A. B. & R. M. SPANSWICK. 1983. Solubilization and reconstitution of an anion-sensitive H⁺-ATPase from corn roots. J. Membr. Biol. **75:** 21–31.

35. CHURCHILL, K. A. & H. SZE. 1984. Anion-sensitive, H⁺-pumping ATPase of oat roots. Plant Physiol. **76:** 490–497.

36. KOLB, H.-A., K. KOHLER & E. MARTINOIA. 1987. Single potassium channels in membranes of isolated mesophyll barley vacuoles. J. Membr. Biol. **95:** 163–169.

37. COLOMBO, R., R. CERANA, P. LADO & A. PERES. 1988. Voltage-dependent channels permeable to K⁺ and Na⁺ in the membrane of *Acer pseudoplatanus* vacuoles. J. Membr. Biol. **193:** 227–236.

38. HEDRICH, R. & E. NEHER. 1987. Cytoplasmic calcium regulates voltage-dependent ion channels in plant vacuoles. Nature **329:** 833–836.

39. SCHUMAKER, K. S. & H. SZE. 1987. Decrease of pH gradients in tonoplast vesicles by NO₃⁻ and Cl⁻: Evidence for H⁺-coupled anion transport. Plant Physiol. **83:** 490–496.

40. DuPONT, F. M., D. L. GIORGI & R. M. SPANSWICK. 1982. Characterization of a proton-translocating ATPase in microsomal vesicles from corn roots. Plant Physiol. **70:** 1694–1699.

41. CHURCHILL, K. A. & H. SZE. 1983. Anion-sensitive, H⁺-pumping ATPase in membrane vesicles from oat roots. Plant Physiol. **71:** 610–617.

42. CHANSON, A., J. FICHMANN, D. SPEAR & L. TAIZ. 1985. Pyrophosphate-driven proton transport by microsomal membranes of corn coleoptiles. Plant Physiol. **79:** 159–164.

43. REA, P. A. & R. J. POOLE. 1986. Chromatographic resolution of H⁺-translocating pyrophosphatase from H⁺-translocating ATPase of higher plant tonoplast. Plant Physiol. **81:** 126–129.

44. WANG, Y., R. A. LEIGH, K. H. KAESTNER & H. SZE. 1986. Electrogenic H⁺-pumping pyrophosphatase in tonoplast vesicles of oat roots. Plant Physiol. **81:** 497–502.

45. POPE, A. J. & R. A. LEIGH. 1987. Some characteristics of anion transport at the tonoplast of oat roots, determined from the effects of anions on pyrophosphate-dependent proton transport. Planta **172:** 91–100.

46. SCHNEIDER, D. L. 1987. The proton pump ATPase of lysosomes and related organelles of the vacuolar apparatus. Biochim. Biophys. Acta **895:** 1–10.

47. PERLIN, D. S. & R. M. SPANSWICK. 1982. Isolation and assay of corn root membrane vesicles with reduced proton permeability. Biochim. Biophys. Acta **690:** 178–186.

48. LEW, R. R. & R. M. SPANSWICK. 1984. Proton-pumping activities of soybean (*Glycine max* L.) root microsomes: Localization and sensitivity to nitrate and vanadate. Plant Sci. Lett. **36:** 187–193.

49. LEW, R. R., N. BUSHUNOW & R. M. SPANSWICK. 1985. ATP-dependent proton-pumping activities of zucchini fruit microsomes. A study of tonoplast and plasma membrane activities. Biochim. Biophys. Acta **821:** 314–347.

50. DE MICHELIS, M. I. & R. M. SPANSWICK. 1986. H⁺-pumping driven by the vanadate-sensitive ATPase in membrane vesicles from corn roots. Plant Physiol. **81:** 542–547.

51. O'NEILL, S. D. & R. M. SPANSWICK. 1984. Solubilization and reconstitution of a vanadate-sensitive H⁺-ATPase from the plasma membrane of *Beta vulgaris*. J. Membr. Biol. **79:** 231–243.

52. VARA, F. & R. SERRANO. 1982. Partial purification and properties of the proton-translocating ATPase of plant plasma membranes. J. Biol. Chem. **257:** 12826–12830.

53. ANTHON, G. E. & R. M. SPANSWICK. 1986. Purification and properties of the H⁺-translocating ATPase from the plasma membrane of tomato roots. Plant Physiol. **81:** 1080–1085.

54. BRISKIN, D. P. & R. T. LEONARD. 1982. Phosphorylation of the adenosine triphosphatase in a deoxycholate-treated plasma membrane fraction from corn roots. Plant Physiol. **70:** 1459–1464.

55. BRISKIN, D. P., W. R. THORNLEY & J. L. ROTI-ROTI. 1985. Target molecular size of the red beet plasma membrane ATPase. Plant Physiol. **78:** 642–644.

56. ANTHON, G. E. & R. M. SPANSWICK. 1989. Kinetic characterization of the purified tomato root plasma membrane ATPase. Proceedings of the 7th International Workshop on Plant Membrane Transport, Sydney. (In press).

57. SPANSWICK, R. M. 1972. Electrical coupling between the cells of higher plants: A direct demonstration of intercellular transport. Planta **102:** 215–227.

58. SPANSWICK, R. M. 1985. The role of $H^+$-ATPases in plant nutrient transport. *In* Frontiers of Membrane Research in Agriculture. J. B. St. John, E. Berlin & P. C. Jackson, eds.: 243–256. Rowman & Allanheld, Totowa, NJ.

59. REINHOLD, L. & A. KAPLAN. 1984. Membrane transport of sugars and amino acids. Ann. Rev. Plant Physiol. **35:** 45–83.

60. LICHTNER, F. T. & R. M. SPANSWICK. 1981. Electrogenic sucrose transport in developing soybean cotyledons. Plant Physiol. **67:** 869–874.

61. RIPP, K. V., P. V. VIITANEN, W. D. HITZ & V. R. FRANCESCHI. 1988. Identification of a membrane protein associated with sucrose transport into cells of developing soybean cotyledons. Plant Physiol. **88:** 1435–1445.

62. BEILBY, M. J. & N. A. WALKER. 1981. Chloride transport in *Chara*. I. Kinetics and current-voltage curves for a probable proton symport. J. Exp. Bot. **32:** 43–54.

63. BOWLING, D. J. F. & J. DUNLOP. 1978. Uptake of phosphate by white clover. I. Evidence for an electrogenic phosphate pump. J. Exp. Bot. **29:** 1139–1146.

64. THIBAUD, J. B. & C. GRIGNON. 1981. Mechanism of nitrate uptake in corn roots. Plant Sci. Lett. **22:** 279–289.

65. RASI-CALDOGNO, F., M. I. DE MICHELIS & M. C. PUGLIARELLO. 1982. Active transport of $Ca^{2+}$ in membrane vesicles from pea. Evidence for a $H^+/Ca^{2+}$ symport. Biochim. Bipohys. Acta **693:** 287–295.

66. SCHUMAKER, K. S. & H. SZE. 1986. Calcium transport into the vacuole of oat roots. Characterization of $H^+/Ca^{2+}$ exchange. J. Biol. Chem. **261:** 12172–12178.

# Renal H$^+$ ATPases[a]

## E. KINNE-SAFFRAN

*Max-Planck-Institut für Systemphysiologie*
*D-4600 Dortmund 1, West Germany*

Luminal membrane H$^+$ ATPases seem to be involved in acid secretion in the proximal tubule as well as the distal tubule of the kidney. This paper summarizes our current knowledge of the existence and the properties of a H$^+$ ATPase related to an ATP-driven proton pump in the luminal membrane of the renal proximal tubule and compares it to the properties of the H$^+$ ATPase present in the distal tubule.

FIGURE 1 depicts a simplified scheme of the renal tubule in which the sites of proton secretion are indicated. The first site is the proximal tubule. Micropuncture studies have shown that the resulting reduction in pH in the primary urine is relatively slight, with a maximum drop to about 6.7.[1] The proton secretion is actually much larger, but it fails to produce a marked decrease in pH mainly because the secreted amounts of protons are used to reabsorb the filtered bicarbonate in the proximal tubule, thus masking the real amount of secreted protons.

The main transport system for protons in the proximal tubule is a sodium/proton exchanger.[3-5] However, micropuncture and microperfusion studies have also shown that about 25–30% of the overall secretion of protons is independent from sodium.[6-8] Moreover, in studies on the regulation of intracellular pH in isolated renal cell lines, proton secretion continues even in the absence of extracellular sodium or when the exchanger is inhibited by amiloride.[9] This process also requires the presence of ATP providing first evidence for the existence of an ATP-driven proton pump.

The second major site of proton secretion in the kidney is the distal part of the nephron. Here, the amount of protons secreted is low compared to that in the proximal tubule, but the ΔpH achieved can be as high as 3 pH units because of the lack in bicarbonate buffering capacity in this segment. Here, ATP-driven proton pumps are most probably the sole mechanism for proton secretion.[10-12]

## PROXIMAL TUBULE

FIGURE 2 is a schematic drawing of the very polarized cell of the initial and middle part of the proximal tubule. The apical membrane, also called brush border membrane, is composed of long, slender, and densely packed microvilli. The main vacuolar apparatus of the cell is found in the subapical region and consists of a few large vacuoles and many small endocytotic vesicles and tubules that originate from the base of the microvilli, the intermicrovillous area.

FIGURE 2 also indicates the cellular location of three H$^+$ ATPases. If the situation found in yeast, fungi, and plants is applied to the proximal tubule, mitochondrial, vacuolar, and plasma membrane-bound H$^+$ ATPases can be distinguished. The inner mitochondrial membrane contains the typical $F_0F_1$-type

---

[a] This work was supported by DFG grant 477/279/89.

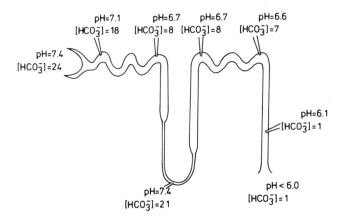

**FIGURE 1.** Schematic representation of renal tubular acidification. (Redrawn from Koeppen *et al.*[2])

ATPase with its high affinity for the inhibitors oligomycin and N,N'-dicyclo-hexylcarbodiimide (DCCD). The vacuolar or endosomal membrane ATPase represents another class of enzymes. As shown by Sabolic and Burckhardt,[13] the sensitivity of this enzyme to inhibitors, in particular N-ethylmaleimide (NEM), is similar to the $H^+$ ATPase of the distal tubule. In the proximal tubule these endocytotic vesicles or endosomes are mainly involved in the reabsorption of proteins present in the primary urine. The $H^+$ ATPase provides for the acidic environment within these vesicles and the lysosomes, which is essential for the breakdown of proteins and dissociation of receptor-ligand complexes.

Evidence for the existence of a $H^+$ ATPase in isolated brush border membranes of the proximal tubule will be discussed in detail. To investigate the properties of a $H^+$ ATPase present in the brush border membrane, these membranes

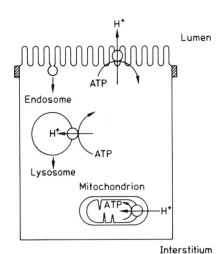

**FIGURE 2.** ATP-dependent proton transport systems in proximal tubule cells.

were isolated and the purity was assessed by enrichment of marker enzymes.[14] Because of the elaborate cytoskeletal core within the microvilli *in situ*, brush border membranes vesiculate to a high degree spontaneously into rightside-out oriented vesicles. Consequently, the former cytoplasmic side of the membranes also forms the inside of the isolated membrane vesicles. Thus, ATP must be incorporated into the vesicles to be able to study ATP-dependent transport processes. Preloading with ATP can be achieved during the initial homogenization of renocortical tissue by including ATP, phosphoenolpyruvate, and pyruvate kinase as an ATP-regenerating system in the homogenization medium.[15] For experiments on proton pump activity in brush border membrane vesicles, either the acidification of the extravesicular medium or the hydrolysis of the intravesicular ATP was monitored. All experiments were performed in the presence of oligomycin and ouabain.

FIGURE 3 shows a representative experiment in which acidification of the extravesicular medium, which occurred in the presence of brush border mem-

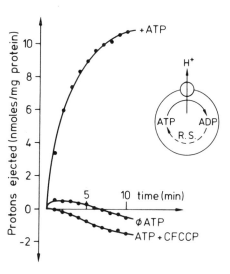

**FIGURE 3.** Proton secretion out of brush border membrane vesicles in the absence or presence of intravesicular ATP and the effect of the protonophor CFCCP (carbonyl-cyanide-p-trifluor-methoxy-phenyl-hydrazone) on proton secretion out of ATP-preloaded membrane vesicles, as indicated by the *symbol*. Five microns CFCCP were added to the extravesicular medium. For further experimental details see reference 15.

brane vesicles preloaded with ATP and the regenerating system, has been recorded. This acidification does not occur when membrane vesicles isolated in the absence of ATP are used. Furthermore, the acidification is abolished in the presence of the protonophor carbonyl-cyanide-p-trifluormethoxyphenylhydrazone (CFCCP) consistent with a recycling of protons into the membrane vesicles.

To investigate whether this ATP-dependent proton secretion is electrogenic or not, the effect of valinomycin was examined. In the presence of potassium and gluconate in the intra- and extravesicular space, the addition of valinomycin immediately stimulated proton secretion from ATP-preloaded vesicles (FIG. 4). This stimulation can be explained by the assumption that initially an electrical potential difference across the membrane is generated by an electrogenic pump, which subsequently inhibits the pump rate. Valinomycin short circuits the membrane and thus diminishes the inhibitory effect of the electrical potential difference. Thus, the proton pump in the brush border membrane seems to be electrogenic. Similar results were recently obtained by Turrini *et al.*[16] They are also

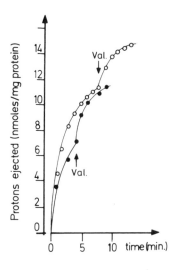

**FIGURE 4.** Effect of valinomycin on proton secretion out of ATP-preloaded brush border membrane vesicles. At different time points, indicated by *arrows*, 10 μg valinomycin were added to the extravesicular medium. (Reprinted from E. Kinne-Saffran *et al.*,[15] with kind permission.)

supported by electrophysiologic measurements of Frömter and Ullrich[17] in the intact proximal tubule.

To further characterize the ATP-driven proton pump present in brush border membrane vesicles we investigated the effect of various proton pump inhibitors. First we tested DCCD, which inhibits all classes of H$^+$ ATPases, but with different affinities. As shown in FIGURE 5, proton movement across the brush border membranes in ATP-preloaded membrane vesicles is also inhibited by DCCD. In this experiment a DCCD concentration of 0.1 mM was used. In FIGURE 6 the concentration dependence of the effect of DCCD and of other inhibitors on proton pump activity in brush border membrane vesicles is summarized. First of all it demonstrates that the affinity of the pump for DCCD is low compared to the mitochondrial ATPase; an apparent $K_i$ of $2 \times 10^{-4}$ M is observed. Besides DCCD the proton pump in brush border membrane vesicles is also inhibited by filipin, an antibiotic that complexes with the cholesterol entity of plasma membranes.[18] Conversely, diethylstilbestrol, a potent inhibitor of H$^+$ ATPases of plant and

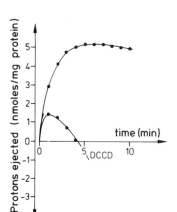

**FIGURE 5.** Effect of 0.1 mM DCCD on proton secretion out of ATP-preloaded brush border membrane vesicles. (Redrawn from Kinne-Saffran *et al.*,[15] with permission.)

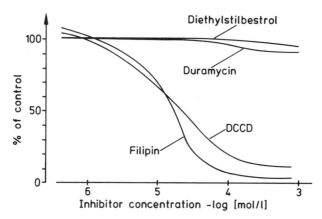

**FIGURE 6.** Effect of various concentrations of DCCD, filipin, diethylstilbestrol, and dura-mycin on the ATP-driven proton pump in ATP-preloaded brush border membrane vesicles. (Redrawn from Kinne-Saffran et al.,[22] with permission.)

fungi,[19,20] and duramycin, which has been reported to inhibit the ATP-driven proton pump of clathrin-coated vesicles,[21] have no effect on the H⁺ pump activity in brush border membrane vesicles.[22]

These results strongly support the view that a proton pump is present in brush border membrane vesicles which is ATP-dependent, electrogenic, and inhibited by DCCD and filipin. To further characterize the biochemical properties of this proton pump, the ATPase activity in permeabilized brush border membrane fragments, rather than sealed vesicles, was investigated. For this purpose brush border membranes were subjected to repeated freezing and thawing, and the activity of the ATPase was measured by the addition of ATP to the incubation medium. In FIGURE 7 the lower symbol at the right depicts the problem encountered in correlating ATPase activity to the proton pump in these membranes. In addition to the

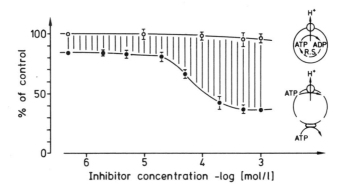

**FIGURE 7.** Effect of various concentrations of diethylstilbestrol on the ATP-driven proton pump in sealed ATP-preloaded brush border membrane vesicles and on the Mg²⁺ ATPase activity in permeabilized brush border membranes, as indicated by the *symbols*. Mean values (± SEM) derived from four experiments are given. (Redrawn from Kinne-Saffran et al.,[22] with permission.)

**TABLE 1.** Effect of Dicyclohexylcarbodiimide (DCCD) and Filipin on Diethylstilbestrol (DESE)-Insensitive $Mg^{2+}$-ATPase Activity in Brush Border Membranes

|                                    | Enzyme Activity[a]        |
| ---------------------------------- | ------------------------- |
| Control                            | $37.9 \pm 2.56$ ($n = 8$) |
| +0.5 mmol DESE                     | $13.2 \pm 1.04$ ($n = 8$) |
| +0.5 mmol DESE and 0.5 mmol DCCD   | $2.1 \pm 0.91$ ($n = 6$)  |
| +0.5 mmol DESE and 0.2 $\mu$g filipin | $1.3 \pm 0.78$ ($n = 6$) |

[a] Enzyme activity is given in $\mu$mol/h/mg protein. Mean values ($\pm$ SEM) are shown. *Numbers in brackets* represent the number of experiments performed.

proton pump these membranes contain other ATPases which are mainly located outside the membrane, but, as already described by Busse *et al.*,[23] are sensitive to diethylstilbestrol. As shown in FIGURE 7, the overall ATPase activity is inhibited by diethylstilbestrol in a dose-dependent manner to a maximum of about 70%. This figure indicates again that at these concentrations of diethylstilbestrol the ATP-driven proton pump is not inhibited. We therefore assumed that the residual ATPase activity observed in the presence of the highest concentration of diethylstilbestrol represents the biochemical equivalent of the ATP-driven proton pump, the $H^+$ ATPase. Further proof of this assumption is provided in TABLE 1. The addition of DCCD as well as filipin inhibits the residual ATPase activity found in the presence of diethylstilbestrol almost completely, results similar to those obtained with DCCD and filipin on the proton pump.

Experiments were also performed to obtain information on the pH dependence of the $H^+$ ATPase. FIGURE 8 shows that the enzyme has a pH optimum at 7.5. It should be noted that this value was obtained in the presence of oligomycin, ouabain, and the specific inhibitor of alkaline phosphatase,[24] levamisole, to prevent ATP breakdown by this enzyme at a higher pH.

The major substrate for the $H^+$ ATPase is ATP (TABLE 2). Besides ATP the enzyme splits also to some extent ITP; GTP and UTP are less effective. This profile is very different from the substrate specificity of the ectoenzymes which hydrolyze all four nucleotides at the same rate.[23]

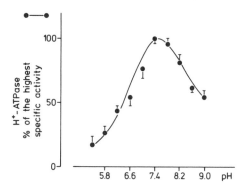

**FIGURE 8.** pH dependence of the diethylstilbestrol-insensitive $Mg^{2+}$ ATPase in permeabilized brush border membranes. Enzyme activity was measured in the presence of 2 mM ouabain, 80 $\mu$g oligomycin, and 200 $\mu$M l-p-bromotetramisole. Enzyme activity is expressed as a percentage of the highest activity measured ($\mu$mol/h/mg protein). Mean values ($\pm$ SEM) derived from four experiments are given.

**TABLE 2.** Activity of H$^+$ ATPase in Brush Border Membranes in the Presence of Different Trinucleotides ($1 \times 10^{-3}$ mol/l) in Relation to the Activity Measured in the Presence of ATP[a]

| Substrate | H$^+$ ATPase (%) | |
|---|---|---|
| ATP | 100 | ($n = 10$) |
| ITP | 42.3 ± 2.09 | ($n = 4$) |
| UTP | 31.3 ± 1.19 | ($n = 4$) |
| GTP | 21.6 ± 1.04 | ($n = 4$) |

[a] Enzyme activity is given in $\mu$mol/h/mg protein. Mean values (± SEM) are shown. *Numbers in brackets* represent the number of experiments performed.

We also attempted to obtain information about the molecular weight of the proton pump in brush border membranes. To this end, radiation inactivation was used, a technique that allows the examination of the molecular weight of membrane-bound proteins without solubilization and dissociation of the membrane components. Isolated brush border membranes were irradiated at −110°C by accelerated electrons, and the H$^+$ ATPase activity was subsequently determined. The activity of the H$^+$ ATPase decreased with increasing doses of radiation. When the logarithm of the remaining activity was plotted against the dose of radiation, a linear correlation was obtained. In parallel experiments we found that inactivation of the enzyme resulted only in a decrease in $V_{max}$, whereas the $K_m$ was not affected. Thus, the single hit criterion appears to apply to the enzyme inactivation, and we therefore conclude that the protein exists in the membrane as a single target or monomer. From these studies a minimum molecular weight for the functional unit of the H$^+$ ATPase in brush border membranes of about 105 kDa can be derived (E. Kinne-Saffran, unpublished observations).

Further evidence for a molecular weight of the H$^+$ ATPase in this range could be derived from $^{14}$C-DCCD labeling studies with isolated brush border membranes, similar to the studies performed recently by Sussman and Slayman[25] with *Neurospora crassa* plasma membranes. As shown in FIGURE 9, $^{14}$C-DCCD labels several protein bands. Strongly labeled bands are present at around 100 and 14

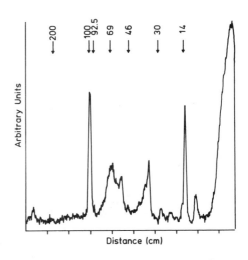

**FIGURE 9.** Densitometric scan of brush border membrane proteins labeled with ($^{14}$C)-DCCD. 500 $\mu$g of membrane protein were incubated for 30 minutes with 2.3 nmol ($^{14}$C)-DCCD/mg protein, 75 mM Glycin/Tris, pH 7.6, 1 mM MgSO$_4$, 2 mM ouabain, and 3 mM Tris-ATP. Membrane proteins were separated by SDS gel electrophoresis with 0.5 mg protein applied per gel slot. Acrylamide concentration of the gradient gel ranged from 5–15%, and an autoradiograph was obtained. The experimental protocol is similar to that described by Friedrich *et al.*[27]

kDa. In rabbit brush border membranes, Igarashi and Aronson[26] recently showed that on the basis of amiloride protection, a 100-kDa protein band is part of the $Na^+/H^+$ exchanger. In the rat kidney this band is not protected by amiloride, suggesting that it may be the $H^+$ ATPase. Further studies are currently performed to substantiate this conclusion.

## DISTAL TUBULE

The luminal membrane of the specialized acid-secreting intercalated cells of the collecting duct contains a $H^+$ ATPase of the vacuolar type, in that it is sensitive to N-ethylmaleimide. The same enzyme activity has been found in endosomal or reserve vesicles present in the apical pole of these cells. There is now suggestive evidence that $H^+$ ATPases can be shuttled extensively between endosomal membranes and the plasma membranes.[28–30] This process is also indicated in the schematic drawing of the lesser polarized distal tubular cell, shown in FIGURE 10. The vacuolar $H^+$ ATPase has been partially purified from bovine kidney medulla.[31] This enzyme is a large molecular mass protein of about 580 kDa and is composed of over 10 subunits. The cDNA of the cytosolic 31-kDa subunit has been cloned, and a higher reactivity to kidney medulla and brain mRNA has been observed.[32]

Recently, we isolated luminal membranes from papillary tissue of bovine kidney.[33] In this membrane fraction a $Mg^{2+}$ ATPase activity, which is insensitive to oligomycin and azide, could be identified. We therefore concluded that this enzyme is a plasma membrane-bound enzyme. Its localization in the luminal membrane of the distal tubule made it a candidate for the biochemical equivalent for the proton pump postulated to be the sole mechanism of acid secretion in this tubular segment.

In contrast to isolated brush border membranes from the proximal tubule, isolated luminal membranes from distal tubules will vesiculate into rightside-out and into inverted membrane vesicles because of the low degree of structural differentiation of the distal tubular cells. In inverted membrane vesicles proton

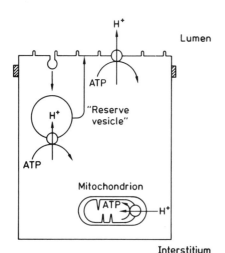

**FIGURE 10.** ATP-dependent proton transport system in distal tubule cells.

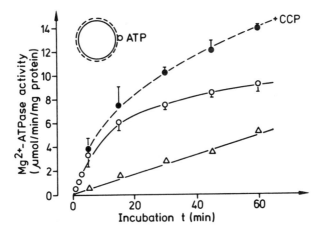

**FIGURE 11.** Effect of the protonophor CFCCP on $Mg^{2+}$ ATPase activity in (luminal) membranes isolated from bovine kidney papilla according to Schwartz *et al.*[33] The *symbol* in this figure represents an inverted membrane vesicle with the catalytic ATP hydrolyzing site facing the medium. Enzyme activities in the absence of CFCCP (O——O) and in the presence of CFCCP (●---●) were determined as described previously[14] and are given in $\mu$mol/h/ mg protein. The activity of the H⁺ ATPase in these membranes (△——△) represents the difference between the two assay conditions. Mean values ($\pm$SEM) derived from four experiments are given.

translocation over the membrane is coupled to an ATPase activity and can be measured as the rate of hydrolysis of extracellular ATP, because the former cytosolic catalytic center of the pump is now exposed to the incubation medium. FIGURE 11 shows that the enzyme reaction is linear only within the first minutes. The rate of ATP hydrolyzed is stimulated markedly when the protonophor CFCCP is added to the incubation medium. Thus, a major portion of the hydrolysis of ATP apparently is coupled to the transport of protons into sealed inverted membrane vesicles. This enzyme activity will therefore be called H⁺ ATPase in the following. With the same experimental setting the pH dependence of the H⁺ ATPase could be determined. In the presence of oligomycin and ouabain the pH optimum was found to be around 6.7 (FIG. 12). From the data given in FIGURE 12 it can also be delineated that at the physiologic pH of distal tubular cells of 7.14 the enzyme is to about 80% active.

As already mentioned, the amount of protons secreted in the distal tubule is low compared to that in the proximal tubule, but the $\Delta$pH achieved can be as high as 3 pH units. This *in vivo* situation could also be demonstrated with isolated rightside-out oriented membrane vesicles that had been preloaded with ATP and an ATP-regenerating system. In these experiments the hydrolysis of intravesicular ATP was monitored, which, in the presence of CFCCP, is linked to proton extrusion from the membrane vesicles.[22] As shown in FIGURE 13, the proton pump is only inhibited to about 20% in the presence of an extravesicular pH of 5.6 or a $\Delta$pH of 2 units across the membrane. Thus, the catalytic center of the ATP-driven proton pump present on the cytoplasmic side (already described) has a much higher pH sensitivity than do those components of the proton pump that are present on the outside of the cell membrane. These data fit well with the physiologic function of the pump in intracellular pH regulation which should be activated

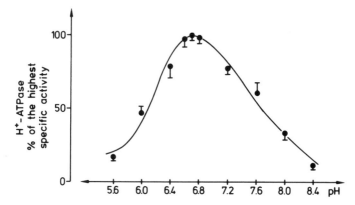

**FIGURE 12.** pH dependence of the H[+] ATPase in (luminal) membranes isolated from bovine kidney papilla according to Schwartz *et al.*[33] Enzyme activity was measured in the presence of 2 mM ouabain and 80 μg oligomycin as described previously.[33] Enzyme activity is expressed as a percentage of the highest activity measured (μmol/h/mg protein). Mean values (± SEM) derived from three experiments are given.

already at minor changes in intracellular pH, whereas the outside of the membrane in this tubular segment has to be capable of coping with much more drastic reductions in urine pH.

TABLE 3 is a compilation of the typical features of proton transport in the proximal and distal tubule and the properties of the H[+] ATPases underlying the ATP-driven proton pumps in these segments. One of the main differences between the proximal and distal tubule is the steepness of the pH gradient that will be generated. As already shown herein, the situation *in vivo* is very close to the results obtained with isolated membranes *in vitro*. In brush border membrane vesicles the pump can work against a maximum of 0.8 pH units, whereas in luminal membranes of the distal tubule the pump is still active when a ΔpH of 2 units is opposing its operation.

Estimates of the relative contribution of ATP-driven proton systems to the overall proton transport capacity of the epithelium obtained in the vesicle studies are in good agreement with the ones observed *in vivo* in the respective tubular

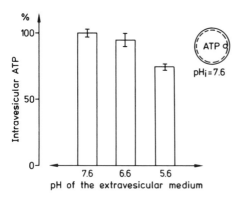

**FIGURE 13.** Effect of extravesicular pH on ATP-driven proton pump in (luminal) membrane vesicles from bovine kidney papilla. Protonophor-induced intravesicular ATP hydrolysis is depicted as a percentage of ATP hydrolysis at an extravesicular pH of 7.6. The *symbol* represents a rightside-out oriented vesicle. Intravesicular ATP hydrolysis was measured as described previously.[15] Mean values (± SEM) derived from three experiments are given.

segments. Taking into account the enrichment factors of the membranes, the protein content of the tubulus, and a stoichiometry of 2 protons pumped per 1 ATP hydrolyzed,[22] the H$^+$ ATPase activities found in the isolated membranes can completely account for the ATP-driven proton transport in the particular segment. The last five lines of TABLE 3 summarize some biochemical properties of the proton pumps and point out the differences. First of all, the proximal and distal proton pumps seem to differ in their pH optimum; their overall molecular weight appears to be different and, lastly, there is a discrepancy in the sensitivity of the pumps to inhibitors, in particular filipin and N-ethylmaleimide. In the proximal tubule using freshly isolated membranes, NEM sensitivity has been confined to the vacuolar and endosomal H$^+$ ATPases. It should be noted, however, that Burckhardt and collaborators most recently showed that the H$^+$ ATPase present in brush border membranes might also be sensitive to NEM after solubilization of the membranes with Triton X100.[16]

TABLE 3. Properties of ATP-Dependent H$^+$ Secretory Mechanisms Present in Luminal Membranes of the Proximal and Distal Tubule

|  | Proximal | Distal |
|---|---|---|
| pH gradient | | |
|    cell : tubular lumen | 7.2 : 6.7 | 7.0 : 4.5 |
| pH gradient generated in luminal membrane vesicles | 0.8 | 2.0 |
| Estimated portion of ATP-dependent H$^+$ secretion as % of overall secretion | 35% | 100% |
| pH optimum of H$^+$ ATPase | 7.5 | 6.7 |
| Apparent molecular weight | 105 kDa | 410–530 kDa |
|  | (?) | (Hetero-oligomer) |
| Inhibitors | | |
|    DCCD (0.5 mM) | ca. 100% | ca. 80% |
|    Filipin (0.2 μg) | ca. 100% | No effect |
|    NEM (1.0 mM) | ? | (100%) |

## CONCLUDING REMARKS

In conclusion, we propose that in the kidney at least three different types of H$^+$ ATPases exist: mitochondrial ATPases, the brush border (plasma membrane) ATPase, and vacuolar ATPases. In the proximal tubule vacuolar ATPase is found in the endosomes and in the area between the microvilli where endocytosis is initiated.

In the $\alpha$ type of intercalated cells[34] of the distal tubule the vacuolar type is present in endosomal (reserve) vesicles and, because of the extensive membrane exchange, also in the luminal plasma membrane. This has also been demonstrated in immunohistologic studies, that is, reserve vesicles and plasma membranes of cells of the distal tubule show reactivity to antibodies raised against a purified distal H$^+$ ATPase[35] or an endosomal H$^+$ ATPase.[36] In the same studies also, reaction with the endocytotic apparatus of the proximal tubule has been observed.[36] It remains to be elucidated whether the H$^+$ ATPase probably present in the basolateral plasma membranes of $\beta$ type intercalated cells is also of the vacuolar type or represents yet another plasma membrane H$^+$ ATPase.

## REFERENCES

1. GOTTSCHALK, L. A., W. E. LASSITER & M. MYLLE. 1960. Am. J. Physiol. **198**: 581–585.
2. KOEPPEN, B., G. GIEBISCH & G. MALNIC. 1985. *In* The Kidney: Physiology and Pathophysiology. D. W. Seldin & G. Giebisch, eds.: 1491–1525. Raven Press. New York, NY.
3. MURER, H., U. HOPFER & R. KINNE. 1976. Biochem. J. **154**: 597–604.
4. KINSELLA, J. L. & P. S. ARONSON. 1980. Am. J. Physiol. **238**: F461–F469.
5. BICHARA, M., M. PAILLARD, F. LEVIEL & J. P. GARDIN. 1980. Am. J. Physiol. **238**: F445–F451.
6. ULLRICH, K. J., G. CAPASSO, G. RUMRICH, F. PAPAVASSILIOU & S. KLOESS. 1977. Pflügers Arch. **368**: 245–252.
7. CHAN, Y. L. & G. GIEBISCH. 1981. Am. J. Physiol. **240**: F220–F230.
8. BICHARA, M., M. PAILLARD, F. LEVIEL, A. PRIGENT & J. P. GARDIN. 1983. Am. J. Physiol. **244**: F165–F171.
9. JANS, A. H. W., K. AMSLER, B. GRIEWEL & R. KINNE. 1986. Biochim. Biophys. Acta **927**: 203–212.
10. ULLRICH, K. J. & F. PAPAVASSILIOU. 1981. Pflügers Arch. **389**: 271–275.
11. PRIGENT, A., M. BICHARA & M. PAILLARD. 1985. Am. J. Physiol. **248**: C241–C245.
12. ZEIDEL, M. L., P. SILVA & J. L. SEIFTER. 1986. J. Clin. Invest. **77**: 113–120.
13. SABOLIC, I. & G. BURCKHARDT. 1986. Am. J. Physiol. **250**: F817–F826.
14. KINNE-SAFFRAN, E. & R. KINNE. 1979. J. Membrane Biol. **49**: 235–251.
15. KINNE-SAFFRAN, E., R. BEAUWENS & R. KINNE. 1982. J. Membrane Biol. **64**: 67–76.
16. TURRINI, F., I. SABOLIC, Z. ZIMOLO, B. MOEWES & G. BURCKHARDT. 1989. J. Membrane Biol. **107**: 1–12.
17. FROMTER, E. & K. J. ULLRICH. 1980. Ann. N.Y. Acad. Sci. **341**: 97–110.
18. DE KRUIFF, B. & R. A. DEMEL. 1974. Biochim. Biophys. Acta **339**: 57–70.
19. GOFFEAU, A. & C. W. SLAYMAN. 1981. Biochim. Biophys. Acta **639**: 197–223.
20. VARA, F. & R. SERRANO. 1982. J. Biol. Chem. **257**: 12826–12830.
21. STONE, D. K., X. S. XIE & E. RACKER. 1983. J. Biol. Chem. **258**: 14834–14838.
22. KINNE-SAFFRAN, E. & R. KINNE. 1986. Pflügers Arch. **407** (Suppl. 2): S180–S185.
23. BUSSE, D., B. POHL, H. BARTEL & F. BUSCHMANN. 1980. J. Membrane Biol. **89**: 147–159.
24. BORGERS, M. & F. THONE. 1975. Histochemistry **44**: 277–280.
25. SUSSMAN, M. R. & C. W. SLAYMAN. 1983. J. Biol. Chem. **258**: 1839–1843.
26. IGARASHI, P. & P. S. ARONSON. 1986. J. Biol. Chem. **262**: 860–868.
27. FRIEDRICH, T., J. SABLOTNI & G. BURCKHARDT. 1986. J. Membrane Biol. **94**: 253–266.
28. STETSON, D. L. & P. R. STEINMETZ. 1983. Am. J. Physiol. **245**: C113–C120.
29. SCHWARTZ, G. J. & Q. AL-AWQATI. 1985. J. Clin. Invest. **75**: 1638–1644.
30. SCHWARTZ, G. J., J. BARASCH & Q. AL-AWQATI. 1985. Nature **318**: 368–371.
31. GLUCK, S. & J. CALDWELL. 1988. Am. J. Physiol. **254**: F71–F79.
32. HIRSCH, S., A. STRAUSS, K. MASOOD, S. LEE, V. SUKHATME & S. GLUCK. 1988. Proc. Natl. Acad. Sci. USA **85**: 3004–3008.
33. SCHWARTZ, I. L., L. J. SHLATZ, E. KINNE-SAFFRAN & R. KINNE. 1974. Proc. Natl. Acad. Sci. USA **71**: 2595–2599.
34. VERLANDER, J. W., K. M. MADSEN & C. C. TISHER. 1987. Am. J. Physiol. **253**: F1124–F1156.
35. BROWN, D., S. HIRSCH & S. GLUCK. 1988. J. Clin. Invest. **82**: 2114–2126.
36. JEHMLICH, K., J. SABLOTNI, W. HAASE & G. BURCKHARDT. 1988. Pflügers Arch. **412** (Suppl. 1): R45 (abstract).

# Simultaneous Binding of Inorganic Phosphate and ATP to Gastric H,K-ATPase

WILLIAM W. REENSTRA

*Department of Physiology-Anatomy*
*University of California*
*Berkeley, California 94720*

Gastric H,K-ATPase, the primary pump for acid secretion by the oxyntic cell, catalyzes an ATP-dependent exchange of 1 cytoplasmic $H^+$ for 1 luminal $K^+$. Hydrolysis of ATP was shown to proceed through an acyl-phosphoenzyme intermediate (E-P) and to show negative cooperativity with respect to ATP concentration.[1] We further characterized the mechanism of ATP hydrolysis and the causes of the negative cooperativity by measuring the steady-state rate of ATP hydrolysis (v) and the steady-state level of E-P.[2] The ATP-dependent phospho-protein, isolated by acid precipitation, was shown to be an acyl-phosphate by demonstrating that (1) greater than 90% of the isolated phospho-protein was subject to hydroxlyamine-catalyzed hydrolysis at pH 6.5, and (2) the observed rate constant was independent of the concentration of ATP used to form the phospho-protein. Based on these results we concluded that the isolated phospho-protein was the intermediate E-P in scheme 1.

As shown in FIGURE 1 the ratio of v to E-P, v/[E-P], increased threefold as the concentration of ATP was increased from 0.1 to 200 $\mu$M. Thus, for the rate law v = [E-P]·k, k—the rate constant for E-P hydrolysis during steady-state hydrolysis of ATP—must be a function of the ATP concentration. One of several mechanisms that are consistent with this finding is shown in scheme 1.

$$K_c^+ + E + ATP \rightleftharpoons E \cdot ATP + H_c^+ \rightarrow E\text{-}P \cdot + ADP$$

$$E \cdot Pi \cdot ATP \cdot K \tag{1}$$

$$Pi + E \cdot K \leftarrow E \cdot Pi \cdot K \rightleftharpoons E\text{-}P \cdot K \rightleftharpoons K_l^+ + E\text{-}P + H_l^+$$

Although data cannot distinguish between this mechanism and mechanisms whereby ATP binds to an E-P form of the ATPase, the data are inconsistent with any mechanism whereby ATP binds to E·K, that is, after the loss of Pi. Although the rate constant for E-P hydrolysis at low ATP concentrations, 320 $min^{-1}$, is similar to a value of 400 $min^{-1}$ obtained with rapid quench techniques, the previous study failed to observe ATP catalysis of E-P hydrolysis.[3] This apparent discrepancy will have to be resolved by future studies.

Additional evidence for the mechanism in scheme 1 was obtained by studying phosphate (Pi) inhibition of ATP hydrolysis.[2] At low (less than 1 $\mu$M) ATP, where hydrolysis proceeds through the free enzyme (E), inhibition by Pi is biphasic, and an apparently Pi-independent rate of ATP hydrolysis is seen at 10 mM Pi (FIG. 2). The reduced rate of hydrolysis at 10 mM Pi is consistent with ATP binding to a phosphate-containing form of the enzyme and hydrolysis proceeding, at a reduced rate, via the E·Pi·ATP·K complex. At high ATP concentrations (100 to 500 $\mu$M)

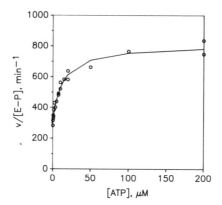

**FIGURE 1.** Dependence of v/[E-P] on the concentration of ATP. Gastric microsomes, purified from hog stomachs by differential centrifugation and lyophilized in 5 mM Pipes-Tris, were assayed for ATP hydrolysis (v) and phosphoenzyme formation (E-P) at 22°C in 10 mM KCl, 1 mM MgSO₄, and 10 mM Pipes-Tris at pH 7.0 with γ³²P-ATP. For v, rates were calculated from the specific activity of the ATP, the concentration of protein, and the ³²Pi in four 200-μl aliquots that were removed at various times and quenched with acid. After ATP was adsorbed to charcoal, Pi was determined by liquid scintillation spectrometry. For E-P, microsomes were assayed under identical conditions. After quenching the reaction with acid the precipitated protein was trapped on filters and E-P was determined from the specific activity of the ATP, the concentration of protein, and the ³²Pi recovered on the filter. (Blanks where E-P was chased with a 100-fold excess cold ATP were subtracted in all cases.) The values of v and E-P were used to calculate v/[E-P]. In all cases less than 15% of the total ATP was hydrolyzed during incubation. (Modified from ref. 2.)

Pi inhibition ($K_I > 100$ mM) is not competitive with ATP and shows no biphasic behavior (data not shown), suggesting that Pi can inhibit ATP hydrolysis by binding to E·ATP. These observations are inconsistent with mechanisms whereby the binding of ATP and Pi is mutually exclusive. The mechanism in scheme 1 is

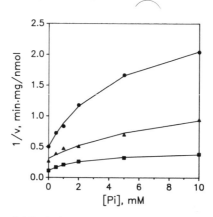

**FIGURE 2.** Inhibition of ATP hydrolysis by inorganic phosphate at low ATP concentrations. Lyophilized membranes were assayed at 22°C in 10 mM KCl, 1 mM MgSO₄, and 10 mM Pipes-Tris at pH 7.0 with γ³²P-ATP. Rates were determined as described in FIGURE 1. ATP concentrations were 0.1 (●), 0.2 (▲), and 0.5 (■) μM ATP.

also consistent with biphasic inhibition of ATP hydrolysis by vanadate, a phosphate analog.[4]

In summary, these studies show that at physiologic ATP concentrations, hydrolysis by the H,K-ATPase proceeds via an E·Pi·ATP·K complex with the simultaneous binding of ATP and Pi. In this respect the mechanism of the H,K-ATPase may differ from the mechanism of ATP hydrolysis by the Na,K-ATPase where most studies suggest that the binding of ATP and Pi are mutually exclusive.[5]

## REFERENCES

1. WALLMARK, B., H. STEWART, E. RABON, G. SACCOMANI & G. SACHS. 1980. J. Biol. Chem. **255:** 7361–7368.
2. REENSTRA, W. W., J. D. BETTENCOURT & J. G. FORTE. 1988. J. Biol. Chem. **263:** 19618–19625.
3. STEWART, H., B. WALLMARK & G. SACHS. 1981. J. Biol. Chem. **256:** 2682–2690.
4. FALLER, L., E. RABON & G. SACHS. 1983. Biochemistry **22:** 4676–4685.
5. GLYNN, I. M. 1988. J. Skou et al., eds. In Progress in Clinical and Biological Research. **268A:** 435–460. Alan Liss, New York.

# Structural Analysis of Gastric ($H^+$ + $K^+$)-ATPase[a]

MASATOMO MAEDA AND MASAMITSU FUTAI

*Department of Organic Chemistry and Biochemistry*
*The Institute of Scientific and Industrial Research*
*Osaka University*
*8-1 Mihogaoka, Ibaraki*
*Osaka 567, Japan*

Gastric ($H^+$ + $K^+$)-ATPase is an intrinsic membrane protein localized in the plasma membranes on the secretory surface of parietal cells and is responsible for acid secretion.[1] As pig enzyme is advantageous in biochemical studies on coupled ion transport and its regulation, information on its primary structure is a prerequisite to further detailed studies.

It is difficult to determine the entire amino acid sequence of this ($H^+$ + $K^+$)-ATPase by protein chemical analysis, because this enzyme is a membrane protein with a large molecular mass. Therefore, complementary DNA to pig gastric mRNA encoding ($H^+$ + $K^+$)-ATPase was cloned, and its amino acid sequence was deduced from the nucleotide sequence.[2] The enzyme was found to consist of 1,034 amino acid residues (Mr. 114,285) including the initiation methionine (FIG. 1.) The amino terminal sequence[3] ([3]KAENYELYQVELGPGP[18]), the aspartic acid residue phosphorylated during a catalytic cycle[4] ([384]CS$D$K[387]), and the fluorescein isothiocyanate (FITC)-reactive lysine residue[5] ([515]LVM$K$GAPE[522]) were identified in the deduced sequence. Potential sites of phosphorylation by cAMP-dependent protein kinase (KRXX$S$ and RRX$S$) and N-linked glycosylation sites ($N$X(S/T)) were also found: residues 363 and 953 for phosphorylation and residues 225, 493, and 773 for glycosylation. The sequence of pig ($H^+$ + $K^+$)-ATPase was highly homologous with that of the corresponding enzyme from rat,[6] with 97.6% of the amino acid residues and 89.1% of the nucleotide residues being identical. There were synonymous codon changes of 28.9% of the amino acid residues. In the $\alpha$ subunit of the closely related ($Na^+$ + $K^+$)-ATPase,[7] 63.0% of the amino acid residues are identical to those of pig ($H^+$ + $K^+$)-ATPase. The hydropathy profile suggests that the enzyme traverses the membrane several times (FIG. 2). The amino acid residues indicated in the figure, located in the same large hydrophilic

[a] This work was supported in part by grants from the Ministry of Education, Science and Culture of Japan, Mitsubishi Foundation, and the Foundation for Promotion of Pharmaceutical Science.

```
  1 MGKAENYELY QVELGPGPSG DMAAKMSKKK AGRGGGKRKE KLENMKKEME INDHQLSVAE   60
    LEQKYQTSAT KGLSASLAAE LLLRDGPNAL RPPRGTPEYV KFARQLAGGL QCLMWVAAAI  120
    CLIAFAIQAS EGDLTTDDNL YLALALIAVV VVTGCFGYYQ EFKSTNIIAS FKNLVPQQAT  180
    VIRDGDKFQI NADQLVVGDL VEMKGGDRVP ADIRILQAQG RKVDNSSLTG ESEPQTRSPE  240
    CTHESPLETR NIAFFSTMCL EGTAQGLVVN TGDRTIIGRI ASLASGVENE KTPIAIEIEH  300
    FVDIIAGLAI LFGATFFIVA MCIGYTFLRA MVFFMAIVVA YVPEGLLATV TVCLSLTAKR  360
    LASKNCVVKN LEAVETLGST SVICSDKTGT LTQNRMTVSH LWFDNHIHSA DTTEDQSGQT  420
    FDQSSETWRA LCRVLTLCNR AAFKSGQDAV PVPKRIVIGD ASETALLKFS ELTLGNAMGY  480
    RERFPKVCEI PFNSTNKFQL SIHTLEDPRD PRHVLVMKGA PERVLERCSS ILIKGQELPL  540
    DEQWREAFQT AYLSLGGLGE RVLGFCQLYL SEKDYPPGYA FDVEAMNFPT SGLSFAGLVS  600
    MIDPPRATVP DAVLKCRTAG IRVIMVTGDH PITAKAIAAS VGIISEGSET VEDIAARLRV  660
    PVDQVNRKDA RACVINGMQL KDMDPSELVE ALRTHPEMVF ARTSPQQKLV IVESCQRLGA  720
    IVAVTGDGVN DSPALKKADI GVAMGIAGSD AAKNAADMIL LDDNFASIVT GVEQGRLIFD  780
    NLKKSIAYTL TKNIPELTPY LIYITVSVPL PLGCITILFI ELCTDIFPSV SLAYEKAESD  840
    IMHLRPRNPK RDRLVNEPLA AYSYFQIGAI QSFAGFTDYF TAMAQEGWFP LLCVGLRPQW  900
    ENHHLQDLQD SYGQEWTFGQ RLYQQYTCYT VFFISIEMCQ IADVLIRKTR RLSAFQQGFF  960
    RNRILVIAIV FQVCIGCFLC YCPGMPNIFN FMPIRFQWWL VPMPFGLLIF VYDEIRKLGV 1020
    RCCPGSWWDQ ELYY*                                                  1034
```

**FIGURE 1.** Amino acid sequence of pig gastric (H⁺ + K⁺)-ATPase deduced from the nucleotide sequence. The consensus sequences of the N-glycosylation site and phosphorylation site of cAMP-dependent protein kinase are indicated by ⌒ and □, respectively. The line above the sequence indicates the lysine-rich sequence. The amino acid sequences determined by protein chemical analyses are underlined.

segment possibly facing the cytoplasm, may form a catalytic site. The amino terminal region contains a lysine-rich sequence similar to that of the α subunit of (Na⁺ + K⁺)-ATPase, although a cluster of glycine residues is inserted into the sequence of the (H⁺ + K⁺)-ATPase (FIG. 1). The lysine-rich sequences in (H⁺ + K⁺)- and (Na⁺ + K⁺)-ATPase may have a common regulatory role(s) in cation binding and its occlusion. The cluster of glycine residues found only in (H⁺ + K⁺)-ATPase may determine the enzyme specificity of H⁺.

Chemical modification experiments often, but not always, provide valuable information about functional residues and regions of the enzymes. During a search for specific modification reagents, we found that pyridoxal 5′-phosphate (PLP) is a unique modifier of (H⁺ + K⁺)-ATPase.[8] Modification of pig gastric (H⁺ + K⁺)-ATPase with PLP resulted in inhibition of K⁺-dependent ATP hydrolysis, phosphoenzyme formation, and H⁺ uptake into vesicles. ATP, ADP, and adenyl 5′-yl imidodiphosphate were protective ligands. The stoichiometry of PLP binding to the enzyme was about 1:1. Limited proteolysis of the enzyme modified with (³H)-PLP indicated that PLP specifically modified a lysine residue located in a 16-kDa fragment of the enzyme cleaved by trypsin. These results suggested that PLP binds to a specific lysine residue in the nucleotide binding site or a region in its vicinity and inhibits the substrate binding or phosphorylation step of (H⁺ + K⁺)-ATPase. Peptides labeled with radioactive PLP could be released from gastric membrane vesicles quantitatively by chymotrypsin treatment, and two peptides were purified by high performance liquid chromatography. From sequence analysis, Lys-497 was concluded to be the binding site of PLP.[9] This residue is conserved in (Na⁺ + K⁺)- and Ca²⁺-ATPases.[7,10,11]

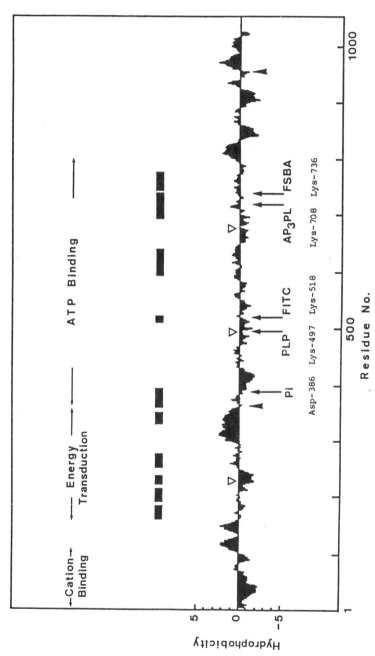

**FIGURE 2.** Hydropathy profile of pig gastric $(H^+ + K^+)$-ATPase. Asp-386 (phosphorylation site), Lys-497 (PLP binding site), Lys-708 (corresponding to the adenosine triphosphopyridoxal $(AP_3$-PL) binding site in $Ca^{2+}$-ATPase[12]), and Lys-736 (corresponding to the 5'-(p-fluorosulfonyl)benzoyl adenosine (FSBA) binding site in the $\alpha$ subunit of $(Na^+ + K^+)$-ATPase[13]) are shown by *arrows*. Potential sites for N-glycosylation and phosphorylation by cAMP-dependent protein kinase are indicated by $\triangledown$ and $\blacktriangle$, respectively. The averaged hydropathic index of a nonadecapeptide is plotted.[15]

# REFERENCES

1. FALLER, L., R. JACKSON, D. MALINOWSKA, E. MUKIDJAM, E. RABON, G. SACCO-
   MANI, G. SACHS & A. SMOLKA. 1982. Ann. N. Y. Acad. Sci. **402:** 146–163.
2. MAEDA, M., J. ISHIZAKI & M. FUTAI. 1988. Biochem. Biophys. Res. Commun. **157:**
   203–209.
3. LANE, L. K., T. L. KIRLEY & W. J. BALL, JR. 1986 Biochem. Biophys. Res. Commun.
   **138:** 185–192.
4. WALDERHAUG, M. O., R. L. POST, G. SACCOMANI, R. T. LEONARD & D. P. BRISKIN.
   1985. J. Biol. Chem. **260:** 3852–3859.
5. FARLEY, R. A. & L. D. FALLER. 1985. J. Biol. Chem. **260:** 3899–3901.
6. SHULL, G. E. & J. B. LINGREL. 1986. J. Biol. Chem. **261:** 16788–16791.
7. OVCHINNIKOV, Y. A., N. N. MODYANOV, N. E. BROUDE, K. E. PETRUKHIN, A. V.
   GRISHIN, N. M. ARZAMAZOVA, N. A. ALDANOVA, G. S. MONASTYRSKAYA & E. D.
   SVERDLOV. 1986. FEBS Lett. **201:** 237–245.
8. MAEDA, M., M. TAGAYA & M. FUTAI. 1988. J. Biol. Chem. **263:** 3625–3656.
9. TAMURA, S., M. TAGAYA, M. MAEDA & M. FUTAI. 1989. J. Biol. Chem. **264:** 8580–
   8584.
10. MACLENNAN, D. H., C. J. BRANDL, B. KORCZAK & N. M. GREEN. 1985. Nature **316:**
    695–700.
11. SHULL, G. E. & J. GREEB. 1988. J. Biol. Chem. **263:** 8646–8657.
12. YAMAMOTO, H., M. TAGAYA, T. FUKUI & M. KAWAKITA. 1988. J. Biochem. (Tokyo)
    **103:** 452–457.
13. OHTA, T., K. NAGANO & M. YOSHIDA. 1986. Proc. Natl. Acad. Sci. USA **83:** 2071–
    2075.
14. SERRANO, R. 1988. Biochim. Biophys. Acta **947:** 1–28.
15. KYTE, J. & R. F. DOOLITTLE. 1982. J. Mol. Biol. **157:** 105–132.

# Models for Gastric Proton Pump in Light of the High Oxyntic Cell Conductance

W. S. REHM,[a] G. CARRASQUER, M. SCHWARTZ,
AND T. L. HOLLOMAN

*Departments of Medicine, Physics, and*
*Engineering Math and Computer Science*
*University of Louisville*
*Louisville, Kentucky 40292*

The secreting *in vitro* frog fundus has a low transmucosal resistance ($R_t$), and the low $R_t$ is due to a low resistance of the lumen-tubular (oxyntic) cell pathway ($R_{LC} = R_L + R_3 + R_4$; see FIG. 1). The parallel paths, that is, the surface cell ($R_{sur} = R_1 + R_2$; see FIG. 1), and the transintercellular (paracellular) $R_{TIC}$ and $R'_{TIC}$ paths have high resistances. These conclusions are based on many lines of evidence that we believe are well accepted by workers in this field.[1-3] Contemporary models[2] for gastric HCl secretion are diagrammed in FIGURE 2. The question arises as to which of these models are compatible with the low resistance of the tubular cells. In model 2A it is postulated that there are two parallel neutral mechanisms in the secretory (luminal) membrane of the tubular cells; KCl moves from cell to lumen via a neutral symport and $K^+$ is then actively exchanged with protons by a second neutral mechanism (one $K^+$ for one proton). With this model a low value for $R_{LC}$ would not be anticipated; it would be necessary to make the *ad hoc* postulate of a low resistance path for some other ion not necessarily essential for secretion. This model has been suggested in the past but has no advocates today, so we will not further comment on it.

In model 2B there are conductive pathways for $K^+$ and $Cl^-$ for transport from cell to lumen and a parallel neutral active $K^+$-$H^+$ exchange mechanism across the membrane. In model 2C there are two high conductance paths, a $Cl^-$ path and an electrogenic proton path. It is postulated that regardless of the complexity of the proton mechanism, it is electrogenic, that is, there is a transfer of $H^+$ from the membrane to the lumen without the involvement of another ion at this site. A completed circuit for the secretion of HCl is necessary, and with $Cl^-$ media, movement of $Cl^-$ from cell to lumen completes the circuit. Model 2B (NP = neutral proton pump) and model 2C (EP = electrogenic proton pump) are both compatible with a low $R_{LC}$.

Our main thrust in this paper is to attempt to distinguish between the NP and EP models on the basis of other experimental evidence.

## METHODS

The fundus of the frog (*Rana pipiens*) with the external muscles removed is mounted between lucite chambers, and a four-electrode system is used for deter-

---

[a] Address for correspondence: Dr. Warren S. Rehm, R. 632, MDR Building, Department of Medicine, University of Louisville, Louisville, KY 40292.

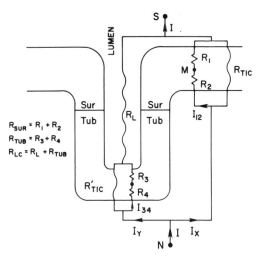

**FIGURE 1.** Scheme to illustrate the transmucosal conductance pathways for the frog gastric fundus. See text.

$$R_{SUR} = R_1 + R_2$$
$$R_{TUB} = R_3 + R_4$$
$$R_{LC} = R_L + R_{TUB}$$

mining $R_t$ and potential difference (PD)[2]. The pH stat method is used for determining the H$^+$ rate. With Cl$^-$ media the nutrient solution contains (mM): 102 Na$^+$, 4 K$^+$, 1 Ca$^{2+}$, 0.8 Mg$^{2+}$, 82.6 Cl$^-$, 25 HCO$_3^-$, 1 phosphate, and 10 glucose; and with Cl-free media it contains: 102 Na$^+$, 4 K$^+$, 1 Ca$^{2+}$, 0.8 Mg$^{2+}$, 25 HCO$_3^-$, 41.3 SO$_4^{2-}$, 41.3 sucrose, 1 phosphate, and 10 glucose. The secretory solution for Cl$^-$ media contained (mM): 156 Na$^+$, 4 K$^+$, and 160 Cl$^-$; and for Cl$^-$-free media it contained 156 Na$^+$, 4 K$^+$, 80 SO$_4^{2-}$, and 80 sucrose. For secretory and nutrient solutions with elevated K$^+$, Na$^+$ was replaced with K$^+$. Histamine (10$^{-4}$ M) was present in the nutrient solution.

A comment on the osmotic pressure of the bathing media is appropriate. We found that the increase in $R_t$ with inhibition was a function of the osmotic pressure

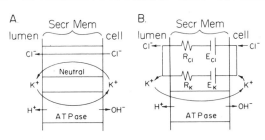

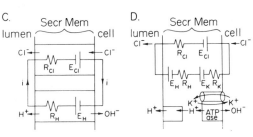

**FIGURE 2.** Proposed models for gastric HCl secretion. See text.

of the secretory solution relative to that of the nutrient one. During secretion, changing from a hypertonic or isotonic secretory solution to a hypotonic one produced only a small increase in $R_t$. On subsequent inhibition of secretion with the hypotonic secretory solution there was a huge increase in $R_t$ (about 1,000 ohm cm$^2$). With increasing osmotic pressures of the secretory solutions, the increase in $R_t$ due to inhibition decreased and reached a lower level (about 100 ohm cm$^2$) with an osmotic pressure definitely less than the one chosen as standard (see above).[1-3] With the standard hypertonic secretory solution the increase in $R_{LC}$ is primarily due to an increase in $R_{TUB}$ (i.e., $R_3 + R_4$ in FIG. 1).

## The $H^+$-$K^+$ ATPase

In recent years primarily on the basis of work by Forte and colleagues[4,5] and Sachs and colleagues,[6,7] evidence for a neutral active $H^+$-$K^+$ ATP-driven exchange mechanism has been obtained from studies on the characteristics of vesicles obtained from the secretory membrane of the tubular cells. These workers have extrapolated the vesicle findings to the intact tissue, and their model is that in FIGURE 2B. It should be pointed out that the FIGURE 2B model is essentially the same as a 1948 scheme proposed by Conway and Brady.[8] According to this model, as already pointed out above, $K^+$ and $Cl^-$ enter the lumen, and $K^+$ and $H^+$ are neutrally exchanged across the lumen-cell border. À propos of the Conway-Brady model, we presented results that we thought made this model untenable. This work involved primarily the effect of changing $K^+$ and $Cl^-$ concentrations and the effect of applied current on the $H^+$ secretory rate. These older findings and an analysis of them are briefly reviewed in reference 2, page G152. However, there was no concrete evidence for the Conway-Brady assumption of a neutral active $K^+$-$H^+$ exchange, and so little further consideration was given to their model. But since the discovery[4] of the $K^+$-$H^+$ ATPase, the implication of using the NP model for the intact tissue needs substantial further examination.

## Adequacy of the NP and EP Models

A challenging finding for testing the adequacy of model 2B is that the *in vitro* frog stomach, with $Cl^-$-free bathing media, continues to secrete acid (about 30% of the rate in $Cl^-$ media), but the PD becomes inverted.[9,10] With $Cl^-$ media the PD is positive (the nutrient side is positive), but after changing to $Cl^-$-free media the PD is inverted, and the nutrient side becomes negative. The PD goes from about +30 mV to about −30 mV.

We interpret the orientation of the PD as the result of two major active ion transport systems in the tubular cells. These systems are an electrogenic $Na^+$-$K^+$-ATPase in the nutrient membrane (with more $Na^+$ moving out of the cell than $K^+$ moving in the opposite direction)[11] and an active proton transport mechanism in the secretory membrane that (even with the NP model in some way) produces an electromotive force (EMF) oriented to make the secretory side positive. As shown in FIGURE 3, $E_S$ represents an equivalent circuit EMF for the secretory membrane due to the proton pump, and $E_N$ the EMF of the nutrient membrane due to the $Na^+$-$K^+$ pump. With $Cl^-$ media $E_N$ is greater than $E_S$ because the PD is positive, but in $Cl^-$-free media $E_S$ is greater than $E_N$ because the PD is negative. Considerable evidence indicates that in the secretory membrane with $Cl^-$ media there is a high $Cl^-$ conductance and a lower $K^+$ conductance. Hence, with $Cl^-$-

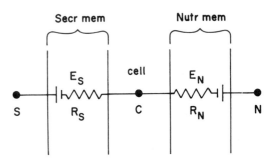

**FIGURE 3.** Equivalent circuit for potential difference (PD) between the secretory (S) and nutrient (N) solutions. $E_S$ and $E_N$ are equivalent circuit EMFs for PD across the secretory and nutrient membranes, respectively.

free media the marked shunting effect of the high $Cl^-$ conductance is absent and therefore $E_S$ is increased (FIG. 4) and the PD is inverted.

With the EP model the foregoing findings are exactly those predicted: $E_S$ becomes greater than $E_N$. In contrast, because the NP model is a neutral proton model, the question arises of how can the inverted PD be explained. Before pursuing this, let us go back to $Cl^-$ media.

It is well established that with standard $Cl^-$ media, inhibition (and vice versa for stimulation) produces an increase in $R_t$ and PD.[12] FIGURE 5 shows an experiment with standard $Cl^-$ media (4 mM $K^+$ on each side) in which the $H^+$ rate was reduced to zero by the addition of thiocyanate (SCN), a well-known inhibitor of acid secretion. The PD increased by about 18 mV and then leveled off at a slightly lower level. $R_t$ increased by about 150 ohm cm$^2$ and then decreased to a level about 65 ohm cm$^2$ above the control level. On the basis of the EP model, inhibition with $Cl^-$ media would result in an increase in $R_H$ and an abolition of $E_H$ (FIG. 4), which explains the increase in PD and $R_t$.

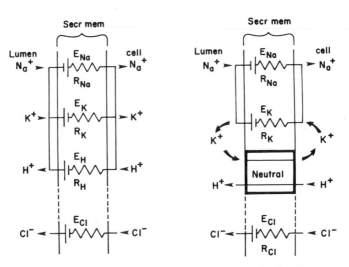

**FIGURE 4.** Equivalent circuits for secretory (Secr) membrane for the electrogenic proton pump model (EP) in FIGURE 2C (*left*) and the neutral pump model (NP) in FIGURE 2B (*right*). With $Cl^-$-free media the equivalent circuits are given by ignoring the $Cl^-$ limbs. See text.

The question arises of how the increase in PD and $R_t$ can be explained on the basis of the NP model. Advocates of the NP model[13,14] suggest that a $K^+$ diffusion potential in the secretory membrane ($E_K$ in FIG. 4) is primarily responsible for $E_S$ in FIGURE 3 and that with inhibition the magnitude of this diffusion potential decreases because $K^+$ continues to enter the lumen from the cell. The increase in the concentration of $K^+$ in the lumen results in a decrease in the ratio of $K^+$ in the cell to that in the lumen, and hence the magnitude of $E_K$ in FIGURE 4 (hence $E_S$ in FIG. 3) would decrease; moreover, because its orientation is to make the secretory positive, it results in the observed increase in PD. We tested this postulate by increasing the $K^+$ on the secretory side from 4 to 80 mM before inhibition.

A typical experiment is shown in FIGURE 6. Increasing $K^+$ from 4 to 80 mM ($K^+$ replaced $Na^+$; $Na^+$ and $K^+$ were both 80 mM) produced a transient increase in PD of about 5 mV and a substantial decrease in $R_t$ from about 135 to about 65 ohm cm². Then the addition of Na SCN to the nutrient solution produced an increase in PD of about 17 mV and an initial increase in $R_t$ of about 65 ohm cm²

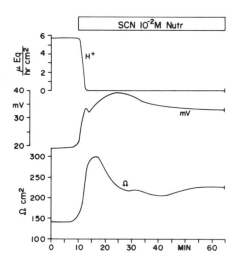

FIGURE 5. Effect on potential difference, resistance, and $H^+$ secretory rate of addition of thiocyanate (SCN) to nutrient solution of frog gastric fundus bathed in standard $Cl^-$ solutions.

with a decline to a lower value of about 40 ohm cm² above the control value.[15,16] The $H^+$ rate decreased to zero in about 5 minutes. The changes in $R_t$, PD, and $H^+$ rate were concurrent. The average increase in PD for the two conditions (FIGS. 5 and 6) was not significantly different and was approximately 15 mV. On the basis of the NP model,[13,14] with 80 mM $K^+$, continued diffusion of $K^+$ following inhibition of $H^+$ secretion will change the $K^+$ concentration by only a small percentage. The ratio of $K^+$ in the cell to that in the lumen will be essentially unchanged. The PD, $R_t$, and $H^+$ rate all changed substantially within a few minutes. The rapidity of the changes in these characteristics in FIGURE 6 makes it improbable that the ratio of $K^+$ in the cell to that in the lumen could change significantly in this short time; this is even true for FIGURE 5 (with 4 mM $K^+$ in the secretory solution).

Results with $Cl^-$-free media show that inhibition of secretion produces a decrease in the magnitude of the PD;[10] in fact, it often returns to the normal orientation. For example, the PD typically goes from $-30$ to about $+5$ mV (FIG. 7). Adding Na SCN (final concentration 20 mM) to the secretory side produced a

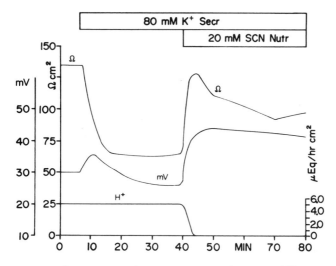

**FIGURE 6.** Effect in Cl⁻ media on potential difference, resistance, and H⁺ secretory rate of increasing K⁺ (4 to 80 mM, K⁺ replacing Na⁺) on the secretory side (*top bar*), and effect of adding SCN to the nutrient side (final concentration 20 mM), with 80 mM K⁺ still present on the secretory side.

decrease in the H⁺ rate to zero and concurrent changes in PD and $R_t$; $R_t$ decreased from about 375 to about 150 ohm cm² and the PD increased from about −30 mV to a slightly positive value. The decrease in $R_t$ we explained as SCN acting like Cl⁻ in the Cl⁻ conductance limb of the secretory membrane (see Discussion). Again, this is what is predicted on the basis of the EP model: the

**FIGURE 7.** Effect on potential difference (PD), resistance, and H⁺ rate with Cl⁻-free solutions of the addition of SCN to the secretory side and subsequent elevation of K⁺ from 4 to 80 mM on the secretory side. NOTE: Before the SCN addition, PD was about −30 mV (the nutrient side negative to secretory side).

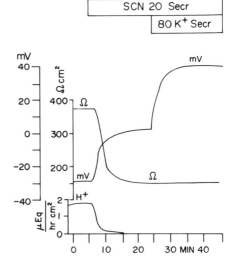

reduction of $E_H$ in FIG. 2C and FIG. 4 to zero would result in an increase in the PD (usually to positive values). Proponents of the NP model[13,14] invoked the $K^+$ diffusion potential to explain the inverted PD. The question arises: How can the PD results be explained on the basis of this model? It is possible that inhibition markedly reduces the $K^+$ conductance of the secretory membrane, which will markedly reduce the contribution of $E_K$ (FIG. 4) to $E_S$ (FIG. 3). This postulate is very easily tested.[17] In fact, the test is presented in FIGURE 7. If inhibition of secretion produces a marked reduction in the magnitude of $K^+$ conductance, then after inhibition the increase in $K^+$ in the secretory membrane from 4 to 80 mM should have essentially no effect on the PD. But as can be seen in FIGURE 7, it causes a typical marked increase in the PD. Hence the postulate that inhibition produces a marked reduction in $K^+$ conductance is contrary to the facts.

We also performed experiments with the reverse sequence to that in FIGURE 7.[16] It might be argued that the expected results could be predicted on the basis of those in FIGURE 7. Nevertheless, we performed the experiments in reverse order. We increased the $K^+$ on the secretory side from 4 to 80 mM and then inhibited secretion (FIG. 8). In going from 4 to 80 mM $K^+$ the PD increased from about $-30$ to $+5$ mM, confirming our previous work.[9] It is pertinent that in $Cl^-$-free media the partial $K^+$ conductance of the secretory membrane is high: increasing $K^+$ from 4 to 80 mM increases the PD by about 40 mV. In contrast, in $Cl^-$ media the partial $K^+$ conductance of the secretory membrane is low.[18–20] The $\Delta$PD, as shown in FIGURE 6, is about 5 mV, an increase in secretory $K^+$ from 4 to 80 mV. In other words, with $Cl^-$ media partial $Cl^-$ conductance of the secretory membrane is higher than that of $K^+$, whereas in the absence of $Cl^-$ the partial $K^+$ conductance of the secretory membrane is relatively high.[21] Now with a high $K^+$ in the secretory side, inhibition produces a large increase in PD, as seen in FIGURE 8. Again these results are predicted by the EP model but not by the NP model.

The conclusion, on the basis of the present and previous evidence (not reviewed herein) and regardless of the complexity of the proton pump, is that it is apparently electrogenic in intact frog tissue.

A comment on the change in $R_t$ shown in FIGURES 6, 7, 8, and 10 is in order.

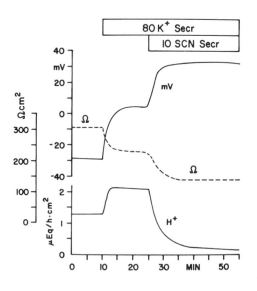

**FIGURE 8.** Effect on potential difference, resistance, and $H^+$ rate with $Cl^-$-free solutions of increasing $K^+$ from 4 to 80 mM on the secretory side, which was followed by the addition of SCN to this side.

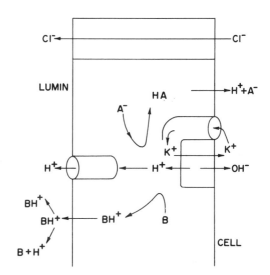

**FIGURE 9.** Model for the secretory membrane in which $H^+$-$K^+$ ATPase is incorporated in this membrane. $A^-$ represents anions like SCN which inhibit the $H^+$ rate, and B represents weak bases that overcome the inhibition by $A^-$. See text.

Elevation of $K^+$ on either side decreases $R_t$. This decrease, we believe, is due to the conductance of the $K^+$ channels being a function of the $K^+$ concentration in the bathing media: a large increase in $K^+$ produces a large increase in conductance. As already pointed out, we explained the large increase in conductance (decrease in $R_t$) caused by addition of SCN in $Cl^-$-free media as due to the anion (usually $Cl^-$) limb being permeable to SCN. In general, in $Cl^-$-free media, SCN produces a greater decrease in $R_t$ than does elevation of $K^+$. For example, in FIGURE 8 the addition of SCN in the presence of 80 mM $K^+$ produces a further decrease in $R_t$, whereas, as shown in FIGURE 7, increasing $K^+$ to 80 mM produces essentially no further decrease in $R_t$ when SCN is present.

In an attempt to reconcile the findings of the neutral $K^+$-$H^+$ exchange mechanism in the vesicles with an electrogenic proton pump in intact tissue, the model presented in FIGURE 2D was devised. FIGURE 9 is an expanded model of that in FIGURE 2D. In this model the neutral $H^+$-$K^+$ exchange occurs within the secretory membrane, and then protons move from the inside of the membrane to the lumen, that is, the overall process is electrogenic: a complete circuit is required. With $Cl^-$ media it is primarily the $Cl^-$ path, whereas in $Cl^-$-free media it is the $K^+$, $Na^+$, and sulfate paths (FIG. 4).[17]

Some may be puzzled as to why the $K^+$ diffusion potential across the nutrient membrane was not considered. Advocates of the $K^+$ diffusion potential with the NP model have not considered it, but we have.[9,17] We found that the partial $K^+$ conductance with $Cl^-$-free media was essentially the same on the two opposing membranes, that is, the magnitude of the change in PD with given changes in the $K^+$ concentration on the two sides is essentially the same.[9] They are obviously oriented in the opposite direction and hence, in the absence of other EMFs, should cancel and yield a PD across the tissue of zero. With the same high $K^+$ concentrations on opposite sides, the PD is essentially the same as with the standard 4 mM $K^+$ on both sides.[9,17] Therefore, there must be another EMF, oriented with its positive pole towards the secretory side, to account for the inverted PD under the conditions of either a low or a high $K^+$ concentration on both sides.

A representative experiment à propos of the foregoing is shown in FIGURE 10; $K^+$ was increased to 80 mM, first on the secretory side and then on the nutrient side. Increasing $K^+$ from 4 to 80 mM on the secretory side resulted in an increase in PD, a decrease in $R_t$, and an increase in the $H^+$ rate. A subsequent increase of $K^+$ to 80 mM on the nutrient side resulted in a return of the PD to the control level and a further decrease in $R_t$. The magnitude of the PD and $H^+$ secretory rate then gradually decreased. At any time during this period inhibitors produce a rapid decrease in $H^+$ rate to zero and an increase in the PD to positive values. As illustrated in FIGURE 10, the addition of the well-known inhibitor omeprazole ($10^{-3}$ M) to the nutrient side produced the typical increase in PD and decrease in $H^+$.

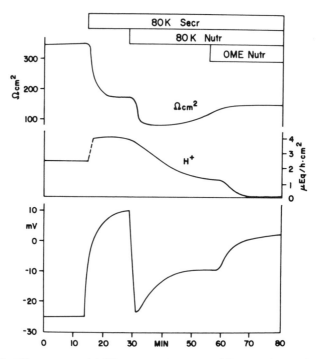

**FIGURE 10.** Effect on potential difference, resistance, and $H^+$ rate with $Cl^-$-free solutions of, first, increasing $K^+$ from 4 to 80 mM on the secretory side, second, of increasing $K^+$ to 80 mM on the nutrient side, and finally, of adding omeprazole (OME) to a final concentration of $10^{-3}$ M to the nutrient side.

## DISCUSSION

The purpose of this study was to evaluate the NP and EP models on the basis of the magnitude and effect of inhibition on the PD and $R_t$ in $Cl^-$ and in $Cl^-$-free media. In $Cl^-$ media the changes in magnitude of the PD and $R_t$ are those predicted by the EP model. In an attempt to explain the changes in PD in $Cl^-$ media on the basis of the NP model, it was proposed that inhibition reduced the magnitude of the $K^+$ diffusion potential in the secretory membrane. Contrary to predictions with 80 mM $K^+$ on the secretory side, the increase in PD with inhibition was the same as that in the controls with 4 mM $K^+$. Hence the increase in PD is not explained on the basis of the NP model. Advocates of the NP model have not presented an adequate explanation for the increase in $R_t$ with inhibition. (For details see references 13 and 15.)

In $Cl^-$-free media the orientation and the effect of inhibition on the PD are those predicted by the EP model. Again on the basis of the NP model the $K^+$ diffusion potential hypothesis fails to account for the changes in PD with a high $K^+$ on the secretory side.

On the basis of the EP model, it would be predicted that $R_t$ should increase with inhibition. Contrary to this prediction, SCN decreases $R_t$ instead of increasing it. As already point out, we interpret the SCN effect on $R_t$ as due to SCN occupying the anion channels in the secretory membrane that are normally occupied by $Cl^-$. Pertinent to this point is that the SCN addition to the secretory side with $Cl^-$ media does not produce a decrease in $R_t$ but an increase: the anion channels are already occupied by $Cl^-$. It is also important to point out that cimetidine produces an increase in $R_t$ with $Cl^-$-free media.[22]

It will be recalled that in $Cl^-$-free media an increase in secretory $K^+$ produces an increase in the $H^+$ secretory rate (FIGS. 8 and 10). This finding can be explained on the basis of either the NP or the EP model. With the NP model the increase in $K^+$ available to the forced $K^+$-$H^+$ exchange mechanism could account for the increase in the $H^+$ secretory rate. With the EP model the increase in PD (increase in the positivity of the nutrient side) also would be expected to increase the $H^+$ rate. This follows, because it has been shown with 4-mM $K^+$ on both sides that in $Cl^-$-free media clamping the voltage of the nutrient side at positive values produces an increase in the $H^+$ secretory rate.[10] In fact, the increase in $H^+$ rate with positive voltage-clamping produces a much greater increase in $H^+$ rate than does increasing the $K^+$ to 80 mM on the secretory side; this large increase in rate occurs even when 80 mM $K^+$ is present on the secretory side. (See references 2, 9, and 10 for interpretations.)

In conclusion, not only the previous work but also the recent work presented herein throws substantial doubt on the NP model. Recall that with the NP model, it is postulated that the $K^+$-$H^+$ exchange occurs across the border between the lumen and oxyntic cells. A model presented in FIGURES 2D and 9 is a tentative attempt to reconcile the vesicle findings with those of the intact tissue. With this model $H^+$ moves across the lumen-cell border, and this movement constitutes an electric current; the circuit is completed at other sites.

## REFERENCES

1. REHM, W. S., M. SCHWARTZ, G. CARRASQUER & M. DINNO. 1988. *In* Membrane Biophysics III: Biological Transport. M. Dinno, ed.:1–22. Alan R. Liss, New York, NY.
2. REHM, W. S., T. C. CHU, M. SCHWARTZ & G. CARRASQUER. 1983. Am. J. Physiol. **245:** G143–G156.
3. REHM, W. S., G. CARRASQUER & M. SCHWARTZ. 1986. Am. J. Physiol. **250:** G639–G647.
4. GANSER, A. L. & J. G. FORTE. 1973. Biochim. Biophys. Acta **307:** 169–172.
5. LEE, H. C., H. BREITBART & J. G. FORTE. 1980. Ann. N.Y. Acad. Sci. **341:** 297–310.
6. SACHS, G., T. BERGLINDH, E. RABON, H. B. STEWART, M. L. BARCELLONA, B. WALLMARK & G. SACCOMANI. 1980. Ann. N.Y. Acad. Sci. **341:** 312–334.

7. SACHS G., H. H. CHANG, E. RABON, R. SCHACKMAN, M. LEWIN & G. SACCOMANI. 1976. J. Biol. Chem. **251:** 7690–7698.
8. CONWAY, E. J. & T. G. BRADY, 1948. Nature (Lond.) **162:** 456–457.
9. DAVIS, T. L., J. R. RUTLEDGE & W. S. REHM. 1963. Am. J. Physiol. **205:** 873–877.
10. REHM, W. S., T. L. DAVIS, C. CHANDLER, E. GOHMANN, JR. & A. BASHIRELAHI. 1963. Am. J. Physiol. **204:** 233–242.
11. SCHWARTZ, M., T. C. CHU, G. CARRASQUER & W. S. REHM. 1981. Biochem. Biophys. Acta **649:** 253–261.
12. REHM, W. S. 1962. Am. J. Physiol. **203:** 63–72.
13. HERSEY, S. J., G. SACHS & D. K. KASBEKAR. 1985. Am. J. Physiol. **248:** G246–G250.
14. RAENSTRA, W. W., J. D. BETTENCOURT & J. G. FORTE. 1987. Am. J. Physiol. **252:** (Gastrointest. Liver Physiol. 15) G1–G5.
15. REHM, W. S., G. CARRASQUER & M. SCHWARTZ. 1986. Am. J. Physiol. **250:** G511–G517.
16. REHM, W. S., M. SCHWARTZ, G. CARRASQUER, E. HAGAN & M. DINNO. 1987. Biochem. Biophys. Acta **899:** 17–24.
17. REHM, W. S., G. CARRASQUER, M. SCHWARTZ & M. DINNO. 1988. *In* Gastrointestinal and Hepatic Secretions: Mechanism and Control. J. S. Davison & E. A. Shaffer, eds.: 121–124. Univ. of Calgary Press, Calgary, Canada.
18. HOLLOMAN, T. L., M. SCHWARTZ, M. DINNO & G. CARRASQUER. 1976. Am. J. Physiol. **231:** 1649–1654.
19. REHM, W. S. 1968. J. G. Physiol. **51:** 250s–260s.
20. SCHWARTZ, M., G. CARRASQUER & W. S. REHM. 1986. Biochem. Biophys. Acta **858:** 301–308.
21. SCHWARTZ, M., G. CARRASQUER, W. S. REHM & M. DINNO. 1987. Biochem. Biophys. Acta **897:** 445–452.
22. SCHWARTZ, M., T. C. CHU, G. CARRASQUER, W. S. REHM & T. L. HOLLOMAN. 1981. Am. J. Physiol. **240:** G267–G273.

# Cytochrome-Mediated Electron Transport in H+-Secreting Gastric Cells

GEORGE W. KIDDER III[a]

*Department of Biological Sciences*
*Illinois State University*
*Normal, Illinois 61761*
*and*
*Mt. Desert Island Biological Laboratory*
*Salsbury Cove, Maine*

In 1974 a vesicle preparation was isolated[1] from dog gastric mucosa which was apparently derived from the mucosal-facing membrane of the parietal cell, was oriented with its cytoplasmic surface outside, and could accumulate $H^+$ in the presence of ATP if preloaded with $K^+$. This finding gave rise to a model for gastric acid secretion in which the active step was provided by an electroneutral exchange of $H^+$ for $K^+$, driven by ATP at a stoichiometry of 1 ATP per $H^+$, and fueled by ATP generated by the mitochondria via oxidative phosphorylation. This model has been the basis for much laboratory work and many publications, which have recently been reviewed,[2,3] and it has been accepted as the mechanism for gastric proton secretion.

A satisfactory model must explain all of the valid observations made on the whole tissue that it purports to model. Inconvenient observations cannot be ignored unless they can be shown to have been in error. We also require that the laws of thermodynamics be obeyed by the model, and that characteristics such as electroneutrality and energetics be consistent with the predictions of the model. I shall attempt to show that in several of these aspects, the current H/K ATPase model is deficient and must therefore be reexamined. Some of these arguments have previously been presented.[4]

The H/K ATPase model postulates an electroneutral pump mechanism whose operation does not, therefore, directly change the electrical potential or resistance across the membrane in which it is placed. However, there are a large number of experiments that directly relate changes in transepithelial potential to changes in gastric acid secretory rate and a similar number of experiments in which rapid changes in electrical resistance of the tissue are noted. Moreover, the postulated pump must be capable of performing under conditions in which there is no $K^+$ in the mucosal bathing solution and no significant $K^+$ transport from the tissue into the mucosal bathing solution. This requires that the $K^+$ necessary for the operation of the H/K ATPase be supplied by the tissue itself, that it be present in sufficient concentrations in the depths of the tubules to satisfy the $K_m$ of the ATPase, and that all the $K^+$ be reabsorbed before the secreted fluid can reach the mucosal bathing solution, to avoid active $K^+$ secretion. These problems have not been satisfactorily addressed by the proponents of this model.

In an accompanying paper, Rehm[5] has undertaken to comment on some of these problems. I shall concentrate on the adequacy of ATP as an energy source

---

[a] Address for correspondence: George W. Kidder III, Dept. of Biological Sciences, Illinois State University, Normal, IL 61761.

for gastric acid secretion and give some indications of the current state of our search for an alternate and more suitable energy source.

First, we must determine the thermodynamic energy requirements for gastric acid secretion across the mucosal-facing membrane of the secretory cell, $\Delta G_H$. For an ion of valance $z$, if we know the concentration on the source ($C_1$) and sink ($C_2$) sides of a membrane and the electrical potential ($\Delta \Psi$) across that membrane, we can calculate the free energy requirement for transport:

$$\Delta G = RT \ln (C_2/C_1) + z\, F \Delta \Psi$$

where R, T, and $F$ have their usual thermodynamic meanings. For gastric $H^+$ secretion, $C_1$ can be derived from the pH of the cytoplasm, and $C_2$ from the pH of the luminal fluid in immediate contact with the secretory cell. For convenience, I shall term the latter the "trans-pH." Because the secretory (oxyntic or parietal) cells are located some distance from the bulk luminal solution and separated from that solution by rather long and thin channels, this trans-pH is not the pH of the bulk solution and may not be the pH of the solution emerging from the tubules. What means do we have for calculating and/or measuring the trans-pH?

This pH can be calculated directly if some simplifying assumptions are made:

1. The tissue secretes only $H^+$ and $Cl^-$.
2. The secretory membrane is very permeable to $H_2O$, so that the solution in the depths of the pits is very close to isotonic equilibrium with the cell cytoplasm.
3. The basolateral surface of these cells is likewise very permeable to water, so the cells are at isotonic equilibrium with the plasma.
4. The ratio of water permeability to HCl permeability at the secretory membrane is sufficiently high that the Staverman coefficient approximates unity.
5. The bulk flow out of the tubules is sufficient to prevent diffusion of molecules from the bulk luminal solution into the depths of the tubules.

Under these conditions, the secretion of HCl into the depths of the tubules will cause an isotonic amount of $H_2O$ to move from cell to tubule. This creates a volume flow which causes this solution to move along the tubules and empty into the bulk luminal solution, while the cell osmolarity is maintained by an equal movement of $H_2O$ from the serosal solution into the cells. The pH will therefore be that of HCl at isotonic equilibrium with the serosal solution and will vary as serosal tonicity varies with species. For mammals this solution will be around pH 0.8. Frogs have less concentrated plasma and will therefore have a pH about 0.97, whereas the concentrated plasma of elasmobranchs, which contains high concentrations of both salts and urea, will yield a trans-pH of 0.57.

Are these reasonable assumptions? Gastric mucosa does secrete some other ions, but as minor components to the major active transports of $H^+$ and $Cl^-$. The permeability assumptions seem well accepted. Only the final assumption regarding the volume flow and its effect on diffusion seems to be in question in some species and under some conditions.

Can we measure the trans-pH? It should be clear that one cannot determine the pH of primary secretion in an Ussing chamber. The rate of acid secretion that would decrease the pH of a small volume of unbuffered mucosal solution from 7 to 6 in 36 seconds (1/100 hour) would require 1,000 hours to move the pH from 2 to 1, which is clearly impractical. Other methods will be required to measure the pH of primary secretion.

In mammals, the secretory rates are sufficiently high that the primary secretion can be collected and measured, and indeed the pH is around 0.8, as predicted.[6] I[7] have measured the pH of the gastric lumen in dogfish by inserting a combination pH electrode through the mouth into the empty stomach of an intact fish and stimulating secretion with injected histamine. FIGURE 1 shows that the measured pH drops to an average of about 1, with one individual reaching a pH of 0.72. This would be the primary secretion pH if there were no movement of acid away from the electrode tip, which is certainly not the case. Thus, although these measurements probably do not succeed in measuring the pH of the actual primary secretion, they set an upper limit for that pH in these species and are in accord with the values calculated for isotonic HCl secretion.

At lower secretory rates, bulk flow may be less important, and diffusion may play a greater role in the delivery of HCl to the bulk luminal fluid, causing a

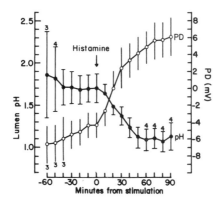

**FIGURE 1.** Changes in the luminal pH of an empty dogfish stomach *in vivo* after histamine stimulation. A combination pH electrode (KCl/calomel reference) was inserted through the mouth and into the stomach of a 2-kg male dogfish, adjusted to the region of lowest pH, and left in that position for the balance of the experiment. Injection of histamine ($5 \times 10^{-6}$ mol in 5 ml dogfish Ringer's solution) into the caudal vein caused a decrease in pH, with the average minimum value being close to pH 1, and pH 0.72 in one case. Average ± SEM for five animals except where noted. The corresponding potential difference (PD) changes (lumen to peritoneal cavity) are also shown.

decrease in the trans-pH under these conditions. Rehm[8] has calculated that in *Rana pipiens* gastric mucosa at moderate secretory rates (3.6 $\mu$Eq/cm$^2 \cdot$ hr), diffusion of H$^+$ from the tubules can account for the rate of appearance of HCl in the lumen without requiring bulk flow, and he infers that the concentration of HCl necessary to account for this diffusion is only 10 mM (pH 2). This would only be true if the bulk solution is isotonic and at pH 7. This rate of diffusion into a bulk solution at a lower pH would require a lower trans-pH, and because the measured $J_H$ in frog is independent of the luminal solution pH down to pH 2.5,[9] such lower pH must exist. Moreover, if the luminal bulk solution contained only water, an osmotic gradient would exist from lumen to cell, which would cause volume flow into the tubules. Water would flow into the cells until the fluid adjacent to the cells was again isotonic. In the absence of any other ions, this isotonicity could only be produced by HCl; the trans-pH would again be set by the tonicity of the cells and would be around pH 1. Secretion continues under hypotonic conditions with a measured decrease in apparent primary secretion pH,[10] so it is clear that the frog cells can produce acid at this concentration.

Therefore, in all species tested it appears that the trans-pH against which acid *can* be secreted can be calculated from the pH of isotonic HCl. Whether or not this is the trans-pH against which acid is *usually* secreted is not our present question.

What experimental means are available to alter the trans-pH in tissues? We cannot alter this pH by lowering the pH of the bulk solution with HCl, although this has been tried.[11] Achieving a pH of less than the isotonic value would require a hypertonic mucosal solution, leading to water flow from tissue to the solution in the depths of the tubules, which would dilute the solution at this point until the trans-pH returned to its isotonic value. Although some increase might in theory be obtained by using an acid with a polyvalent anion (such as $H_2SO_4$), the acid chosen would have to be completely ionized at this pH, and the amount of pH decrease would be small even in this case. To my knowledge such experiments have not been attempted. But there is another way to increase the trans-pH, which is to increase the osmolarity of the cell cytoplasm.

If the cell cytoplasm is made hypertonic, the trans-pH would be expected to rise for the reasons just given, which would decrease its pH and therefore increase the gradient for $H^+$ up which acid must be secreted. This experiment has already been done for us, because various organisms have various plasma tonicities. We have no reason to believe that the fundamental mechanism of gastric acid secretion differs between species, and we must therefore produce a model that can account for the most extreme conditions. In the elasmobranch, the plasma contains 511 mOsM of salts and 350 mOsM of urea under normal conditions, and should therefore produce a trans-pH of 0.57 if it secreted only $H^+$ and $Cl^-$, as it seems to do. In non-elasmobranchs, there may be other reasons for not achieving this pH, such as the inability of the membranes to withstand such high acidity, but we do not expect the fundamental mechanism to be different. We therefore require that the energy source for gastric acid secretion be capable of transporting acid against the maximum gradient that can be shown to exist in any species, even though in some species it is never required to do so.

Can we experimentally alter the gradient in a tissue? If we could increase the osmolarity of the cell cytoplasm without other effects, we expect that the trans-pH would decrease, and therefore that the gradient up which acid is secreted would increase. As the cell osmolarity is increased, the "load" on the pump would be expected to slow its rate, and at some point the gradient up which acid is secreted would reach the maximum up which the pump can secrete, bringing acid secretion to a halt.

The osmotic agent for this purpose must be chosen with care. If an impermeable molecule were added to the serosal solution, it would cause cell shrinkage and might inhibit acid secretion by this means. Therefore the ideal molecule would be quite permeable to the basolateral membrane of the cell, but impermeable to the apical membrane. Clearly the osmotic molecule should not have any specific toxic effects or alter the electrical properties of the system. In many ways, urea would seem to be a suitable choice. It is uncharged and is a normal constituent of plasma, being present in high concentrations in some plasmas. Although it breaks hydrogen bonds, the concentrations used for this purpose are very high (4 M) compared to our anticipated use. One would not expect it to be very permeable to the mucosal membrane, because if it were, species such as elasmobranchs would loose large quantities of urea by this route, which would be a drain on their nitrogen metabolism.

We have therefore used urea to alter the osmolarity of the cells of frog and dogfish gastric mucosa. For the former species, the results are shown in FIGURE 2. The addition of urea to the serosal solution is inhibitory of gastric acid secretion, with the measured rate decreasing smoothly to near zero with added urea. Under short-circuit conditions, the inhibition by small amounts of urea is somewhat greater, and there is a small but not statistically significant decrease in secretory rate due to short circuiting in the absence of urea. Because the normal potential difference is about 40 mV (serosal positive), this potential difference is in a direction to decrease the total electrochemical potential gradient up which $H^+$ must be transported, and removal of this electrical "load" would be expected to

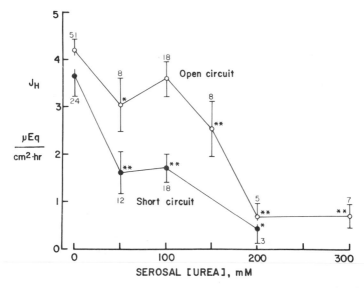

**FIGURE 2.** Effect of added serosal urea on the rate of gastric acid secretion in *Rana catesbeiana*. Rates are for the last 15 minutes of the 1-hour exposure period and are expressed relative to that tissue in the absence of added urea. Control osmolarity, 239 mOsM. Error bars are ± SEM, for the indicated number of replicates. Significance of the difference between experimental points and the appropriate control: *, $0.05 > p > 0.01$; **, $p < 0.01$, by $t$ test.

enhance the operation of the pump, as observed. While short circuiting does alter many electrical properties of the tissue, probably including the cellular ion compositions,[12,13] the effects seen here are consistent with the view that the decreased pump activity is due to the increased electrical component of the electrochemical gradient up which acid is secreted (the Rehm[14] effect), and that changes in this gradient by current passing and by urea addition are additive. One would not predict this effect for an electroneutral pump such as the $H^+/K^+$ antiport.

In some systems, urea is an inhibitor of biochemical activity, and this inhibition can be reversed by trimethylamine oxide (TMAO) at concentrations about half that of the urea.[15] To examine this effect, TMAO was applied either with or without urea. As shown in TABLE 1, TMAO behaves like urea, as if the effect of

**TABLE 1.** Effects of Urea and TMAO on Acid Secretion in Bullfrog[a]

| Serosal Addition (mM) | | Total mOsM | $J_H$, % of Control |
|---|---|---|---|
| None | (control) | 239.0 | 100.00 |
| TMAO | 37.5 | 276.5 | 95.75 ± 4.07 (ns) |
| Urea | 75.0 | 314.0 | 88.43 ± 10.83 (ns) |
| TMAO | 75.0 | 314.0 | 30.78 ± 12.99 (**) |
| TMAO | 37.5 + urea 75.0 | 351.5 | 60.86 ± 4.83 (**) |
| Urea | 150.0 | 389.0 | 43.92 ± 7.15 (**) |
| TMAO | 75.0 + urea 75.0 | 389.0 | 17.33 ± 7.29 (**) |
| TMAO | 75 + urea 150.0 | 464.0 | 7.60 ± 3.16 (**) |

[a] Effects of urea and trimethylamine oxide (TMAO) added to the serosal solution for 1 hour. Control is preceding hour without additions. Means ± SEM, ns indicates no significant difference from control (100%), and ** indicates $p < 0.01$. The inhibition roughly follows total osmolarity regardless of the compound used as osmotic agent. There is no reversal of urea inhibition by TMAO; if anything, TMAO is a more effective inhibitor, but the differences are not significant.

this agent is also a simple osmotic one. We find no evidence that the urea effect is biochemical rather than biophysical.

Similar experiments have been performed in the dogfish.[16] In dogfish, the normal urea content of the plasma is 350 mOsM, and the total osmolarity 847 mOsM (salts and urea). As shown in FIGURE 3, when the urea is omitted from the bathing solutions, the secretory rate drops, apparently for reasons not connected with the additional load on the pump. Increasing the urea concentration causes a decrease in secretory rate, with the rate going to zero at 1,475 mOsM. HCl at this concentration is calculated to have a pH of 0.13. If the cytoplasmic pH is 7.4, this is a gradient of 7.27 pH units, or $1.9 \times 10^7$-fold. Microelectrode studies[17] suggest that the apical membrane potential is about 40 mV (cell negative) under these conditions, which gives a total electrochemical gradient equivalent to nearly 8 pH units. From the foregoing equation, $G_H$ can be calculated to be 10,873 cal/mol.

In both frog and elasmobranchs, it is possible that urea is having additional inhibitory effects beyond those predicted from isotonic HCl secretion. Although the dogfish is normally exposed to high concentrations of urea, the frog is not. The gradients calculated are those against which the tissue can be shown to secrete, and are not necessarily the maximum gradients that could be produced in the absence of possible inhibitory effects of urea. The values for electrochemical gradients just given are thus to be taken as minimum ones.

If the dogfish gastric mucosa can transport protons against such a gradient, it must have a mechanism that can provide this amount of energy. We know that the

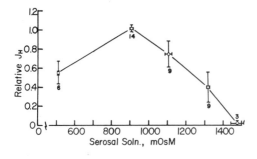

**FIGURE 3.** Effect of serosal urea on acid secretion in dogfish gastric mucosa relative to control in 350 mM urea (plasma level). Tissues exposed to solutions for 2 hours, of which last half hour is reported. Osmolarity determined by freezing point depression. Means ± SEM for both variables, with number of tissues.

ultimate source of energy is aerobic metabolism, and many have assumed that the proximate energy source is ATP hydrolysis. If the ratio of $H^+$ secreted to ATP hydrolyzed were $1:1$, as usually assumed, each hydrolysis must provide enough energy to overcome this adverse gradient. In theory any spontaneous chemical reaction can provide an infinite amount of energy per mole of reactants, although the required concentration ratio (products/reactants) can become extreme. If we accept that $\Delta G_{ATP}^{0'}$ is $-7300$ cal/mol and that at pH 7.4, $\Delta G_{ATP}$ may be $-7,600$ cal/mol, we require that the ratio ATP/ADP be over 200 for the reaction to provide enough energy to support the maximum gradient in elasmobranchs. $\Delta G_{ATP}^{0}$ would have to be $-9,900$ cal/mol for there to be enough energy at the usual ATP/ADP of about 5 to cause acid secretion to proceed. There does not seem to be enough energy available from the hydrolysis of ATP under physiologic conditions to meet the minimum energy requirements for gastric acid secretion, let alone drive acid secretion at a measurable rate.

Neither can one escape this conclusion by noting that $\Delta G_{ATP}^{0}$ rises with rising pH,[18] and that at a pH of 9 (which could conceivably occur in the cell) one might approach a value of 10,000. This pH increase would add to $\Delta G_H$ by increasing the pH gradient up which secretion must take place, opposing the gain in $\Delta G_{ATP}^{0}$. Thus, at a cytoplasmic pH of 9, the calculated $\Delta G$ for acid secretion in the urea-inhibited dogfish gastric mucosa is 13,065, and even the increased $\Delta G_{ATP}$ at this pH is still insufficient.

Therefore, if it is true that the mechanism of urea inhibition of the dogfish gastric mucosa is as just stated, acid secretion cannot be driven at a $H^+/ATP$ of unity in this species unless the concentration of ADP in the cytoplasm is vanishingly low. In light of the other uses for ATP in cells, this seems unlikely. Although it can always be argued[3] that species are different, unless one is prepared to postulate totally different secretory mechanisms to explain conflicting data, this demonstration of a high $\Delta G_H$ in elasmobranchs implies that there is a problem in other species as well.

This is only one of the problems with the identification of ATP as the proximal energy-donating compound in gastric acid secretion. A second major problem is the kinetics of the process. When oxygen is removed from both surfaces of the chambered gastric mucosa, acid secretion drops promptly to zero, accompanied by a decrease in cellular ATP which follows a similar time course.[19] This is consistent with the idea that ATP is the driving compound for gastric acid secretion, and that the removal of oxygen and the consequent removal of the ATP-regenerating system allows ATP concentration to fall, inhibiting acid secretion. However, when oxygen is readmitted, ATP levels rise promptly to their control values, and if the absence of ATP were preventing acid secretion, $J_H$ should rise at a similar rate. The facts are otherwise. Following the readmission of oxygen, there is a lag period of many minutes before *any* acid secretion is seen; thereafter, it rises quickly to near control values.[20] This lag period can be increased to as long as 50 minutes under some conditions. Thus, although the *decrease* in acid secretion and the decrease in ATP concentration have kinetics consistent with those of an ATP-driven pump, the *increase* in ATP and the increase in $J_H$ do not. By contrast, the relation between secretory lag and cytochrome redox lag is excellent, as shown in FIGURE 4.

Another problem appears when attempts are made to reverse the anaerobic inhibition of acid secretion by exogenous ATP. The gastric mucosa is impermeable to ATP from its mucosal surface, but an extensive series of experiments[20,21] indicate that the permeability of cells from the serosal surface, although small, is finite. When ATP is added to anoxic gastric mucosae, there is absolutely no indication of an increase in secretory rate. The permeability is such that the added ATP should have been entering the cells at a rate of between 0.13 and 0.8 $\mu$M/cm$^2$ · hr (depending on the assumptions),[20] which should have supported acid secretion at 0.13 to 0.8 $\mu$Eq/cm$^2$ · hr, which are low but easily measured rates. The actual measured rate is zero.

It is possible to construct rather elaborate hypotheses to explain these observations, and it is common to construct alternative models by ignoring these observations completely. The simple explanation that ATP is not the proximal energy source for gastric acid secretion must remain a viable one.

It should not be inferred from the foregoing that I do not believe in the existence of a proton-translocating ATPase. There seems to be little question that

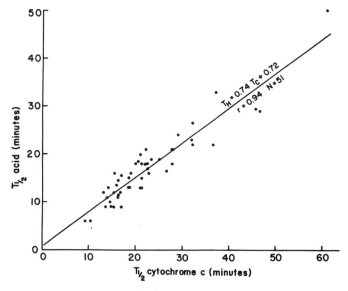

**FIGURE 4.** Secretory lag versus cytochrome c reoxidation following reoxygenation of bullfrog gastric mucosa. In each case, the half-time for response has been taken. Secretory lag is spontaneously variable and has been prolonged by a variety of means in obtaining this figure. The 51 points shown are tightly clustered around the least squares straight line.

such a mechanism exists and is capable of moving H$^+$ against moderate gradients. The gradients that have been demonstrated to date, however, fall at least two orders of magnitude short of the requirements for gastric acid secretion. The gradients in vesicles are in line with that which ATP is capable of supplying, as expected. This mechanism clearly has a role in the transport properties of gastric cells and perhaps in other tissues.[3] The problems just noted, however, would seem to indicate that the H/K ATPase is not the primary, high potency H$^+$ pump of the gastric mucosa.

If ATP is not the energy source for gastric acid secretion, we need to find another candidate, because gastric acid secretion does in fact occur against steep gradients. From the physiology of the system, we know that the energy source is closely linked to aerobic metabolism, because acid secretion is immediately and reversibly inhibited by the absence of oxygen. This mechanism needs to have rather more energy per mole than ATP can supply, and it must be closely associated with the apical plasma membrane of the secretory cells, because this is the site of gastric acid secretion. Many years ago, we[22] suggested, on the basis of spectrophotometry of the secreting gastric mucosa of the bullfrog, that cytochrome c was involved in this process as more than just a component in the ATP generating system, and that a portion of cytochrome c might well be extramitochondrial.

If such a cytochrome system exists and operates in a manner similar to that of mitochondrial cytochromes, it could provide the energy necessary for acid secretion against considerable gradients. Mitchell[23] many years ago proposed a proton-translocating scheme based on the operation of the cytochrome system, and that scheme has generally been accepted as operating in mitochondria, chloroplasts, and prokaryotes. In this model, the transport of electrons down the cytochrome chain in the mitochondrial membrane causes the translocation of $H^+$ across the mitochondrial membrane, creating a "chemiosmotic" gradient that is then dissipated by driving the phosphorylation of ADP to ATP. If such a cytochrome system were present in the apical membrane of the secretory cell, $H^+$ could be secreted into the lumen by the operation of this system. The "oriented ATPase" responsible for converting the energy of the $H^+$ gradient into phosphate bond energy would be absent, however, because a $H^+$ gradient is the net output.

This scheme would explain the absolute requirement for oxygen displayed by gastric acid secretion. Because it is an electrogenic transport system, it also explains the changes in potential difference and resistance observed with changes in secretory rate. It has no specific transport requirement for $K^+$, although a $K^+$ requirement would be expected for the tissue as a whole. Moreover, because the energy from such a proton translocation site is considerably greater than that ultimately recovered in ATP, it can satisfy the thermodynamic demands of the acid secretory system. The problem with such a scheme is in being able to test it.

Tissue spectrophotometry gives a picture of the reactions of all cytochromes in the cell. If there are mitochondrial cytochromes as well as plasma membrane cytochromes, as there certainly are, the spectrophotometric signal will be an addition of similar cytochromes at both sites. Indeed, if the cytochromes from these two systems are identical, it is very difficult to design experiments that distinguish between them. However, if it can be shown that some spectral difference exists between members of these two chains, it should be possible to proceed. I believe I have now found such a difference.

This story starts with the observation that acid secretion in the bullfrog gastric mucosa is resistant to inhibition by carbon monoxide,[24] an observation subsequently extended to the skate gastric mucosa.[25] Because CO is one of the classical cytochrome oxidase inhibitors, this might mean that this tissue can interact with oxygen by a different route. However, cyanide and azide, the other classical inhibitors of the terminal oxidase, are very effective inhibitors of secretion. It was also known that an azide-inhibited tissue was not the same as an anoxic tissue (there was a demonstrable difference in potential difference), so we suspected that azide was having some effect other than (or in addition to) the inhibition of cytochrome oxidase. This is not unprecedented.[26]

To investigate this situation, an oxygen electrode system was constructed, and the respiration of the isolated bullfrog gastric mucosa was examined.[27] Because hyperbaric conditions are necessary for the CO experiments, azide was used for these studies, accepting the fact that gastric acid secretion would be inhibited. A portion of the respiration was normally sensitive to azide, whereas a large fraction

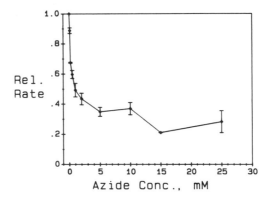

**FIGURE 5.** Effect of azide on oxygen consumption by bullfrog gastric mucosa. The tissue was mounted as a flat sheet (7.02 cm²) between buffered solutions of minimal (6.25 ml) volume, with an oxygen electrode in each chamber half. Gassing and circulation were supplied by a reservoir and air lift system, using 10% $CO_2$ in $O_2$. When the hoses to the reservoir were clamped, respiration decreased the oxygen concentration in the solutions, which is recorded against time. Oxygen concentration was never allowed to fall below 0.7 atm, the point at which $J_H$ is inhibited;[29] the clamps were removed and the tissue reequilibrated with oxygen. Azide was added to both solutions at least a half hour before the measurement period. A portion of respiration is rather sensitive to azide, whereas a significant portion is hardly affected. (From ref. 18, by permission.)

was very insensitive, as seen in FIGURE 5. A Dixon plot of these data gives two components, as shown in FIGURE 6, for which the x-intercepts give $K_I$s of 0.55 and 29.81 mM, respectively. Similar results were obtained in the skate gastric mucosa.[28] This observation strongly implies that oxygen reacts with these tissues through two different oxidases. One of these is probably the usual cytochrome oxidase, cytochrome $a_3$, which is sensitive to $N_3^-$. The other is much less sensitive to azide and supports respiration when cytochrome $a_3$ is inhibited. This alternate

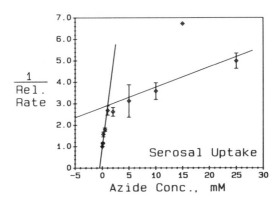

**FIGURE 6.** Dixon plot of oxygen consumption from the serosal surface of frog gastric mucosa versus azide concentration. Two components are resolved, one with a $K_I$ appropriate to conventional cytochrome oxidase and one with a much higher $K_I$. (From ref. 18, by permission.)

oxidase may be the means by which acid secretion is supported in the presence of CO. The inhibition of acid secretion by azide would appear to be due to some additional effects of this compound, perhaps related to its activities as an uncoupling agent.[26]

If an alternate oxidase exists, is it a cytochrome? Spectra taken in the absence of azide show a shoulder on the cytochrome oxidase peak that can be resolved by deconvolution into a peak at 590 $\mu$m, as shown in FIGURE 7. This shoulder has been observed for years and was always identified as the carbon monoxide complex of cytochrome $a_3$, because in the usual experiment a low concentration of CO was included to prevent observation of residual hemoglobin. This cannot be the explanation in the spectrum presented. Moreover, as seen in FIGURE 8, the reaction of the authentic a + $a_3$ peak at 605 was inhibited by the presence of $N_3^-$, leaving a peak at 590 which remains free to interact with oxygen. Because considerable proportions of the other cytochromes are also capable of being oxidized by

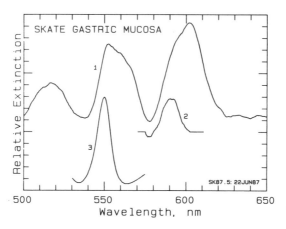

**FIGURE 7.** Spectra of the skate gastric mucosa in the complete absence of CO, taken by the dual-differential technique. Spectrum #1 is the $N_2$-$O_2$ difference spectrum, showing the asymmetric cytochrome oxidase peak near 605 nm. Spectrum #2 is the result of deconvoluting this peak by removal of the 602.5-nm component reconstructed from the long wavelength side, to leave the 590-nm peak responsible for the shoulder. Spectrum #3 is the alpha-peak of authentic cytochrome c, for calibration purposes. (From ref. 18, by permission.)

oxygen under these conditions, it would appear that this 590 peak may represent the alternate oxidase, which is capable of interacting with cytochromes further down the chain. At present this identification is a tentative one, but it is clear that an unusual cytochrome pigment has been uncovered in this tissue.

To pursue this further, cells have been isolated from the gastric mucosa of frog and skate, and their oxygen uptake has been tested for sensitivity to azide. Essentially the same results obtain with these cells as with the intact tissue. In addition, it has been possible to use cyanide as an additional inhibitor, which is prohibited in intact tissues by the requirements for frequent gassing. To our surprise, the respiration of these cells is quite sensitive to $CN^-$, which may indicate that our alternate oxidase is $CN^-$ sensitive although $N_3^-$ and CO insensitive. Work is proceeding along these lines.

The goal is to be able to use the spectral properties of the new pigment to provide an assay for its purification and characterization, along with the other cytochrome components of the putative secretory cytochrome chain. If this system can be demonstrated to exist, it will provide a model for gastric acid secretion which is free of the problems that surround the ATPase model.

In conclusion, there are a number of ways in which the H/K ATPase model for gastric acid secretion fails to account for the data from intact tissues. Its mucosal solution $K^+$ requirement and the difficulties such a model has in explaining the electrical events associated with acid secretion are one such problem. A second problem is the high energy requirement for acid secretion, which it seems unlikely that ATP can supply. I have presented data on the effects of raising the serosal osmolarity which seem to imply acid secretion into a very low pH, with a very high energy requirement. The kinetics of the changes in ATP concentration in the cells, and the failure of added ATP to reverse anoxic inhibition of acid secretion

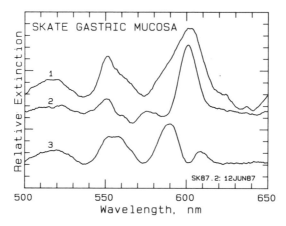

**FIGURE 8.** Difference spectra of skate gastric mucosa in the presence of azide. Spectrum #1 is the $N_2$-$O_2$ control, without azide. Spectrum #2 is the same tissue in the presence of 5 mM $N_3^-$ and oxygen, minus the oxygen control, and shows those components reduced in the presence of azide. Spectrum #3 is the $N_2$-$O_2$ difference spectrum in the presence of 5 mM $N_3^-$, showing those components that can respond to $O_2$ in the presence of azide. Azide is seen to cause the reduction of most of the cytochrome oxidase (602.5) and about half of cytochrome c (550), but little of cytochrome b (564). In the presence of azide, the components not reduced in the presence of azide are free to react with $O_2$. (From ref. 18, by permission.)

are also problems that must be faced. Finally, experiments with the H/K ATPase system fail to show pH gradients that approach those required to explain gastric acid secretion. On balance, it would appear that this model can only be made to serve the purposes of gastric acid secretion by some rather tortured explanations or by ignoring inconvenient data.

The alternative model suggests that proton secretion is driven directly by a cytochrome system located in or closely associated with the apical plasma membrane, without the intervention of ATP. It proposes an electrogenic secretion of $H^+$, which is consistent with the electrical data. It has no counterion requirement, although it does require $Cl^-$ movement, as is observed. It is an absolutely aerobic system, in accord with the observations. The difficulty until now has been to demonstrate its existence. The demonstration of a large fraction of inhibitor-insensitive respiration in this tissue, along with a cytochrome peak that seems to be unique to this system, gives us a means to pursue this problem by fractionation and characterization of this pigment. Hopefully these studies will be fruitful.

## REFERENCES

1. LEE, S., G. SIMPSON & P. SCHOLES. 1974. ATPase of dog gastric microsomes. Changes of outer pH in suspensions of membrane vesicles. Biochem. Biophys. Res. Commun. **60:** 825–864.
2. FORTE, J. G. & J. M. WOLOSIN. 1987. HCl secretion by the gastric oxyntic cell. *In* Johnson, L. R., ed.: 853–863. Physiology of the Gastrointestinal Tract, 2nd Edition. Raven Press, New York.
3. SACHS, G. 1987. The gastric proton pump: The $H^+,K^+$-ATPase. *In* Johnson, L. R., ed.: 865–881. Physiology of the Gastrointestinal Tract, 2nd Edition. Raven Press, New York.
4. KIDDER, G. W. III. 1980. Theories on gastric acid secretion. Ann. N. Y. Acad. Sci. **341:** 259–273.
5. REHM, W. S., G. CARRASQUER, M. SCHWARTZ & T. L. HOLLOMAN. 1989. Models for gastric proton pump in light of the high oxyntic cell conductance. Ann. N. Y. Acad. Sci. (This volume).
6. MOODY, F. G. & R. P. DURBIN. 1965. Effects of glycine and other instillates on concentration of gastric acid. Am. J. Physiol. **209:** 122–126.
7. KIDDER, G. W. III. 1977. Effect of mucosal pH on PD in dogfish gastric mucosa. Bull. Mt. Desert Isl. Biol. Lab. **17:** 21–23.
8. REHM, W. S. Personal communication.
9. KIDDER, G. W. III & J. T. BLANKEMEYER. 1978. Cytochromes and gastric acid secretion. A reevaluation of mucosal acidification experiments. Biochim. Biophys. Acta **512:** 192–198.
10. DURBIN, R. P. 1979. Osmotic flow of water in isolated frog gastric mucosa. Am. J. Physiol. **236:** E63–E69.
11. HERSEY, S. J. 1974. Interactions between oxidative metabolism and acid secretion in gastric mucosa. Biochem Biophys. Acta **344:** 157–203.
12. KIDDER, G. W. III & W. S. REHM. 1970. A model for the long time-constant transient voltage response to current in epithelial tissues. Biophys. J. **10:** 215–236.
13. KIDDER, G. W. III & M. G. ELROD. 1984. Voltage clamping changes resistance and current-voltage plot of frog gastric mucosa. Am. J. Physiol. **246:** G574–G579.
14. REHM, W. S. 1945. The effect of electric current on gastric secretion and potential. Am. J. Physiol. **144:** 115–125.
15. YANCEY, P. H. & G. N. SOMERO. 1980. Methylamine osmoregulatory solutes of elasmobranch fishes counteract urea inhibition of enzymes. J. Exp. Zool. **212:** 205–213.
16. KIDDER, G. W. III. 1980. Effect of urea hyperosmolarity on acid secretion in dogfish gastric mucosa. Bull. Mt. Desert Isl. Biol. Lab. **20:** 39–42.
17. KIDDER, G. W. III & E. L. KIDDER. 1982. Microelectrode studies of skate gastric mucosa. Bull Mt. Desert Isl. Biol. Lab. **22:** 30–31.
18. ALBERTY, R. A. 1968. Effect of pH and metal ion concentration on the equilibrium hydrolysis of adenosine triphosphate to adenosine diphosphate. J. Biol. Chem. **243:** 1337–1373.
19. DURBIN, R. P. 1968. Utilization of high-energy phosphate compounds by stomach. J. Gen. Physiol. **51:** 233s–239s.
20. KIDDER, G. W. III. 1973. Purine nucleotide entry, exit and interconversions in bullfrog gastric mucosa. Am. J. Physiol. **224:** 809–817.
21. KIDDER, G. W. III. 1973. Effects of 5'-adenylyl methylenediphosphonate (AMP-PCP) on gastric acid secretion. Biochim. Biophys. Acta **298:** 732–742.
22. KIDDER, G. W. III, P. F. CURRAN & W. S. REHM. 1966. Interactions between the cytochrome system and H ion secretion in bullfrog gastric mucosa. Am. J. Physiol. **211:** 513–519.

23. MITCHELL, P. 1966. Chemiosmotic coupling in oxidative and photosynthetic phosphorylation. Biol. Rev. **41:** 445–502.
24. KIDDER, G. W. III. 1980. Carbon monoxide insensitivity of gastric acid secretion. Am. J. Physiol. **238:** G197–G202.
25. KIDDER, G. W. III & E. L. KIDDER. 1986. Carbon monoxide insensitivity of the gastric mucosa of *Raja erinacea*. Bull. Mt. Desert Isl. Biol. Lab. **26:** 43–46.
26. ELLIASSON, L. & I. MACHIESEN. 1956. The effect of 2,4-dinitrophenol and some oxidase inhibitors on the oxygen uptake in different parts of wheat roots. Physiol. Plant. **9:** 265–279.
27. KIDDER, G. W. III & M. S. AWAYDA. 1989. Effects of azide on gastric mucosa. Biochim. Biophys. Acta **973:** 59–66.
28. KIDDER, G. W. III & A. T. MILLER. 1987. Azide insensitivity of oxygen consumption and some cytochromes of the gastric mucosa of *Raja erinacea*. Bull. Mt. Desert Isl. Biol. Lab. **27:** 106–107.
29. KIDDER, G. W. III & C. W. MONTGOMERY. 1975. Oxygenation of the frog gastric mucosa *in vitro*. Am. J. Physiol. **229:** 1510–1513.

# Differential Function Properties of a P-Type ATPase/Proton Pump[a]

CLIFFORD L. SLAYMAN AND GERALD R. ZUCKIER

*Department of Cellular and Molecular Physiology*
*Yale School of Medicine*
*New Haven, Connecticut 06510*

A very widespread biologic correlate of the acquisition of cell walls as osmotic controllers was the development (or retention) of proton pumps as the primary agents of active transport. The majority of plant, fungal, and bacterial cells eject a stream of protons, coupled to the capture of photons, the oxidation of pyridine nucleotides, or the hydrolysis of ATP, and they use the resultant transmembrane difference of potential to drive the uptake of organic nutrients (e.g., amino acids, sugars, nucleotides) and critical ions (K[+], phosphate, etc.) and the extrusion of catabolic products and superfluous ions. Plasma-membrane proton pumps also contribute essentially to the background conditions for cytoplasmic pH control, just as in wall-less cells plasma-membrane sodium pumps establish the background conditions for osmotic control.

Because the physiologic roles of these two species of ion pumps are so similar, it is not surprising, as described in a previous paper at this symposium,[1] that plasma-membrane *proton* pumps closely resemble the pumps for alkali metal cations in animal-cell membranes in general physical properties, in reaction chemistry, in secondary structure, and appreciably in primary structure. But against the impressive similarities between these two species of ion pumps, and among several others as well (e.g., plasma membrane and SR calcium pumps, the gastric proton pump, and the bacterial potassium pump), certain differences stand out sharply and probably will turn out to be correspondingly important to our eventual understanding of the molecular mechanism(s) of ion pumping.

For economy, I shall consider only four of the differences: stoichiometry, apparent substrate affinity, intrinsic velocity for one particular reaction step, and a preponderant localization of acidic residues near the N-terminus of the polypeptide sequence. All of these have been most clearly defined in the plasma-membrane proton pumps of fungi which, like the alkali cation pumps, are $M_r \sim 100{,}000$ proteins that become phosphorylated during the reaction cycle and are blocked by orthovanadate ions. This general class of membrane enzymes has been dubbed $E_1$-$E_2$ ATPases[2] or P-type ATPases.[3]

## PUMP STOICHIOMETRY

The fungal plasma-membrane proton pump carries out the simplest overall reaction yet discovered for P-type ATPases: it ejects one proton for each ATP molecule split,[4] whereas the related animal-cell enzymes simultaneously pump multiple ions, usually with counterflow. In the sodium pump,[5] three Na[+] are

[a] This work was supported by research grant GM-15858 from the National Institute of General Medical Sciences.

exchanged for two $K^+$; in the gastric proton pump,[6,7] one $H^+$ for one $K^+$ or perhaps two $H^+$ for two $K^+$; in the calcium pump of sarcoplasmic reticulum,[8] two $Ca^{2+}$ for two $H^+$; and in the plasma-membrane calcium pump,[9] one $Ca^{2+}$ for two $H^+$.

Unitary stoichiometry for the *Neurospora* proton pump has been known from thermodynamic data for more than 10 years. Total cytoplasmic concentrations of ATP, ADP, and inorganic phosphate are 2.7, 0.8, and 10 mM, respectively,[10] and the standard free energy for ATP hydrolysis can be estimated at 8 kcal/mol, so the operant free energy for ATP hydrolysis is approximately 500 mV. The usual pH difference across the plasma membrane is 1.4 units (80 mV) outside acid,[11] the measured membrane potential ($V_m$) can be as high as 350 mV (cell interior negative), and pump stalling potentials can exceed 400 mV.[12,13] Because the operating free energy of the pump can thus exceed 400 mV, while the energy from ATP hydrolysis is near 500 mV, the only possible integral stoichiometry is one net charge per cycle. (This calculation is only slightly influenced by the possibility of cytoplasmic ADP binding, which could reduce the free ADP concentration by 10- to 100-fold, thereby increasing the free energy of ATP hydrolysis only to about 600 mV.) Furthermore, the ATPase in broken membranes shows no specific cation coactivation,[14] so there is no reason to invoke participation of other cations.

Independent confirmation of one $H^+$ transported/one ATP split by fungal proton pumps has come from proton-counting experiments on isolated membranes. Vesicles of fungal plasma membranes are easily prepared which are morphologically everted,[15] so the catalytic region of the enzyme is exposed to the medium; such vesicles can use ATP to pump the vesicle interior more than 2 pH units acidic to the suspending medium.[15,16] Steady state is achieved in such experiments when the rate of inward pumping just balances leaks through the vesicle membranes; then after pumping is suddenly halted (as by the addition of 10 $\mu m$ orthovanadate), the pumped acid leaks out. When the initial net leak in such experiments was compared with the steady-state consumption of ATP, a stoichiometry of 0.96 : 1 was observed with *Neurospora* vesicles.[17] Similar numbers have also been obtained with yeast vesicles.[18]

Chemical determination of stoichiometry for proton pumps in intact, metabolically active microorganisms is very difficult, because of numerous alternative pathways both for proton transport and for ATP hydrolysis. A useful indication of stoichiometry, however, is often found in the kinetic relationship between transport velocity and substrate concentration. In *Neurospora* this can be assessed by determining how pump current varies with changes of cytoplasmic pH ($pH_i$), provided the current is measured during short circuiting of the membrane, so there is no voltage to oppose the pump. For such experiments, changes of $pH_i$ have been forced by extracellular addition of a weak acid, such as butyric acid,[11] and by direct electrophoretic injection of protons.[19] Pump current then changes almost in direct proportion to the *measured* cytoplasmic $H^+$ concentration.[20,21] That is, the kinetic response of the pump to changing $[H^+]_i$ is the same as the response to a first-order substrate.

## APPARENT AFFINITY FOR ATP

Another conspicuous difference from most P-type ATPase/cation pumps is the high apparent $K_m$ (low affinity) for ATP seen in fungal and plant proton pumps. This property was first observed in studies of the ATP dependence of membrane potential in *Neurospora*[22]: cyanide blockade of respiration results in rapid decay (time constant $\approx 5$ seconds) of cytoplasmic ATP and similarly rapid decay of membrane potential, following a brief lag period. Plots of $V_m$ versus ATP concentration in such experiments are essentially hyperbolic, with $K_m$ values of 1.5 to 2.5 mM ATP. For times up to ~30 seconds, membrane resistance is almost constant, so changes of potential can be translated directly to changes of current (i.e., pump velocity). The $K_m$ value has been confirmed by measurements of ATPase activity in membrane fragments of *Neurospora*[14] and *Schizosaccharomyces*,[23] whereas smaller values (0.1–1.0 mM) were found in *Saccharomyces*[24] and in tobacco and beet tissue.[25,26] Additional support has come from measurement of acid uptake by everted plasma-membrane vesicles from *Neurospora*,[15] tobacco,[27] and corn.[28]

The finding thus of millimolar $K_m$'s was very surprising in view of well-established micromolar $K_m$'s for ATP reactivity in the familiar P-type ATPases from animal cells. During slow hydrolysis by electroplax or brain $Na^+,K^+$-ATPase, operating (for example) in the presence of sodium but the *absence* of potassium, that enzyme displays a $K_m$ below 1 $\mu M$.[29] Speeding up hydrolysis with added $K^+$ increases the ATP-$K_m$ to the neighborhood of 100 $\mu M$,[29,30] which is the lower range of values for the $H^+$-ATPases. Rapid-phosphorylation studies on the $Ca^{2+}$-ATPase of rabbit sarcoplasmic reticulum[31,32] have yielded ATP-$K_m$'s of ~10 $\mu M$ for that enzyme, whereas "physiologic" transport and ATPase measurements[33,34] show two components: a large one at micromolar $K_m$, and a small one at millimolar $K_m$. The $H^+,K^+$-ATPase from hog gastric mucosa likewise displays two kinetic components for ATP hydrolysis: one with $K_m$ near 2 $\mu M$, and the other with $K_m$ near 50 $\mu M$,[35] with some variation depending on the test $K^+$ concentration and the assay temperature.[36] It seems clear for all three of these enzymes that the lowest $K_m$ values, ~1 $\mu M$ and below, obtain under conditions of retarded or blocked transport, when the phosphoprotein intermediate is allowed to accumulate. This is the condition in which Michaelis enzymes *should* display the smallest $K_m$ values, approaching the dissociation constant for enzyme-bound ATP, which may be as small as 0.1 $\mu M$ in the $Na^+,K^+$-ATPase.[37]

No inhibitory maneuver or agent comparable to potassium withdrawal from the sodium pump is yet known for fungal protons pumps, but measurement of the dissociation constant ($K_d$) for ATP in the absence of $Mg^{2+}$ ions and imposition of large membrane potentials to oppose pumping should yield useful (if only partial) information. Despite some exaggeration due to methodologic differences, it seems likely that the disparity of $K_m$ values between the fungal or plant proton ATPases and other cation ATPases is real. One possible simple interpretation is that the high ATP-$K_m$ for fungal ATPases might reflect obligatory activation by nucleotide binding at a nonhydrolyzing, low-affinity binding site (masking much higher affinity at the catalytic site), much as has been proposed for the $Ca^{2+}$-ATPase of sarcoplasmic reticulum.[8] That suggestion is circumstantially supported by a slight sigmoidicity in plots of membrane potential, ATPase activity, or vesicle $H^+$ transport, against ATP concentration.[14,15,22] The suggestion remains controversial, however, and in our judgment is unlikely to account for the whole discrepancy.

## INTRINSIC REACTION RATE FOR $E_1 \sim P \rightleftharpoons E_2 \cdot P$

A more subtle feature of the fungal proton ATPase, which differentiates it from most of its congeners, is its transition rate at the voltage-dependent step, that is, at the transformation between $H^+ \cdot E_1 \sim P$ and $H^+ \cdot E_2 \cdot P$, shown in the reaction diagram in FIGURE 1A. Because the partial reaction chemistry of P-type proton ATPases has not yet been worked out, FIGURE 1A has necessarily been drawn by reduction of an accepted reaction scheme for the $Na^+,K^+$-ATPase[38]: $H^+$ has been substituted for $Na^+$ as the first ion bound, and all steps concerned *only* with multiple ion binding or $K^+$ countermovement have been deleted. For the sodium pump, the voltage-dependent step was recently identified as the $E_1 \sim P$-to-$E_2 \cdot P$ transition by (1) the effect of chemically clamped membrane potentials upon sodium flux in kidney vesicle preparations[40] and (2) the effect of voltage amplitude, for electrical clamp-steps, upon transient currents in perfused cardiac cells which were blocked by zero extracellular potassium.[41]

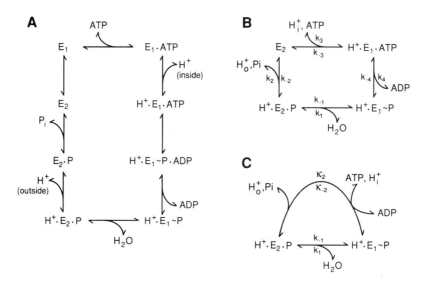

**FIGURE 1.** Kinetic models for the *Neurospora* plasma-membrane ATPase. **(A)** Hypothetical reaction diagram[39] obtained by simplification of a standard model for the $Na^+,K^+$-ATPase.[38] All multiple ion and ion-return steps have been deleted, because the $H^+$-ATPase only ejects a single $H^+$ for each ATP molecule split. **(B)** Reduced version of **A** emphasizing phosphorylation/dephosphorylation reactions relevant to $^{18}O$-exchange experiments. **(C)** Reduced version of **A** emphasizing voltage-dependent reactions ($k_1$, $k_{-1}$) versus ensemble of voltage-independent reactions ($K_2$, $K_{-2}$), relevant to current-voltage analysis. Correct parameter evaluation in such reduced models always *under*estimates corresponding or constituent rate constants in the full model.

The same conclusion had been reached previously for the *Neurospora* proton pump on both à priori and experimental grounds.[13] First, this pump runs at nearly maximal velocity against membrane potentials approaching 200 mV and at substantial velocities even beyond 350 mV, and the electrical work thereby done *on each charge* is too large to come from any source other than the phosphate bond. Second, steady-state kinetic analysis, applied to pump currents measured over a wide range of clamped voltages, always shows the ratio of rate constants, $k_1/k_{-1}$, at $E_1 \sim P \rightleftharpoons E_2 \cdot P$ to be large, on the order of $10^6$, which *again* represents too much energy for any source other than $\sim P$.[10,21]

*Modeling of Current-Voltage Data.* In steady-state voltage-clamp experiments, current through the *Neurospora* proton pump (and also through all examples of the sodium pump that have been tested) rises monotonically with the clamped membrane potential. This fact allows the electrical data *per se* to be described by a much simpler reaction model than that in FIGURE 1A, indeed by the two-state model in FIGURE 1C. That comprises a single voltage-dependent step (with the rate parameters $k_1$ and $k_{-1}$) and a lumped voltage-independent step ($K_2$ and $K_{-2}$) which contains all of the scalar chemical reactions plus electroneutral transit of the membrane.[10,42] (We assume the voltage dependence to be described by charge transit across an Eyring barrier,[43] because that is closely related to the definition of electrochemical potential, but we *do not* assume it to be symmetric.) Correctly fitted values for the reaction parameters in such a "reduced" model must be adjusted in two ways to give physically satisfactory numbers. First, because the lumping procedure incorporates unevaluated reaction states into the explicit states $E_1$ and $E_2$, thus making their *apparent* values too large, it follows that conjugate rate parameters $k_1$ and $k_{-1}$ must necessarily be *too small*. The appropriate corrections or *reserve factors* $r_1$ and $r_2$ can be written as algebraic combinations of the concealed reaction parameters from FIGURE 1A.[13] Second, as with all steady-state measurements, the absolute value of a rate constant cannot be known without independent knowledge of the number of transporter molecules participating: that is, the site density in molecules/$\mu m^2$, $E_T = E_1 + E_2$.

For practical analyses, the lumped two-state parameter $K_2$ has an especially simple and useful form: $K_2 = i_{sat+}/zFE_T$, in which z is the transport stoichiometry in charges per cycle, F is the Faraday constant, and $i_{sat+}$ is the positive saturation current for the pump (measured during strong clamped depolarization). Model fitting has been done with all parameters scaled so that $K_2$ (in $s^{-1}$) is *numerically* equal to $10\times i_{sat+}$, expressed in $\mu A/cm^2$, so that the implied value of $E_T$ is 6,200 sites/$\mu m^2$. In our experiments then, fully corrected values for the forward voltage-dependent rate parameter, $k_1$ in FIGURE 1A, are given by this equation (1): $k_{1(actual)} = k_{1(apparent)} \cdot r_1 \cdot 6200/E_{T(actual)}$.

To estimate the absolute magnitude of $k_1$, we examined the steady-state current-voltage relationship of the *Neurospora* proton pump under a wide range of conditions: normal metabolism, energetic downshifting by cyanide blockade, and lowered and elevated cytoplasmic pH and extracellular pH. The data have been fitted in groups to the two-state model of FIGURE 1C, allowing the fewest parameters possible to change from curve to curve within each group.[10,21] In general, the voltage-dependent parameters $k_1$ and $k_{-1}$ have not been required to change from curve to curve, and essentially all of the data can be described by values of $k_1$ down to 3,000 $s^{-1}$, but no lower. Corrections that must be applied to this value are discussed in the foregoing section (Modeling of Current-Voltage Data) and summarized in equation 1. The actual density of proton pumps in the *Neurospora* plasma membrane, estimated from freeze-fracture pictures displaying 80- to 120-A particles,[15,44] is in the neighborhood of 3,000/$\mu m^2$ and certainly does not exceed 6,200/$\mu m^2$. In addition, trial calculations with more elaborate models (explicitly incorporating, e.g., $pH_i$ and $[ATP]_i$, and $pH_o$) indicate the reserve factor $r_1$ to be in the range of 3–10. Thus, from equation 1, the actual value of $k_1$ must lie above 3,000 $s^{-1}$ and could be as high as 60,000 $s^{-1}$. From a geometric point of view, the lower limit was set by a combination of observational constraints on the modeling: the demonstrably large reversal potential (ca. 400 mV), the high operating $K_m$ for ATP (>1 mM), the proportionality between pump current and cytoplasmic proton concentration, and a generally convex shape to the pump's overall current-voltage relationship in intact membranes.

Anyone familiar with the myriad of partial reaction studies on the other P-type ATPases is likely to be skeptical of a number like 3,000–60,000/s, because it is 2–3 orders of magnitude larger than the corresponding values obtained (for example) on the $Na^+,K^+$-ATPase or the $Ca^{2+}$-ATPase of sarcoplasmic reticulum (TABLE 1). For those pumps, the numbers range under 300/s for all temperatures (≤37°C) and

**TABLE 1.** Estimated Rate Constants for the Transition $M_x^+ \cdot E_1 \sim P \rightleftharpoons M_x^+ \cdot E_2 \cdot P$ in P-Type Membrane ATPases[a]

| Case No. | Value ($s^{-1}$) | Enzyme | Species | Method | Temp.[b] (°C) | Exp. $pH_i$[c] | Ref. Nos. |
|---|---|---|---|---|---|---|---|
| 1 | 3,000–30,000 | PM $H^+$-Pump | *N. crassa* | Steady-state I-V analysis in intact cells | 22 | 7.2 | 46 and this report |
| 2 | 9,200 | PM $H^+$-ATPase | *S. pombe* | Steady-state $^{18}O$ exchange in $PO_4$; PM vesicles ± lysolecithin, 0 ATP | 30 | 6.0 | 45 |
| 3[d] | 51 | SR $Ca^{2+}$-ATPase | Rabbit skeletal muscle | Steady-state $^{18}O$ exchange in $PO_4$; SR vesicles, zero ATP, and $Ca^{2+}$ | 25 | 6.5 | 47 |
| 4 | 60 | SR $Ca^{2+}$-ATPase | Rabbit skeletal muscle | Decay of chem. P-E formed from $PO_4$ in zero $Ca^{2+}$; SR vesicles | 25 | 6.8 | 48 |
| 5 | 170 | PM $Na^+$-Pump | General | Best fit for multiple features of electrogenic transport | 37 | . . . | 49 |
| 6 | 270 | PM $Na^+$-Pump | Guinea pig cardiac cells | Ouabain sens. charge relaxation; patch-perfused cells in zero $K^+$ | 35 | 7.4 | 41 |
| 7[e] | 24 | $Na^+,K^+$-ATPase | Rabbit kidney | Bilayer current from attached PM fragments; flash-liberated ATP | 20 | 7.0 | 50 |
| 8[e] | 100 | $Na^+,K^+$-ATPase | Pig kidney | Bilayer current from attached PM fragments; flash-liberated ATP | 22 | 6.2 | 51 |
| 9[d] | 6 | $Na^+,K^+$-ATPase | Pig kidney | Flash-activated single-turnover $Na^+$ release from PM vesicles | 15 | 7.2 | 52 |
| 10 | 460 | $H^+,K^+$-ATPase | Hog stomach | Steady-state $^{18}O$ exchange in broken PM vesicles | 37 | 7.4 | 53 |

[a] The symbol $M_x^+ \cdot$ designates x cations bound.
[b] A rough correction for different temperatures can be obtained by assuming a $Q_{10}$ of 3.
[c] Intracellular pH ($pH_i$) here refers to the solution bathing the physiologic cytoplasmic side of the membrane or pump molecules.
[d] Since these reactions contain more steps than simply the $E_1$-$E_2$ transition, estimated rate constants are necessarily *minimal* values for the single step.
[e] Results represent similar data, but different physical interpretations of the data.

all pH values (6.2–7.4) tested. It is important, therefore, that a large figure, 9,200/s, has also been calculated by Amory et al.[45] from $^{18}O$-exchange measurements on the plasma-membrane $H^+$-ATPase of Schizosaccharomyces pombe.

*Analysis of Phosphate/Medium Oxygen Exchange.* The $^{18}O$ experiments can most easily be recounted by reference to the four-state reaction model of FIGURE 1B. In the experiments of Amory et al.,[45] ATP was absent, and inorganic phosphate fully substituted with $^{18}O$ was added to the medium. Even though neither synthesis of ATP (from ADP and inorganic phosphate) nor formation of the phosphoprotein from inorganic phosphate[54] has been chemically demonstrated in the fungal $H^+$-ATPase, the enzyme does bind tracer-detectable amounts of phosphate[45] and does catalyze exchange of the phosphate-$^{18}O$'s with water oxygen, as if binding were to the active site (via $k_{-2}$; FIG. 1B) and as if tumbling occurred within the active site prior to water entry (via $k_1$) and rerelease of the phosphate ($k_2$). Theoretical analysis[55] posits several parameters for efficient description and interpretation of phosphate/medium-$^{18}O$ exchange data: (i) $V_{ex}$, the absolute rate of $^{18}O$ release (efflux) from fully labeled phosphate (corrected for nonenzymatic release), equals $k_1[E_1 \sim P]$ in FIGURE 1B; $[E_1 \sim P]$ varies hyperbolically with the concentration of ($^{18}O$-labeled) inorganic phosphate, such that (ii) $V_{ex(max)} = E_T k_{-1}/(k_1 + k_{-1})$; (iii) $K_4$ is the measured rate constant for efflux of the first $^{18}O$ ($-P^{18}O_4$ going to $-P^{18}O_3{}^{16}O$); (iv) $K_t$ is the rate constant for efflux of total $^{18}O$ from phosphate; and (v) $P_c$, the partition coefficient for reaction of $E_2 \cdot P$, $= k_{-1}/(k_{-1} + k_2) = 0.33(4 - K_4/K_t)$. (In the circumstance in which multiple water entries can occur before phosphate release ($k_{-1}$ is large relative to $k_2$, FIG. 1B), $K_4$ and $K_t$ should be nearly equal, but when each water entry leads to phosphate release, $K_4$ should be large (up to $4\times$) with respect to $K_t$.)

The following remarkable result has emerged from $^{18}O$-exchange studies on the $Ca^{2+}$-pump of sarcoplasmic reticulum,[47,56] on the $H^+$-ATPase of S. pombe,[45] and, as far as have been determined, on the $Na^+,K^+$-ATPase:[57] provided that the level of ($P^{18}O_4$-labeled) phosphate in the medium is poised *at the same fraction of the reaction* $K_m$ for all three enzymes, then each of the measurable parameters ($V_{ex}$, $K_4$, and $K_t$) is roughly the same for all three enzymes. Conspicuous differences between the $H^+$-ATPase and the other two lie in the magnitude of phosphate-$K_m$ for $^{18}O$ exchange: 177 mM (!) for the $H^+$-ATPase,[45,54] compared with 1–5 mM for the other two,[58,59] and in a very low apparent value for $[E_1 \sim P]$ in the absence of ATP.[45]

Simple calculation of $k_1$, $k_{-1}$, and $k_2$ requires independent estimates of $E_T$ (total ATPase present) and $[E_1 \sim P]$. The latter has come from direct measurements for the $Na^+,K^+$- and $Ca^{2+}$-ATPases[58,59] and from enzyme-bound tracer phosphate for the $H^+$-ATPase.[45] Of the three rate constants, $k_1$ is evaluated from equation i (see preceding section on Analysis of Phosphate/Medium Oxygen Exchange); $k_{-1}$, from equation ii, and $k_2$ from equation v. The resultant values are shown in TABLE 2 for three different ATPases assayed under comparable conditions.

The fact that the $E_1 \leftrightharpoons E_2$ transition is about two orders of magnitude faster in the proton pumps than in P-type enzymes pumping other cations raises a number of interesting questions. For example, does it indicate a close relationship between the mode of proton movement within the protein and that through water or certain clathrates, where apparent proton mobility exceeds that for other cations by more than an order of magnitude? Or does the high value obtain in plant plasma membrane ATPases, which are closer to the fungal enzymes in both structure and function than are most of the animal enzymes? Or does the higher rate constant occur in the gastric $H^+,K^+$-ATPase, which is structurally closely related to the $Na^+,K^+$-ATPase? Diaz and Faller[53] recently examined this question by $^{18}O$-exchange in hog gastric ATPase (TABLE 1, case 10). Their calculated value of 460/s

**TABLE 2.** Critical Phosphorylation/Dephosphorylation Parameters for Three P-Type ATPases[a]

| Case No. | Enzyme | $k_1$ | $k_{-1}$ | $k_2$ | $K_m$ (mM) | Temp. (°C) | Ref. Nos. |
|---|---|---|---|---|---|---|---|
| | | (all units of 1/s) | | | | | |
| 2 | PM H$^+$-ATPase from Schizosaccharomyces | 9,200 | 57 | 68 | 177 | 30 | 45 |
| 3 | SR Ca$^{2+}$-ATPase from rabbit muscle | 51 | 32 | 380 | 2[b] | 25 | 47,59 |
| 10 | H$^+$,K$^+$-ATPase from hog stomach | 460 | 310 | 1,100 | 2 | 37 | 53 |

[a] Calculated from $^{18}$O-exchange data; k's defined as in FIGURE 1B.
[b] $K_m$ experiment at 15°C.

(at 37°C) would come much closer to values for the Na$^+$,K$^+$- and Ca$^{2+}$-ATPases than to the fungal H$^+$-ATPase when the likely temperature coefficient ($Q_{10} \geq 3$) is considered.

## THE SUPERCHARGED N-TERMINUS

As already reviewed in this symposium,[1] the fungal PM-ATPase is closely similar to other P-type cation ATPases in its secondary structure and also has extensive identity of primary sequence in certain regions, such as the nucleotide-binding area (including Lys628 to Ala650 in the *Neurospora* enzyme) and the flanks of the phosphorylation site (Asp378). There is even a strong similarity in distribution of charged amino acids within the hydrophobic, membrane-spanning helices of all of the P-type ATPases,[1] despite widely differing transport specificities for the different enzymes.

However, one respect in which the fungal enzyme clearly does differ from the others is in the amino acid composition of, and charge distribution along, the N-terminal segment (Ala2 to Lys115 in *Neurospora*). Its primary structure is summarized in FIGURE 2 and TABLE 3. For the three main species of fungal ATPase

**TABLE 3.** Composition and Charge Distribution in N-Terminal Segments of P-Type ATPases[a]

| Source | Total AAs | Arg | Lys | His | Asp | Glu | Sum + | Sum − | Charge |
|---|---|---|---|---|---|---|---|---|---|
| *N. crassa* | 115 | 5 | 8 | 4 | 12 | 20 | 17 | 32 | −15 |
| Seg. 38–66 | 29 | 0 | 0 | 2 | 8 | 11 | 2 | 19 | −17 |
| *S. cerevisiae* | 115 | 3 | 7 | 2 | 15 | 11 | 12 | 26 | −14 |
| Seg. 36–64 | 29 | 0 | 0 | 1 | 10 | 4 | 1 | 14 | −13 |
| *S. pombe* | 113 | 2 | 6 | 3 | 13 | 19 | 11 | 32 | −21 |
| Seg. 35–63 | 29 | 0 | 0 | 0 | 7 | 10 | 0 | 17 | −17 |
| Rat stomach | 103 | 4 | 15 | 1 | 3 | 11 | 20 | 14 | +6 |
| Sheep kidney | 92 | 7 | 12 | 3 | 9 | 8 | 22 | 17 | +5 |
| *L. donovani* | 92 | 5 | 10 | 2 | 5 | 12 | 17 | 17 | 0 |
| *A. thaliana* | 60 | 2 | 8 | 0 | 3 | 11 | 10 | 14 | −4 |
| Rabbit muscle | 59 | 1 | 4 | 3 | 2 | 11 | 8 | 13 | −5 |
| *Escherichia coli* | 32 | 2 | 4 | 0 | 0 | 2 | 6 | 2 | +4 |

[a] N-terminal segment defined by the common start-point for the first hydrophobic helix.[1]

```
                  1 - -                -    +1 --+   -     1  ++  ------   --  ------          1+ --    -  1 - +     +++1  +++-+-1- +
1) MADHSASGAPALSTNIESGKF-DEKAAEAAAYQPKPKVEDDEDEDIDALIEDLESHDGHDAEEEEEATP--GGGRVVPEDMLQTDPRVGLTSEEVVQRRRKYGLNQMKEEKENHFLK       ELGFEVGPIOFVMEGAAVLAAGL

                  1 - -                +  -- -   -    1  -  ------   --  ----ll -- -          1 - +     -   1-      +++11  -+-1 +
2) MTDTSSSSSSASSVSAHQPTQEK-P-AKTYDDAAS-ESSDDDDIDALIEELQSNHGVDDEDSDNDGPVAAGEARPVPEEYLQTDPSYGLTSDEVLKRRKYGLNQMADEKESLVVK        FVMFEVGPIOFVMEAAAILAAGL

                  1 - -    -+- -       1              -- ----ll  --  ----l--           .  1 --- 1  -  + +       -1--++++   1 +-- --1 +
3) MADNAGEYHDAEKHAPEQQAPPPQQ-P-A-HAAAPAQ-DDEPDDDIDALIEELFSEDVQEEQDNDDAP-AAGEAKAVPEELLQTDMNTGLTMSEVEERRKYGLNQMKEELENPFLK      EIMFEVGPIOFVMEMAAALAAGL

                  1 +-     1-        -1 +  +++ 1  +++-+1- +++ - 1-.    -+        +1 +-     -1 +-    +1 + - +1 1+
4) MGKENYELYSVELGTGPGGDMAAKMSKKKAGGGGKKKEKLENMKKEMEMNDHQLSVSELEQKYQTSATKGLKASLAAELLLRDGPNALRPPRGTPEYVKFAR                   OLAGGLQCLMHVAAAICLIAFAIOAS

                  1 +       +-+- -  -+++ ++++-- -- ++-   --+ +   -++   - + 1  +-    1+-         1 - + 1+
5) MGKVGRDKYEPAAVSEHGDKKAKKERDMDELKKEVSMDDHKLSLDELHRKYGTDLNRGLTTARAAEILARDGPNALTPPTTPEWVKFCR                                OLFGGESMLLWIGAVLCFLAYGIOAAT

                  1 ++ -   1 --+ -   .1- -   + 1+ +   +   +1 -++1 +      +1  - ---++         -++
6) MSSKKYELDAAFEDKPESHSDAEMTPQKPQRRQSVLSKAVSEHDERATGPATDPVPPSKGLTTEEAEELLKKYGRNELPEKKTPSWLIYVR                             GLWGEMPAALWIIAILIEFAL

                1   1 - -+ -1 - -+       - -     +   +-    - --+            + --+++-+ + +
7) MSGLEDIKNETVDLEKIPIEEVFQQLKCTREGLITTQEGEDRIVIFGPNKLEKKESKILK                                                           FLGFMWNPLSWVMEAAALMAIALA

               1- - + --         -1        +++ -+  - 1  -- +            -- --
8) MEAAHSKSTEECLAYFGVSETTCLIPDQVKRHLEKYGHNELPAEEGKSLWELVIEQFED                                                            LLVRILLLAACISFVLAWE

                      1 ++        -1     +-1 ++        ++
9) MSRKQLALFEPTLVVQALKEAVVKKLNPQAQWR                                                                                       NEVMFLVWIGSLLTTCISIAMASGAMPGN
```

**SYMBOL KEY**

Amino acids:   R = arginine      K = lysine

• = histidine (+ at physiological pH)

- Amino acids:   D = aspartate     E = glutamate

1 Sequence counter: 1, 11, 21, etc., from the

N-terminal methionine

**SOURCE KEY**

1) _Neurospora crassa_ H⁺-ATPase

2) _Saccharomyces cerevisiae_ H⁺-ATPase

3) _Schizosaccharomyces pombe_ H⁺-ATPase

4) Rat (_Rattus norvegicus_) gastric H⁺,K⁺-ATPase

5) Sheep (_Ovis_) kidney Na⁺,K⁺-ATPase (α -subunit)

6) _Leishmania donovani_ ?H⁺-ATPase

7) _Arabidopsis thaliana_ ?H⁺-ATPase

8) Rabbit (_Oryctolagus cuniculus_) SR Ca⁺⁺-ATPase from fast-twitch muscle

9) _Escherichia coli_ K⁺-ATPase (B-subunit)

| Line # | 1 | 2 | 3 | 4 | 5 | 6 | 7 | 8 | 9 |
|--------|---|---|---|---|---|---|---|---|---|
| Ref. # | 60,61 | 62 | 63 | 64 | 65 | 66 | 67 | 68 | 69 |

**FIGURE 2.** Comparison of N-terminal segments of several P-type membrane ATPases. The top three sequences, for the fungal enzymes, are aligned to maximize coincident amino acids. Charged amino acids are designated by symbols directly above. Note the remarkable strings of negative amino acids near the middle of the fungal sequences. Nothing comparable is found in the other enzymes. Reference key:

whose genes have been sequenced, the N terminus is both long and highly charged, bearing 15–21 excess *negative* residues in its span of 113–115 amino acids. By comparison, the N-terminal segment of rat gastric $H^+$-ATPase has an excess of six *positive* charges in a span of 103 amino acids; that of sheep kidney $Na^+,K^+$-ATPase has five excess positive charges in 92 amino acids; and that of muscle $Ca^{2+}$-ATPase has five excess negative charges in its 59-amino acid span. Even more remarkably, essentially the entire surplus of negative charges for the fungal enzymes resides in a single 29-amino acid stretch (Glu38 to Glu66 in *Neurospora*), which contains 14–19 negative residues and only 0–2 positive residues.

Furthermore, recent studies on trypsin digestion of the *Neurospora* ATPase have shown the "supercharged" segment to be *required* for hydrolytic activity. It has been known for several years[70] that controlled trypsin degradation creates large, well-defined fragments of the *Neurospora* enzyme, whose distribution and stability are strongly influenced by the presence or absence of nucleotides and the enzyme-specific inhibitor, orthovanadate. Mandala and Slayman[71] refined earlier experiments by blotting the large fragments with antibodies specific for the N-terminus (amino acids 3–46) or C-terminus (amino acids 886–920) and by sequencing the fragments via Edman degredation. Three trypsin cleavage sites were thus identified near the N-terminus: Lys24, Lys36, and Arg73. (Two sites, Arg900 and Arg911, were also found near the C-terminus.) Removal of the first 36 residues had little effect on hydrolytic activity of the enzyme, but removal of the next 37 residues (cleavage at Arg73) completely inactivated the enzyme.

Several facts concerning this peculiar negatively charged region now stand out: it is present exclusively in the three P-ATPases which are pure $H^+$ pumps (rather than coupled ion pumps); its overall charge is maintained even though the detailed amino-acid sequence is not as well conserved as the average structures among the three enzymes; mainly by virtue of its charge the region is strongly hydrophilic and therefore certainly not within the membrane-spanning domain of the enzyme; antibody studies[72] have verified its *intracellular* location, which was first deduced from hydropathy plots.[60,62] Although in the primary sequence, it is remote from the phosphorylation site (Asp378 in *Neurospora*) and the nucleotide-binding area (Near Lys628–Ala650), it might be imagined to cluster tightly against a positively charged region: Lys385-Lys474, with 10 excess (+) residues lying between.

Against the welter of similarities[1,73] between the fungal plasma-membrane $H^+$-ATPases and a variety of other P-type transport ATPases, it is interesting to speculate on possible functional meanings for this peculiarly charged or acidic region.

Is it particularly indicative of "pure" $H^+$ pumps? This possibility is suggested by the facts that the gastric proton pump is specifically and tightly $K^+$ coupled and that the *plant* plasma-membrane proton pump remains disputed as a partially $K^+$-coupled enzyme. In that regard, the structural differences between the cloned genes for a *Leishmania* membrane ATPase and especially the *Arabidopsis* enzyme raise the obvious question of whether these indeed are $H^+$-ATPases, as has been tacitly assumed. Finally, could the structural and functional properties of the fungal plasma-membrane $H^+$-ATPases mean that a different reaction pathway is followed by that enzyme than by the other P-type ATPases? This is improbable for the reactions of nucleotides and phosphate, but worth considering for the transported ions. The general lability of hydrogen bonds, the ready mobility of conjugated $C=C$ double bonds, and the potential abundance of protons from solvent water all conjure additional mechanisms for proton movement[74–76] that are not available for sodium, potassium, or calcium.

It is ironic that primary sequences of transport proteins are now easier to obtain than are precise functional data. Proton transport by the newly sequenced enzymes from *Leishmania* and *Arabidopsis,* for example, are thus far unproven. But it is clear in consequence that careful comparative functional studies on the new proteins, as they are sequenced, could yield much information about the molecular mechanisms of transport, in the manner long anticipated for site-directed mutagenesis studies.

## ACKNOWLEDGMENTS

The authors are indebted to Dr. L. D. Faller for use of results (TABLE 1, case 10) prior to publication, and to Dr. R. K. Nakamoto for helpful discussion.

## REFERENCES

1. NAKAMOTO, R. K., R. RAO & C. W. SLAYMAN. 1989. Ann. N.Y. Acad. Sci. **574:** 165.
2. MITCHELL, P. & W. H. KOPPENOL. 1982. Ann. N.Y. Acad. Sci. **402:** 584–601.
3. PEDERSEN, P. L. & E. CARAFOLI. 1987. TIBS **12:** 146–150.
4. SLAYMAN, C. L. 1987. J. Bioener. Biomembr. **19:** 1–20.
5. POST, R. L. & P. C. JOLLY. 1957. Biochim. Biophys. Acta **25:** 118–128.
6. REENSTRA, W. W. & J. G. FORTE. 1981. J. Membr. Biol. **61:** 55–60.
7. RABON, E. C., T. L. MCFALL & G. SACHS. 1982. J. Biol. Chem. **257:** 6296–6299.
8. INESI, G. & L. DE MEIS. 1985. *In* The Enzymes of Biological Membranes. A. Martonosi, ed. Vol. **3:** 157–191. Plenum, New York.
9. NIGGLI, V., E. SIGEL & E. CARAFOLI. 1982. J. Biol. Chem. **257:** 2350–2356.
10. SLAYMAN, C. L. 1973. J. Bacteriol. **114:** 752–766.
11. SANDERS, D. & C. L. SLAYMAN. 1982. J. Gen. Physiol. **80:** 377–402.
12. GRADMANN, D., U.-P. HANSEN, W. S. LONG, C. L. SLAYMAN & J. WARNCKE. 1978. J. Membr. Biol. **39:** 333–367.
13. GRADMANN, D., U.-P. HANSEN & C. L. SLAYMAN. 1982. Curr. Top. Membr. Transp. **16:** 257–276.
14. BOWMAN, B. J. & C. W. SLAYMAN. 1977. J. Biol. Chem. **252:** 3357–3363.
15. PERLIN, D. S., K. KASAMO, R. J. BROOKER & C. W. SLAYMAN. 1984. J. Biol. Chem. **259:** 7884–7892.
16. DUFOUR, J.-P., A. GOFFEAU & T. Y. TSONG. 1982. J. Biol. Chem. **257:** 9365–9371.
17. PERLIN, D. S., M. J. D. SAN FRANCISCO, C. W. SLAYMAN & B. P. ROSEN. 1986. Arch. Biochem. Biophys. **248:** 53–61.
18. MALPARTIDA, F. & R. SERRANO. 1981. FEBS Lett. **313:** 351–354.
19. BLATT, M. R. & C. L. SLAYMAN. 1987. Proc. Natl. Acad. Sci. USA **84:** 2737–2741.
20. SANDERS, D., U.-P. HANSEN & C. L. SLAYMAN. 1981. Proc. Natl. Acad. Sci. USA **78:** 5903–5907.
21. SLAYMAN, C. L. & D. SANDERS. 1984. *In* Hydrogen Ion Transport in Epithelia. J. G. Forte, D. G. Warnock & F. C. Rector, eds.: 47–56. Wiley & Sons, New York.
22. SLAYMAN, C. L., W. S. LONG & C. Y.-H. LU. 1973. J. Membr. Biol. **14:** 305–338.
23. DUFOUR, J.-P. & A. GOFFEAU. 1980. Eur. J. Biochem. **105:** 145–154.
24. SERRANO, R. 1978. Mol. & Cell. Biochem. **22:** 51–63.
25. KASAMO, K. 1979. Plant & Cell Physiol. **20:** 281–292.
26. O'NEILL, S. D. & R. M. SPANSWICK. 1984. J. Membr. Biol. **9:** 245–256.
27. SZE, H. & K. A. CHURCHILL. 1981. Proc. Natl. Acad. Sci. USA **78:** 5578–5582.
28. DE MICHAELIS, M. I. & R. M. SPANSWICK. 1986. Plant Physiol. **81:** 542–547.
29. PLESNER, L. & I. W. PLESNER. 1981. Biochim. Biophys. Acta **643:** 449–462.
30. ROBINSON, J. D. 1974. FEBS Lett. **47:** 352–355.

31. FERNANDEZ-BELDA, F., M. KURZMACK & G. INESI. 1984. J. Biol. Chem. **259:** 9687–9698.
32. FROEHLICH, J. P. & P. F. HELLER. 1985. Biochemistry **24:** 126–136.
33. INESI, G., J. J. GOODMAN & S. WATANABE. 1967. J. Biol. Chem. **242:** 4637–4643.
34. KANAZAWA, T., S. YAMADA, T. YAMAMOTO & Y. TONOMURA. 1971. J. Biochem. **70:** 95–123.
35. WALLMARK, B., H. B. STEWART, E. RABON, G. SACCOMANI & G. SACHS. 1980. J. Biol. Chem. **255:** 5313–5319.
36. FALLER, L., R. JACKSON, D. MALINOWSKA, E. MUKIDJAM, E. RABON, G. SACCOMANI, G. SACHS & A. SMOLKA. 1982. Ann. N.Y. Acad. Sci. **402:** 146–163.
37. NØRBY, J. G. & J. JENSEN. 1971. BBA **233:** 104–116.
38. KARLISH, S. J. D., D. W. YATES & I. M. GLYNN. 1978. Biochim. Biophys. Acta **525:** 252–264.
39. SLAYMAN, C. L. & D. SANDERS. 1985. Biochem. Soc. Symp. **50:** 11–29.
40. REPHAELI, A., D. E. RICHARDS & S. J. D. KARLISH. 1986. J. Biol. Chem. **261:** 12437–12440.
41. NAKAO, M. & D. C. GADSBY. 1986. Nature **323:** 628–630.
42. HANSEN U.-P., D. GRADMANN, D. SANDERS & C. L. SLAYMAN. 1981. J. Membr. Biol. **63:** 165–190.
43. LAUGER, P. & G. STARK. 1970. Biochim. Bipohys. Acta **211:** 458–466.
44. SLAYMAN, C. L., P. KAMINSKI & D. STETSON. 1989. *In* Biochemistry of Cell Walls and Membranes of Fungi. P. Kuhn & T. Trinci, eds. :295–312. Springer-Verlag, Berlin.
45. ARMORY, A., A. GOFFEAU, D. B. MCINTOSH & P. D. BOYER. 1982. J. Biol. Chem. **257:** 12509–12516.
46. SLAYMAN, C. L., D. SANDERS, J. WARNCKE, G. ZUCKIER & M. BLATT. 1989. Steady-state electrical kinetics of the plasma-membrane proton pump in *Neurospora*. I. Optimization of 2-, 3-, and 4-state models during cyanide-induced, metabolic downshifts. In preparation.
47. MCINTOSH, D. B. & P. D. BOYER. 1983. Biochemistry **22:** 2867–2875.
48. INESI, G., M. KURZMACK, D. KOSK-KOSICA, D. LEWIS, H. SCOFANO & H. GUIMARAES-MOTTA. 1982. Zeit. Naturforsch. **37C:** 685–691.
49. CHAPMAN, J. B., E. A. JOHNSON & J. M. KOOTSEY. 1983. J. Membr. Biol. **74:** 139–153.
50. BORLINGHAUS, R., H.-J. APELL & P. LAUGER. 1987. J. Membr. Biol. **97:** 161–178.
51. FENDLER, K., E. GRELL & E. BAMBERG. 1987. FEBS Lett. **224:** 83–88.
52. FORBUSH, B., III. 1984. Proc. Natl. Acad. Sci. USA **81:** 5310–5314.
53. DIAZ, R. A. & L. D. FALLER. 1989. Biochemistry **28:** 6908–6914.
54. AMORY, A. & A. GOFFEAU. 1982. J. Biol. Chem. **257:** 4723–4730.
55. HACKNEY, D. 1983. J. Biol. Chem. **255:** 5320–5328.
56. ARIKI, M. & P. D. BOYER. 1980. Biochemistry **19:** 2001–2004.
57. DAHMS, A. S. & P. D. BOYER. 1973. J. Biol. Chem. **248:** 3155–3162.
58. KANAZAWA, T. & P. D. BOYER. 1973. J. Biol. Chem. **248:** 3163–3172.
59. DE MEIS, L., A. DE SOUZA OTERO, O. B. MARTINS, E. W. ALVES, G. INESI & R. NAKAMOTO. 1982. J. Biol. Chem. **257:** 4993–4998.
60. HAGER, K. M., S. M. MANDALA, J. W. DAVENPORT, D. W. SPEICHER, E. J. BENZ & C. W. SLAYMAN. 1986. Proc. Natl. Acad. Sci. USA **83:** 7693–7697.
61. ADDISON, R. 1986. J. Biol. Chem. **261:** 14896–14901.
62. SERRANO, R., M. C. KIELLAND-BRANDT & G. R. FINK. 1986. Nature **319:** 689–693.
63. GHISLAIN, M., A. SCHLESSER & A. GOFFEAU. 1987. J. Biol. Chem. **262:** 17549–17555.
64. SHULL, G. E. & J. B. LINGREL. 1986. J. Biol. Chem. **261:** 16788–16791.
65. SHULL, G. E., A. SCHWARTZ & J. B. LINGREL. 1985. Nature **316:** 691–695.
66. MEADE, J. C., J. SHAW, S. LEMASTER, G. GALLAGHER & J. R. STRINGER. 1987. Mol. Cell. Biol. **7:** 3937–3946.
67. HARPER, J. F., T. K. SUROWY & M. R. SUSSMAN. 1989. Proc. Natl. Acad. Sci. USA **86:** 11234–11238.
68. MACLENNAN, D. H., C. J. BRANDL, B. KORCZAK & N. M. GREEN. 1985. Nature **316:** 696–700.

69. HESSE, J. E., L. WIECZOREK, K. ALTENDORF, A. S. REICIN, E. DORUS & W. EPSTEIN. 1984. Proc. Natl. Acad. Sci. USA **81:** 4746–4750.
70. ADDISON, R. & G. A. SCARBOROUGH. 1982. Biol. Chem. **257:** 10421–10426.
71. MANDALA, S. M. & C. W. SLAYMAN. 1988. J. Biol. Chem. **263:** 15122–15128.
72. MANDALA, S. M. & C. W. SLAYMAN. 1989. J. Biol. Chem. **264:** 16276–16281.
73. GOFFEAU, A. & C. W. SLAYMAN. 1981. Biochim. Biophys. Acta **639:** 197–223.
74. NAGLE, J. F. & S. TRISTRAM-NAGLE. 1983. J. Membr. Biol. **74:** 1–14.
75. FREUND, F. 1981. TIBS **6:** 142–145.
76. SCARBOROUGH, G. A. 1985. Microbiol. Rev. **49:** 214–231.

# Chloroplast Thylakoid Membrane-Bound Ca²⁺ Acts in a Gating Mechanism to Regulate Energy-Coupled Proton Fluxes [a]

RICHARD A. DILLEY AND GISELA G. CHIANG

*Department of Biological Sciences*
*Purdue University*
*West Lafayette, Indiana 47907*

All cellular organisms, from primitive unicellular procaryotes to the higher organisms, make ATP via a $H^+$-ATP synthase. The implication, supported by a considerable body of experimental evidence,[1] is that electron transport reactions generate the proton gradients, in or across a membrane diffusion barrier, required to drive ATP formation. Energy coupling is the process wherein $H^+$ flux (downhill energetically) through the $F_0F_1$ complex is coupled to the endergonic reaction ADP + Pi = ATP, whether in oxidative phosphorylation (bacteria and mitochondria) or photosynthetic phosphorylation. Beyond the general statement that the process occurs, few details are known as to how the protons move through the $F_0$ membrane channel subunits or how the protons interact with the $F_1$ subunits in the phosphorylation reaction. Another fascinating question is how different $H^+$-ATPase complexes apparently evolved from common primitive ancestral genes to develop membrane-specific and function-specific $H^+$-ATPases which are known to occur in various organisms with the function not of ATP synthesis, but as $H^+$ transporting ATPases that develop membrane potentials and $H^+$ gradients that are linked to a variety of ion or sugar or amino acid transport processes.[2]

In chloroplast thylakoid membranes the two photochemical reaction centers, with their linked electron and proton transport reaction steps, cooperate to accumulate protons from the external (stroma) phase into the membrane or across the membrane into the lumen. FIGURE 1A shows a "cartoon" model depicting a thylakoid membrane with the redox-linked proton release reactions from the photosystem II (PSII) water oxidizing system and the $H^+$ translocating plastoquinone oxidation by PSI. Shown there is a model of delocalized energy coupling, that is, the protons are shown as being directly deposited into the lumen phase by the redox reactions, from where the $H^+$ efflux through the $CF_0$-$CF_1$ drives ATP formation. The $H^+$ gradient is shown as obligatorily fully delocalized. Considerable evidence has been published in support of the delocalized chemiosmotic energy coupling hypothesis (cf. Ferguson's review[1]). Gräber and colleagues[3,4] have made a systematic study of light-driven, $\Delta\psi$-driven and $\Delta pH$ (in the dark)-driven phosphorylation with thylakoids, and their data fit very well the expectations of the delocalized chemiosmotic hypothesis. Other photophosphorylation results[5,11,12] also were consistent with delocalized energy coupling. However, recent evidence, to be reviewed herein, indicates that this relatively simple delocalized gradient energy coupling model cannot be the sole correct formulation of the molecular events occurring between proton release and ATP formation in thylakoids.

[a] This work was supported in part by grants from the National Science Foundation and the U.S.D.A.

## LOCALIZED AND DELOCALIZED ENERGY COUPLING H⁺ GRADIENTS OCCUR

In the course of testing the chemiosmotic hypothesis, several laboratories working with chloroplast thylakoids demonstrated that the predictions of the delocalized energy coupling model are routinely violated.[6–10,13,14] Considerable uncertainty existed as to why different conclusions were reported, until Horner and Moudrianakis[9] and Sigalat *et al.*[10] suggested that different findings were

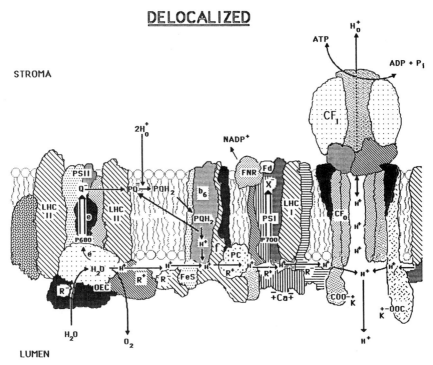

**FIGURE 1A.** A delocalized energy coupling model for chloroplast thylakoid membrane structure depicting the proton gradient freely equilibrating with the aqueous lumen volumen. (See ref. 37 for details of the model.)

caused by the different levels of KCl used in the thylakoid preparations. Our group followed up that lead and found clear evidence that changes in ionic conditions can cause a *reversible switching* of the energy coupling patterns between localized and delocalized responses.[15–17] As will be described, the work led to a model for localized energy coupling expressed in Figure 1B. A Ca²⁺ regulated gating function is suggested to control H⁺ fluxes between localized and delocalized pathways.[17]

The technique that most clearly shows the dual coupling patterns is the effect (the delocalized case), or lack of effect (the localized case), of a permeable buffer such as pyridine or hydroxyethanemorpholine (HEM) on the number of single-turnover light flashes (each flash gives roughly 3 H⁺ per redox chain or about 3.5

## LOCALIZED

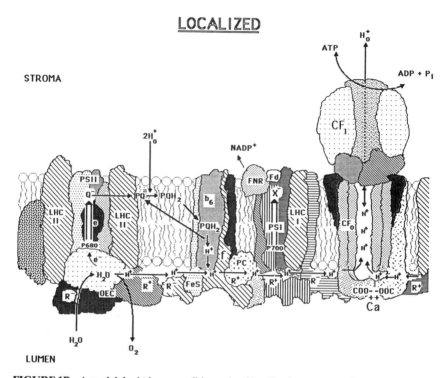

**FIGURE 1B.** A model depicting a possible mode of localized energy coupling. As discussed in the text, $Ca^{2+}$ ions are hypothesized to regulate a proton flux gate between a membrane-localized domain and the lumen.

nmol $H^+$ [mg chl · flash]$^{-1}$) required to pump up the proton gradient to initiate ATP formation. Under conditions commonly used (no electric field contribution), a $\Delta pH$ of 2.3 units is required to energize ATP formation. FIGURE 2 shows the ATP detection assay we routinely use, which is the luminescence of luciferin-luciferase detected in real-time when flashing a thylakoid suspension containing the phosphorylation reaction components plus the luciferin-luciferase components.[18] The figure indicates the expected effect of the electric field component (no valinomycin present, left-hand trace, FIG. 2) in giving the threshold energization of ATP formation after 2–3 flashes, whereas with valinomycin and $K^+$ when only the $\Delta pH$ component is available, more flashes are required to initiate ATP formation. When the $\Delta pH$ builds up in thylakoids experiencing delocalized energy coupling, theory predicts and experiments verify that 5 mM pyridine causes a *further* 10–15 flash extension of the ATP formation onset lag. FIGURE 3 shows exactly that effect of 5 mM pyridine when thylakoids were *stored* in a medium containing 100 mM KCl. Such thylakoid storage will be referred to as the *high salt* condition to distinguish that case from *low salt* storage conditions wherein 200 mM sorbitol or sucrose replaced the 100 mM KCl.[15,16]

The low salt-stored thylakoids did not show the pyridine-dependent extension of the ATP formation onset lag (FIG. 3), clearly inconsistent with the predictions of the delocalized energy coupling model. In agreement with the earlier results with permeable buffers[6,7] and other studies using somewhat different protocols[8,9] (in all those cases the thylakoids were stored in a low salt medium), the thylakoid proton gradient did not interact with the pyridine that we know had equilibrated with the lumen volume before the beginning of the flash train.[16] Various control measurements were done with the low and high salt-stored thylakoids to test for

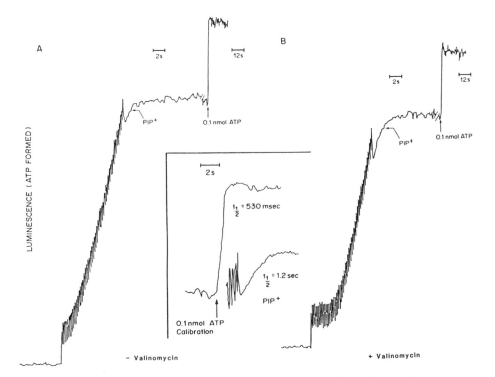

**FIGURE 2.** Single-turnover flash-initiated phosphorylation measured by luciferin-luciferase luminescence. Flashes were delivered at a rate of 5 Hz to thylakoids containing 10 μg of Chl suspended at 10°C in 1 ml of reaction mixture containing 50 mM Tricine-KOH (pH 8.0), 10 mM sorbitol, 3 mM MgCl₂, 2 mM KH₂PO₄, 0.1 mM ADP, 0.1 methylviologen, and 10 μM diadenosine pentaphosphate. DTT, 5 mM, was included to protect critical luciferase sulfhydryls. The vertical spike was the result of a light leak and served as a useful event marker. The flash lag (actual) for the onset of ATP formation was determined by the first detectable rise in luciferin-luciferase luminescence, whereas the nanomole ATP yield per flash was calculated from the linear rise in bioluminescence. The extrapolated lag for the onset of ATP formation was ascertained from where the linear rise in luminescence would intersect a baseline drawn through the initial, nonphosphorylating flashes. Valinomycin was omitted in **A** and 400 nM was included in **B**. The lag for the onset of ATP formation increased from about 3 to 15 flashes after the addition of valinomycin, whereas the ATP yield per flash was about 0.75 nmol ATP/mg Chl, with or without valinomycin. After 50 flashes, phosphorylation continues to yield 2.7 nmol ATP/mg Chl (PIP⁺) in the absence of valinomycin and 5.8 nmol ATP/mg Chl in its presence. PIP⁺ was extrapolated to 200–300 ms after the last flash. **Inset:** Comparison of the kinetics of luminescence due to the addition of standard ATP (*top trace*) to the kinetics of the PIP⁺ ATP yield (*bottom trace*). Just the last five flashes of the flash sequence are shown for the PIP⁺ experiment. The data were taken from a different experiment than that used for FIGURE 1A and B, but similar conditions to those in FIGURE 1B were used. The indicated time scale was the same for both traces. Three separate experiments were performed, all of which gave very similar data. (From Beard and Dilley.[19])

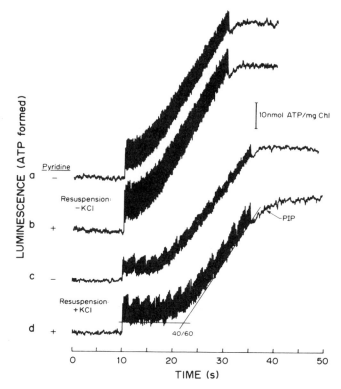

**FIGURE 3.** The effect of pyridine at pH 8.0 on single-turnover flash-initiated phosphory-lation with thylakoids stored in the absence or presence of 100 mM KCl. Thylakoids were washed and resuspended in 5 mM Hepes (pH 7.5), 2 mM $MgCl_2$, 0.5 mg/ml BSA, and 200 mM sucrose (**a,b:** − KCl) or 100 mM KCl in place of the sucrose (**c,d:** +KCl). Flashes (100 in **a,b;** 125 in **c,d**) were delivered at a rate of 5 Hz to thylakoids containing 14 μg chlorophyll suspended at 10° in 1-ml reaction mixture as specified in FIGURE 2B. (See ref. 15 for details.)

different responses of the two thylakoid preparations, but Beard and Dilley[15,16] found no significant differences in thylakoid volume, $H^+$, or $e^-$ transport capac-ity, electric field generation (totally collapsed when valinomycin, $K^+$ were present). Chiang and Dilley (in preparation) found no differences in thylakoid lumen volumes assayed by electron microscopic measurements with and without pyridine and in the dark or in the light with phosphorylating conditions.

A critical point in evaluating the results is whether, in the flash train with the low salt-stored thylakoids, there could have been a deficiency of $K^+$ (relative to the high salt-stored case) to act with the valinomycin to keep the $\Delta\psi$ component suppressed. That is, either $H^+$ ions or Pyr-$H^+$ forming in the lumen could contrib-ute to a positive $\Delta\psi$ if not enough $K^+$ were available in the lumen to collapse the $\Delta\psi$. This was tested by Beard and Dilley[16] and found not to be a problem (i.e., doubling the $K^+$ content in the phosphorylation medium had no effect on the ± pyridine effects with the high and low salt-stored thylakoids).

Two quite different ATP formation experiments support the conclusion that the low salt-stored thylakoids have a localized, and high salt-stored membranes a delocalized proton gradient energy coupling mechanism. One is the effect of the two storage treatments on postillumination phosphorylation (PIP) following the flash train,[15,16] and the other is the effect of lipid-soluble Ca²⁺ chelators given to low salt-stored thylakoids.[17] Pyridine stimulation of postillumination phosphorylation has been used as convincing evidence for delocalized energy coupling.[12] The traditional way to do the PIP experiment is to add ADP and Pi in the dark immediately after the illumination, thus trapping some of the accumulated proton gradient.[12] We refer to this as PIP⁻, the minus sign implying that ADP,Pi were added after the light phase. With the very sensitive luciferin-luciferase ATP assay we use, we can readily detect not only that PIP⁻ mode of ATP formation, but also the smaller PIP⁺ (the plus meaning ADP,Pi were present during the flash train) ATP yield that occurs after the last flash when ADP and Pi were present during the energizing flash train (FIG. 2, right side). Our low and high salt-stored thylakoids routinely give an easily detectable PIP⁺ ATP yield, but only the high salt case showed an increased ATP yield in the presence of pyridine (FIG. 4). The two thylakoid preparations in that experiment, as in most, gave similar efficiencies of ATP formation during the flash train either plus or minus pyridine. Thus, the PIP⁺ ATP yield differences are consistent with the low salt case having the energetic proton gradient develop in a domain not in equilibrium with the lumen, the same conclusion drawn from the ATP onset lag experiments (FIG. 3).

However, both the low and high salt-stored thylakoids show considerable proton delivery to the lumen (shown by similar, large pyridine buffering effects in

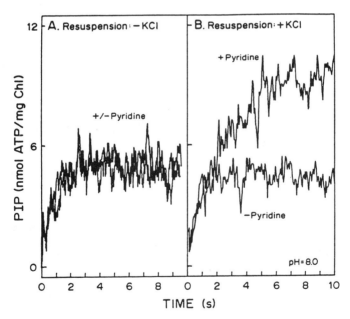

**FIGURE 4.** Effect of pyridine on the luminescence resulting from postillumination phosphorylation with thylakoids resuspended in the absence or presence of KCl at pH 8. The PIP⁺ portions of the luminescent signals (cf. FIG. 2B) are presented: **(A)** thylakoids resuspended in low-salt medium; and **(B)** thylakoids resuspended in high-salt medium. (From Beard and Dilley.[15])

the lumen), *provided* that the flashes were given in the absence of ADP and Pi (cf., TABLE III, ref. 27). That was shown by the direct $H^+$ uptake measurements and by the similar pyridine stimulation of the (traditional) PIP$^-$ ATP yield. TABLE 1 shows the results of such an experiment comparing PIP$^+$ and PIP$^-$ ATP yields using 100 flashes of excitation. Both low or high salt-stored thylakoids gave a significantly greater PIP$^-$ ATP yield if pyridine was present, but similar, and much smaller, PIP$^+$ yields with or without pyridine. TABLE 1 also shows that low salt-stored thylakoids can be induced to show a pyridine-dependent PIP$^+$ ATP yield increase (and a pyridine-dependent ATP onset lag increase), provided the thylakoid storage stages are supplemented with $Ca^{2+}$ chelators EGTA (at pH 5.5 which apparently allows the EGTA to cross the membrane into the lumen) or TMB-8 (trimethoxybenzoic acid 8-(dimethylamino) octylester).[17] The chelator ef-

**TABLE 1.** Effect of Pyridine and $Ca^{2+}$ Chelators on Two Types of Postillumination Phosphorylation (PIP) ATP Yield

| Treatment | Pyridine | Traditional PIP$^-$ ATP Yield[a] (nmol ATP/mg Chl) | PIP$^+$ ATP Yield[b] (nmol ATP/mg Chl) |
|---|---|---|---|
| Low salt-stored thylakoids | — | 20 ± 0.4 | 4.3 ± 0 |
| | +5 mM | 30 ± 0.6 | 4.3 ± 0.4 |
| High salt-stored thylakoids | — | 22 ± 2 | 5.4 ± 0.5 |
| | +5 mM | 32 ± 2 | 9.5 ± 0.4 |
| Low salt-stored[c] thylakoids + 25 $\mu$M TMB-8 | — | — | 3.9 |
| | +5 mM | — | 6.2 |
| Low salt-stored,[c] + 25 $\mu$M TMB-8, + 5 mM CaCl$_2$ | — | — | 3.9 |
| | +5 mM | — | 3.8 |
| Low salt,[c] + 2 mM EGTA at pH 5.5 | — | — | 3.8 |
| | +5 mM | — | 4.2 |

[a] For the traditional (PIP$^-$) ATP yield assay, the ADP and Pi were added after the last flash in a 100-flash sequence. The flash rate was 5 Hz. The values reported are the means ± SE of three determinations (cf. ref. 27 for details).

[b] For the PIP$^+$ ATP yield, the reaction was carried out as specified in the legend of FIGURE 2, and ADP and Pi were present during the flash sequence. The PIP$^+$ ATP yield is that observed in the dark after the last flash (cf. refs. 19 and 27 for details).

[c] See ref. 17 for details of using TMB-8 and EGTA.

fects occurred with little or no deleterious effects on the efficiency of ATP formation during the flash train. (The $Ca^{2+}$ effects will be discussed in more detail below.) It is clear, therefore, that the pyridine effects signaling delocalized energy coupling do not require the KCl treatment *per se,* but must be due to another factor(s) that is sensitive to the KCl treatment, as well as treatments that perturb $Ca^{2+}$. The reversibility of the TMB-8 $Ca^{2+}$ chelator effect on the pyridine-induced PIP$^+$ ATP yield increase by adding 5 mM CaCl$_2$ indicates that the TMB-8 is probably exerting its $H^+$ gradient delocalizing action through its $Ca^{2+}$ chelating action, rather than some other mechanism. In accord with the $Ca^{2+}$ chelator effect on the pyridine response of the PIP$^+$ ATP yield for the low salt-stored thylakoids, the ATP formation onset flash lag was extended about 10 flashes by 5 mM pyridine given to thylakoids treated with the $Ca^{2+}$ chelators.[17]

The simplest interpretation consistent with all the facts is that the low salt-stored thylakoids can keep the energizing proton gradient localized within some membrane phase domain and that the localized proton gradient does not equilibrate fully with the lumen as long as conditions allow ATP formation to occur during the flash train (thus dissipating to some extent the developing proton gradient). Following this hypothesis, basal ($-$ADP,Pi) conditions during the flash train, by not permitting energy-linked dissipation of the H$^+$ gradient through the CF$_1$, are predicted to lead to a greater H$^+$ accumulation in the localized domains, leading to a spillover into the lumen, where the H$^+$ ions can interact with pyridine leading to the increased PIP$^-$ ATP yield. An essential part of this interpretation is that considerable H$^+$ accumulation can occur in membrane domains. There is, in fact, direct evidence for that sort of membrane phase H$^+$ buffering, to be considered next.

## MEMBRANE PHASE SEQUESTERED H$^+$ BUFFERING DOMAINS AND ENERGY COUPLING

If localized proton gradient coupling is a correct view for certain circumstances—and it seems to be—it is necessary to discover how charge pairing or charge movement in the localized domains compensates for the electrical effect of depositing positive protons into the putative membrane domain(s). Recent developments have given quantitative insights into the existence of fixed cationic and anionic groups in sequestered thylakoid domains. First, we should specify what magnitude of the sequestered proton accumulation is required to reach the energization threshold for ATP formation. When the electric field can contribute to energization, ATP formation can begin on the second or third flash[18,19] in dark-adapted thylakoids that have no ATPase activity in the dark. At about 3H$^+$ released per redox chain per flash,[20] only 6–9 protons per redox chain or per CF$_1$ are released by 2–3 flashes, enough to build up the energetic threshold when $\Delta\psi$ can contribute. When $\Delta\psi$ cannot contribute to energization, as when valinomycin, K$^+$ are present, usually about 12–15 flashes are required to initiate ATP formation.[7,19] This means that about 50–60 nmol H$^+$ (mg Chl)$^{-1}$ are generated, an enormous charge effect that requires compensating charge movement. Hangarter and Ort[21] have shown that the first 60 nmol H$^+$ (mg Chl)$^{-1}$ or so of proton accumulation into dark-adapted thylakoids occurs with a very slow H$^+$ efflux. The accumulation of that amount of H$^+$ ions can be electrically balanced through either cation out-exchange or anion cotransport. It has been known since 1965 that the large scale proton uptake (300–1,000 nmol H$^+$ (mg Chl)$^{-1}$) is generally balanced by cation (Mg$^{2+}$,K$^+$) efflux,[22] although anion uptake into thylakoids is possible.[23] Thylakoid carboxyl groups provide the fixed negative charge sites to buffer the accumulated protons and are charged-paired with Mg$^{2+}$ and K$^+$ in the resting state.[22,23,28,29] No data are yet available on what compensating charge movements actually occur during the attainment of the 2–3 flash or the 12–15 flash threshold energization. But there is recent information on fixed charge groups in the membrane domains and how they relate to membrane energization.

There are two sequestered membrane phase H$^+$ buffering arrays, an amine array of pK$_a$ = 7.5 of magnitude near 30–40 nmol per milligram Chl,[24-26] and a much larger magnitude, lower pKa carboxylic group array, the evidence for which is clear even if it is less direct than that for the amine array.[27-29]

*Domain Protons in the Amine Array are on the Main Path of H$^+$ Flux Driving ATP Formation.* Protons associated with pK$_a$ = 7.5 amine groups are not at sufficient energy for a $\Delta$pH alone to drive ATP formation unless the external pH were close to pH 10.5. However, a $\Delta\psi$ could electrophorese domain protons or lumen protons near pH 7 to 8 into the $CF_0$-$CF_1$ and effectively concentrate them to a $\Delta$pH energetically competent to energize ATP formation.[33] This is probably what happens in ATP formation that begins after two or three flashes in the absence of valinomycin, $K^+$. Interesting results were found in comparing the threshold ATP formation energization lag dependence on the protonation status of the amine array with or without the $K^+$ ionophore present. The buried amine array $H^+$ buffering groups are normally protonated and metastable, held out of equilibration with pH 8.5 aqueous phases (either the external or the lumen phases).[25,30] The domain protons can be released into pH 8.5 suspending buffer by a low concentration of Cl-CCP, whereupon added bovine serum albumin (BSA) will absorb the Cl-CCP and the thylakoids will be capable of good energy coupling.[18] Using this reversible uncoupling technique, it was shown that depleting the amine buffering array in the pH 8.5 suspending phase prior to a flash phosphorylation regime caused an extension of the ATP energization lag of about 10–12 flashes.[18] Given that there are initially about 30–40 nmol $H^+$ (mg Chl)$^{-1}$ in the protonated sequestered amine domain which are lost upon Cl-CCP addition and that each flash delivers about 3–3.5 nmol $H^+$ (mg Chl)$^{-1}$, it is clear that 10–12 flashes could replenish about 30–40 nmol $H^+$ (mg Chl)$^{-1}$ to the buried domain. If those amine groups in the $-NH_2$ form were directly in the domain pathway of $H^+$ ions traversing from the redox release sites ($H_2O$ and $PQH_2$ oxidation), then it is possible that the $NH_2$ groups would bind 30–40 nmol $H^+$ (mg Chl)$^{-1}$ as the energetic $H^+$ gradient builds up to the threshold $\Delta$pH of = 2.3 units, hence accounting for an increase of 10–12 flashes in the onset lag. A second light flash train given to the thylakoids (treated as just described with the Cl-CCP reversible uncoupling treatment) gave a flash lag about 10 flashes shorter than that seen on the first flash train, consistent with the notion that the proton gradient, built up in the first flash train, reprotonated the amine groups so that on the second flash train the protons released by the redox reactions did not get absorbed by the 30–40 nmol ($-NH_2$ groups) per milligram chl, and thus the critical $\Delta$pH for energization built up with fewer flashes.[18] This interpretation, while reasonable, did not rule out the possibility that the amine array may be not on the "main path" of $H^+$ ions to the $CF_0$-$CF_1$, but on a side pathway, a type of dead-end buffering array that can compete for protons with the main pathway when only a $\Delta$pH builds up to drive ATP formation. One way to test this is to allow the $\Delta\psi$ component to contribute to the protonmotive force, which we know results in ATP formation onset in as little as 2–3 flashes. In that circumstance, and following the hypothesis that the energy coupling system only uses delocalized (lumenal) protons in the energization, with the corollary that protons from the redox reactions must traverse the lumen on their way to the $CF_0$-$CF_1$, then it follows that a $\Delta\psi$ component should be able to drive the lumenal $H^+$ ions, assumed to be near the lumenal side of the $CF_0$, through the $CF_0$-$CF_1$ regardless of the protonation state of the buried domain protons. Such experiments were done with low salt-stored thylakoids, which by other criteria show localized energy coupling, before and after depleting the buried domain metastable $H^+$ pool. FIGURE 5 (from Theg et al.[31]) shows that with a $\Delta\psi$ contribution, there is still a near 10-flash longer ATP onset lag after the reversible Cl-CCP domain depletion treatment. Those results are clearly contradictory to the hypothesis that lumenal protons are obligatorily the ones having the only (and first) access to the $CF_0$ $H^+$ channel. The data are consistent with the concept that, in low salt-stored thylakoids, protons in the membrane domains are on the main path to the $CF_0$ and are used in lieu of lumenal protons.

This interpretation was supported by the observation that when high salt-stored thylakoids, which show delocalized energy coupling, were used in the foregoing reversible uncoupler experiment (no valinomycin, K⁺), the data indicated that lumenal protons *were* the first to be driven through the $CF_0$ by the $\Delta\psi$ pulses.[31] TABLE 2 shows that for high salt-stored membranes, the ATP formation onset lag was the same whether or not the reversible Cl-CCP proton depletion treatment was given. A complicating factor for interpreting these data is that Allnutt *et al.*[32] showed that the metastable domain protons are present in the dark

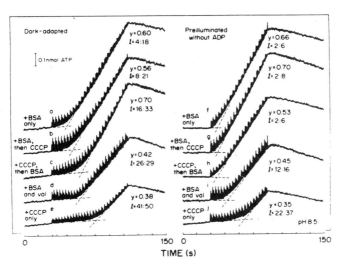

**FIGURE 5.** The effect of reversible uncoupling on the ATP onset flash lag in alkaline media. The assay medium (pH 8.5) was as described in ref. 31, "Materials and Methods," except that ADP was initially omitted. Unless an order of addition is indicated, the listed compounds were present at the time of chloroplast addition ($t = 0$); otherwise, the second compound was added at $t = 30$ seconds. The protocol was as follows: $t = 0$, chloroplasts added to assay medium containing indicated compounds; $t = 0.5$ minutes, where indicated, the second compound was added; $t = 3.5$ minutes, ADP added; $t = 4$ minutes, luciferin-luciferase added; and $t = 5$ minutes, the sequence of 90 flashes at 1 Hz started. For the traces on the right (f–j), 10 seconds of red light were given at $t = 1$ minute (val, valinomycin); $y =$ ATP flash yield in nanomoles of ATP formed per milligram of Chl flash; $l =$ ATP onset flash lag (first number corresponds to first detectable rise of signal from baseline, and second number to the point of crossing of extrapolated steady-state flash yield to baseline). (From Theg *et al.*[31])

when either low or high salt-stored thylakoids were used in the H⁺ depletion experiments. Our present interpretation is that the high salt storage must not disrupt the membrane domains in the dark sufficiently to allow the domain H⁺ to equilibrate with the pH 8.5 aqueous phases, but upon energization with flash-driven redox turnovers, some membrane change occurs, such that the protons released must gain access to and equilibrate with the lumen in times very short compared to the time required to build up the energetic proton gradient.[32]

That experiment stands as a separate, but supportive, data set to the permeable buffer experiments, and together we feel that they give broadly based support

TABLE 2. Effect of Sequestered Domain $H^+$ Depletion Treatment on ATP Onset Lag Parameters in 100 mM KCl-Treated (High-Salt) Compared to Low-Salt-Treated Thylakoids in the Absence of Valinomycin[a]

| Treatments (Additions before Flashes) | Number of Flashes to ATP Formation Onset | |
|---|---|---|
| | 1st Cycle | 2nd Cycle |
| A. High-salt storage | | |
| 1. BSA only | 7/20 | |
| 2. CCCP only | 19/41 | |
| 3. BSA, then CCCP | 13/29 ± 1 | 10/22 ± 1 |
| 4. CCCP, then BSA | 13/31 ± 2 | 13/27 ± 1 |
| 5. BSA, then CCCP[b] | 13/28 ± 1 | 10/27 |
| 6. CCCP, then BSA[b] | 13/33 ± 1 | 10/25 ± 3 |
| 7. BSA, valinomycin[c] | 39/49 ± 1 | 36/43 ± 2 |
| B. Low-salt storage | | |
| 1. BSA only | 8/20 | |
| 2. CCCP only | 27/44 | |
| 3. BSA, then CCCP | 11/25 | |
| 4. CCCP, then BSA | 19/37 ± 1 | |
| 5. BSA, then CCCP[a] | 14/26 ± 1 | 12/18 ± 2 |
| 6. CCCP, then BSA[a] | 22/38 ± 4 | 11/16 ± 2 |

[a] The effect of Cl-CCP and BSA on the ATP onset flash lag. Conditions are described as in FIGURE 5 (without valinomycin except where noted). See ref. 31 for details.

[b] Experiments from a different day.

[c] 400 nM valinomycin added.

to the concept of there being two pathways for energy-coupled proton diffusion into the $CF_0$-$CF_1$, one path utilizing a localized domain that is not in equilibration with the aqueous lumen and the other pathway utilizing proton diffusion through the lumen. These concepts are embodied in the localized and delocalized energy coupling models shown in FIGURE 1A and B (cf. refs. 9 and 13 for related comments on dual energy coupling pathways). A different line of experiments, implicating $Ca^{2+}$ ions as the factor reversibly regulating the observation of either the localized or delocalized energy coupling, provides a compelling argument that the two modes of $H^+$ flux coupling have physiologic significance in chloroplast energy coupling.

## CA$^{2+}$ IONS REGULATE LOCALIZED-DELOCALIZED ENERGY COUPLING RESPONSES

Upon observing that the 100 mM KCl storage effects could be reversed by centrifuging the thylakoids and resuspending the pellet in the low salt buffer,[16] it was concluded that the KCl effects were not caused, for example, by salt extraction of a thylakoid protein or some other permanent membrane perturbation. Rather, the salt effects appeared to be causing a type of reversible switching effect. Subsequent experiments[17] showed that 1 mM $CaCl_2$ or 30 mM $MgCl_2$ added to the 100 mM KCl blocked the delocalizing tendency of the KCl (FIG. 6). Moreover, either EGTA (at pH 5.5) or 25 $\mu$M TMB-8 (3,4,5 trimethoxybenzoic acid 8-(dimethylamino) octylester), a lipid-soluble $Ca^{2+}$ chelator, given to the low salt-stored thylakoids (only present in the storage medium, being diluted $\simeq$ 150-fold

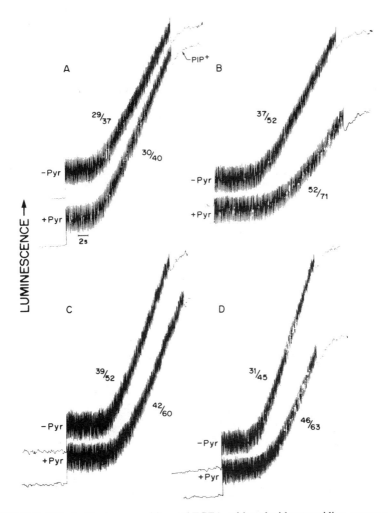

**FIGURE 6.** Effect of ionic composition and EGTA, with and without pyridine, on energization lag for ATP formation. Thylakoid storage treatment, phosphorylation medium, and luciferin-luciferase ATP assay were as described in ref. 17. Energization lag parameters are listed by each trace. In assay **A–D**, the *top trace* was from a sample without pyridine and the *bottom trace* was from a sample with 5 mM pyridine present, 3.5 minutes before beginning the flashes. Thylakoid samples were *stored* in the following media prior to dilution (approximately 5 μl added to 800 μl of reaction medium) into the phosphorylation medium (identical for all samples): **(A)** Low-salt (control): 200 mM sucrose, 5 mM Hepes-KOH pH 7.5, 2 mM MgCl₂, and 0.5 mg ml⁻¹ bovine serum albumin; **(B)** High-salt: 100 mM KCl and 30 mM sucrose replaced the 200 mM sucrose used in part A. **(C)** High-salt + 1 mM CaCl₂: storage medium was as in B plus 1 mM CaCl₂; **(D)** High salt + 30 mM MgCl₂ + 2 mM EGTA: the medium was as in B plus 30 mM MgCl₂ and 2 mM EGTA. In **A**, the *bottom trace* identifies the postillumination phosphorylation ATP yield (PIP⁺). (From Chiang and Dilley.[17])

for the ATP formation assay) caused them to respond to 5 mM pyridine in the ATP onset lag and postillumination ATP yield assays the same as high salt-stored membranes (without such treatments); that is, the $Ca^{2+}$ chelators caused a delocalized $H^+$ gradient coupling response to occur without the necessity for high KCl treatment. Adding 1–2 mM $CaCl_2$ blocked those chelator effects with low salt-stored thylakoids. We were thus led to the hypothesis, embodied in the $Ca^{2+}$ gating structures drawn in FIGURE 1B, that $Ca^{2+}$ ions interacting with lumenal-facing protein carboxyl groups function as a gating structure in regulating $H^+$ flux into the lumen from the membrane domains.

The $Ca^{2+}$ gating hypothesis suggested testing the effects of some of the drugs known to perturb $Ca^{2+}$ transport and/or $Ca^{2+}$ binding in other systems. A $Ca^{2+}$ channel blocker, verapamil, had no effect on any of the phosphorylation parameters tested (data not shown), but antagonists that block $Ca^{2+}$ binding to calmodulin and other calcium binding sites and block the action of the $Ca^{2+}$-calmodulin complex, showed interesting effects. Calcium binding antagonists[34,35] trifluoperazine (TF) and chlorpromazine (CPZ) were given to thylakoids only in the storage stage, not in the ATP formation assay step.

TABLE 3 shows that 5 $\mu$M CPZ in the storage phase had no effect on the ATP formation onset parameters ($\pm$ pyridine) and no effect on the ATP yield per flash (i.e., there was no uncoupling action of the CPZ pretreatment; CPZ at 5 $\mu$M in a phosphorylation assay is a known uncoupler, but the dilution to near 0.05 $\mu$M rendered it ineffective as an uncoupler). Interestingly, CPZ *blocked* the pyridine stimulation of the $PIP^-$ ATP yield in the postillumination ATP formation where ADP, Pi were added *after* the flash train. The CPZ had no effect on the $PIP^+$ ATP yield, where ADP, Pi were present during the flash train. With low salt-stored thylakoids, the model based on past results posits that localized proton fluxes drive ATP formation which occurs in the flash train and in the $PIP^+$ phase, but *lumenal* protons, such as those associated with protonated pyridine in the lumen, can be involved in the $PIP^-$ mode of postillumination ATP formation.[27] The TABLE 3 data suggest that for low salt-stored thylakoids CPZ blocks the component of lumenal protons derived from pyridine buffering from gaining access to the $CF_0$-$CF_1$. However, in the low salt-stored case CPZ caused an increase in the $PIP^-$ ATP yield from 20 (control) to 32 (+CPZ). That increase is puzzling and seems to suggest that just as CPZ blocks lumenal protons associated with pyridine buffering to contribute to $PIP^-$ ATP yield, so it may block domain protons from going to the lumen. That is, the CPZ caused a greater accumulation of protons into the membrane phase buffering groups. Such effects were not seen with the high salt-stored thylakoids. That interpretation requires that for the CPZ-treated thylakoids, the near 30 nmol ATP (mg Chl)$^{-1}$ $PIP^-$ yield must be driven by protons buffered by membrane-phase buffering groups during the flash train. We have established that the membrane phase buffering capacity, independent of buffering in the lumen,[27] is near 150 nmol $H^+$ (mg Chl)$^{-1}$, so if 3 $H^+$ are needed to efflux through the $CF_0$-$CF_1$ per ATP formed, then an ATP yield of 30 nmol (mg Chl)$^{-1}$ could reasonably be expected to result from the efflux of protons buffered in the membrane phase at the acidic conditions needed to protonate carboxyl groups.[27,29]

A different test for whether CPZ pretreatment blocks $H^+$ entry into the lumen is needed and is available by measuring total $H^+$ uptake into the membranes and lumen with and without pyridine. Pyridine buffering action in the lumen is observed by an additional $H^+$ uptake, easily observable with the pH electrode method or with the cresol red (external) pH indicator dye.[27] TABLE 4 indicates that with low salt-stored membranes 5 $\mu$M CPZ in the storage stage largely blocked, in a subsequent H$^+$ uptake assay, the expected pyridine-dependent increase in H$^+$ uptake during a flash train. However, this was the case only for low salt-stored thylakoids, for in the high salt-storage case, the same extent of pyridine stimulation of H$^+$ uptake occurred with or without CPZ.

**TABLE 3.** Effect of Chlorpromazine on Phosphorylation Parameters at pH 7.5 Using Low-Salt-Stored Thylakoids

| Treatments | Pyridine | Energization Lag: Number of Flashes | Yield/Flash (nmol ATP/mg Chl) | nmol ATP/mg Chl | |
|---|---|---|---|---|---|
| | | | | PIP$^+$ | PIP$^-$ |
| A. Low salt | − | 45/51 | 0.52 | 4.0 | 20 ± 0.4 |
| | + | 49/58 | 0.44 | 4.0 | 30 ± 0.6 |
| Low salt plus | − | 47/55 | 0.52 | 4.3 | 32 ± 2 |
| 5 μM CPZ | + | 50/61 | 0.50 | 4.7 | 28 ± 2 |
| B. High salt | − | . . . | . . . | 3.6 | 22 ± 2 |
| | + | . . . | . . . | 5.3 | 32 ± 2 |
| High salt plus | − | . . . | . . . | 3.6 | 21 ± 1 |
| 5 μM CPZ | + | . . . | . . . | 5.4 | 32 ± 2 |

A. Thylakoids were washed and resuspended in a low-salt medium containing 2 mM MgCl$_2$, 0.5 mg/ml BSA, and 200 mM sucrose or in the same medium with 5 μM chlorpromazine (CPZ). Flashes (100) were delivered at 5 Hz to thylakoids containing 15 μg Chl suspended at 10°C in 800 μl reaction mixture consisting of 50 mM Tricine-KOH pH 7.5, 10 mM sorbitol, 3 mM MgCl$_2$, 1 mM KH$_2$PO$_4$, 5 mM DTT, 0.1 mM methylviologen, 400 mM valinomycin, 5 μM diadenosine pentaphosphate, and 10 μl luciferin-luciferase prepared from the LKB ATP assay kit. 0.1 mM ADP and 1 mM KH$_2$PO$_4$ were included in the foregoing reaction mixture before flash excitation to ascertain the energization lag, yield/flash, and PIP$^+$. The energization lag was determined by the number of flashes required before the first detectable rise in luminescence (number given before the slash) and by back extrapolation of the steady-state rise in luminescence to the baseline (number after the slash).

The standard deviation for the onset lag was not more than ± 2, for the yield/flash ± 0.08, for PIP$^+$ ± 0.9, and for PIP$^-$ ± 2.0.

PIP$^+$ = postillumination phosphorylation occurring after a flash train having ADP, Pi present during the flash excitation.

PIP$^-$ = postillumination phosphorylation in the traditional mode, with ADP, Pi added after the last flash.

B. Thylakoids were washed and resuspended in a high-salt medium in which 30 mM sucrose and 100 mM KCl replaced 200 mM sucrose. Either chlorpromazine (CPZ) or trifluoperazine (TF) was added to the foregoing resuspension mixtures as treatments during thylakoid storage, prior to dilution into the assay medium. PIP$^+$ and PIP$^-$ ATP yields were determined as described in ref. 27.

In PIP$^-$ experiments 100 flashes were delivered at 5 Hz.

This is a fascinating, but not unexpected finding when considered in light of other work with CPZ, which shows that CPZ only binds to the Ca$^{2+}$-bound form of calmodulin.[35] Our working hypothesis, using the Ca$^{2+}$-calmodulin analogy, is that Ca$^{2+}$ is bound to the H$^+$ flux gating structure in the low salt-stored thylakoids, but the Ca$^{2+}$ is displaced from that site in the high salt-storage case; hence CPZ may not bind to that site. That is not to say a calmodulin protein is the Ca$^{2+}$ binding site, for CPZ can interact with Ca$^{2+}$-noncalmodulin protein complexes.[42]

**TABLE 4.** Effect of Chlorpromazine on Flash-Induced Proton Uptake in the Absence and Presence of Pyridine at pH 7.5 under Basal Conditions (no ADP or Pi)

| Treatment | nmol $H^+$ (mg Chl)$^{-1}$ | | Difference +Pyr-(−)Pyr |
|---|---|---|---|
| | −Pyr | +Pyr | |
| Low salt | 211 ± 4 | 338 ± 15 | 127 |
| Low salt + 5 μM CPZ | 226 ± 19 | 258 ± 9 | 32 |
| Low salt + 1 mM CaCl$_2$ | 191 ± 11 | 303 ± 18 | 112 |
| Low salt + 5 μM CPZ + 1 mM CaCl$_2$ | 183 ± 11 | 227 ± 3 | 44 |
| High salt | 192 ± 15 | 320 ± 14 | 128 |
| High salt +5 μM CPZ | 187 ± 10 | 310 ± 10 | 123 |

Thylakoids were washed and resuspended in either a low-salt or a high-salt medium described in previous tables. CaCl$_2$ and/or chlorpromazine was added as indicated. 100 flashes were delivered at 5 Hz to thylakoids diluted in 2 ml (20 μg/ml) of reaction mixture consisting of 50 mM sorbitol, 0.3 mM Tricine pH 7.5, 0.15 mM KCl, 3 mM MgCl$_2$, 0.1 mM MV, and 400 nM valinomycin in the presence and absence of 5 mM pyridine. The reaction solution, thermostatted at 10°C, was adjusted to pH 7.5 prior to each assay, and continuous recordings of pH were obtained using a strip chart recorder connected to a Corning Model 12 pH meter. (See ref. 41 for details.)

Calmodulin has not been reported as occurring in thylakoid membranes, but it may occur in the chloroplast stroma or outer envelope,[34,42] although further work remains to be done to be certain of these points.

## THYLAKOID PROTEINS ASSOCIATED WITH CPZ BINDING

The CPZ effects just reported are consistent with there being a calcium binding site involved in the $H^+$ flux gating action also just discussed. This stimulated a search for thylakoid proteins that bind CPZ in the $Ca^{2+}$-occupied but not in the $Ca^{2+}$-free form.

CPZ can be UV-activated to a reactive species that forms covalent adducts with calmodulin but only when $Ca^{2+}$ is bound to the protein.[34] We UV-exposed low and high salt-stored thylakoids with 5 μM [$^3$H] CPZ present, followed by SDS-PAGE analysis of the protein labeling. FIGURE 7 shows that more CPZ labeling occurred in the low salt-stored (dashed line) than in the high salt-stored (solid line) membranes and that the main labeled bands were in the region if 8–14 kDa. This was suggestive of CF$_0$ labeling, so a butanol extraction, ether precipitation procedure was followed to isolate the 8-kDa CF$_0$ subunit in high purity.[36] That procedure yielded a highly enriched 8-kDa band (with a large amount of 6-kDa protein as well) that was heavily labeled with [$^3$H] CPZ (FIG. 8). The 8-kDa CF$_0$ subunit reacts avidly with DCCD[36] (dicyclohexylcarbodiimide), so thylakoids were double labeled with [$^{14}$C] DCCD and [$^3$H] CPZ, followed by SDS-PAGE analysis. FIGURE 9 shows that both labels were concentrated in the 8-kDa region, leading us to conclude that the 8-kDa CF$_0$ protein is likely to be involved in the $Ca^{2+}$ binding events that regulate $H^+$ flux gating between thylakoid localized domains and the lumen. The cartoon representation in FIGURE 1A and B is one way of modeling this concept. One or more of the thylakoid proteins contributing

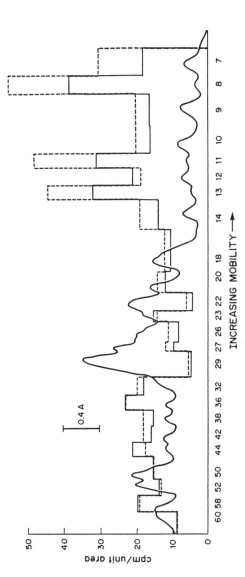

**FIGURE 7.** Coomassie stain band pattern (——) of an SDS-PAGE gel of low salt-stored thylakoids, after photoaffinity labeling with [³H]chlorpromazine. Thylakoids from low or high salt-stored conditions were suspended in either of those media at 25 g Chl ml⁻¹ with 5 M [³H]chlorpromazine. Exposure to a Mineralit UV source was for 20 minutes at 4°C. Gels were run on the SDS solubilized thylakoid proteins with the standard techniques. Radioactivity of gel slices, expressed as CPM per unit area of stained band; —— data from high salt-stored thylakoids, ---- low salt-stored sample. Count data are averages from these gel lanes. (From Chiang and Dilley.[41])

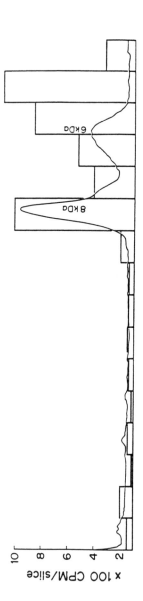

**FIGURE 8.** Coomassie stained SDS-PAGE gel of an 8-kDa $CF_0$ preparation from photoaffinity-labeled thylakoids. Thylakoids (33 g Chl ml$^{-1}$) were photoaffinity labeled with 5 M [$^3$H]chlorpromazine for 20 minutes in UV light, followed by butanol extraction and ether precipitation to obtain the 8-kDa $CF_0$ protein. The main band is at 8 kDa. Radioactivity was counted in 5 mM gel slices and is given by the bar graph. (See ref. 41 for details.)

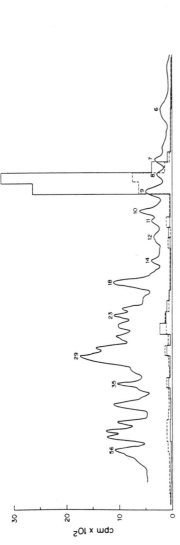

**FIGURE 9.** Optical density profile and dual labeling pattern for [¹⁴C]DCCD and [³H]chlorpromazine labeled low salt-stored whole thylakoid proteins. Low salt-stored chloroplast membranes (8 mg chlorophyll) were irradiated for 1 hour at 3°C with 254 nm UV light in the presence of 6 μM (22.3 Gi/mmol) [³H]chlorpromazine (---) in the modification media containing 0.2 M sucrose, 5 mM Hepes, pH 7.5, 2 mM MgCl₂, and 0.5 mM CaCl₂. The chloroplasts were then washed once and resuspended in the same modification media as above with 500 nmol (54.8 mCi/mmol) [¹⁴C]DCCD (——). After 1-hour incubation at 25°C the chloroplasts were washed and resuspended in 1 ml cold water. Samples of the whole thylakoid proteins were delipidated by washing once in 95% acetone prior to loading approximately 150 μg protein/lane on 15% SDS-polyacrylamide slab gels. Gel lanes were sliced in 2-mm slices, digested by 5 hours of incubation at 80°C in 0.6 ml 30% (v/v) H₂O₂, and counted after the addition of 10 ml tritosol scintillation cocktail. (See ref. 41 for details.)

$-COO^-$ groups to the putative $Ca^{2+}$ binding and $H^+$ flux gating site may be $CF_0$ components. The CPZ labeling of polypeptides in the 13–15 kDa region shown in FIGURE 7 may indicate that one of the other $CF_0$ subunits known to occur in the 13–15 kDa range could be involved, but that remains to be tested.

## PHYSIOLOGIC IMPLICATIONS OF $CA^{2+}$-GATED $H^+$ FLUXES

Inquiry into the possible physiologic ramifications of localized or delocalized $H^+$ gradient energy coupling leads to some interesting possibilities, some of which were discussed in a recent review.[37] The main points that suggest why localized coupling might be advantageous to plants are:

1. Localized gradients can be built up to the threshold protonmotive force more quickly than can a delocalized gradient owing to the lesser buffer capacity of the localized domains.[16,27] In natural environments where transient light flecks are a frequent occurrence and/or the main light source, this could be advantageous. Our estimate of the relative buffer capacities of the localized and delocalized pathways, while somewhat variable,[16,27] indicates about a factor of two more buffering capacity by the lumen, which is reflected in about a twofold longer ATP formation onset lag with high salt-stored thylakoids (TABLE III of ref. 16).

2. Localized energy coupling may avoid harmful osmotic swelling which can occur following salt uptake into the lumen driven by putative $H^+_{in} = K^+_{out} (Na^+)$ exchange. It is known that thylakoids can experience large amplitude salt-dependent swelling in the light, probably driven by the lumenal acidic condition through $H^+_{in} = K^+_{out} (Na^+)$ exchange, with $Cl^-$ and $H_2O$ uptake following the $K^+ (Na^+)$ uptake (cf. discussion in ref. 37). Interestingly, the light-dependent swelling does not always occur,[38] being regulated perhaps by the release of C18:3 fatty acids and/or other factors yet to be elucidated.[39] In any event, by maintaining localized gradient energy coupling, thylakoids could even more easily avoid the stressful effects of lumenal swelling.

3. Related to this, we found that intact chloroplasts,[43] having the outer envelope present and retaining the complex and ion-rich electrolyte phase of proteins, nucleic acids, and metabolites in the stromal contents, responded to high KCl incubation just as isolated thylakoids, that is, they were induced by the 100 mM KCl storage to respond to 5 mM pyridine in the ATP onset lag experiment as though they had a delocalized gradient coupling mode. The intact chloroplasts were salt-treated while intact and osmotically burst in the phosphorylation reaction cuvette just before the assay. However, with the normal levels of near 30 mM $Mg^{2+}$ and near 15 mM $Ca^{2+}$ reported to be in the stromal phase,[40] we predict that the intact organelles would tend to show localized coupling responses because the high divalent cation content will allow for sufficient $Ca^{2+}$ to be available to keep the $CF_0$ gating site in the $Ca^{2+}$ (closed gate) form.

4. Under conditions of limiting the $CF_1$ energy coupling reactions (ATP formation), the build-up of the protonmotive force in the domains and in the vicinity of $CF_0$ gate could protonate the putative carboxyl groups of the gate, allowing the $Ca^{2+}$ to drift away and the gate to open, letting protons flow into the lumen. The additional carboxyl groups of the thylakoid proteins facing the lumen are suggested to buffer the pH in the lumen, providing a reservoir of acidic-buffered protons to be used for driving ATP formation in subsequent moments if coupling conditions at the $CF_1$ become operative. In this way the energy of proton pumping can be transiently stored as a $H^+$ gradient in the lumen carboxylic buffering array

as well as in the membrane domains rather than have the protons flow out of the membrane in a wasteful efflux.

The foregoing speculations are just that and are not final answers. It is hoped that the new developments surrounding the suggested $Ca^{2+}$ gating of $H^+$ fluxes will stimulate others to probe into what appears to be a very fascinating aspect of energy coupling.

## ACKNOWLEDGMENT

The excellent help of Janet Hollister in manuscript preparation is gratefully noted.

## REFERENCES

1. FERGUSON, S. J. 1985. Fully delocalized chemiosmotic or localized proton flow pathways in energy coupling? A scrutiny of experimental evidence. Biochim. Biophys. Acta **811:** 47–95.
2. NELSON, N. 1988. Structure, function and evolution of proton ATPases. Plant Physiol. **86:** 1–3.
3. GRABER, P. 1982. Phosphorylation in chloroplasts: ATP synthesis driven by $\Delta\psi$ and $\Delta$pH of artificial or light-generated origin. Curr. Top. Membr. & Transport **16:** 215–245.
4. JUNESCH, U. & GRABER, P. 1987. Influence of the redox state and the activation of the chloroplast ATP synthase on proton-transport-coupled ATP synthesis/hydrolysis. Biochim. Biophys. Acta **893:** 275–288.
5. JUNGE, W. 1987. Complete tracking of transient proton flow through active chloroplast ATP synthase. Proc. Natl. Acad. Sci. (USA) **84:** 7084–7088.
6. ORT, D. R., DILLEY, R. A. & GOOD, N. E. 1976. Photophosphorylation as a function of illumination time. II. Effects of permeant buffers. Biochim. Biophys. Acta **449:** 108–124.
7. GRAAN, T., FORES, S. & ORT, D. R. 1981. The nature of ATP formation associated with single turnovers of the electron transport carriers in chloroplasts. *In* Energy Coupling in Photosynthesis. B. R. Selman & S. Selman-Reiner, eds. Vol. **20:** 25–34. Elsevier-North Holland, New York. *Developments in Biochemistry.*
8. HORNER, R. D. & E. N. MOUDRIANAKIS. 1983. The effect of permeant buffers on initial ATP synthesis by chloroplasts using rapid mix-quench techniques. J. Biol. Chem. **258:** 11643–11647.
9. HORNER, R. D. & E. N. MOUDRIANAKIS. 1986. Effect of permeant buffers on the initiation of photosynchronous phosphorylation and postillumination phosphorylation in chloroplasts. J. Biol. Chem. **261:** 13408–13414.
10. SIGALAT, C., F. HARAUX, F. DE KOUCHKOVSKY, S. P. N. HUNG, & Y. DE KOUCHKOVSKY. 1985. Adjustable microchemiosmotic character of the proton gradient generated by systems I and II for photosynthetic phosphorylation in thylakoids. Biochim. Biophys. Acta **809:** 403–413.
11. DAVENPORT, J. W. & R. F. McCARTY. 1980. The onset of photophorphorylation correlates with the rise in transmembrane electrochemical proton gradients. Biochim. Biophys. Acta **589:** 353–357.
12. VINKLER, C., M. AVRON & P. D. BOYER. 1980. Effects of permeant buffers on the initial time course of photophosphorylation and post-illumination phosphorylation. J. Biol. Chem. **255:** 2263–2266.
13. PICK, U., M. WEISS & H. ROTTENBERG. 1987. Anomalous uncoupling of photophosphorylation by palmitic acid and by gramicidin D. Biochemistry **26:** 8295–8302.
14. OPANASENKO, V. K., T. P. RED'KO, V. P. KUZ'MINA & L. E. YAGUZHINSKY. 1985.

The effect of gramicidin on ATP synthesis in pea chloroplasts: Two modes of phosphorylation. FEBS Lett. **187:** 257–269.

15. BEARD, W. A. & R. A. DILLEY. 1986. A shift in chloroplast energy coupling by KCl from localized to bulk phase delocalized proton gradients. FEBS Lett. **201:** 57–62.

16. BEARD, W. A. & R. A. DILLEY. 1988. ATP formation onset lag and postillumination phosphorylation initiated with single-turnover flashes. III. Characterization of the ATP formation onset lag and postillumination phosphorylation for thylakoids exhibiting localized or bulk-phase delocalized energy coupling. J. Bioenerg. Biomembr. **20:** 129–154.

17. CHIANG, G. & R. A. DILLEY. 1987. Evidence for $Ca^{2+}$-gated proton fluxes in chloroplast thylakoid membranes: $Ca^{2+}$ controls a localized to delocalized proton gradient switch. Biochemistry **26:** 4911–4916.

18. DILLEY, R. A. & U. SCHREIBER. 1984. Correlation between membrane-localized protons and flash-driven ATP formation in chloroplast thylakoids. J. Bioenerg. Biomembr. **16:** 173–193.

19. BEARD, W. A. & R. A. DILLEY. 1988. ATP formation onset lag and postillumination phosphorylation initiated with single-turnover flashes. I. An assay using luciferin-luciferase luminescence. J. Bioenerg. Biomembr. **20:** 85–106.

20. GRAAN, T. & D. R. ORT. 1982. Photophosphorylation associated with synchronous turnovers of the electron transport carriers in chloroplasts. Biochim. Biophys. Acta **682:** 395–403.

21. HANGARTER, R. & D. R. ORT. The relationship between light-induced increase in $H^+$ conductivity of thylakoid membranes and the activity of the coupling factor. Eur. J. Biochem. **158:** 7–12.

22. DILLEY, R. A. & L. P. VERNON. 1965. Ion and water transport processes related to the light-dependent shrinkage of chloroplasts. Arch. Biochem. Biophys. **11:** 365–375.

23. HIND, G., H. Y. NAKATANI & S. IZAWA. 1974. Light-dependent redistribution of ions in suspensions of chloroplast thylakoid membranes. Proc. Natl. Acad. Sci. USA **71:** 1484–1488.

24. BAKER, G. B., D. BHATNAGER & R. A. DILLEY. 1981. Proton release in photosynthetic water oxidation. Evidence for proton movement in a restricted domain. Biochemistry **20:** 2307–2315.

25. LASZLO, J. A., G. M. BAKER & R. A. DILLEY. 1984. Chloroplast thylakoid proteins having buried amine buffering groups. Biochim. Biophys. Acta **764:** 160–169.

26. THEG, S. M., J. D. JOHNSON & P. H. HOMANN. 1982. Proton efflux from thylakoids induced in darkness and its effect on photosystem II. FEBS Lett. **145:** 25–29.

27. BEARD, W. A., G. CHIANG & R. A. DILLEY. 1988. ATP formation onset lag and postillumination phosphorylation initiated with single-turnover flashes. II. Two modes of postillumination phosphorylation driven by either delocalized or localized proton gradient coupling. J. Bioenerg. Biomembr. **20:** 107–128.

28. MURAKAMI, S. & L. PACKER. 1970. Protonation and chloroplast membrane structure. J. Cell. Biol. **47:** 332–351.

29. WALZ, D. V., L. GOLDSTEIN & M. AVRON. 1974. Determination and analysis of the buffer capacity of isolated chloroplasts in the light and in the dark. Eur. J. Biochem. **47:** 403–407.

30. BAKER, G. M., D. BHATNAGAR & R. A. DILLEY. 1982. Site-specific interaction of ATPase-pumped protons with photosystem II in chloroplast thylakoid membranes. J. Bioener. Biomembr. **14:** 249–264.

31. THEG, S. M., G. CHIANG & R. A. DILLEY. 1988. Protons in the thylakoid membrane-sequestered domains can directly pass through the coupling factor during ATP synthesis in flashing light. J. Biol. Chem. **263:** 673–681.

32. ALLNUTT, F. C., R. A. DILLEY & T. KELLY. 1988. Effect of high KCL concentrations on membrane-localized metastable proton buffering domains in thylakoids. Photosynth. Res. **20:** 161–172.

33. MITCHELL, P. 1966. Chemiosmotic coupling in oxidative and photosynthetic phosphorylation. Biol. Rev. Cambridge Phil. Soc. **41:** 445–540.

34.  KLEE, C.B., T. H. CROUCH & P. G. RICHMAN. 1980. Calmodulin. Ann. Rev. Biochem.
     **49:** 489–515.
35.  PROZIALECK, W. C., M. CIMINO & B. WEISS. 1981. Photoaffinity labeling of calmodu-
     lin by phenothiazine antipsychotics. Mol. Pharmacol. **19:** 264–269.
36.  NELSON, N., E. EYTAN, B. NOTSANI, H. SIGRIST, K. SIGRIST-NELSON & C. GITLER.
     1977. Isolation of a chloroplast DCCD-binding proteolipid active in proton transloca-
     tion. Proc. Natl. Acad. Sci. USA **74:** 2375–2378.
37.  DILLEY, R. A., S. M. THEG & W. A. BEARD. 1987. Membrane-proton interactions in
     chloroplast bioenergetics: Localized proton domains. Ann. Rev. Plant Physiol. **38:**
     343–389.
38.  DILLEY, R. A. & D. DEAMER. 1971. Light-dependent chloroplast volume changes in
     chloride media. Bioenergetics **2:** 33–38.
39.  SIEGENTHALER, P. A. 1972. Aging of the photosynthetic apparatus. IV. Similarity
     between the effects of aging and unsaturated fatty acids on isolated spinach chloro-
     plasts as expressed by volume changes. Biochim. Bipohys. Acta **275:** 182–191.
40.  NAKATANI, H., J. BARBER & M. J. MINSKI. 1979. The influence of the thylakoid
     membrane surface properties on the distribution of ions in chloroplasts. Biochim.
     Biophys. Acta **545:** 24–35.
41.  CHIANG, G. & R. A. DILLEY. 1989. Calcium regulation of localized to delocalized
     proton gradient switching in thylakoids. The 8 kDa CF₀ subunit is part of the Ca²⁺
     gating structure. *In* Photosynthesis. W. Briggs, ed. Vol. **8:** 437–455. Plant Biol.
     Series, Alan R. Liss, New York.
42.  ROBERTS, D. M., T. J. LUKAS & D. M. WATTERSON. 1986. Structure, function and
     mechanism of action of calmodulin. C.R.C. Rev. **4:** 311–339.
43.  CHIANG, G. & R. A. DILLEY. 1989. Intact chloroplasts show Ca²⁺-gated switching
     between localized and delocalized proton gradient energy coupling. Plant Physiol.
     **90:** in press.

# Protons, the Thylakoid Membrane, and the Chloroplast ATP Synthase[a]

WOLFGANG JUNGE

*Biophysik, FB Biologie/Chemie*
*Universität Osnabrück*
*D-4500 Osnabrück, FRG*

Light-driven proton pumps and proton-translocating ATP synthases stand very close to the beginning of photoautotrophic life. Whereas nature found different solutions to proton pumping (e.g., retinal-based bacteriorhodopsin and (bacterio)chlorophyll-based reaction centers from eubacteria to green plants), the proton-translocating ATP synthases belong to one superfamily[1] of bipartite enzymes that are formed from a channel portion, which is embedded in the coupling membrane, and a peripheral, catalytic portion (reviewed in ref. 2). The mechanism of coupling between proton flow and ATP synthesis is usually discussed in the framework of the chemiosmotic theory.[3]

In the thylakoid membrane of chloroplasts a transmembrane protonmotive force is generated by light-driven electron transport[4] which involves three big protein-complexes, two photosystems, and the cytochrome $b_6$,f-complex (FIG. 1). Thylakoid membranes of shade plants are stacked (see upper insert in FIG. 1). While the ATP synthase $CF_oCF_1$ resides exclusively in exposed membrane regions, the proton pumping activity of photosystem II is exclusively located in appressed regions,[5] that is, up to 300 nm away from an ATP synthase. The lower insert in FIGURE 1 shows a simple equivalent circuit for the cyclic proton current, consisting of pumps, P, two line resistors, $R_{L1}$ and $R_{L2}$ (note the relatively large lateral distance between pumps and ATP synthases), an access resistor in the enzyme, $R_A$, and finally the largest resistor, $R_C$, representing the coupling site. Here protons are forced to do useful work, whereas proton flow over the other resistors only produces ohmic heat.

On the long way to understanding the coupling mechanism between proton flow and ATP synthesis we have studied the nature of the pathways for protons between pumps and ATP synthases, the magnitude of the losses of protonmotive force along these paths, the conductance of the channel portion of the ATP synthase, which is supposed to act as a proton well,[6] and, finally, elements of the protonic coupling site in the enzyme.

Thylakoids are a favorable object for two reasons: (1) Excitation of thylakoids with a short flash of light generates a voltage transient across the membrane (in nanoseconds) and a transient pH difference (microseconds to milliseconds). (2) The rise in the transient protonmotive force and its subsequent decay via leak conductances or via the ATP synthase can be followed at high time resolution by appropriate indicator dyes: intrinsic pigments respond to the transmembrane voltage by electrochromism,[7] the surface adsorbed neutral red (in the presence of a nonpermeant buffer) reports small pH transients at the lumen side of the thylakoid membrane,[8-11] and any hydrophilic pH indicator is practically selective for pH transients in the medium.[12] Taken together, the spectrophotometric techniques have allowed "complete tracking of proton flow."[13,14]

[a] This work was supported by the DFG (SFB-171-A2/B3).

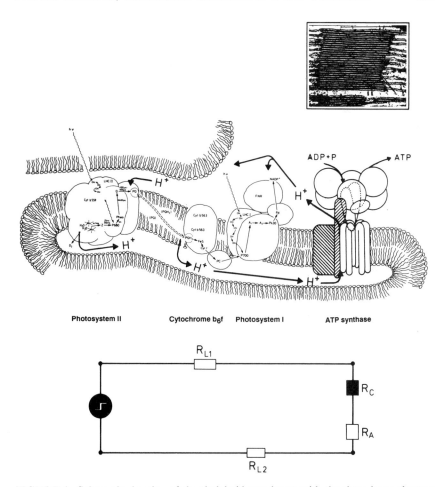

**FIGURE 1.** Schematic drawing of the thylakoid membrane with the three large electron transfer complexes and the ATP synthase. Apposed membrane portions with photosystem II and exposed portions with photosystem I and the synthase are apparent. Sites of light driven proton binding from the outer medium (the stroma) and of proton release into the lumen, lateral proton flow, and transmembrane proton backflow over the ATP synthase are indicated. The insert in the upper right shows an electron micrograph of stacked thylakoid membranes. Each of the disk-like structures contains at least 100 pieces of any protein complex, but the true function unit is larger as disks are interconnected with each other. The lower portion shows a simplistic equivalent circuit for cyclic proton flow between pumps and the ATP synthase.

## ON THE PATHWAY OF PROTONS BETWEEN PUMPS AND ATP SYNTHASES AND THE DICHOTOMY BETWEEN LOCALIZED AND DELOCALIZED COUPLING

There is a long-standing debate on whether very many proton pumps are coupled with very many ATP synthases simply by proton flow through aqueous compartments, as postulated by Mitchell[3] (delocalized coupling), or whether there is a preferential interaction with pumps feeding protons directly into neighboring ATP synthases (localized coupling). Detailed theories for the latter case (e.g., ref. 15) have been put forward mainly to account for certain observations in mitochondria and photosynthetic bacteria.[16] The dichotomy between localized and delocalized coupling has two different aspects, one related to the dimension in plane of the membrane and the other normal to the membrane. Over the lateral dimension and in thylakoids coupling is *per se* delocalized simply because a large fraction of proton pumps can be far from ATP synthases. For the normal dimension, however, it is still under debate whether protons *obligatorily* move through the aqueous volumes that are separated by the coupling membrane, thereby interacting with many buffering groups,[3] or whether there are special ducts for protons in the membrane[17,18] or along the membrane surface.[19,20] It is evident that the nature of the pathway for protons into $F_0F_1$ ATP synthases [e.g., through non-aqueous environment, as proposed[17]] strongly bears on the energetics and on the molecular mechanism of photophosphorylation. Four facets of this issue are dealt with in the following: (a) The properties of the very narrow (5 nm wide) spaces between membranes. (b) The possibility of enhanced proton diffusion at the surface of the membrane. (c) The magnitude of ohmic losses due to lateral proton flow. (d) The existence of intramembrane proton ducts that are not always in equilibrium with the adjacent water spaces.

(a) *Properties of the narrow aqueous phases between thylakoid membranes.* Both the lumen of thylakoids and the gap between adjacent thylakoids in a stack are only about 5 nm wide,[21] a distance comparable to the Debye length which describes the range of electrostatic interactions in electrolyte solutions (e.g., 3 nm at 10 mM of a $1:1$ electrolyte[22]). These narrow spaces are not adequately described as an aqueous bulk phase but rather as charged surface or Donnan-matrix.[23] They are filled with charged groups on lipid heads and on the surface of membrane proteins.

Thylakoids, which appear as stacked disks in the electron micrograph inserted in FIGURE 1 (radius about 300 nm, repetition period about 20 nm), are truly interconnected and able to form large spherical blebs (radius about 1 $\mu$m) when suspended in distilled water. In the following discussion we neglect this complicated connectivity, and we use the term thylakoid in a loose way, denoting stacked, disk-shaped membranes. The internal volume of a thylakoid disk is so small ($1.4 \cdot 10^6$ nm$^3$) that one might wonder whether the concept of pH is still valid. At pH 7 this volume contains an average of less than 0.1 free protons. In other words, a snapshot reveals only one free proton in one disk and no free proton in nine others. That the concept of pH is valid is owed to many buffering groups (ensemble average) and to the rapidity of protonation/deprotonation reactions (time average). The specific buffering capacity of the thylakoid lumen, approximately 100 mmol/mol chlorophyll around neutral pH, has been determined by optical[9] and by spin probe techniques.[24] The number of buffering groups in the aforementioned small volume can be calculated from the definition of the buffering capacity; see Eq. 1a:[9]

$$\beta = -\Delta[H^+_{total}]/\Delta pH \qquad (1a)$$

wherein $\beta$ denotes the buffering capacity, which is the ratio between the change in total concentration of protons (bound plus free) and the resulting pH change. If only one type of buffering group is present (dissociation constant K and concentration $c_{buffer}$), the buffering capacity depends on the concentration of free protons, $[H^+]$, as given in Eq. 1b.

$$\beta = 2.3 \cdot c_{buffer} \cdot [H^+] \cdot K/([H^+] + K)^2 \qquad \textbf{(1b)}$$

If the measured buffering capacity of the thylakoid lumen[9,24] was not attributable to a host of groups with different pK (as in reality), but to a homogeneous set with uniform pK 7, the number of such buffering groups in this small volume was about $3 \cdot 10^4$.

According to Eigen,[25] and for buffering groups at the surface of lipid structures and proteins experimentally established by Gutman,[26] each of these groups undergoes rapid protonation/deprotonation with a relaxation time in the order of 100 $\mu$s. As the diffusion of protons over aqueous distances of 300 nm occurs in a few microseconds, these buffering groups can be conceived of as one common pool. Then, the concentration of free protons in this small volume is the time average over protolytic events in the pool. On a time scale of 100 $\mu$s there will be more than $10^4$ events, that is, the concentration of free protons will be sharply defined with less than 1% standard deviation.

The observation[10] that certain hydrophilic buffers, added perhaps at 10 mM, do not affect the magnitude of flash-induced pH transients in the lumen of unswollen thylakoids is at least in part a consequence of the tiny internal volume (3 l/mol chlorophyll). A concentration of 10 mM of a hydrophilic but neutral compound implies less than $10^4$ molecules in the aforementioned small volume. This is much less than the amount of intrinsic buffers. The same buffers, however, are effective in swollen thylakoids[9] with the specific internal volume expanded more than 10-fold higher from the original 3 l/mol chlorophyll,[21,24] which means that now more than $10^5$ buffer molecules are present, more than intrinsically. Supporting this argument, lipid soluble or amphiphilic buffers like imidazole also act on unswollen thylakoids, probably because they are enriched by adsorption to the membrane surface and not restricted to the small aqueous volume. On the same line, they are still more effective than hydrophilic buffers even in swollen thylakoids.[9]

In summary, although the behavior of the narrow spaces between thylakoid membranes in some respects differs from one of an extended aqueous bulk phase, there is no reason to doubt that the pH is well defined and that these spaces can serve to couple proton pumps with ATP synthases.

*(b) Enhanced proton diffusion at the surface of the membrane?* One aspect of the old but ongoing debate over the validity of Mitchell's chemiosmotic theory[3] is the question of whether the diffusion of protons at the surface of the membrane is enhanced over the one in bulk water. Originally a matter of speculation,[19] it has gained thrust by the experiments of Prats, Tocanne, and Teissie[20] on proton diffusion along the surface of phospholipid monolayers deposited on an aqueous subphase. Their results have allowed them to claim 20-fold enhancement of the surface diffusion coefficient over the one in bulk water, and furthermore that this result is relevant for the protonic energy coupling in real biomembranes.

Experiments with stacked thylakoid membranes do not support this notion. The layout is illustrated in FIGURE 2. Excitation with one short flash of light causes one turnover of each photosystem II (FIG. 1) with comcomitant proton uptake from the outer phase (and proton release into the lumen). This generates an alkalinization jump in the narrow space between appressed membranes (hatched in FIG. 2). In the absence of ADP and P the membrane is rather proton

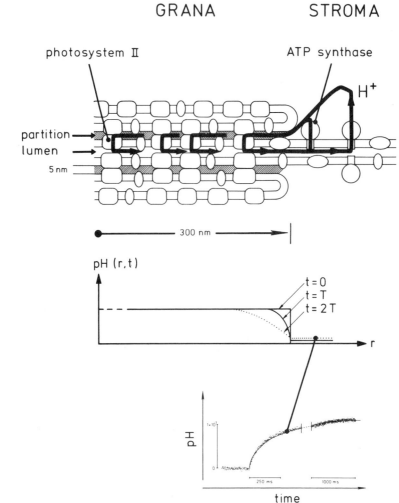

**FIGURE 2.** Schematic side view of stacked thylakoid membranes with appressed grana lamellae and interconnecting stroma lamellae. The narrow space between the outer surface of thylakoid membranes in a granum, called partition, is hatched. The *arrows* illustrate (a) light-driven proton pumping by photosystem II which is directed from the partition into the lumen, (b) opposite directed proton flow which is coupled to ATP synthesis, (c) lateral proton flow through the lumen, and (d) lateral backflow through partitions. The drawing in the middle illustrates the box shaped pH profile induced by excitation with a single flash of light at time zero and its evolution in time (t = T, t = 2T). The bottom trace is a reproduction of original data on the flash induced, and photosystem II-related pH rise in the stroma compartment and a theoretical curve (see text and refs. 27–29).

tight. (The relaxation time of a transmembrane pH difference then is 15 s.[11]) Thus, alkalinization relaxes by *lateral* diffusion of protons, hydroxyl anions, and mobile buffers along the surface to and from the medium. The concomitant rise of pH in the medium (data points and fit curve illustrated at the bottom of FIG. 2) is measured spectrophotometrically. At first glance the slowness of the observed slow rise is surprising. The half-rise time of about 100 ms[12,27] is $10^5$ times larger than expected by solving Fick's equation and assuming the same diffusion coefficients for protons and hydroxyl ions as in water. But this can be understood by taking into account the presence of fixed buffers in this domain. The rise velocity of the medium pH is a function of the "effective diffusion coefficient," $D^{eff}$ (see refs. 27 and 28):

$$D^{eff} = (2.3/\beta_{tot}) \cdot \{D_{H^+} \cdot [H^+] + D_{OH^-} \cdot [OH^-] + \Sigma(D_i \cdot \beta_i/2.3)\} \qquad (2)$$

depends on the buffering capacity, $\beta$, of the mobile buffers (suffix i) and of all buffers (fixed and mobile = *tot*al), respectively. D denotes the diffusion coefficient of protons, hydroxyl anions, and mobile buffers,[28] as specified by the suffix. The observed dependence of the relaxation on the medium pH and on added mobile buffers follows the expectation based on Eq. 1. The predicted minimum of the rise velocity around neutral pH and the enhancement of the velocity by added mobile buffers have been observed.[29] The $10^5$-fold delay has been found compatible with the amount of fixed buffers present.[27] This has led to the conclusion *that protons adjacent to the outer side of stacked thylakoid membranes are, if anything, less mobile than in bulk water.*[29] It may be argued that enhanced surface diffusion is absent in protein-loaded biomembranes and only exists in membranes formed from pure lipid, as stated in ref. 20. This has become highly questionable because of recent experiments by Gutman *et al.*[30] and Menger *et al.*[31] The former find no evidence for enhanced proton diffusion in the ultrathin layer between the osmotically compressed bilayers in multilamellar lipid vesicles. The latter arrive at the same conclusion in their study on proton diffusion along the surface of a lipid monolayer spread on an aqueous subphase.

Does the narrowness of the aqueous phases and their granularity (by protruding proteins) invoke the critical diffusion phenomena described by percolation theory?[32] A tentative answer may be inferred from studies on proton conduction on lyophilized and lightly rehydrated purple membranes.[33] The threshold for the onset of conduction as a function of relative hydration is observed at a water-to-protein ratio (w/w) of 0.045 g/g, far below the point for the full water coverage, 0.25 g/g.[33] We calculated a water-to-protein ratio of 0.6 g/g for thylakoid membranes, which suggests that critical threshold phenomena that are predicted by percolation theory are not present in normally hydrated thylakoid membranes.

(c) *Losses of protonmotive force by lateral proton flow.* Comparing the efficiency of photosystem II to drive ATP synthesis with that of photosystem I, Haraux and de Kouchkowsky[34] found slightly higher figures for the latter. This might be understood in terms of the losses of the protonmotive force during lateral flow of protons and hydroxyl anions between photosystems II in the appressed membrane portions and the ATP synthases in the exposed ones (see FIG. 2). A total flux of protons, I (moles/s), over the boundary of a thylakoid disk requires a drop in proton concentration between the center (suffix c) and the fringe (suffix f) of the disk. For a disk that is homogeneously filled with pumps the concentration drop has been calculated:[27]

$$[H^+]_c - [H^+]_f = I/4\pi hD \qquad (3)$$

It depends on the thickness of the disk-shaped slab between membranes, h, and on the diffusion coefficient, D. A proton flux, which is equivalent to the highest rate of ATP synthesis in a model thylakoid (radius 300 nm, area per chlorophyll molecule 2.2 nm), namely, $1.3 \cdot 10^{-19}$ mol $s^{-1}$, and assuming a thickness of 5 nm and the diffusion coefficient as in bulk water, $D_{H^+} = 9.3 \cdot 10^{-9}$ m$^2$ s$^{-1}$, implies a drop in proton concentration of 0.23 $\mu$M.[27] The magnitude of the corresponding pH drop, $\Delta pH = -\Delta[H^+]/2.3 \cdot [H^+]$, depends on the pH in the medium. It decreases toward more acid pH. A similar relation holds in the alkaline pH domain, where the diffusion of $OH^-$ dominates. Taking these results together and assuming the same diffusion coefficients as in bulk water the pH drop has been calculated.[27] At the outer surface of stacked thylakoids it amounts to 0.14 pH units, if the outer side is kept at pH 8. It is below 0.01 pH units at the internal side and at pH 4.[27] If the diffusion coefficients in these narrow spaces are lower than in bulk water, greater losses are expected. *It is probable that ohmic losses of protonmotive force are small, but they are not negligible in tightly stacked thylakoids.*

(*d*) *Are there localized proton ducts in the membrane?* This question is not easy to solve. It has been speculated that the respiratory chain and photosynthetic electron transport inject protons into the nonaqueous environment of the respective coupling membrane from where they are used for ATP synthesis.[17] For thylakoid membranes it has indeed been established by Dilley and coworkers (see ref. 18 for review) and also by others[35,36] that the membrane contains certain buffering groups that are not normally in equilibrium with the adjacent aqueous bulk phases. Amine groups with an unusually high pK (around 7.8) are involved.[18] These groups can transiently trap protons that are released, for example, by photosystem II as a consequence of water oxidation. The pool size is limited to about 6 protons per photosystem II, and proton liberation into the thylakoid lumen is again detectable after the pool is filled.[36] Under site-specific blocking of every second photosystem II by a herbicide (DCMU), the relative pool size per active reaction center II is doubled.[37] This has proved that the pool is delocalized and not, as one might expect, restricted to the particular protein molecule, such as photosystem II, from which protons are released. The question is whether or not these proton-trapping domains are relevant for photophosphorylation. There is ample evidence that a transmembrane and bulk-to-bulk pH difference[38] and an electric potential difference[39] (reviewed in ref. 40) can drive ATP synthesis in thylakoid membranes or in lipid vesicles with $CF_oCF_1$.[41] However, there is also evidence that the depletion from protons of the intramembrane buffering pool delays the onset of photophosphorylation in thylakoids that are excited by a series of light flashes.[42] This has been interpreted to indicate that the intramembranous groups may be on the pathway of protons into the ATP synthase. With the limited proton storage capacity of these groups,[35-37] this is relevant only for the abrupt onset of illumination from the dark but not for continuous illumination as experienced by plants in daytime. From the physiologic standpoint it is not much to worry about.

Another challenge to a chemiosmotic mechanism is also related to the onset lag of photophosphorylation. For thylakoids receiving a series of light flashes starting from the dark and with valinomycin added to minimize the electric component of the protonmotive force, the chemiosmotic theory predicts prolongation by added buffers of the onset lag of photophosphorylation. Buffers simply slow down the building up of a sufficiently large pH difference. On the contrary, in thylakoids that are prepared under low salt, added pyridin fails to enhance the onset lag of photophosphorylation[43] (and see Dilley, this volume). It behaves as expected only in high salt thylakoids. The former behavior has been interpreted as evidence for the passage of protons through localized proton ducts that are inaccessible to the added buffer[43] (see ref. 18 for review). The stringency of this interpretation depends on the accessibility of the thylakoid lumen for these buffers. We checked the buffering power of added pyridin by measuring its effect on the extent of the flash-induced absorption changes of neutral red, which are indicative of pH transients in the lumen (when the external phase is strongly buffered by bovine serum albumin, see ref. 9). In low salt thylakoids, pyridin (10 mM, pH 7.8) failed to prolong the onset lag of photophosphorylation (as in ref. 43), and it also failed to quench the extent of the pH-indicating absorption changes. Conversely, in high salt thylakoids it prolonged the time lag (as in ref. 43) but concomitantly it decreased (by buffering) the extent of pH transients in the lumen. Thus enhancement/nonenhancement of the onset lag of photophosphorylation was paralleled by buffering/nonbuffering of pH transients in the lumen. The same parallel was observed for other buffers, one of which acted on both prepara-

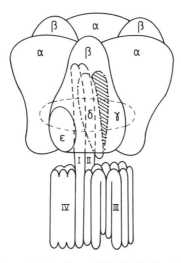

**FIGURE 3.** Schematic drawing of $CF_oCF_1$.[75] (For details, see text.)

tions (tris), others failed to act on both (HEPES, MES, tricine), and imidazole revealed a selective behavior as pyridin (A. Borchard & W. Junge, unpublished). These phenomena may be caused by different accessibility in different preparations of the thylakoid lumen to certain added buffers. *One should be careful to take the failure of pyridin to enhance the onset lag of photophosphorylation as proof of a role in photophosphorylation of intra-membrane proton ducts.*

## PROTON FLOW THROUGH CFOCF1

The structure of the chloroplast ATP synthase is schematically illustrated in FIGURE 3. $CF_1$, the extrinsic portion, contains the catalytic sites of ATP synthesis and hydrolysis, and $CF_o$, a membrane-spanning complex, acts as a proton conductor.[2] $CF_1$ is composed of five different subunits, $\alpha$, $\beta$, $\gamma$, $\delta$, and $\varepsilon$, in the order of decreasing molecular mass, in stoichiometric proportion $3:3:1:1:1$.[44] The large subunits, $\alpha$ and $\beta$, are arranged alternatingly to form a pseudohexagon[45] and they interact with nucleotides and phosphate. The positions of the smaller subunits of $CF_1$ are less well defined; their arrangement at the interface between $CF_0$ and $CF_1$, as illustrated in FIGURE 3, may be close to the truth. Their role in regulating proton flow through the synthase and in energy transduction will be discussed later on. In $CF_0$, four different subunits have been identified and named I to IV. Their stoichiometry is still under debate. Subunit III, the proteolipid, has a hairpin structure[46] and it is present in 6 to 12 copies. By sequence similarity with the subunits of the homologous enzyme of *Escherichia coli,* it is known that subunit I contains only one membrane-spanning helix with a bulky hydrophilic headpiece and that subunit IV is made from 5–7 membrane-spanning helices.[47]

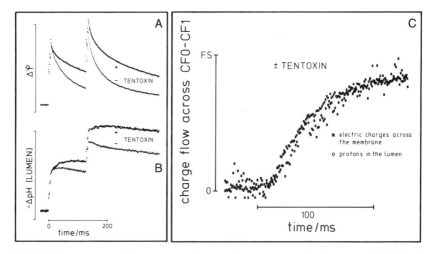

**FIGURE 4.** Transients of the transmembrane voltage (**A**) and of the lumenal pH (**B**) in the presence of 20 $\mu$M ADP and 60 $\mu$M $P_i$ and with and without tentoxin (specific blocker of ATP synthase). Reproduced from ref. 14. Suspensions of thylakoids have been excited with two groups of three short flashes each. The transmembrane voltage was recorded by electrochromism[7] and the pH transient by neutral red.[8–11] Part **C** gives a superimposition of two data sets, namely, the number of protons entering $CF_0CF_1$ from the lumen and the number of charges crossing $CF_0CF_1$ in the time interval between the first and the second flash group (Normalization and procedure to extract these data from the traces in **A** and in **B** in ref. 14). Broadly speaking the two sets of data points in **C** coincided (see text).

Proton conduction by the integral enzyme is controlled by the presence of nucleotides which interact with $CF_1$. FIGURE 4 shows transients of the transmembrane voltage (FIG. 4A) and of the lumenal pH (FIG. 4B) under excitation of thylakoids with two groups of three short flashes each. The traces have been obtained with ADP and $P_i$! present without ($-$) and with added tentoxin ($+$) to block ATP synthesis.[14] If the ATP synthase is blocked, the voltage that is generated by two sets of three flashes decays more rapidly (some 100 ms) than does acidification of the lumen (some 10 s). This is indicative of an electric leak conductance which is higher for other ions than for the proton. It is responsible for the electric balance of the greater part of the inwardly directed proton pumping in thylakoids.[4] It is noteworthy that specific ion channels that might account for the nonprotonic leak conductance of thylakoid membranes have not been characterized up to now, except for a voltage-dependent anion channel that has been detected by patch clamping of thylakoids from giant chloroplasts of *Peperomia metallica*.[48]

Only when the ATP synthase is active are the electric decay and the decay of the acidification accelerated concomitantly. This clearly shows the dominance of the protonic conductance through $CF_oCF_1$ over the leak conductances. FIGURE 4C shows a comparison of the extra charge displacement through active $CF_oCF_1$ (as extracted from FIG. 4A, see ref. 14) and of the extra proton displacement (from FIG. 4B) during the interval between the first and the second flash group. The two sets are coincident within noise limits. *Hence, every charge crossing the active ATP synthase is evident as a proton entering the enzyme from the lumen.*[14]

## PROTON FLOW THROUGH $CF_o$, THE CHANNEL PORTION OF THE ATP SYNTHASE

Turnover numbers of $CF_oCF_1$ range up to more than 400 ATP molecules formed per second.[41] With a proton-to-ATP stoichiometry of 3 this implies a turnover number of 1,200 protons per second. In most experiments aiming at proton conduction by the channel portion, $F_O$ or several of its subunits have been isolated and incorporated into lipid vesicles (see ref. 47 for review). Then proton leakage across the membrane has been monitored by pH electrodes. This approach has produced proton conductance that was sensitive to DCCD, as in $F_O$, but with turnover numbers of $10 \text{ s}^{-1}$, falling short by orders of magnitude from the required one. This was incommensurate with the proposed function as a low-impedance access for protons to the coupling site in the ATP synthase.[6] The shortcoming may have been due to the survival of only a small proportion of $F_O$ channels in reconstitution experiments (safe guard mechanism?) and/or to insufficient time resolution of pH electrodes.

In an alternative approach to determine the time-averaged single channel conductance of $CF_o$, a fraction of $CF_1$ has been removed by EDTA treatment of thylakoid membranes. Relaxation of the flash light-induced transmembrane voltage and of the pH transients in the lumen and in the medium have been monitored.[13,49,50] FIGURE 5 illustrates the situation with $CF_1$ removed and with ferricyanide added as terminal electron acceptor, so that for any two protons released into the thylakoid lumen, only one proton is taken up from the suspending medium. FIGURE 6 shows the transients of the external pH (top), of the transmembrane voltage (middle), and of the lumenal pH (bottom) with both $CF_o$ exposed (−DCCD) and $CF_o$ blocked by DCCD. The two traces in the upper left show alkalinization of the suspending medium when the proton channel is blocked and a net acidification of same extent when the channel is open. (The efflux of two protons overcompensates the alkalinization by one proton.) Correlation of the electric and protonic decay processes (left column in FIG. 6) with the amount of *exposed $CF_o$* (determined by immunoelectrodiffusion) and application of the capacitor equation (assuming the usual $1 \text{ }\mu\text{F cm}^{-2}$ for the thylakoid membrane) revealed that: *Under the assumption that every exposed $CF_o$ is actually conducting, the average proton conductance of exposed $CF_o$ is about 10 fS.* This is equivalent to the translocation of 6,000 protons at 100 mV electric driving force.[13]

The traces in the right show the same set of experiments except that the electric potential difference has been shunted by the addition of gramicidin (with $K^+$ present). The short circuiting of the transmembrane voltage is immediately

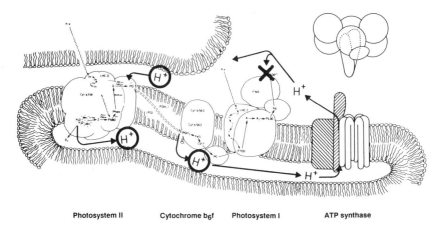

Photosystem II          Cytochrome b₆f      Photosystem I              ATP synthase

**FIGURE 5.** Schematic drawing of protolytic reaction sites when ferricyanide is added as terminal electron acceptor and with $CF_1$ removed by EDTA. This figure serves to illustrate the meaning of the original traces in Figure 6.

apparent from the middle traces. Under these conditions the pH difference across the membrane relaxes more slowly. It discharges the larger proton buffering capacity of the lumen rather than the electric capacitance of the membrane (as in the left column of FIG. 6). Around neutral pH the ratio between the specific electric capacitance and the equivalent term derived from the buffering capacity of the lumen is larger than 10, as determined in reference 9. This is the reason for the larger extent of transmembrane voltage after excitation of thylakoids with a single-turnover flash of light (namely, 30–50 mV[4]) as contrasted with the extent of transmembrane pH difference (namely, 0.06 units, only[9]). Conversely, comparison of the traces in the left and right columns of FIGURE 6 clearly demonstrates that in the absence of gramicidin there is no leak conductance to other cations, that is, $CF_0$ is highly selective for protons (but see below).

CF₁ can be removed from thylakoid membranes without leaving behind proton conducting $CF_0$.[13,51] If the electric relaxation (see left column in FIG. 6) is attributable to only a few active channels out of many exposed ones, then the conductance of the open channels is higher than the average 10 fS. This has been subjected to a systematic study.[49,50] A lucky circumstance as well as a prerequisite for the experimental approach has been that EDTA treatment of thylakoid membranes, which is standard to remove the $CF_1$ counterpart of $CF_0$, also causes destacking of thylakoids and finally their fragmentation into smaller spherical vesicles containing a total of about 100 $CF_0CF_1$ or $10^5$ chlorophyll molecules. Only the small size of these vesicles opens the way to a situation where among the about $10^{11}$ partially $CF_1$-depleted vesicles in the absorption cell of a typical experiment, some may have no conducting $CF_0$ and others 1, 2, or more. Indeed, the relaxation of the flash light-induced voltage and of the pH transients has revealed a biphasic decay. This has been subjected to a statistical analysis based on Poisson's distribution of active channels over vesicles. Assuming ohmic behavior of $CF_0$ (in the 50 mV range of these experiments), the decay of the voltage indicating electrochromic absorption changes is expected to follow Eq. 4:[49,52]

$$U_{app}(t) = U_0 \exp(-\bar{n}) \exp(\bar{n} \exp(-Gt/A \cdot \hat{c})). \qquad (4)$$

This equation has only two fit parameters, namely, ñ, the average number of open channels per vesicle, and G, the time-averaged conductance of one active $CF_0$. A, the area of one vesicle, can be inferred from similar experiments with the channel-forming antibiotic gramicidin (see below and ref. 49) and ĉ, the specific membrane capacitance, is taken as usual, namely, 1 $\mu Fcm^{-2}$. The experimentally observed biphasic decay of the electrochromic absorption changes and its relation to the parameters ñ and G is illustrated in FIGURE 7. Eq. 4 is a good approximation even if a vesicle area distribution with 30% standard deviation from the mean is

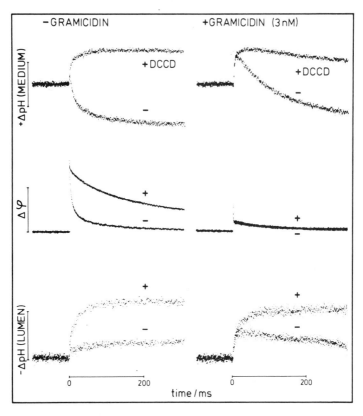

$\Rightarrow$ 10 fS per exposed $CF_0$
$\Rightarrow$ 6000 $H^+$ $s^{-1}$ @ 100 mV
$\Rightarrow$ monospecific for protons

**FIGURE 6.** Complete tracking of proton flow through $CF_0$ (adapted from ref. 13). Traces in the upper part show pH transients in the suspending medium of thylakoids, those in the middle transients of the transmembrane voltage, and traces at the bottom show pH transients in the lumen, all of them induced by one short flash of light. They are observed in partially $CF_1$-depleted thylakoids, with the covalent channel blocker DCCD present and absent, respectively. In the right column the transmembrane voltage is shunted by addition of the alkali cation pore former gramicidin. For details, see text.

taken into account.[49] Application of this analysis to $CF_1$-depleted vesicles has yielded a *time-averaged conductance of active $CF_o$ of about 1 pS*. At 100 mV electric driving force this is equivalent to the passage of $6 \cdot 10^5$ protons per second. However, only a few percent of exposed $CF_o$ have been highly conducting under these circumstances.

As a test for this statistical analysis of electrochromic absorption changes the conductance of gramicidin has been analyzed in the same vesicles as before. With $CF_o$ blocked by DCCD, gramicidin has been added at pM concentration. The analysis has yielded a conductance of 2.5 pS, which is in the range of published figures for the open state of gramicidin under a given ionic milieu (see ref. 49 for

$$U_{app}(t) = U_0 \cdot exp(-\bar{n}) \cdot exp\left(\bar{n} \cdot exp\left(-\frac{G}{A \cdot \hat{c}} \cdot t\right)\right)$$

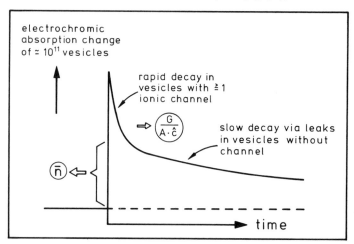

**FIGURE 7.** Schematic drawing of the flash-induced electrochromic absorption changes (ordinate) as function of time in a suspension with small thylakoid vesicles that are partially $CF_1$ depleted, so that each vesicle out of the $10^{11}$ vesicles that are contained in an optical absorption cell carried zero, one, two, . . . active $CF_o$ channels. This has enabled a poisson statistical fit of measured decay curves with only two essential fit parameters, namely, the time-averaged single-channel conductance, G, and the average number of open channels, $\bar{n}$.[49,50]

details). Is the gramicidin channel always open in thylakoid membranes, though? This is indeed suggested by a linear concentration dependence of the electric relaxation rate in thylakoids for gramicidin concentrations above some 10 pS and a quadratic one for lower concentrations (Schönknecht *et al.*, unpublished). The rather high dimerization constant of gramicidin in thylakoids points to an open probability close to 1.

The conductance of $CF_o$ has been investigated as a function of pH, pD (isotopic substitution), the concentration of other cations, addition of glycerol to alter the water structure, and under variations of temperature.[50] The technique has been the same as that already described. The results are as follows: (1) The proton is the major charge carrier through $CF_o$ even at pH 8 and against a background of 300 mM NaCl or KCl or 30 mM $MgCl_2$ in the medium. No other interpretation is possible, as time-resolved pH transients are one basis for this conclusion. The apparent discrimination by a factor of $10^7$, say, against $Na^+$ is surprising, as the homologous $F_oF_1$ ATP synthase from *P. modestum* seems to be able to operate facultatively on $H^+$ or on $Na^+$.[53] (2) Between pH 5.6 and pH 8 the conductance of $CF_o$ does not vary as a function of the medium pH, independent of the ionic strength. This shows that the conductance is not rate limited by the protonation/deprotonation of one particular group. (3) The conductance is lower in $D_2O$ than in $H_2O$ by a constant factor of 1.7. The isotope effect may be secondary. (4) Addition of glycerol decreases the conductance of $CF_o$ and abolishes the hydrogen/deuterium isotope effect. This suggests that the conduction of the channel may then be governed by events related to the water structure in the channel mouth. (5) With 42 kJ/mol the Arrhenius activation energy of proton conduction by $CF_o$ in the thylakoid membrane is intermediate between that of a pore (e.g., 30 kJ/mol for gramicidin) and that of a carrier (e.g., 65 kJ/mol for valinomycin) in the same membrane.

A protonic conductance of 1 pS exceeds by orders of magnitude the calculated convergence value that is limited by diffusive supply to a pore mouth of perhaps 1 nm diameter in a medium at neutral pH. I also greatly exceeds calculated rates of proton transfer through a short hydrogen bonded chain[55] and measured conductances (neutral pH) of gramicidin or of channels formed from leucine- and serine-containing synthetic polypeptides.[61] Protolytic reaction rates in excess of the diffusion limit have been reported for certain uncouplers in biomolecular lipid membranes.[54] Hydrolysis, the reaction of water with an acceptor group in the channel mouth, has been discussed to account for the seeming discrepancy.[54,55] This, however, is expected to reveal a pH dependence that has not been observed in the foregoing experiments with $CF_o$. At the moment, one can only speculate about the origin of the extremely high proton conductance of $CF_o$. Channel properties as large mouth and a short selectivity filter, which have been discussed in the context of $K^+$ maxi channels,[56] the drag force in the coulomb cage that surrounds the channel mouth,[57] the transient supply of protons from fixed buffers in its vicinity and/or mobile buffers may lift the supply rate of protons to the observed figure.

The foregoing experiments have established that *$CF_o$ is a kinetically competent access channel for protons in the integral ATP synthase, $CF_oCF_1$, and the site of proton selectivity of $CF_oCF_1$ is the $CF_o$ portion of the enzyme.*

$CF_oCF_1$ has recently been addressed by electrophysiologic techniques. In one approach, Wagner *et al.*[58] reconstituted purified $CF_oCF_1$ into azolectin vesicles, these were fused into a lipid monolayer on aqueous subphase, and a $CF_oCF_1$-doped bilayer was formed at the dip of a micropipette (dipstick technique). At voltages greater than 100 mV, there has been a steep rise in the open probability of channels with a conductance of 1–5 pS. Evidence has been presented for the protonic nature of the observed gated currents. That these are attributable to $CF_oCF_1$ and not to $CF_o$ alone has become evident, as added ADP (3 $\mu$M) and $P_i$! (5 $\mu$M), both of which interact with the $CF_1$ portion, have decreased the open probability of conduction events.[58] These concentrations of ADP and $P_i$ are known to affect tight-binding sites on $CF_1$, but they are too low for efficient ATP synthesis. Higher concentrations of ADP and $P_i$, however, lead to enhanced proton flow over $CF_oCF_1$ (see FIG. 4). This suggests that the aforementioned effects may represent a (over-voltage) valve reaction of $CF_oCF_1$, with $CF_1$ shifted in a voltage-dependent reaction to transiently expose the path of protons through

$CF_0$. This was compatible with a similar magnitude of conductance and with the same pH independence as in the previous studies on exposed $CF_0$ (free of $CF_1$) in thylakoid membranes. It is noteworthy that the large turnover number of these channels ($>10^6 s^{-1}$ at 180 mV) in some circumstances has been sustained over several 100 ms (i.e., transport of more than $10^5$ protons).

Another approach has been fusion of small $CF_0CF_1$-containing vesicles to form large liposomes (by dehydration/rehydration[59]). Patch clamping of these large liposomes has revealed cation channels with several conductance levels in the range of 10 pS (at 100 mM KCl or NaCl) but without a pronounced selectivity for protons.[60] Addition of venturicidin, an inhibitor of proton flow through $CF_0$, that is supposed to interact with the proteolipid, has decreased the open probability. Similar channels have been observed with subunit III of $CF_0$ alone. This has been interpreted to indicate that some of the $CF_0CF_1$ molecules are deranged or disintegrated, as during dehydration/rehydration, and that the products, mainly subunit III, are capable of forming cation channels[60] as certain synthetic polypeptides.[61] The general cation conductance is obviously suppressed in intact $CF_0$. The sharp selectivity filter for protons may be brought about by interaction of subunit III with the other subunits of $CF_0$.

## ON THE ROLE OF THE SMALLER SUBUNITS OF $CF_0CF_1$

With $CF_0$ acting as a proton channel and the hexagon formed by the $\alpha$ and the $\beta$ subunits interacting with the nucleotides, one may ask for the coupling site where protons are forced to do useful work. The $\gamma$ subunit regulates proton flow through the enzyme. Its thiol groups[62] are involved in the diurnal redox modulation of the enzyme activity which is switched off at night to prevent the ATP produced by mitochondria to be dissipated by chloroplasts (see ref. 63). The $\varepsilon$ subunit seems also involved in the regulation of proton conduction by the enzyme.[64] The $\delta$ subunit is necessary for efficient coupling between $CF_0$ and $CF_1$,[65] but in contrast to the other two, which exert their respective roles only in conjunction with the other small subunits present, it seems to have some function of its own. After hints that $\delta$ may remain back on $CF_0$ after removal of $CF_1$, where it keeps $CF_0$ nonconducting,[66] we have found that *isolated $\delta$,*[67] *when added back to $CF_1$-depleted thylakoids, blocks proton flow through open $CF_0$*[68] and thereby restores photophosphorylation[69] by those $CF_0CF_1$ that have remained on the $CF_1$-depleted membrane. The stopcock action of $\delta$ on $CF_0$ has to be relieved in the intact, ATP-synthesizing enzyme. It is probable that $\delta$ *then either acts as (part of) the valve*, which admits protons from $CF_0$ further up into the enzyme to the coupling site, *or as (part of) the conformational transducer* between protons and ATP. It is compatible with this role that $\delta$, which is not necessary for the binding of $CF_1$ to $CF_0$,[65,70] can bind not only to $CF_0$[67-69] but also to $CF_1$ with one high (100 nm) and one or two low affinity sites.[71]

There is good evidence of a similarity between the quarternary structures of $F_0F_1$ ATP synthases from different sources. Yet there is only limited sequence homology, for example, between the $\delta$ subunits of *E. coli*[72] and spinach chloroplasts (36% including conservative replacements).[73] Nevertheless an attempt to construct functional hybrids between the *E. coli* $EF_1$ (minus $\delta$) and chloroplast $\delta$ and vice versa has been successful. The hybrid constructs plugged proton conduction through $CF_0$ or $EF_0$.[74] This is highly suggestive of a mechanical role of subunit $\delta$ in a conformational coupling mechanism between protons and ATP formation,[75] perhaps by a rotating binding site mechanism as proposed.[76]

## SUMMARY AND OUTLOOK

According to the chemiosmotic theory,[3] proton pumps and ATP synthases are coupled by lateral proton flow through aqueous phases. Three long-standing challenges to this concept, all of which have been loosely subsumed under 'localized coupling' in the literature, were examined in the light of experiments carried out with thylakoids: (1) *Nearest neighbor interaction between pumps and ATP synthases.* Considering the large distances between photosystem II and $CF_oCF_1$, in stacked thylakoids this is *a priori* absent. (2) *Enhanced proton diffusion along the surface of the membrane.* This could not be substantiated for the outer side of the thylakoid membrane. Even for the interface between pure lipid and water, two laboratories have reported the absence of enhanced diffusion. (3) *Localized proton ducts in the membrane.* Intramembrane domains that can transiently trap protons do exist in thylakoid membranes, but because of their limited storage capacity for protons, they probably do not matter for photophosphorylation under continuous light. Seemingly in favor of localized proton ducts is the failure of a supposedly permeant buffer to enhance the onset lag of photophosphorylation. However, it was found that failure of some buffers and the ability of others in this respect were correlated with their failure/ability to quench pH transients in the thylakoid lumen, as predicted by the chemiosmotic theory. It was shown that the chemiosmotic concept is a fair approximation, even for narrow aqueous phases, as in stacked thylakoids. These are approximately isopotential, and protons are taken in by the ATP synthase straight from the lumen.

The molecular mechanism by which $F_oF_1$ ATPases couple proton flow to ATP synthesis is still unknown. The threefold structural symmetry of the headpiece that, probably, finds a corollary in the channel portion of these enzymes appeals to the common wisdom that structural symmetry causes functional symmetry. "Rotation catalysis" has been proposed.[76] It is of heuristic value to visualize $CF_oCF_1$ as a mechanical coupling device. Its maximum turnover number ranges up to $400 \ s^{-1}$ for ATP and $1200 \ s^{-1}$ for protons. At about 200 mV electric driving force this implied a conductance of about 1 fS. Its channel portion ($CF_o$), however, has revealed a very large protonic conductance of 1 pS (three orders of magnitude greater than the protonic conductance of gramicidin around neutral pH). This was seemingly pH independent (between 5.8 and 8). The passage of other cations through $CF_o$ is strictly suppressed (even at pH 8 and with 300 mM NaCl in the medium). Components of $CF_o$, on the other hand, mainly the proteolipid subunit, can form $Na^+$-permeable cation channels in lipid bilayers. The magnitude and specificity of the proton conductance of $CF_o$ is not well understood physicochemically. In physiologic terms there is no need to supply protons at such high rate to a $10^3$-fold slower enzyme.

The pathway of protons after their entry into $CF_o$ is unknown, although several critical residues have been pinned down, for example, by site-directed mutagenesis (reviewed in refs. 3, 46, and 47). The entry of protons into the $CF_1$ portion with direct chemical action in the catalytic process has been proposed,[77] so far without any experimental evidence. If protons left the enzyme already at the interface between $CF_o$ and $CF_1$, a mechanical mode of energy transduction to the catalytic site had to be faced. Attempts have been made to identify the $CF_1$ subunits that throttle the very high rate of proton passage through $CF_0$ to the slow coupled rate. These pointed to a role for subunit $\delta$, an elongated protein (MW 20 k) that can block proton flow through $CF_0$. The nature of blocking is unknown, but again it is heuristically appealing to view $\delta$ as part of a conformational transducer.[75] The construction of hybrids from $F_1(-\delta)$, $F_O$, and $\delta$ taken from chloroplasts and from *E. coli* has revealed some functionality of the constructs. It is expected that further hybrids between even more remote enzymes of the $F_0F_1$ family may give a clue to a discrimination between mechanical and chemical coupling. Of course, the X-ray crystal structure analysis of this large enzyme (MW 550 k) is badly needed for any thorough understanding. Equally needed is a higher kinetic resolution of partial reactions and, if possible, of mechanical transients in the operating enzyme.

## ACKNOWLEDGMENTS

This article largely relies on the valuable contributions of my coworkers and colleagues, G. Althoff, A. Borchard, Dr. S. Engelbrecht, Dr. H. Lill, Dr. A. Polle, G. Schönknecht, and Dr. R. Wagner.

## REFERENCES

1. NELSON, N. & L. TAIZ. 1989. Trends Biochem. Sci. **3:** 113–116.
2. SENIOR, A. E. 1988. Physiol. Rev. **68:** 177–231.
3. MITCHELL, P. 1982. Biol. Rev. **41:** 445–502.
4. JUNGE, W. 1982. Curr. Top. Membr. Transp. **16:** 431–465.
5. ANDERSSON, B. & J. M. ANDERSON. 1980. Biochim. Biophys. Acta **593:** 427–440.
6. MITCHELL, P. 1977. FEBS. Lett. **78:** 1–20.
7. JUNGE, W. & H. T. WITT. 1987. Z. Naturforsch. **23b:** 244–254.
8. AUSLÄNDER, W. & W. JUNGE. 1975. FEBS. Lett. **59:** 310–315.
9. JUNGE, W., A. G. McGEER, W. AUSLÄNDER & I. RUNGE. 1979. Biochim. Biophys. Acta **546:** 121–141.
10. HONG, Y. Q. & W. JUNGE. 1983. Biochim. Biophys. Acta **722:** 197–208.
11. JUNGE, W., G. SCHÖNKNECHT & V. FÖRSTER. 1986. Biochim. Biophys. Acta **852:** 93–99.
12. POLLE, A. & W. JUNGE. 1986. Biochim. Biophys. Acta **848:** 257–264.
13. SCHÖNKNECHT, G., W. JUNGE, H. LILL & S. ENGELBRECHT. 1986 FEBS Lett. **203:** 289–294.
14. JUNGE, W. 1987. Proc. Natl. Acad. Sci. USA **48:** 7084–7088.
15. WESTHOFF, H. V. & K. VAN DAM. 1986. Mosaic Nonequilibrium Thermodynamics and the Control of Biological Free Energy Transduction. Elsevier, Amsterdam.
16. WESTHOFF, H. V., B. A. MELANDRI, G. F. VENTUROLI & D. B. KELL. 1984. Biochim. Biophys. Acta **768:** 257–292.
17. WILLIAMS, R. P. J. 1969. J. Theor. Biol. **1:** 1–13.
18. DILLEY, R. A., S. M. THEG & W. A. BEARD. 1987. Ann. Rev. Plant Physiol. **38:** 348–389.
19. HAINES, T. 1983. Proc. Natl. Acad. Sci. USA **80:** 160–164.
20. PRATS, M., J. TEISSIE & J. F. TOCANNE. 1986. Nature **322:** 756–758.
21. MURPHY, D. J. 1986. Biochim. Biophys. Acta **593:** 33–94.
22. McLAUGHLIN, S. 1977. Curr. Top. Membr. Transport **9:** 71–144.
23. SIGGEL, U. 1981. Bioelectrochem. Bioenerg. **8:** 327–346.
24. WILLE, B. 1988. Biochim. Biophys. Acta **936:** 513–530.
25. EIGEN, M. 1963. Angew, Chem. **75:** 489–588.

26. GUTMAN, M. & E. NACHLIEL. 1985. Biochemistry **24:** 2941–2946.
27. JUNGE, W. & A. POLLE. 1986. Biochim. Biophys. Acta **848:** 265–273.
28. POLLE, A. & W. JUNGE. 1989. Biophys. J. **56:** 27–31.
29. JUNGE, W. & S. MCLAUGHLIN. 1987. Biochim. Biophys. Acta **890:** 1–5.
30. GUTMAN, M., E. NACHLIEL & S. MOSHIACH. 1989. Biochemistry **28:** 2936–2941.
31. MENGER, F. M., S. D. RICHARDSON & G. R. BROMLEY. 1989. J. Am. Chem. Soc. **111:** 6893–6894.
32. STAUFFER, D. & A. CONIGLIO. 1987. Physica A. **143A:** 326–330.
33. RUPLEY, J. A., L. SIEMANKOWSKI, G. CARERI & F. BRUNI. 1988. Proc. Natl. Acad. Sci. USA **85:** 9022–9025.
34. HARAUX, F. & Y. DE KOUCHKOWSKY. 1982. Biochim. Biophys. Acta **679:** 235–247.
35. THEG, S. M. & P. H. HOMANN. 1982. Biochim. Biophys. Acta **679:** 221–234.
36. THEG, S. M. & W. JUNGE. 1983. Biochim. Biophys. Acta **723:** 294–307.
37. POLLE, A. & W. JUNGE. 1986. FEBS Lett. **198:** 263–267.
38. NEUMANN, J. & A. T. JAGENDORF. 1964. Arch. Biochem. Biophys. **107:** 109–119.
39. WITT, H. T., E. SCHLODDER & P. GRÄBER. 1976. FEBS Lett. **69:** 272–276.
40. SCHLODDER, E., P. GRÄBER & H. T. WITT. 1982. *In* Electron Transport and Phosphorylation. J. Barber, ed.: 105–175, Elsevier, Amsterdam.
41. SCHMIDT, G. & P. GRÄBER. 1985. Z. Naturforsch. **42C:** 231–236.
42. THEG, S. M., G. CHIANG & R. A. DILLEY. 1988. J. Biol. Chem. **263:** 673–681.
43. BEARD, W. A. & R. A. DILLEY. 1986. FEBS Lett. **201:** 57–62.
44. MORONEY, J. V., L. LOPRESTI, B. F. MCEVEN, R. E. MCCARTY & G. HAMMES. 1983. FEBS Lett. **158:** 58–62.
45. TIEDGE, H., H., LÜNSDORF, G. SCHÄFER & H. U. SCHAIRER. 1985. Proc. Natl. Acad. Sci. USA **82:** 7874–7878.
46. HOPPE, J. & W. SEBALD. 1984. Biochim. Biophys. Acta **768:** 1–27.
47. SCHNEIDER, E. & K. ALTENDORF. 1987. Microbiol. Rev. **51:** 477–497.
48. SCHÖNKNECHT, G., R. HEDRICH, W. JUNGE & K. RASCHKE. 1988. Nature **336:** 589–592.
49. LILL, H., G. ALTHOFF & W. JUNGE. 1987. J. Membr. Biol. **98:** 69–78.
50. ALTHOFF, G., H. LILL & W. JUNGE. 1989. J. Membr. Biol. **108:** 263–271.
51. LILL, H., S. ENGELBRECHT, G. SCHÖNKNECHT & W. JUNGE. 1986. Eur. J. Biochem. **160:** 627–634.
52. SCHMID, R. & W. JUNGE. 1975. Biochim. Biophys. Acta **394:** 76–92.
53. LAUBINGER, W. & P. DIMROTH. 1987. Eur. J. Biochem. **168:** 475–480.
54. KASIANOWICZ, J., R. BENZ & S. MCLAUGHLIN. 1987. J. Membr. Biol. **95:** 73–89.
55. BRÜNGER, A., Z. SCHULTEN & K. SCHULTEN. 1983. Z. Phys. Chem. **136:** 1–63.
56. HILLE, B. 1984. Ionic Channels of Excitable Membranes. Sinauer Associates, Inc., Sunderland, Mass.
57. PESKOFF, A. & D. M. BERS. 1988. Biophys. J. **53:** 863–875.
58. WAGNER, R., E. C. APLEY & W. HANKE. 1989. EMBO J. **8:** 2827–2834.
59. KELLER, B. U., R. HEDRICH, W. L. C. VAZ & M. CRIADO. 1988. Pflügers Arch. **411:** 94–100.
60. SCHÖNKNECHT, G., G. ALTHOFF, A. C. APLEY, R. WAGNER & W. JUNGE. 1989. FEBS Lett. In press.
61. LEAR, J. D., Z. R. WASSERMANN & W. F. DEGRADO. 1988. Science **240:** 1177–1181.
62. WEISS, M. A. & R. E. MCCARTY. 1977. J. Biol. Chem. **252:** 8007–8012.
63. JUNESCH, U. & P. GRÄBER. 1987. Biochim. Biophys. Acta **893:** 275–288.
64. RICHTER, M. L. & R. E. MCCARTY. 1987. J. Biol. Chem. **262:** 15037–15040.
65. ANDREO, C. S., W. J. PATRIE & R. E. MCCARTY. 1982. J. Biol. Chem. **257:** 9968–9975.
66. JUNGE, W., Y. Q. HONG, L. P. QIAN & A. VIALE. 1984. Proc. Natl. Acad. Sci. USA **81:** 3078–3082.
67. ENGELBRECHT, S. & W. JUNGE. 1987. FEBS Lett. **219:** 321–325.
68. LILL, H., S. ENGELBRECHT & W. JUNGE. 1988. J. Biol. Chem. **263:** 14518–14523.
69. ENGELBRECHT, S. & W. JUNGE. 1987. Eur. J. Biochem. **172:** 213–218.
70. ENGELBRECHT, S., K. SCHÜRMANN & W. JUNGE. 1989. Eur. J. Biochem. **1179:** 117–122.

71. WAGNER, R., E. C. APLEY, S. ENGELBRECHT & W. JUNGE. 1988. FEBS Lett. **230:** 109–115.
72. GAY, N. J. & J. WALKER. 1981. Nucleic Acids Res. **9:** 3919–3926.
73. HERMANS, J., C. H. ROTHER, J. BICHLER, J. STEPPUHN & R. G. HERRMANN. 1988. Plant Mol. Biol. **10:** 323–330.
74. ENGELBRECHT, S., G. DECKERS-HEBESTREIT, K. ALTENDORF & W. JUNGE. 1989. Eur. J. Biochem. **181:** 485–491.
75. ENGELBRECHT, S. & W. JUNGE. 1989. Biochem. Biophys. Acta, in press.
76. BOYER, P. 1987. Biochemistry **26:** 8503–8507.
77. MITCHELL, P. 1985. FEBS Lett. **182:** 1–7.

# Proton Channels in Snail Neurones

## Does Calcium Entry Mimic the Effects of Proton Influx?[a]

ROGER C. THOMAS

*Department of Physiology*
*School of Medical Sciences*
*University of Bristol*
*Bristol, BS8 1TD, UK*

Hydrogen ions are normally found at very low free concentrations (less than $10^{-7}$ molar) both inside and outside neurones.[1,2] Thus a significant $H^+$ or proton current was hard to imagine 10 years ago when I discovered[3] that snail neurones depolarized by isotonic KCl rapidly lost injected $H^+$. At the time I thought K/H exchange might be the explanation. Meanwhile, Robert Meech had discovered that intracellular pH rapidly changed when he voltage-clamped snail neurones to positive membrane potentials. We subsequently collaborated[4] to show that the membrane became highly permeable to $H^+$ when depolarized, that $pH_i$ in depolarized cells was determined by $pH_o$ and the membrane potential ($E_m$), and that a current was carried by $H^+$ leaving after an injection into a depolarized cell. The $H^+$ appeared to leave via channels blocked by cadmium, cobalt, and other heavy metals, but not blocked by $Cs^+$ or inhibitors of $pH_i$ regulation. More recently outward currents carried by $H^+$ have been found in axolotl oocytes[5] and in pond snail neurones.[6]

No clear inward $H^+$ current has so far been detected; however, measurements on surface $pH$[7] recently led me to believe that $H^+$ could enter snail neurones when $pH_i$ was relatively high and $E_m$ was depolarized to around zero. I concluded that the putative channels were perhaps not invariably rectifying. Related $pH_i$ decreases with modest depolarizations had, however, been previously attributed[8] to $Ca^{2+}$ influx, as shown in FIGURE 1. That $pH_i$ is decreased by Ca entry or injection, via interaction with intracellular binding sites and subcellular organelles, has been known for many years.[9,10]

The experiments described in this contribution were designed to establish the cause of the rather variable $pH_i$ decrease seen with modest depolarizations of voltage-clamped snail neurones. This $pH_i$ increase is only seen when cells have near-normal $pH_i$, a condition not often obtained during earlier experiments. I have found that it is relatively easy to block the $pH_i$ increase, but the $pH_i$ decrease can be blocked only by complete removal of external calcium. I conclude that the $pH_i$ decrease is caused by a surprisingly large $Ca^{2+}$ entry, which is anomalously reduced by raising external Ca, and very variable from cell to cell. This confirms earlier conclusions from current measurements that the proton channels appear to be strongly rectifying.

[a] This work was supported in part by the Medical Research Council.

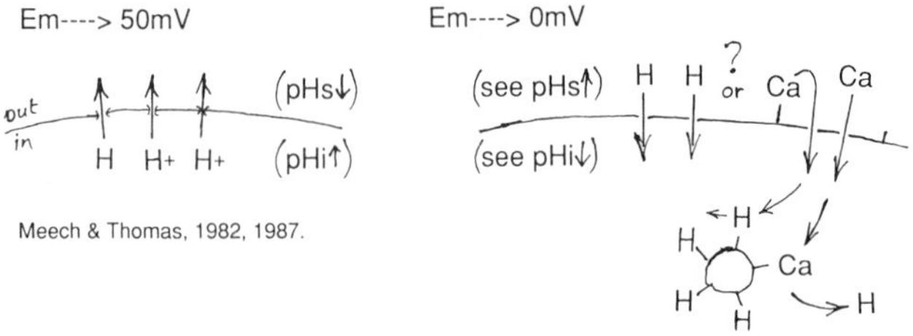

**FIGURE 1.** Schematic diagram showing proposed ion movements changing surface and intracellular pH ($pH_s$ and $pH_i$) of neurones clamped at +50 mV or about 0 mV. Calcium ions are suggested to vacate surface binding sites which subsequently bind $H^+$ and to displace $H^+$ from intracellular binding sites.

## METHODS

Experiments were done on exposed 70–150 $\mu$m diameter cells in the brain of the common snail, *Helix aspersa*, dissected as previously described.[7] The normal snail Ringer's solution was $CO_2$ free and contained (mM): NaCl 80, KCl 4, CaCl$_2$ 7, MgCl$_2$ 5, and Hepes (2-N-2-hydroxyethylpiperazine-N′-2-ethanesulfonic acid) 20 mM, the pH being adjusted by the addition of NaOH. Solutions with lower concentrations of Hepes had additional NaCl to compensate; the pH electrodes were calibrated in a Ringer's solution buffered to pH 6.5 with 20 mM Pipes (piperazine-N,N-bis-2 ethanesulfonic acid).

The pH measurements were done using eccentric double-barrelled microelectrodes that were made as previously described[7] from aluminosilicate glass, except that they were silanized under vacuum. They were usually used the same day as they were filled and always required careful tip breakage before the reference or pH side had a reasonable resistance or pH response, respectively. An acceptable $pH_i$ electrode had a reference side resistance of 20–40 megohms and a pH response of over 50 mV per pH unit. Microelectrodes for voltage clamping were filled with 3 M CsCl.

Voltages from the amplifiers connected to the various electrodes and current and voltage clamp outputs were displayed on an oscilloscope and potentiometric pen recorder, and stored on floppy discs by an Apple IIe microcomputer for later analysis. The measured values of $pH_i$ used to plot graphs were all taken from the data stored on discs rather than from the pen recordings.

## RESULTS

Most previous experiments[4-6,8] on proton channels and currents were done with acid-loaded cells to maximize the currents or pH changes seen on depolarization. With a near-normal $pH_i$, the pH changes are relatively small, as shown in FIGURE 2. In this experiment I recorded both surface and intracellular pH ($pH_s$ and $pH_i$) while clamping in 10 mV for 5s steps from −50 to +40 mV. Ignoring the first series, marred by electrode problems, the second series shows typical $pH_i$ and $pH_s$ changes. From −40 to about zero membrane potential, $pH_i$ decreased and $pH_s$ slightly increased. From zero to +40, $pH_i$ increased and $pH_s$ decreased, the latter much more than it had increased earlier. As shown before,[4,7,8] there is much evidence that the $pH_i$ increase and $pH_s$ decrease are due to $H^+$ efflux through channels or possibly a rheogenic carrier. No other ions are required for $H^+$ efflux to occur.[8]

The $pH_s$ increase and $pH_i$ decrease could be caused directly by $H^+$ influx or indirectly by $Ca^{2+}$ influx. I have investigated the role of calcium by raising and lowering the superfusate Ca concentration from its normal 7 mM (without any Ca buffers) in the experiment of FIGURE 2. Both changes in Ca reduced both $pH_s$ and $pH_i$ changes on depolarization, the increase in Ca being more effective than its removal. This result is hard to explain with a simple Ca entry model.

I therefore tested various channel blockers. A typical experiment is illustrated in FIGURE 3. This neurone was unusually sensitive to modest depolarizations, exhibiting a remarkably fast $pH_i$ decrease and a large subsequent $pH_i$ increase. The Ca-channel blocker nicardipine (10 $\mu$M) had only a small reversible effect on the $pH_i$ changes. However, zinc, shown by Mahaut-Smith[11] to be the most potent blocker of $H^+$ currents in snail neurone, almost completely blocked any $pH_i$ increase. Zinc had little effect on the $pH_i$ *decrease,* suggesting that the two processes are indeed different. Zinc also slowed the $pH_i$ recovery from the previous increase.

The effect of nicardipine is analyzed in more detail in FIGURE 4 which is a plot of the change in $pH_i$ (referred to its value just before the voltage clamp series) against clamp potential. I also tested nicardipine at 100 $\mu$M and nimodipine at 10 $\mu$M, with equally unimpressive results.

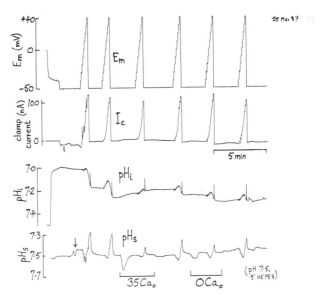

**FIGURE 2.** Records of membrane potential ($E_m$) clamp current ($I_c$) intracellular pH ($pH_i$) and surface pH ($pH_s$) plotted against time from an experiment to show the effects of raising to 35 mM or removing extracellular Ca on the pH changes induced by depolarizing in steps each of 10 mV for 5 seconds. Throughout, the superfusate was buffered at pH 7.5 with 5 mM Hepes. The records start at the point where the combination $E_m$ and $pH_i$ microelectrode was inserted. The *arrow* marks the point where the surface electrode was pushed against the cell.

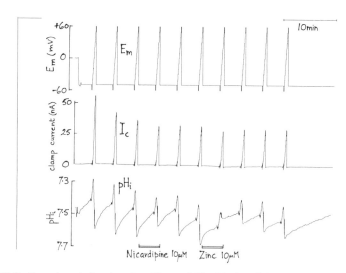

**FIGURE 3.** Experiment showing the effects of 10 $\mu$M nicardipine and $ZnCl_2$ on the $pH_i$ changes induced by voltage clamp steps from $-60$ to $+60$ mV, 3.75 s at each potential. Superfusate was buffered at pH 7.5 with 20 mM Hepes.

A similar $pH_i$ change against clamp potential plot for two levels of zinc is shown in FIGURE 5. It appears that 10 $\mu$M zinc caused over 90% inhibition of the $H^+$ efflux at positive potentials, although it was not so potent in the earlier experiments on $H^+$ currents.[11]

I have also attempted to block $pH_i$ decreases by injecting decamethonium.[12] It had no detectable effect. Nevertheless it seemed essential to repeat the earlier Ca-

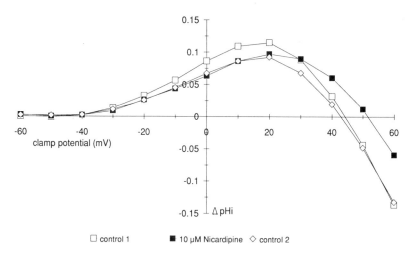

**FIGURE 4.** Graph of the change of $pH_i$ (from its value just before $E_m$ was increased to $-60$ mV) against clamp potential before, during, and after superfusion with 10 $\mu$M nicardipine, taken from the experiment shown in FIGURE 3.

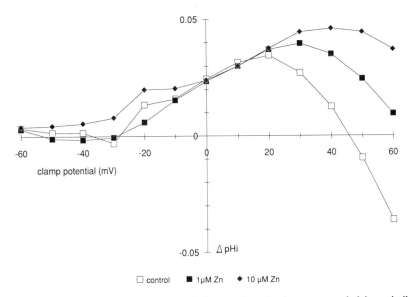

**FIGURE 5.** Graph showing the effect of zinc on the $pH_i$ changes recorded in a similar experiment to that in FIGURE 3, except that zinc was applied at two different concentrations.

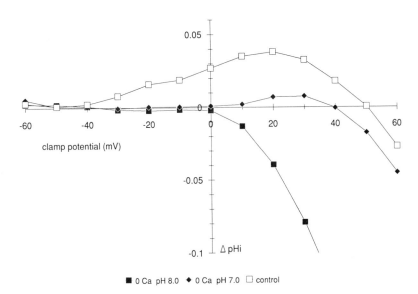

**FIGURE 6.** The effect of Ca removal at two different pH values on the $pH_i$ changes induced by voltage clamp steps. Control pH 7.5. The OCa solutions both contained 1 mM EGTA to minimize free $Ca^{2+}$. Experimental procedure similar to that in FIGURE 3 was used.

free experiment, but with added 1 mM EGTA to reduce Ca levels even further than in simple zero Ca (Mg increased to keep total divalents constant).

FIGURE 6 shows data from an experiment in which OCa EGTA was tested at two pH levels, in case the EGTA somehow mimicked a change in $pH_o$. The test solutions were superfused for 3 minutes before each depolarization series. In this and several other similar experiments OCa EGTA essentially abolished the $pH_i$ increases. When both control and OCa EGTA were applied at the same pH (not shown), the $pH_i$ increase at positive potentials was much larger and faster in OCa. It is also noteworthy that the shift of the OCa lines along the clamp potential axis with $pH_o$ is larger than previously seen[13] with normal Ca levels. Other ways of modifying the apparent Ca influx have been tested. Substituting barium for calcium caused a small decrease in $pH_i$, but had little effect on the $pH_i$ response to a series of voltage clamp steps. Similarly, substituting Li for Na, which should interfere with possible Na:Ca exchange, had little significant effect.

## CONCLUSIONS

The foregoing results confirm earlier reports that $H^+$ currents are unidirectional, only flowing out of the cell. The $pH_i$ and $pH_s$ changes seen with modest depolarizations appear to be indirectly due to $Ca^{2+}$ entry. The entering $Ca^{2+}$ presumably exposes surface $H^+$ binding sites which raise $pH_s$, and inside the cell displace $H^+$ from internal sites and organelles.[9]

The effects of raising external Ca fivefold, as shown in FIGURE 2, are difficult to explain. Why does increasing the number of external $Ca^{2+}$ ions seem to allow fewer to enter? Also puzzling is the lack of effect of what must be at least a low external Ca. Complex interaction between $H^+$ and $Ca^{2+}$ and the channel properties must be occurring. A specific Ca entry blocker may be difficult to find if, as seems likely, Ca entry occurs by a variety of pathways, not necessarily all channels. It will, however, be difficult to study the pH effects of $H^+$ movement in isolation except at very low levels of external Ca.

The physiologic significance of the apparent proton channels is very hard to assess. Perhaps they provide a pathway that helps prevent the buildup near the inside of the cell membrane of too many $H^+$ displaced from binding sites by entering $Ca^{2+}$. It is still uncertain how large or widely distributed are the proposed channels. Recent patch clamp studies by Byerly and Suen[14] have revealed that the unitary proton current is very small, too small (less than 0.004 pA) to resolve with present methods.

## ACKNOWLEDGMENTS

Many thanks to Mike Rickard for programming various computers to record, analyze, and plot my data, and to Sue Maskell for exceptional secretarial speed. I am also grateful to Lou Byerly for sending me unpublished work.

## REFERENCES

1. ROOS, A. & W. F. BORON. 1981. Intracellular pH. Physiol. Rev. **61:** 296–434.
2. THOMAS, R. C. 1984. Review lecture. Experimental displacement of intracellular pH and the mechanism of its subsequent recovery. J. Physiol. (Lond) **354:** 3–22P.

3. THOMAS, R. C. 1979. Recovery of pH$_i$ in snail neurones exposed to high external potassium. J. Physiol. (Lond) **296:** 77P.

4. THOMAS, R. C. & R. W. MEECH. 1982. Hydrogen ion currents and intracellular pH in depolarised voltage-clamped snail neurones. Nature **299:** 826–828.

5. BARISH, M. E. & C. BAUD. 1984. A voltage-gated hydrogen ion current in the oocyte membrane of the axolotl, Ambystoma. J. Physiol. (Lond) **352:** 243–263.

6. BYERLY, L., R. W. MEECH & W. MOODY. 1984. Rapidly-activating hydrogen ion currents in perfused neurones of the snail *Lymnaea stagnalis*. J. Physiol. (Lond) **351:** 199–216.

7. THOMAS, R. C. 1988. Changes in the surface pH of voltage-clamped snail neurones apparently caused by H$^+$ fluxes through a channel. J. Physiol. (Lond) **398:** 313–327.

8. MEECH, R. W. & R. C. THOMAS. 1987. Voltage-dependent intracellular pH in Helix aspersa neurones. J. Physiol. (Lond) **390:** 433–452.

9. MEECH, R. W. & R. C. THOMAS. 1977. The effect of calcium injection on the intracellular sodium and pH of snail neurones. J. Physiol. (Lond) **265:** 867–879.

10. AHMED, Z. & J. A. CONNOR. 1980. Intracellular pH changes induced by calcium influx during electrical activity in molluscan neurons. J. Gen. Physiol. **75:** 403–426.

11. MAHAUT-SMITH, M. 1987. The effect of zinc on calcium and hydrogen ion currents in snail neurones. J. Physiol. (Lond) **382:** 129P.

12. AUGUSTINE, G. J., M. P. CHARLTON & R. HORN. 1988. Role of calcium-activated potassium channels in transmitter release at the squid giant synapse. J. Physiol. (Lond) **398:** 149–164.

13. THOMAS, R. C. 1988. Proton channels in snail neurons studied with surface pH glass microelectrodes. *In* Proton Passage Across Cell Membranes. :168–183. Wiley, Chichester (Ciba Foundation Symposium 139).

14. BYERLY, L. & Y. SUEN. 1989. Characterisation of proton currents in neurones of the snail, *Lymnaea stagnalis*. J. Physiol. (Lond) **413:** 75–89.

# Cell Volume Changes and the Activity of the Chloride Conductance Path

ASER ROTHSTEIN AND CHRISTINE BEAR

*Research Institute*
*Hospital for Sick Children*
*Toronto, Canada M5G1X8*

Water is the main volume-filling component of the cell, occupying some 70% of cell space. Control of volume is therefore almost synonomous with control of water content. In animal cells water, because it can pass relatively rapidly through the cell membrane, is usually in virtual osmotic equilibrium.[1-4] Consequently, water content and size are largely determined by the cell's total osmotic content. In turn, because much of the osmotic content is $K^+$, $Na^+$, and $Cl^-$, the control of these ions is of primary importance in cell volume regulation. Other osmolytes, particularly small nonelectrolytes, may also contribute, particularly in cells of organisms that must face large changes in the osmolarity of their environments.[5]

The control of ion content depends on ion pumps that generate ion gradients, balanced against pathways that allow slow dissipation of gradients. The balance between pumps and dissipating systems results in steady-state levels of ions and is therefore a prime determinant of cell volume. Maintenance of steady-state gradients is essential. If ions were allowed to approach equilibrium, cells would swell continuously because of uncompensated osmotic pressure of their nondiffusible components, primarily charged macromolecules. In its early formulation the concept of volume control by ion steady-states was known as the "pump and leak hypothesis,"[6] the most important pump being the $Na^+/K^+$ ATPase, responsible for maintaining $Na^+/K^+$ gradients. Initially, "leaks" were considered to be permeation paths for individual ion species. Because ions are charged, their movements via such pathways would be conductive, involving current flow, generation of membrane potentials, and electrical constraints on ion movements. For salt to be gained or lost, with corresponding changes in volume, cations and anions would have to permeate at the same time (they would be electrically coupled).

In recent years it has become evident that the pump and leak hypothesis was based on an oversimplified view of the leak, because recent information indicates that two additional classes of transport pathways are particularly important in control of ion content (and cell volume), the ion cotransporters and ion exchangers. In both classes the movement of one ion is tightly coupled to the movements of one or more other ions. Consequently the downhill gradient of one ion can drive another ion in the uphill direction, a process called secondary active transport. In these coupled flows no net charge is necessarily moved, so in contrast to leak pathways the processes are usually electroneutral. Exchanger activities, for example $Na^+/H^+$, or $Cl^-/HCO_3^-$ do not result directly in a net gain in osmolyte, but their parallel operation can result in large net inflows of NaCl driven by the preexisting $Na^+$ gradient, resulting in cell swelling ($H^+$ and $HCO_3^-$ neutralize each other).[3,7] The cotransport systems, $Na^+/Cl^-$, $K^+/Cl^-$, and $Na^+/K^+/2Cl^-$, are essentially salt moving systems that directly modulate volume changes.[7-10]

Although the present paper is especially concerned with the role of conductive $Cl^-$ channels in cell volume regulation, it is important to recognize that $Cl^-$ flow and associated volume changes can also be mediated by cotransport or ion exchange systems. Experimentally it is essential to differentiate between the various $Cl^-$ pathways, conductive, exchange, and cotransport. This is not always an easy task.

No formal model has been proposed to accommodate all of the pathways that contribute to ion steady states. It would be a formidable task because of the number of contributing pathways, changes in the pathways due to intervention of regulatory phenomena, and the different "mixes" of pathway activities in different cell types. Under normal circumstances, cell volume adjustments would usually be modest and difficult to distinguish experimentally in terms of small increases in activity of particular pathways. To overcome this difficulty, the usual strategy is to perturb volume dramatically and to examine the cell's capacity to accommodate. Usually the perturbation is imposed by changing the tonicity of the medium in order to osmotically swell or shrink the cells.[2,3,7,8,10] In shrunken cells, depending on the cell type, either cotransport systems or ion exchange systems are activated, allowing NaCl gain (and volume compensation), driven by the preexisting $Na^+$ gradient.[2,3,7,10] In swollen cells, volume compensation results primarily from KCl loss driven by the outward $K^+$ gradient. Two patterns have also been observed. In red blood cells, $Cl^-$ loss occurs via an electroneutral $K^+/Cl^-$ cotransport,[2,7,8] whereas $Cl^-$ loss in most cell types, KCl loss is achieved by opening of independent, conductive, $K^+$ and $Cl^-$ channels.[2,3,7,10] It is the latter phenomenon, particularly the volume-activated $Cl^-$ channel, that will be discussed here. Because more is known about the phenomenon in lymphocytes and Ehrlich ascites cells, these cell types will generally, but not exclusively, be used to illustrate the underlying phenomena.

## REGULATORY VOLUME DECREASE: VOLUME-ACTIVATED $Cl^-$ PERMEABILITY

Most cells, when osmotically swollen by exposure to diluted medium, will soon shrink back toward normal size, a phenomenon usually called regulatory volume decrease or RVD.[2,3,7,10] The phenomenon is not triggered by low tonicity per se, because it will also occur if cells are swollen under isotonic conditions by uptake of a permeant osmolyte such as urea[11] or of substrates such as sugars or amino acids.[12,13]

A typical RVD for human peripheral blood lymphocytes[14] is illustrated in FIGURE 1 (lower curve). Because the cells are more permeable to water than to osmolytes, the initial response to diluted medium is an osmotic swelling due to water equilibration. The somewhat slower reshrinking phase is the RVD. Early studies with mouse lymphoblasts demonstrated that RVD resulted from a loss of KCl driven by the preexisting outward $K^+$ gradient via volume activation of a $K^+$ permeability pathway.[15] The importance of $K^+$ permeability is illustrated in the upper curve of FIGURE 1. Cells suspended in KCl medium do not undergo RVD. Instead, after the initial osmotic response, they undergo a large secondary swelling. The increased $K^+$ permeability induced by osmotic swelling allows uptake of KCl driven by the KCl gradient, which is in the inward direction under these circumstances. If $Cl^-$ is replaced by the impermeant anion, $SO_4^{--}$, the secondary swelling is blocked (middle curve). The latter data demonstrate the importance of permeant anions in volume responses.

Earlier studies were focused almost exclusively on the large changes in $K^+$ permeability triggered by cell swelling, and it was implicitly assumed that $Cl^-$ also escaped the cell in some undefined manner. In red blood cells, one of the first types used to study RVD[16] (nucleated cells of birds), no special $Cl^-$ mechanism was required because anion permeability is normally exceptionally high. In other cells, however, conductive $Cl^-$ permeability may be relatively low, limiting salt outflow. For example, membrane potentials of many cell types are dominated by the $K^+$ gradient, indicating that $K^+$ permeability substantially exceeds $Cl^-$ permeability. This circumstance clearly holds in the case of lymphocytes based on evaluation of relative conductive permeabilities to $K^+$ and $Cl^-$ measured experimentally by the flourescent dye technique.[17,18] The potential is highly dependent on the external $K^+$ concentration (FIG. 2), but much less dependent on the $Cl^-$

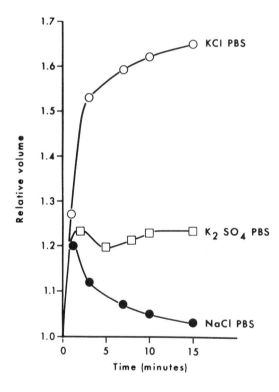

**FIGURE 1.** Volume adjustments of peripheral blood monocytes exposed to hypotonic medium: effect of the medium composition. Cells were suspended in isotonic phosphate-buffered media of the indicated salts. At zero time the cells were subjected to hypotonic challenge (0.67 × isotonic), and volumes were measured by the Coulter counter. The figure is reproduced from ref. 14.

concentration (and the $Na^+$ concentration). These findings suggest that in lymphocytes, conductive permeability to $K^+$ is relatively high compared to conductive permeability to $Cl^-$. Another kind of experiment leads to the same conclusion. If lymphocytes are treated with valinomycin or gramicidin to substantially increase cation permeability, little volume change occurs[17,19] despite large ion gradients. Salt gains or losses and associated volume changes must therefore be limited by low conductive $Cl^-$ permeability in isotonic cells. Paradoxically, $Cl^-$ fluxes are considerably higher[3] than those for $K^+$ (compare control values of Fig. 3A and B). Clearly, a nonconductive pathway (anion exchange) allows $Cl^-$ to pass rapidly through the membrane, and consequently, the conductive component constitutes only a small fraction of the total flux.

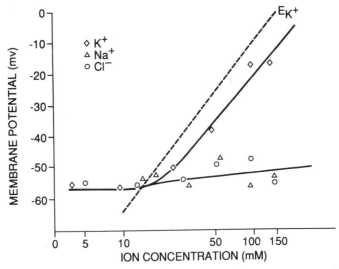

**FIGURE 2.** The membrane potentials of peripheral blood lymphocytes as a function of external ion concentrations. The potentials were determined using the fluorescent dye, 3,3'-dipropylthiadicarbocyanine, calibrated by the valinomycin procedure. $K^+$, $Na^+$, and $Cl^-$ were isosmotically replaced by $Na^+$, choline$^+$ and isothionate$^-$, respectively. The dotted line represents the theoretical Nernst potential for $K^+$. The figure is redrawn from ref. 17 (Fig. 5).

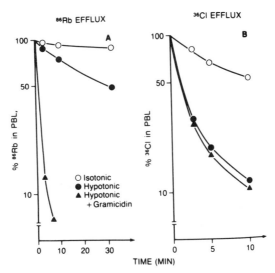

**FIGURE 3.** Effect of hypotonic challenge and gramicidin on $K^+$ ($^{86}Rb^+$) and $^{36}Cl^-$ fluxes of peripheral blood lymphocytes. Cells were loaded with isotope, washed, and resuspended in isotonic or hypotonic (0.67 × isotonic) medium with or without 0.5 $\mu$M gramicidin. Aliquots were taken at intervals, and the isotope remaining in the cells was determined after centrifugation through oil. The figure is reproduced from reference 3.

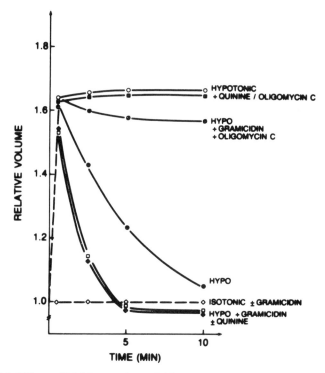

**FIGURE 4.** Effects of inhibitors and gramicidin on RVD of peripheral blood lymphocytes. Cells were suspended in isotonic choline-Cl. At zero time they were subjected to hypotonic challenge (0.67 × isotonic) in the presence of the indicated substances, and volumes were measured by the Coulter counter. Concentrations were: quinine, 100 $\mu$M; oligomycin, 5 $\mu$g/ml; gramicidin, 0.5 $\mu$M. The figure is taken from unpublished data associated with reference 20.

Given that the conductive Cl$^-$ permeability of isotonic cells is relatively low compared to that of K$^+$, rapid RVD could not occur if K$^+$ permeability alone were activated. Swelling must therefore activate a large conductive anion permeability as well. In fact, osmotic swelling induces large increases in both K$^+$ and Cl$^-$ fluxes,[3] as is illustrated in FIGURE 3. The fluxes become extraordinarily large, allowing escape of a large fraction of cell KCl in a relatively short time. Because the fluxes of Cl$^-$ are, in fact, increased to a greater degree than are those of K$^+$, it would be expected that in swollen cells Cl$^-$ conductance would become dominant in determining membrane potential. As expected, the membranes became depolarized[17] and the membrane potentials became responsive to the Cl$^-$ concentration gradient, indicating that a large part of the volume-activated Cl$^-$ flux must be conductive in nature and that during RVD Cl$^-$ conductance exceeds that of K$^+$.

Measurements of membrane potential have only limited value in assessing conductive Cl$^-$ permeability. The changes reflect shifts in the relative sizes of K$^+$ and Cl$^-$ conductances, but not the magnitudes of their fluxes. Flux measurements, as already noted, include nonconductive as well as conductive components. A simple method to assess conductive Cl$^-$ fluxes is based on volume changes in the presence of cation ionophores. The ionophore ensures that cation flow is so rapid that Cl⁻ permeability is limiting to salt movement and, therefore, to volume change. In practice, with lymphocytes, it was found that gramicidin was particularly useful. Its striking and cation-specific stimulation of $K^+$-flux is illustrated in FIGURE 3. Because it also permits rapid permeation of $Na^+$, leading to cell swelling, measurement of RVD (shrinkage) requires substitution of an impermeant cation, such as $choline^+$ or N-methyl-glucamine⁺. Cl⁻ permeability can also be calculated from the rate of cell swelling of cells suspended in KCl medium.

As already noted, addition of gramicidin to lymphocytes suspended in isotonic medium causes little or no change in volume, confirming that under isotonic conditions conductive Cl⁻ permeability is relatively low. In contrast, the effect of the ionophore on osmotically swollen cells is substantial.[19] RVD is speeded up (FIG. 4), as is the secondary swelling of cells suspended in KCl medium (FIG. 5), indicating that in swollen cells $K^+$ permeability must have been rate limiting to salt loss or gain. These results are entirely consistent with those just cited, based on flux and potential data. They indicate that RVD results from an increase in conductive $K^+$ permeability and an even greater increase in Cl⁻ permeability, with reversal in their relative magnitudes.

In the experiment in FIGURE 4, it can be noted that RVD is inhibited by quinine, but that the block is overcome by gramicidin. Quinine, in general, is a blocker of $K^+$ pathways and it appears to act in the same manner in volume activation. The fact that the ionophore can bypass the inhibition indicates that

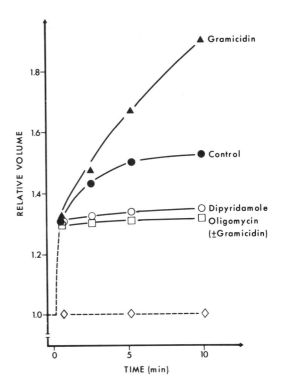

**FIGURE 5.** Effects of inhibitors and gramicidin on secondary swelling of peripheral blood lymphocytes exposed to hypotonic KCl. Cells were suspended in isotonic KCl. At zero time they were subjected to hypotonic challenge (0.67 × isotonic) in the presence of the indicated substances, and volumes were measured by use of the Coulter counter. Concentrations were: dipyridamol, 100 $\mu$M; oligomycin, 5 $\mu$g/ml; gramicidin, 0.5 $\mu$M. The figure is reproduced from ref. 19.

quinine must act by blocking the conductive $K^+$, but not the $Cl^-$ pathway. In contrast, inhibition of RVD by oligomycin C is not bypassed by gramicidin. Its effect must therefore be exerted on the $Cl^-$ pathway. Similarly, FIGURE 5 illustrates that oligomycin and dipyridamol inhibit secondary swelling in KCl medium, with no unblocking by gramicidin, a clear indication that these inhibitors inhibit the $Cl^-$ pathway.

In addition to being pharmacologically distinct with substantially different sensitivities to a variety of inhibitors,[20] the $K^+$ and $Cl^-$ pathways differ in other characteristics. They are not present, for example, in any fixed relationship to each other, and they are distinctly different in degree of volume activation, time course, and extent, resulting in large changes in membrane potential (normally depolarization). The distinction between the pathways is exaggerated in some cells. Thus, in B lymphocytes, volume activation of the $K^+$ channel is minimal, whereas activation of the $Cl^-$ pathway is considerable.[21] Consequently, RVD is slow or minimal unless gramicidin is added to overcome the deficit in cation permeability. In Ehrlich ascites cells, RVD also involves distinct volume-activated $K^+$ and $Cl^-$ channels,[10,22] with many properties similar to those of lymphocytes.

## PROPERTIES OF THE VOLUME-ACTIVATED Cl⁻ PATHWAY

Certain properties of the volume-activated $Cl^-$ pathway, as already noted, are: (1) it is electrogenic; (2) it is distinct from the $K^+$ pathway, but is electrically coupled to it, as would be expected, so that when both pathways are activated by cell swelling, KCl and osmotically obliged water will leave the cell; (3) when activated, it can allow exceptionally rapid fluxes, so that more than 30% of the cell's KCl can escape in a few minutes; and (4) it has a specific susceptibility to inhibitors, one of the most potent being oligomycin C.[20] Other properties have largely been elucidated by measuring volume changes of gramicidin-treated cells. The anion specificity is $SCN^-/I^-/NO_3^-/Br^-/Cl^-/acetate^-/SO_4^{--}/gluconate^-$. A surprising finding is that volume activation, at least in lymphocytes, appears to be an all-or-none phenomenon for individual cells.[23] Activation is not evident until cell swelling is about 10%; activation increases with swelling up to about 20%; and greater degrees of swelling do not result in further increases. In the intermediate range of 15% swelling, two populations of cell size are observed, those that have and those that have not responded (FIG. 6). At 20%, all cells have apparently responded, and the population is homogeneous, with all the cells enlarged. This behavior can be explained by assuming that each cell has a finite threshold of response in the range of 10–18% swelling. Once activation occurs, each cell appears to respond maximally. The threshold operates in both directions. Cells swollen in diluted KCl medium, with $Cl^-$ pathways open, when suddenly reduced in size by the addition of concentrated KCl to a value just below the threshold (10%), respond by shutting down the $Cl^-$ pathway.

Activation appears to be related to cell size per se. If cells are allowed to go through a cycle of RVD, thereby becoming partially depleted of KCl, another cycle of RVD can be initiated by further dilution of the medium. The threshold for activation of the new cycle occurs at exactly the same size as the first. Thus, osmolarity and ionic concentrations per se are not factors. This all-or-none behavior is in contrast to that of the volume-activated $K^+$ pathway, in which the response is clearly proportional to the degree of swelling.[24] The $Cl^-$ threshold phenomenon has not been explored to any degree in other cell types.

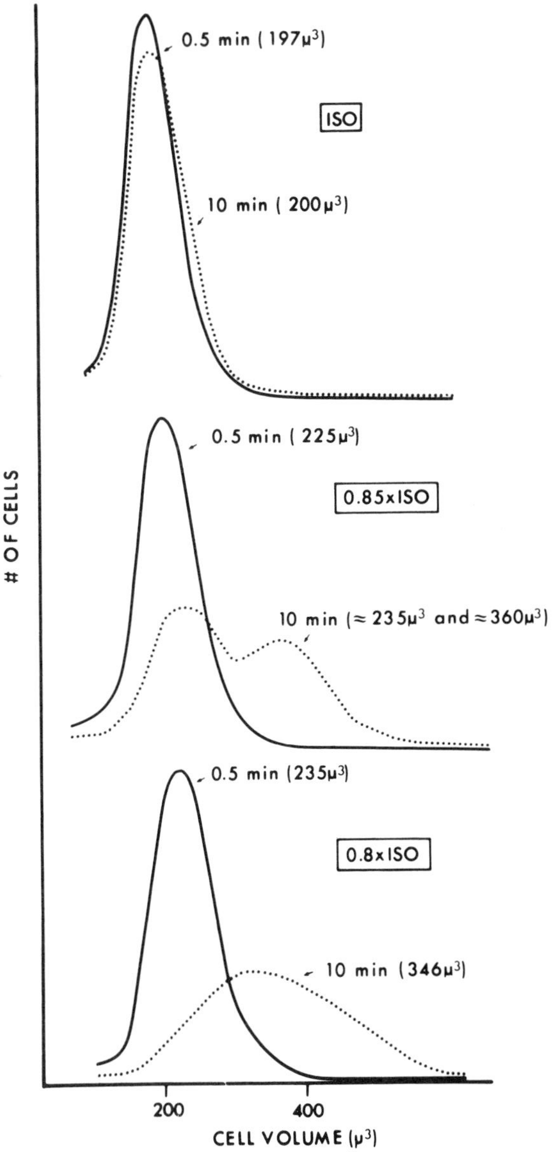

FIGURE 6. Changes in cell volume distributions of peripheral blood lymphocytes as depicted by the Coulter channel analyzer system. Cells were suspended in isotonic KCl medium and challenged with the hypotonicities indicated, in the presence of 0.5 $\mu$M gramicidin. Readings were obtained after 0.5 and 10 minutes. The mean cell size for the populations or subpopulations is indicated. The figure is reproduced from ref. 23.

## DEPENDENCE ON Ca$^{2+}$

Ca$^{2+}$ activation of Cl$^-$ pathways has been reported in the basolateral membranes of epithelial cells in isotonic medium.[25] The role of Ca$^{2+}$ in the behavior of the volume-activated Cl$^-$ pathway is, however, not entirely clear. It varies somewhat in different cell types. In lymphocytes, extracellular Ca$^{2+}$ is not required for

RVD,[14] and osmotic swelling does not alter the concentration of $Ca^{2+}$ in the cytoplasm, measured by the fluorescent technique, using Quin-2.[26] Conversely, $Ca^{2+}$ depletion blocks RVD,[14] but the effect in this case is due to inhibition of the $K^+$ pathway. The volume-activated $Cl^-$ pathway of the depleted cells is functional, as demonstrated by a rapid RVD after the addition of gramicidin. No effects of $Ca^{2+}$ on volume-activated $Cl^-$ permeability have been noted. However, indirect evidence suggests that in isotonic cells, $Cl^-$ permeability can be influenced by $Ca^{2+}$. For example, the addition of the $Ca^{2+}$ ionophore A23187 plus $Ca^{2+}$ results in volume changes.[14] Given that the $Cl^-$ permeability of isotonic cells is known to be low, restrictive to salt loss and volume changes (gramicidin does not

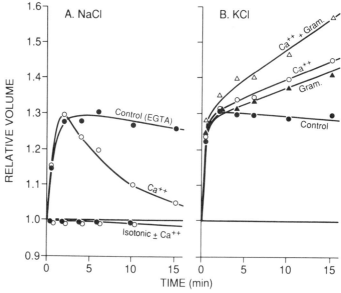

**FIGURE 7.** Volume adjustments of dissociated MDCK cells after exposure to hypotonic challenge (0.7 × isotonic) **(A)** or KCl **(B)** medium. The effects of $Ca^{2+}$ and gramicidin. Confluent cells were dissociated with trypsin, washed, and suspended in isotonic NaCl or KCl. At zero time they were subjected to hypotonic challenge (0.7 × isotonic) in the presence of the indicated substances, and volumes were measured by the Coulter counter. Concentrations were: $Ca^{2+}$, 1.5 $\mu M$; gramicidin, 0.5 $\mu M$.

cause volume changes; see above), it can be concluded that $Cl^-$ permeability must have been increased by ionophore-induced uptake of $Ca^{2+}$ by the cell. It is not known, however, whether this $Ca^{2+}$ modulated pathway is the same as the volume-activated pathway.

In Ehrlich ascites cells, responsiveness to $Ca^{2+}$ is more definitely established. In isotonic cells, A23187 induces volume changes, which in this case can be attributed directly to an increase in $Cl^-$ permeability[22] based on the gramicidin procedure. As in lymphocytes, however, it is not certain that the $Cl^-$ pathways activated by $Ca^{2+}$ are identical to those activated by swelling. External $Ca^{2+}$ is not essential for RVD, but it distinctly increases the response.[27] Another interesting connection between $Ca^{2+}$ and the volume-activated pathway has been established. As in lymphocytes,[23] the $Cl^-$ pathway, once activated, stays open for only about 10 minutes and then closes spontaneously. In ascites cells,[22] however, it has been demonstrated that the pathway does not shut down in the presence of $Ca^{2+}$ plus A23187. Thus elevated, $Ca^{2+}$ appears necessary to keep the channels open.

In epithelial cells, RVD requires the presence of $Ca^{2+}$ in the medium,[11,28,29] and increases in cellular $Ca^{2+}$ take place. It was not clear, however, if $Ca^{2+}$ exerted its effect only on the $K^+$ pathway or on $Cl^-$ permeability as well. Recently we addressed this question using dissociated MDCK cells (dog renal epithelial cell origin). From FIGURE 7A it is evident that RVD requires the presence of $Ca^{2+}$. This dependence is also evident in swelling of cells suspended in KCl medium (FIG. 7B). No secondary swelling occurs in the absence of $Ca^{2+}$ unless gramicidin is added. These data suggest that $Ca^{2+}$ is required to open the $K^+$ pathway, and that in the absence of $Ca^{2+}$ the deficiency can be bypassed by providing an alternative $K^+$ pathway (gramicidin). The data also indicate that volume activation of the $Cl^-$ pathway, in the absence of $Ca^{2+}$, is sufficient to support a considerable swelling rate. It is of interest, however, that $Ca^{2+}$ significantly increases the swelling rate of gramicidin-treated cells. This finding supports the view that $Ca^{2+}$ stimulates the volume-activated $Cl^-$ pathway in these cells.

## TRIGGERING AND CONTROL OF RVD

The events that lead from cell swelling to the opening of $Cl^-$ pathways are not well delineated. Clearly, the process must be initiated by a triggering event, perhaps followed by intervention of a "messenger" system, followed, after RVD is completed, by deactivation. Some factors can be excluded from consideration. For example, cell swelling per se, rather than changes in ionic strength or ionic concentrations, appears to be important. Thus it has already been noted that isotonic swelling associated with uptake of substrates or other nonelectrolytes[11-13] is as effective as hypotonic swelling in initiating RVD. In lymphocytes the triggering always appears to take place when cells reach a particular size regardless of prior manipulations.[23] One possibility is that the unfolding of the membrane extensions associated with cell enlargement[3] results in a triggering event, perhaps due to disturbances of the underlying cytoskeleton. Indeed, swelling has been noted to result in a reversible dissolution of the cytoskeletal fibrous system,[3] and several investigators have proposed a direct role for the cytoskeleton in volume control.[29-32] Conversely, many cell types apparently follow simple osmotic behavior; with no evidence that mechanical intervention plays a direct role in size determination. Furthermore, changes in volume during RVD can be quantitatively accounted for by changes in permeability and loss or gain of salts. Attempts to find meaningful correlations between the effects of agents that influence the cytoskeleton and their effects on RVD in lymphocytes have been unsuccessful.[23] Finally, RVD can be triggered in red blood cells that do not possess an intracellular cytoskeletal network. It is quite possible, however, that swelling stress on the network of extrinsic proteins on the cytoplasmic side of the bilayer (such as the spectrin system of red blood cells) plays a triggering role. The fact that pipette pressure applied to the membrane during the "patch clamp" procedure can induce channel activity (stress channels) may be relevant to the question. (See subsequent section on channels.)

The question of linkage between the triggering event and the opening of the $Cl^-$ pathway has some possible answers, but much remains to be investigated.

Cell swelling, for example, increases $Ca^{2+}$ permeability,[14] and it has been noted herein that $Ca^{2+}$ plays a direct activating role in epithelial cells and in preventing channel closing in Ehrlich ascites cells. It is well known that agonists related to the cyclic AMP system are activators of $Cl^-$ channels in epithelial cells,[33] but cyclic AMP analogs and relevant agonists do not influence RVD in lymphocytes[23] or in dissociated MDCK cells (unpublished observations).

Recently, interesting studies on Ehrlich ascites cells have focused attention on lipid substances in the leukotriene pathway. To briefly review the background, arachidonic acid, a substituent of phospholipids, inhibits the $Cl^-$ pathway and increases $Na^+$ permeability.[34] It gives rise, by a series of biochemical reactions, to either prostaglandins or leukotrienes. Cell swelling favors the latter pathway,[35] with synthesis increased by as much as 18-fold, whereas the prostaglandin pathway is repressed. Furthermore, if the leukotriene pathway is blocked by a specific lipoxygenase inhibitor, RVD in the presence of gramicidin (the $Cl^-$ channel) is blocked. Of four leukotrienes tested, only one, LTD4, directly increases the rate of cell shrinkage, suggesting that it may be the channel activator. In many cells, leukotriene synthesis is stimulated by an increase in cell $Ca^{2+}$, providing a basis for a tentative series of events: triggering$\rightarrow$ $Ca^{2+}\rightarrow$ $LTD_4\rightarrow$ activation of $Cl^-$ channels.

Once activated, the $Cl^-$ channel does not influence the course of RVD until cell volume has returned to almost normal. As just noted, conductance of the $Cl^-$ pathway greatly exceeds that of the $K^+$ pathway, so the RVD is usually $K^+$ limited.[17] Because activation of the $K^+$ pathway involves a graded response with respect to the degree of swelling, RVD follows the same graded response, with half-times of the process relatively independent of the amount of swelling.[24]

Termination of RVD, on the other hand, appears to be due to closure of the $Cl^-$ pathway. RVD ceases when lymphocytes are about 110% of normal size.[23] This, as already noted, is about the threshold value for the opening or closing of the $Cl^-$ channels. The activity of the $K^+$ pathway, however, although it decreases somewhat as RVD proceeds, may remain above normal for a time after RVD is complete.[15]

## NATURE OF THE VOLUME-ACTIVATED $Cl^-$ PATHWAY

Given that conductive $Cl^-$ pathways generally appear to involve ion flow through channels,[33] it has usually been assumed that the volume-activated pathway was also a channel. Until recently, however, direct evidence using patch-clamp technique was lacking. In fact, in the recent review just cited,[33] only one abstract related to this topic was listed. In the last year, however, several relevant studies have appeared. In Ehrlich ascites cells swollen either by uptake of glycine or by exposure to hypotonic medium, $Cl^-$ channel activity in cell-attached patches is substantially increased.[13] Biophysical characteristics were determined after excision of the patch in the inside-out configuration. Unitary conductance is 26 pS at 0 mV, and with exposure to symmetrical KCl solutions (150 mM) pronounced outward rectification is evident. Selectivity for $Cl^-$ is considerably (11-fold) greater than that for $K^+$.

In oppossum kidney epithelial cells, channels are activated by exposure to hypotonicity, but they are relatively nonselective between anions and cations.[36] The channel size is 26 pS. It is of interest that similar channels can be activated under isotonic conditions by stretching the membrane with applied negative pressure, suggesting that RVD may be triggered by membrane stretch.

In thymic lymphocytes, whole cell $Cl^-$ currents are activated by imposition of osmotic gradients (cell interior hypertonic).[37] The conductance starts after 40 seconds, reaching maximum current in 3 minutes. The whole-cell current-voltage relationship showed outward rectification. The channels are very small (minichannels) with an estimated single channel conductance of 2.6 pS. Two interesting observations are reported. First, osmotic channel opening requires the presence of ATP in the intracellular compartment, indicating involvement of a phosphorylation process. Second, the osmotically "opened" channels can be "closed" by application of negative pressure to the cell interior, suggesting a mechanical linkage to channel opening and closing.

In human skin fibroblasts, we have obtained preliminary evidence that $Cl^-$ minichannels are also activated by hypotonic shock. In cell-attached patches,

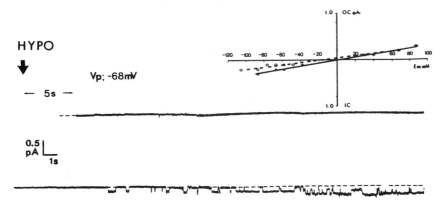

**FIGURE 8.** Activation of $Cl^-$ channels in human skin fibroblasts by hypotonic swelling. *Lower panel: Arrow* indicates 50% dilution of medium (NaCl, 140 mM; CaCl$_2$, 2 mM; MgCl$_2$, 1 mM; glucose, 10 mM; and HEPES, 10 mM; pH, 7.2). The recording is in the cell-attached mode, with pipette potential held at $-68$ mV. Low pass filter at 100 Hz. *Upper panel:* The I-V relationship of the evoked channel in excised inside-out membrane patches. Closed symbols = symmetric NaCl (140 mM); open symbols = asymmetric NaCl (bath, 70 mM NaCl, and pipette, 140 mM).

current steps were evident after 40 seconds and sustained for several minutes in four of eight trials (FIG. 8). A unitary conductance of 3 pS was estimated in excised inside-out patches similar in magnitude to those found in lymphocytes, as noted above. The channel appears to be highly selective for $Cl^-$ over $Na^+$, based on ion substitution experiments.

To summarize, in each of the cell types studied, volume-activated $Cl^-$ channels were observed, but their properties were variable. In two cases the conductances were moderately large, about 25 pS, whereas in the other two the conductances were very small, about 3 pS. In the latter, given the large size of the volume-activated fluxes, the number of channels must be quite large. Other variations were noted, such as the degree of $Cl^-$ specificity and the presence of rectification. Possible triggering by mechanical stretch was noted in two cases. The possible role of $Ca^{2+}$ was not investigated.

## RELATION OF THE VOLUME-ACTIVATED Cl⁻ PATHWAY TO OTHER Cl⁻ PATHWAYS

It is clear from studies at the cell and single channel level[33] that many species of Cl⁻ channels are present in the same and in different cells, and as noted herein, at least two kinds of channels are activated by cell swelling. Insufficient data are available at this time to make definitive statements concerning uniqueness or relatedness between volume-activated and other channels. A few observations are pertinent. At the macro (cellular) level, fluxes of the volume-activated pathways are exceedingly large, allowing escape of a large fraction of cell Cl⁻ in a few

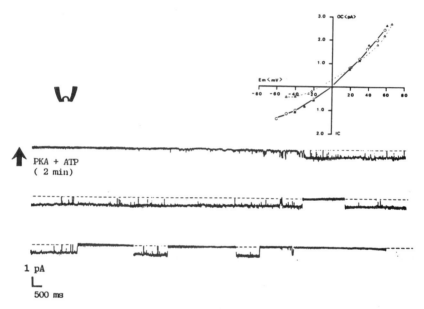

**FIGURE 9.** Activation of Cl⁻ channels in human fibroblasts by the catalytic subunit of kinase A. *Lower panel: Arrow* indicates the addition of the catalytic subunit (0.3 μM) plus ATP (1 mM) to the cytoplasmic surface of the excised patch. The pipette potential was held at −40 mV. Low pass filter at 1,000 Hz. *Upper panel:* The I-V relationship. *Closed triangles* = symmetric NaCl (140 mM); open circles = N-methyl-glucamine-chloride (bath, 140 mM) and NaCl (pipette, 140 mM); open triangles = asymmetric NaCl (bath, 70 mM; and pipette, 140 mM).

minutes, far faster than fluxes reported for cAMP-activated fluxes, for example. Direct comparisons in MDCK cells indicate a difference of over 20-fold. Another difference is that the volume-activated systems, at least those of lymphocytes[19] and MDCK cells (unpublished observations), are not influenced by agonists such as cAMP derivatives that "turn on" Cl⁻ channels involved in transepithelial fluid transport.

At the micro level the most ubiquitous channel is one that is activated by β-agonists and that is functionally important in secretion and absorption processes.[33] This class of channels is characterized by outward rectification in symmetrical NaCl solutions, has a conductance of 30 pS at 0 mV, and can be activated by depolarization. In Ehrlich ascites cells,[13] the volume-activated channel has a similar current-voltage relationship, but activation by cAMP has not been reported. In oppossum kidney epithelial cells,[36] channel conductance is in the same range, but little anion-cation specificity is evident. In fibroblasts, a cAMP-activated channel has been found[38] that can be activated by the catalytic subunit of kinase A plus ATP in inside-out membrane patches, as illustrated in FIGURE 9. In preliminary experiments this type of channel does not appear to be activated by osmotic swelling. Only the minichannels are observed (FIG. 8). Evidence so far suggests that volume-activated Cl⁻ efflux and β-agonist-induced Cl⁻ secretion occur through distinct conductive pathways.

## DISCUSSION

The capacity to regulate volume is an essential function of unwalled cells. In most cells, this capacity can be demonstrated by sudden exposure to hypotonic medium. The osmotic swelling is followed by a rapid return toward normal size, accomplished largely by the loss of KCl and osmotically obligated water, the driving force being the preexisting KCl gradient. At least two mechanisms have been described that account for KCl loss, depending on the cell type. In general, blood cells lose salt via volume activation of an electroneutral KCl cotransport system, whereas in other cell types, the loss is via independent, conductive, volume-activated $K^+$ and Cl⁻ channels. The reason for two distinct mechanisms is not currently apparent.

Although the presence of volume-activated Cl⁻ pathways has been reported in many cell types, they have only been characterized in detail in a small number, so generalizations are tentative. Even in the few cells in which the pathway has been characterized, differences in behavior are evident, such as sensitivity to inhibitors and dependence on $Ca^{2+}$. At the micro level, using the patch-clamp technique, recent studies have demonstrated volume activation of Cl⁻ channels, but in some cells the conductance per channel is very low (about 3 pS), whereas in others it is much higher (26 pS). It can be tentatively concluded that no single type of Cl⁻ channel accounts in general, for volume regulation in cells.

Many types of Cl⁻ channels have been reported, the most ubiquitous being those that are cAMP activated, important in fluid secretion and absorption. Current evidence suggests that the latter are not involved in cell volume regulation. The relationship of volume-activated channels to other Cl⁻ channels is not yet clear.

The nature of the triggering action that recognizes an increase in volume is not clear, although a mechanical effect on the membrane structure is a natural candidate. In Ehrlich ascites cells the messenger that signals channel opening appears to be an increased level of leukotriene ($LTD_4$) that is induced by cell swelling. $Ca^{2+}$ clearly plays a role in some cells, although in lymphocytes the evidence so far is negative. One possibility is that it activates the leukotriene pathway.

The tentative nature of the conclusions just outlined indicates that much further research on volume-activated Cl⁻ channels is necessary.

## REFERENCES

1. MACKNIGHT, A. D. C. & A. LEAF. 1977. Physiol. Rev. **57:** 510–573.
2. ROTHSTEIN, A. 1989. *In* Monovalent Cations in Biological Systems. C. A. Pasternak, ed. CRC Press, Boca Raton, FL (in press).

3. GRINSTEIN, S., A. ROTHSTEIN, B. SARKADI & E. W. GELFAND. 1984. Am. J. Physiol. **246:** C204–C215.
4. SPRING, K. R. 1985. Fed. Proc. **44:** 2526–2529.
5. GILLIES, R. 1983. Mol. Physiol. **4:** 3–16.
6. TOSTESON, D. C. & J. F. HOFFMAN. 1960. J. Gen. Physiol. **44:** 169–194.
7. EVELOFF, J. L. & D. G. WARNOCK. 1987. Am. J. Physiol. **252:** F1–F10.
8. HOFFMAN, E. K. 1986. Biochim. Biophys. Acta **864:** 1–31.
9. GECK, P. & B. PFEIFFER. 1985. Ann. N. Y. Acad. Sci. **456:** 166–182.
10. HOFFMANN, E. K. 1985. Fed. Proc. **44:** 2513–2519.
11. WONG, S. M. E. & H. S. CHASE, JR. 1986. Am. J. Physiol. **250:** C841–C852.
12. LARIS, P. C. & G. V. HENIUS. 1982. Am. J. Physiol. **242:** C326–C332.
13. HUDSON, R. L. & S. G. SCHULTZ. 1988. Proc. Natl. Acad. Sci. USA **85:** 279–283.
14. GRINSTEIN, S. A., DUPRE & A. ROTHSTEIN. 1982. J. Gen. Physiol. **79:** 849–868.
15. ROTI ROTI, L. W. & A. ROTHSTEIN. 1973. Exp. Cell Res. **79:** 295–310.
16. KREGENOW, F. M. 1981. Ann. Rev. Physiol. **43:** 493–505.
17. GRINSTEIN, S., C. A. CLARKE, A. DUPRE & A. ROTHSTEIN. 1982. J. Gen. Physiol. **80:** 801–823.
18. GRINSTEIN, S., J. D. GOETZ & A. ROTHSTEIN. 1984. J. Gen. Physiol. **84:** 565–584.
19. SARKADI, B., E. MACK & A. ROTHSTEIN. 1984. J. Gen. Physiol. **83:** 497–512.
20. SARKADI, B., R. CHEUNG, E. MACK, S. GRINSTEIN, E. W. GELFAND & A. ROTHSTEIN. 1985. Am. J. Physiol. **248:** C480–C487.
21. CHEUNG, R. K., S. GRINSTEIN & E. W. GELFAND. 1982. J. Clin. Invest. **70:** 632–638.
22. HOFFMAN, E. K., I. H. LAMBERT & L. O. SIMONSON. 1986. J. Membr. Biol. **91:** 227–244.
23. SARKADI, B., E. MACK & A. ROTHSTEIN, A. 1984. J. Gen. Physiol. **83:** 513–527.
24. CHEUNG, R. K., S. GRINSTEIN, H.-M. DOSCH & E. W. GELFAND. 1982. J. Cell. Physiol. **112:** 189–196.
25. CHANG, D. & C. DAWSON. 1988. J. Gen. Physiol. **92:** 281–306.
26. RINK, T. J., A. SANCHEZ, S. GRINSTEIN & A. ROTHSTEIN. 1983. Biochim. Biophys. Acta **762:** 593–596.
27. HOFFMAN, E. K., L. O. SIMONSON & I. H. LAMBERT. 1984. J. Membr. Biol. **78:** 211–222.
28. DAVIS, C. W. & L. A. FINN. 1987. Fed. Proc. **44:** 2520–2525.
29. FOSKETT, J. K. & K. R. SPRING. 1985. Am. J. Physiol. **248:** C27–C36.
30. MELMED, R. N., P. J. KARANIAN & R. D. BERLIN. 1981. J. Cell. Biol. **90:** 761–768.
31. MILLS, J. W. & M. LUBIN. 1986. Am. J. Physiol. **250:** C319–C324.
32. VAN ROSSUM, G. D. V. & M. A. RUSSO. 1981. J. Membr. Biol. **59:** 191–209.
33. GOGOLEIN, H. 1988. Biochim. Biophys. Acta **947:** 521–547.
34. LAMBERT, I. H. 1987. J. Membr. Biol. **98:** 207–221.
35. LAMBERT, I. H., E. K. HOFFMANN & P. CHRISTENSEN. 1987. J. Membr. Biol. **98:** 247–256.
36. UBL, J., H. MURER & H. A. KOLB. 1988. Biophys J. **53:** 57a.
37. CAHALAN, M. D. & R. S. LEWIS. 1988. *In*. Cell Physiology of Blood. R. Gunn & J. C. Parker, eds.  : 282–301. Rockefeller University Press, New York.
38. BEAR, C. E. 1988. FEBS Lett. **237:** 145–149.

# Role of the Cytoskeleton in Regulation of Gastric HCl Secretion[a]

JOHN CUPPOLETTI AND DANUTA H. MALINOWSKA

*Department of Physiology and Biophysics*
*University of Cincinnati College of Medicine*
*Cincinnati, Ohio 45267–0576*

A multifunctional and multicomponent transport system is responsible for the ATP-dependent secretion of HCl across the apical membrane of the gastric parietal cell. The components of the gastric proton pump include the $(H^+ + K^+)$ ATPase and mechanisms for the net transport of $K^+$ and $Cl^-$ into the lumen.[1-6] $K^+$ exits the cell through the apical membrane, down its electrochemical gradient, accompanied by equivalents of $Cl^-$, and is then available for $K^+/H^+$ exchange by the $(H^+ + K^+)$ATPase. The net result is secretion of HCl into the lumen, with $K^+$ recycling between the cytosolic and the luminal compartments. HCl secretion thus depends on the presence of both the $(H^+ + K^+)$ATPase and the $K^+$ and $Cl^-$ transport mechanisms in the same membrane and in an active form.[5,6] The $(H^+ + K^+)$ATPase has been identified[1,2] and isolated in functional form,[5,6] and the primary sequence determined,[7] but much less is known of the transport of $K^+$ and $Cl^-$. Measurements of biionic diffusion potential[6] driven by $K^+$ and $Cl^-$, isotope exchange,[6,8] diffusion of $K^+$ or $Cl^-$ driven by electrochemical gradients of these ions,[6,8] and recent preliminary results of patch clamp studies[9,10] suggest that $K^+$ and $Cl^-$ transport occurs through channels. Heavy metals and divalent cations inhibit $H^+$ transport in gastric vesicles,[8,11] but the site of action is not known. $Ba^{2+}$ and $TEA^+$ inhibit the $K^+$ channel,[12] carboxylic acids inhibit the $Cl^-$ channel,[13,14] and all of these agents inhibit HCl accumulation. Inhibition of either channel results in loss of net $K^+$ and $Cl^-$ transport due to electrical coupling considerations, but the transport functions can be measured and inhibited separately. It is not known if the $K^+$ and $Cl^-$ transport function resides on a single protein or on separate proteins.

HCl secretion by the gastric parietal cell is stimulated by histamine, gastrin, and carbachol (a cholinergic agonist). The total $(H^+ + K^+)$ATPase activity as measured in the presence of the ionophore nigericin in vesicles prepared from secreting and nonsecreting stomachs is similar.[6] However, $H^+$ accumulation and $(H^+ + K^+)$ATPase activity in sealed vesicles from secreting and nonsecreting stomachs are vastly different, reflecting the $H^+$ transport properties of the stomach from which they were derived.[5,6] $K^+$ and $Cl^-$ transport is also different in these vesicles. $K^+$ and $Cl^-$ transport occurs across vesicles from secreting stomachs and is inactive across vesicles from nonsecreting stomachs.[6,8] Regulation of HCl secretion thus appears to involve regulation of the $K^+$ and/or $Cl^-$ channels, because the specific activity of the $(H^+ + K^+)$ATPase is equal in both resting and stimulated vesicles if the KCl pathway is bypassed by preloading with $K^+$ salts, by using ionophores such as nigericin, or using the permeant analog of $K^+$, ammonia.[6,15,16]

Increases in HCl secretion rate and $K^+$ and $Cl^-$ transport are accompanied by

(

[a] This work was supported by National Institutes of Health Grant DK38808.

striking morphologic changes of the parietal cells.[17,18] The size of the secretory canaliculus and the length of the microvilli lining the surface of the canaliculus increase, thereby increasing the space for acid accumulation within the cells. In resting cells, the tubulovesicles do not appear to communicate with the lumen of the gastric glands, whereas in stimulated cells, the large canalicular space is contiguous with the lumen. Thus, the morphologic change also provides access of HCl to the lumen. The increase in membrane area of the secretory canaliculus results from the recruitment of intracellular vesicles[17] (tubulovesicles) containing the $(H^+ + K^+)$ATPase.[19] Thus, recruitment leads to an increase in the number of $(H^+ + K^+)$ATPase molecules in the canalicular membrane per unit area as determined by freeze fracture.[20] Cytochalasin B (CB) and cytochalasin E (CE), which block elongation of microfilaments, inhibit stimulated HCl secretion,[21–23] and CB has been shown to block the morphologic change associated with stimulation.[21,22] This finding suggests that microfilaments, which are found in close association with the secretory canalicular membrane of the stimulated parietal cell, are intimately involved in the regulation of HCl secretion.

Membrane reorganization may also be important to the organization of the functional $H^+$ pump in the stimulated secretory membrane. It has been suggested by others that the $(H^+ + K^+)$ATPase containing tubulovesicles and the $K^+$ and $Cl^-$ channels may exist in separate membranes in the resting cell.[8,11] In this case, stimulation could result from fusion of $(H^+ + K^+)$ATPase containing vesicles with other vesicles or apical membranes containing $K^+$ and/or $Cl^-$ channels (insertion mechanism). Alternatively, $K^+$ and $Cl^-$ channels may preexist in the parietal cell tubulovesicle membrane together with the $(H^+ + K^+)$ATPase, and stimulus-dependent activation of $K^+$ and/or $Cl^-$ channels by phosphorylation by cAMP-dependent protein kinase and other mechanisms may occur (activation mechanism).

This study was designed to distinguish between the suggested possible mechanisms of insertion and activation of the $K^+$ and $Cl^-$ transport system(s) in the regulation of HCl secretion. We used cytochalasin E in gastric glands to prevent the membrane reorganization on which the formation of the expanded canalicular membrane depends. We then measured HCl accumulation both in gastric glands and in vesicles isolated from the glands and looked for evidence of $K^+$ and $Cl^-$ channel activity after treatment with secretagogues. If the $(H^+ + K^+)$ATPase and the $K^+$ and $Cl^-$ channels exist in separate membranes in resting cells, and if stimulation brings about fusion of these separate membranes, cytochalasin E would be expected to block the stimulatory effects of secretagogues by preventing membrane fusion. Thus, vesicles prepared from gastric glands that were treated with cytochalasin E and then secretagogues, should exhibit $(H^+ + K^+)$ATPase activity and $H^+$ transport properties of vesicles from resting glands, because of the absence of the $K^+$ and $Cl^-$ transport systems. If, however, activation of $K^+$ and/or $Cl^-$ channels occurs, cytochalasin-E-treated, secretagogue-stimulated glands would yield vesicles that continued to show stimulated characteristics.

The results of this study argue against the hypothesis that the $(H^+ + K^+)$-ATPase and $K^+$ and $Cl^-$ transport systems reside in different membranes. The evidence presented rather suggests that the $K^+$ and $Cl^-$ transport system(s) coexist with the $(H^+ + K^+)$ATPase in resting tubulovesicle membranes and that activation of $K^+$ and/or $Cl^-$ channels occurs upon secretagogue stimulation by modification of the existing transport system(s). The approach taken in the present study may generally be applicable to the understanding of mechanisms of regulation of secretion by other epithelia where recruitment, recycling, and activation of transport processes occur simultaneously.

## RESULTS

### Effect of Cytochalasin E (CE) on Histamine-Stimulated $H^+$ Accumulation in Gastric Glands

The glands were incubated without or with different concentrations of CE for 15 minutes at 37°C. $10^{-4}$ M histamine or $10^{-4}$ M cimetidine was then added for a

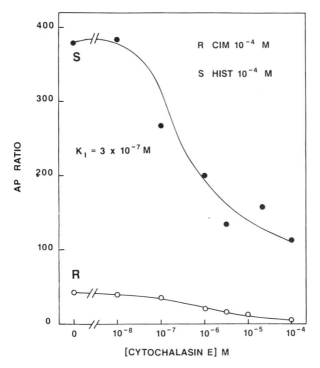

**FIGURE 1.** Effect of CE on acid accumulation in resting and stimulated gastric glands. Rabbit gastric glands from New Zealand White rabbits were obtained by collagenase digestion,[24] and acid secretion was measured by following the uptake of the weak base, [$^{14}$C]-aminopyrine (AP).[25] The glands were first treated without or with different concentrations of CE alone (15 minutes at 37°C) and then $10^{-4}$ M histamine (S, ●—●) or $10^{-4}$ M cimetidine (R, ○—○) were added for 30 minutes for stimulation and inhibition of HCl secretion, respectively. CE was prepared fresh daily in DMSO, and DMSO was added to controls.

further 30 minutes to obtain stimulated and resting glands, respectively. At the end of the incubation, $H^+$ accumulation was measured by the cell:medium ratio of the radioactive weak base, [$^{14}$C]-aminopyrine (AP ratio).[25] The results in FIGURE 1 show a typical experiment. The AP ratio was 43 in resting glands and increased to 379 in stimulated glands. CE inhibited AP accumulation in a dose-dependent manner. The AP ratio of stimulated glands was reduced by 70% at the highest concentration of CE used ($10^{-4}$ M). The CE-sensitive portion of AP accumulation was inhibited with an apparent $K_I$ of $3 \times 10^{-7}$ M. The AP ratio in resting glands was reduced by about 80% over the same range of concentration of CE. We next investigated whether CE treatment inhibited microfilament elongation in the parietal cell.

### *Effect of Cytochalasin E on the State of Elongation of the Microfilaments*

The gastric parietal cell is highly enriched in actin, and it has previously been shown that gastric glands stain with a fluorescent derivative of phallicidin, a fungal metabolite that reacts with F- but not G-actin.[26] Staining of gastric glands, indicating the presence of F-actin, was observed to be distributed largely within stimulated parietal cells of the gastric glands with most intense staining at the apex of the cell, where the canalicular membrane is largely localized. We employed a related F-actin stain, fluorescein phalloidin (Molecular Probes), to measure the extent of organization of the microfilament network in isolated gastric gland suspensions. With fluorescence microscopy of stimulated gastric glands (not shown), we found that fluorescein phalloidin indeed differentially stained the apical pole of stimulated parietal cells in histamine-stimulated glands. The staining in resting glands was diffuse and much less intense. Staining of peptic cells was minimal. Measurements of fluorescence were facilitated by the availability of a front face (reflectance) attachment (SPEX Industries) on the spectrofluorimeter which allows fluorescence measurement of turbid suspensions. As shown in FIGURE 2, when glands were digitonin permeabilized[3] and stained with fluorescein phalloidin, the fluorescence of stimulated glands was 44% more intense than was that of resting glands. CE ($3 \times 10^{-6}$ M) reduced F-actin staining by 31% and 38% in resting and stimulated glands, respectively, as compared to controls (without CE). Thus, CE was effective in reducing the level of F-actin in the parietal cell. Because the bulk of F-actin in the parietal cell is associated with the secretory membrane of the stimulated parietal cell, these results indicate that CE treatment effectively interfered with the microfilament-mediated fusion of tubulovesicles to form the expanded secretory canaliculus.

### *Effect of Cytochalasin E on ATP-Driven $H^+$ Accumulation in Permeabilized Parietal Cells*

As already shown, CE disrupts the microfilaments of parietal cells, which would prevent the fusion of tubulovesicles with other membranes. Unlike CB, CE is without effects on D-glucose transport.[28] CE (as also shown with CB[21]) has no direct effect on vesicular $H^+$ transport or ATP hydrolysis at the highest concentrations used here (data not shown). However, it is difficult to rule out all nonspecific effects on the parietal cell which might lead to inhibition of HCl secretion. We therefore employed permeable parietal cells to examine the specificity of the effects of CE in inhibition of HCl secretion. Digitonin selectively permeabilizes the basolateral membrane of the parietal cell and leaves the secretory membrane intact.[3] Permeabilized glands in the presence of mitochondrial inhibitors have been used to establish that ATP, $K^+$, and $Cl^-$ are necessary and sufficient for HCl secretion.[3,4] With such a permeable system metabolic effects on the parietal cells can thus be bypassed by supporting acid secretion with exogenous ATP, $K^+$, and $Cl^-$. In addition, permeable glands and parietal cells have been shown to retain their functional state, resting or stimulated, after permeabilization.[4,29,30] Parietal cells were treated first with CE ($3 \times 10^{-6}$ M) for 15 minutes, then with histamine or cimetidine for 15 minutes. The cells were then permeabilized with digitonin, and oligomycin, N', N'-dicyclohexylcarbodiimide (DCCD), and ouabain were added to inhibit mitochondrial function and ($Na^+ + K^+$)ATPase activity. AP accumulation was measured after MgATP addition. The effect of CE pretreatment on the ATP-dependent AP ratios is shown in FIGURE 3. The $\Delta$ATP-AP ratio of resting parietal cells was 10, and this figure doubled to 22 upon stimulation. Treatment of parietal cells with $3 \times 10^{-6}$ M CE before cimetidine or histamine

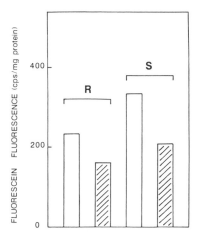

**FIGURE 2.** F-actin staining of gastric glands. Gastric glands, prepared and treated as described in FIGURE 1 without and with $3 \times 10^{-6}$ M CE, followed by cimetidine (R, resting) or histamine (S, stimulated) were permeabilized by digitonin (20 $\mu$g/ml of gland suspension)[3] for 5 minutes at 37°C and then incubated with 4 units of fluorescein phalloidin per milliliter of gland suspension for 10 minutes at 37°C. The glands were then washed 3 times by centrifugation with detergent-free media containing 0.3 M sucrose, 4 mM PIPES/8 mM Tris (pH 7.4). Glandular fluorescence was determined using front face (reflectance) spectrofluorimetry (excitation and emission wavelengths of 492 and 530 nm, respectively) with a SPEX Industries dual-excitation, dual-emission spectrofluorimeter. Fluorescence of unstained glands (autofluorescence plus light scatter) was subtracted from the fluorescence of the stained glands. Protein was determined on each of the samples[27] and fluorescence was corrected for protein. The fluorescence was linear with protein concentration over the range employed in these studies. *Hatched columns*, $+ 3 \times 10^{-6}$ M CE; *unhatched clear columns*, $+$ DMSO (control).

addition reduced the $\Delta$ATP-AP ratio of resting cells to 5 and dramatically reduced the $\Delta$ATP-AP ratio of histamine-stimulated cells to 6.

One possible explanation of the CE-induced reduction in the ATP-dependent AP ratio in stimulated permeabilized cells arises from our expected finding that CE reduces the level of F-actin in the cells. Reduced cellular F-actin would lead to failure to elaborate the intracellular canaliculus (secretory membrane), leading to a smaller space within which acid can accumulate and thus a smaller AP ratio. Accumulation of AP in an acid space is dependent not only on the pH of the membrane-bounded space, but also on the size of the space. Thus, in CE-treated cells, AP ratios could be low regardless of the activity of the transport proteins present in the intracellular vesicles. Alternately, the reduction in the AP ratio

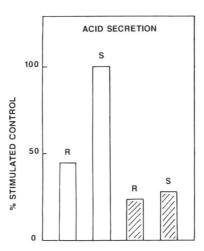

**FIGURE 3.** Effect of CE on ATP-driven AP accumulation in permeable parietal cells. Crude gastric cells were prepared by collagenase and pronase digestion, and parietal cells (85% pure) were purified using Nycodenz (Accurate Chemical Company) buoyant density gradients.[31] The cells in medium containing 60 mM KCl, 1.2 mM $MgSo_4$, 20 mM HEPES/9.6 mM Tris (pH 7.4), 0.2 mM EGTA, and 130 mM sucrose were treated with CE ($3 \times 10^{-6}$ M) for 15 minutes and then incubated with $10^{-4}$ M cimetidine (R, resting) or $10^{-4}$ M histamine (S, stimulated) for 15 minutes. Digitonin (20 $\mu$g/ml cell suspension) was then added for 5 minutes to permeabilize the cells,[4,30] followed by oligomycin (10 $\mu$g/ml), N'N'-dicyclohexylcarbodiimide (1 $\mu$M) and ouabain ($10^{-4}$ M). MgATP (5 mM, pH 7.4) was added and the AP ratio was measured after 5 minutes. Resting $\Delta$ATP-AP ratio = 10; stimulated $\Delta$ATP-AP ratio = 22. *Hatched columns,* + $3 \times 10^{-6}$ M CE; *unhatched clear columns* + DMSO (control).

could also result from inhibition of fusion of membranes containing components necessary for the function of the transport systems operative in the stimulated state. To distinguish between these possible explanations, we determined the transport characteristics of the intracellular vesicles after CE treatment by examining ATP driven, KCl-dependent $H^+$ uptake by vesicles derived from CE-treated gastric glands.

### Effect of Cytochalasin E on the Intrinsic Activity of the Secretory Apparatus

CE prevents microfilament-mediated fusion of tubulovesicles, which leads to the elaborated secretory membrane. The insertion model suggests that this would prevent the fusion of resting tubulovesicles containing only the ($H^+$ + $K^+$)-ATPase with membranes containing the $K^+$ and $Cl^-$ ion pathways on which $H^+$ pump activity depends. Thus, according to the insertion model, vesicles isolated from glands that were pretreated with CE and then treated with histamine should not exhibit stimulated $H^+$ transport properties. The activation model, however, would predict that vesicles isolated from CE-treated and histamine-stimulated glands would exhibit stimulated $H^+$ transport even in the absence of a morphologic change. Gastric glands, rather than parietal cells, were used for vesicle isolation, because of the large quantity of starting material. Resting and stimulated gastric glands have previously been used for vesicle isolation, and such vesicles retain the functional characteristics of the glands from which they were derived.[32] The acid secretory state was measured in the same glands in parallel. Control glands were preincubated with $3 \times 10^{-6}$ M CE for 15 minutes and then either stimulated with $10^{-4}$ M histamine or placed in the resting state by treatment with $10^{-4}$ M cimetidine (30 minutes). The resting glands used in these experiments had an AP ratio of $36 \pm 7$ (SEM, $n = 3$), and the stimulated glands had an AP ratio of $217 \pm 64$ (SEM, $n = 3$). CE treatment reduced AP ratios by 56% to $16 \pm 3$ (SEM, $n = 3$) in resting and by 83% to $38 \pm 2$ (SEM, $n = 3$) in stimulated glands. Vesicles were then prepared from each treatment group as previously described.[6] A representative sodium dodecyl sulfate polyacrylamide gel electrophoretogram (SDS-PAGE) of the vesicles thus obtained and stained with Coomassie blue is shown in FIGURE 4a. Several faint minor bands and a prominent band at about 95 kD corresponding to the $(H^+ + K^+)$ATPase were present in all treatment groups, with no major differences between the groups. The same gel after silver staining (FIG. 4b) shows additional bands at about 70 and 50 kD. There were no major differences in the silver staining pattern between treatment groups.

Despite the similarity in the polypeptide patterns of these vesicles, the characteristics of $H^+$ transport were dramatically different between treatment groups (FIG. 5). Gastric vesicles from cimetidine-treated glands (resting) exhibited virtu-

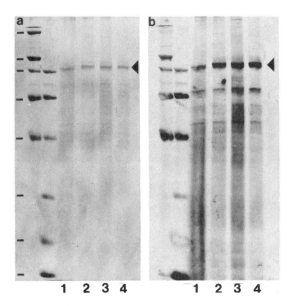

**FIGURE 4.** SDS-PAGE of gastric vesicles isolated from gastric glands. Gastric glands were isolated and treated as indicated in FIGURE 1 without and with $3 \times 10^{-6}$ M CE (15 minutes), followed by cimetidine or histamine (30 minutes). Gastric vesicles were isolated from the glands as previously described.[6] SDS-PAGE of the gastric vesicles (100 μg protein/lane) was carried out using the method of Laemmli.[33] Protein was determined by the method of Lowry.[27] **(a)** The gel stained with Coomassie blue and **(b)** the same gel stained with silver. *Lane 1*, cimetidine-treated glands; *lane 2*, CE- and cimetidine-treated glands; *lane 3*, histamine-treated glands; *lane 4*, CE- and histamine-treated glands. *Arrowhead* = about 95 kD protein, the $(H^+ + K^+)$ATPase. The molecular weight standards indicated at the right (*from top to bottom*) correspond to: 200, 116.25, 92.5, 66.2, 45, 31, 21.5, and 14.4 kD.

ally no H⁺ transport in the presence of 20 mM KCl and 0.5 mM ATP-Mg unless valinomycin (10 $\mu$g/ml) plus 100 mM TMACl were also added. The dependence of H⁺ transport on the exogenous ionophore valinomycin and a large concentration gradient of Cl⁻ indicates that the K⁺ and/or Cl⁻ ion channels are not active, whereas the (H⁺ + K⁺)ATPase *per se* is active because H⁺ pump activity was elicited by valinomycin and TMACl.[6] It should be noted that the extent of H⁺ transport using vesicles from cimetidine-treated glands in the absence of valinomycin over the entire time course of the experiment (15 minutes) was less than 10% of that obtained with valinomycin. In contrast, vesicles obtained from histamine-stimulated glands showed rapid H⁺ transport without the need for valinomycin and a large Cl⁻ gradient. This indicates that the K⁺ and Cl⁻ channels and the (H⁺ + K⁺)ATPase were active in vesicles isolated presumably from the secretory

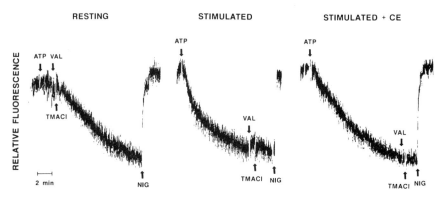

**FIGURE 5.** H⁺ transport in vesicles derived from resting, stimulated, and CE-treated stimulated gastric glands. KCl-dependent and ATP-dependent H⁺ transport in vesicles obtained from cimetidine-treated (resting), histamine-treated (stimulated), and CE-treated, histamine-stimulated (stimulated + CE) gastric glands was determined by measuring the rate and extent of fluorescence quenching of acridine orange as previously described.[6] The medium contained 20 mM KCl, 0.3 M sucrose, 4 mM PIPES/8 mM Tris (pH 7.4). MgATP (0.5 mM, pH 7.4) was added to start the reaction. The extent of H⁺ transport stimulated by the addition of valinomycin (VAL, 10 $\mu$g/ml) plus 100 mM tetramethyl ammonium chloride (TMACl) was taken as a measure of the transport capacity of resting vesicles.[6] Nigericin (NIG), 10 $\mu$g/ml, rapidly dissipated the H⁺ gradients formed.

membrane of histamine-stimulated gastric glands. CE treatment followed by histamine treatment of glands did not prevent stimulation of H⁺ transport in the vesicles. H⁺ transport was rapid and did not need the presence of valinomycin and a large Cl⁻ gradient. This indicates that in vesicles isolated from CE-treated, histamine-stimulated glands, the K⁺ and Cl⁻ channels and the (H⁺ + K⁺)ATPase were active despite the fact that the AP ratio was about 80% inhibited and despite the lack of fusion of membranes.

## DISCUSSION

The present study was undertaken to delineate whether activation or insertion of apical membrane K⁺ and Cl⁻ ion channels occurs on stimulation of HCl secretion in the gastric parietal cell. CE was employed as an inhibitor of HCl secretion and as a probe to investigate the role of the cytoskeleton in stimulation of HCl secretion. The more specific CE was used in the present studies, because we had previously shown that it was approximately 100 times more potent than CB in inhibiting acid secretion (apparent $K_I = 3 \times 10^{-7}$ M for CE versus about $10^{-4}$ M for CB[23]) because it required much less time for maximal effect (5–15 minutes versus several hours),[21,23] and because CE leads to greater inhibition than does CB at all concentrations.[23] It was therefore expected that $H^+$ secretion as measured by AP ratio in intact cells and glands would be potently inhibited by CE because of interference with the fusion of tubulovesicles required for elaboration of the secretory membrane.

It has long been known that the parietal cell undergoes a morphologic change on stimulation. The surface area of the apical membrane increases 3- to 10-fold, with recruitment of membrane vesicles (tubulovesicles) from intracellular stores in order to elaborate by fusion the microvillar structure characteristic of the stimulated parietal cell.[17,18,20] Approximately 3.5% of the parietal cell protein is actin.[26] Approximately 60% of this is in the F-(polymerized) form, and these polymers (microfilaments) are associated with the parietal cell apical membrane.[26] Staining patterns with fluorescent phallotoxins[26] and our fluorescent measurements indicated that the number of microfilaments (F-actin) associated with apical membrane increases with stimulation.

CE reduced the level of F-actin in resting and stimulated parietal cells as measured by the use of fluorescent phallotoxins and inhibited HCl accumulation in stimulated intact gastric glands. However, vesicles derived from CE-treated stimulated glands exhibited the characteristics of $H^+$ transport of vesicles from stimulated (not resting) glands, that is, no need for valinomycin and a large $Cl^-$ gradient. Thus, despite CE treatment, which prevents membrane fusion, the $(H^+ + K^+)$ATPase and the $K^+$ and $Cl^-$ channels were active in the membrane vesicles. This study therefore shows that fusion of vesicles containing the $(H^+ + K^+)$ATPase with membranes containing the $K^+$ and $Cl^-$ channels (insertion mechanism) does not occur, but rather activation of $K^+$ and/or $Cl^-$ channels, present in the same membrane as the $(H^+ + K^+)$ATPase, is occurring on stimulation of HCl secretion. Because possible metabolic effects of CE were eliminated by our experiments in permeable parietal cells, the inhibitory effect of CE on HCl accumulation in intact gastric glands must therefore be interpreted to result from lack of fusion of tubulovesicles to form the expanded secretory canalicular space into which HCl accumulates. Thus, the morphologic change is also essential for elaboration of stimulated HCl secretion in the intact parietal cell, but it is a separable process from that of channel activation. The use of the cytochalasins to inhibit gastric HCl secretion is not unique to this study. However, the use of cytochalasins to dissociate the characteristic morphologic change of the parietal cell from changes in the activity of the transport functions of the multicomponent secretory apparatus sheds new light on the mechanisms of regulation of HCl secretion.

The major mechanisms of acute regulation of plasma membrane transport processes can be classified into two general categories. (1) Transport proteins can undergo covalent or noncovalent modification, thereby altering the maximal rates or affinities for substrate. (2) The level of functional membrane transport proteins in the membrane can be altered by recruitment of the proteins from intracellular stores, thus increasing the number of functional transport units. The results of the present study suggest that both of these general mechanisms operate simultaneously to regulate HCl secretion. Thus, activation of $K^+$ and/or $Cl^-$ channels, present in the intracellular tubulovesicles, together with the $(H^+ + K^+)$ATPase occurs, and fusion of the tubulovesicles with the apical membrane (recruitment) occurs to form the expanded secretory membrane.

In most cases, removal of secretagogues or application of antagonists leads to a return of the morphology to the resting state: shortening of the microfilaments and reappearance of tubulovesicles in the cytosol and concomitant loss of HCl secretion. However, two general lines of evidence suggest that expansion of the secretory membrane, although a necessary prerequisite, is insufficient in itself for HCl secretion. (1) Thiocyanate, anoxia, substrate removal, and the inhibitor picoprazole inhibit HCl secretion but do not inhibit the stimulated morphology.[18,34,35] (2) Stimulation of HCl secretion with carbachol and pentagastrin followed by treatment with the $H_2$ receptor antagonist cimetidine (which lowers cAMP) leads to inhibition of HCl secretion without a concomitant loss of secretory membrane surface area.[34,36] Dissociation of HCl secretion from the morphologic change has thus been accomplished previously. However, these experiments did not address the question of whether the multicomponent $H^+$ pump requires the morphologic change in order to become active.

The effect of anoxia or substrate removal in stopping HCl secretion and membrane retrieval has been interpreted as a metabolic requirement for these processes. Picoprazole inhibits the $(H^+ + K^+)$ATPase directly,[37] and thiocyanate dissipates the $H^+$ gradient formed without inhibiting the $(H^+ + K^+)$ATPase.[16,38] Cimetidine inhibition of carbachol- and pentagastrin-stimulated HCl secretion without concomitant change in the morphology of the parietal cell is enlightening, because cimetidine acts primarily to lower intracellular cAMP levels of parietal cells, thus supporting the view that activation of the $H^+$ pump components by cAMP-dependent protein kinase is essential to maintenance of HCl secretion. Because treatment with a cholinergic receptor antagonist, 1-hyoscymamine, after carbachol and pentagastrin stimulation did however result in inhibition of both HCl secretion and the stimulated morphology,[34,36] these authors concluded that cAMP effects were limited to the $K^+$ and $Cl^-$ channels and that $Ca^{2+}$ effects were limited to the cytoskeleton. We disagree that cAMP effects must be limited to the $K^+$ and $Cl^-$ channels, because we now know that several putative cytoskeletal proteins[23] of the parietal cell show cAMP-mediated protein phosphorylation.[39,40] Some of these appear to be associated with the secretory membrane of the stimulated parietal cell. However, it is possible that reduction of intracellular cAMP by cimetidine adversely affects/inhibits the membrane retrieval process mediated by cytoskeletal proteins, thus leading to maintenance of the stimulated morphology. Additionally, cimetidine, by reducing intracellular cAMP, may abolish other unknown cAMP-dependent protein-kinase–mediated events, some of which may be essential to maintenance of the stimulated state of secretion.

In contrast to the aforedescribed studies, we directly addressed the question whether the morphologic change is required for activation of the multicomponent $H^+$ pump. The persistence of stimulus-dependent activation of the $H^+$ pump in vesicles isolated from CE-treated gastric glands, despite inhibition of membrane fusion (the morphologic change) and acid secretion (AP accumulation), is the central experimental finding of this study. This finding localizes the $K^+$ and $Cl^-$ channels to the tubulovesicles of the resting membrane together with the $(H^+ + K^+)$ATPase and indicates that activation and not insertion of $K^+$ and/or $Cl^-$ channels occurs on stimulation. Additionally, our experiments show that $H^+$ pump activation and the morphologic change (recruitment) are two separable processes that occur on stimulation of HCl secretion. Nevertheless, both processes are essential for the full expression of stimulated HCl secretion. It remains, however, to be clearly delineated whether both $K^+$ and $Cl^-$ channels or only the

$K^+$ or only the $Cl^-$ channel are activated on stimulation. Our finding that $H^+$ pump activation (which occurs as a result of channel activation) is a separate process from the morphologic change suggests that it may be possible to study channel regulation *in vitro* using vesicles from nonsecreting and secreting tissues.

These studies were facilitated by the fact that the parietal cell is an excellent model for the study of secretory mechanisms. Isolated functional gastric glands and parietal cells, and isolated vesicle systems have been developed. Hormone and drug effects can be studied in these systems. Secretory and morphologic changes in response to secretagogues are large, and $H^+$ secretion is easily measured. The presence of active channels in the $(H^+ + K^+)$ATPase-containing membrane is evident from the characteristics of $H^+$ transport in isolated vesicles, providing a convenient assay for the relevant population of channels that are active and in association with the $(H^+ + K^+)$ATPase. The approach taken in this study has allowed us to overcome some major obstacles in understanding the mechanisms of regulation of HCl secretion that arise from the lack of convenient markers for the presence of inactive $K^+$ and/or $Cl^-$ channels.

The approaches and findings of these studies are highly relevant to regulation of secretion by other epithelia. The morphologic change that accompanies stimulation of secretion and regulation of secretion by altering channel activity is not unique to the parietal cell. NaCl transport by a colonic tumor cell line is stimulated by treatments that elevate intracellular cAMP,[41] and this stimulation may be due in part to recruitment of vesicles containing the $Cl^-$ channels and in part to the effects of cAMP in mediating the activation of the intrinsic activity of the channel.[41,42] Similar mechanisms appear to be operative in activation of zymogen granule secretion in the pancreatic acini.[43] The approaches taken in the present study may be useful in delineating the mechanisms that regulate secretion in these systems.

In summary, we noted that in the gastric parietal cell, the $(H^+ + K^+)$ATPase and $K^+$ and $Cl^-$ channels of the secretory apparatus are found together in the same membranes, the tubulovesicles, at rest and that stimulation of HCl secretion involves activation, rather than insertion, of the ion channels.

## ACKNOWLEDGMENTS

We would like to thank Dr. Ulrich Hopfer for discussions and encouragement in these studies and Lisa Cannon for technical assistance.

## REFERENCES

1. GANSER, A. L. & J. G. FORTE. 1973. Biochim. Biophys. Acta **307:** 169–180.
2. SACHS, G., H. H. CHANG, E. RABON, R. SCHACKMAN, M. LEWIN & G. SACCOMANI. 1976. J. Biol. Chem. **251:** 7690–7698.
3. MALINOWSKA, D. H., H. R. KOELZ, S. J. HERSEY & G. SACHS. 1981. Proc. Natl. Acad. Sci. USA **78:** 5908–5912.
4. MALINOWSKA, D.H., J. CUPPOLETTI & G. SACHS. 1983. Am. J. Physiol. **245:** G573–G581.
5. WOLOSIN, J. M. & J. G. FORTE. 1981. FEBS Lett. **125:** 208–212.
6. CUPPOLETTI, J. & G. SACHS. 1984. J. Biol. Chem. **259:** 14592–14959.
7. SHULL, G. E. & J. B. LINGREL. 1986. J. Biol. Chem. **261:** 16788–16791.
8. WOLOSIN, J. M. & J. G. FORTE. 1985. J. Membr. Biol. **83:** 261–272.
9. DEMAREST, J. R., D. D. F. LOO & G. SACHS. 1988. Biophys. J. **53:** 58a.

10. SHOEMAKER, R. L., P. J. VELDKAMP & G. SACCOMANI. 1988. Biophys. J. **53:** 525a.
11. IM, W. B., D. P. BLAKEMAN & J. P. DAVIS. 1985. J. Biol. Chem. **260:** 9452–9460.
12. CUPPOLETTI, J. & D. H. MALINOWSKA. 1988. Gastroenterology **94(5pt2):** A82.
13. CUPPOLETTI, J. 1988. Faseb J. **2(4):** A718.
14. MALINOWSKA, D. H. & J. CUPPOLETTI. 1988. Faseb J. **2(4):** A718.
15. MALINOWSKA, D. H. 1988. Fed. Proc. **46:** 363.
16. HERSEY, S. J., L. STEINER, S. MATHERAVIDATHU & G. SACHS. 1988. Am. J. Physiol. **254:** G856–G863.
17. HELANDER, H. F. & B. I. HIRSHOWITZ. 1972. Gastroenterology **63:** 951–961.
18. BLACK, J. A., T. M. FORTE & J. G. FORTE. 1981. Gastroenterology **81:** 509–519.
19. SMOLKA, A., H. F. HELANDER & G. SACHS. 1983. Am. J. Physiol. **245:** G589–G596.
20. FORTE, J. G., J. A. BLACK, T. M. FORTE, T. E. MACHEN & J. M. WOLOSIN. 1981. Am. J. Physiol. **241:** G349–G358.
21. BLACK, J. A., T. M. FORTE & J. G. FORTE. 1982. Gastroenterology **83:** 595–604.
22. CASSIDY, M. M. & M. A. DINNO. 1983. *In* membrane Biophysics II: Physical Methods in the Study of Epithelia. M. A. Dinno & A. Callahan, eds.: 343–364. Alan R. Liss, New York.
23. CUPPOLETTI, J. & D. H. MALINOWSKA. 1988. *In* Progress in Clinical and Biological Research M. A. Dinno & W. M. D. Armstrong, eds. **258:** 23–35. Alan R. Liss, New York.
24. BERGLINDH, T. & K. J. OBRINK. 1976. Acta Physiol. Scand. **96:** 150–159.
25. BERGLINDH, T., H. F. HELANDER & K. J. OBRINK. 1976. Acta Physiol. Scand. **97:** 401–414.
26. WOLOSIN, J. M., C. OKAMOTO, T. M. FORTE & J. G. FORTE. 1983. Biochim. Biophys. Acta **761:** 171–182.
27. LOWRY, O. H., N. J. ROSEBROUGH, A. L. FARR & A. J. RANDALL. 1951. J. Biol. Chem. **193:** 265–275.
28. CUPPOLETTI, J., E. MAYHEW & C. Y. JUNG. 1981. Biochim. Biophys. Acta **642:** 392–404.
29. HERSEY, S. J. & L. STEINER. 1985. Am. J. Physiol. **248:** G561–G568.
30. MALINOWSKA, D. H. 1989. *In* Methods of Enzymology. Vol 5(II), Chap. 44. S. Fleischer, ed. Academic Press, New York, NY.
31. BERGLINDH, T. 1985. Fed. Proc. **44:** A616.
32. URUSHIDANI, T. & J. G. FORTE. 1987. Am. J. Physiol. **252:** G458–G465.
33. LAEMMLI, U.K. 1970. Nature (Lond.) **227:** 680–685.
34. HELANDER, H. F. & G. W. SUNDELL. 1984. Gastroenterology **87:** 1064–1071.
35. FORTE, T. M., T. E. MACHEN & J. G. FORTE. 1975. Gastroenterology **69:** 1208–1222.
36. HELANDER, H. F., H. BLOM & G. SACHS. 1988. *In* Gastrointestinal and Hepatic Secretions—Mechanisms of Control. J. S. Davison & E. A. Shaffer, eds. Chapter 15: 144–146. University of Calgary Press, Calgary, Canada.
37. FELLENIUS, E., T. BERGLINDH, G. SACHS, L. OLBE, B. ELANDER, S.-E. SJOSTRAND & B. WALLMARK. 1981. Nature (Lond.) **290:** 159–161.
38. REHM, W. S., G. CARRASQUAR & M. SCHWARTZ. 1981. *In* Membrane Biophysics: Structure and Function in Epithelia. M. A. Dinno & A. Callahan, eds. :229–246. Alan R. Liss, New York.
39. MALINOWSKA, D. H., G. SACHS & J. CUPPOLETTI. 1988. Biochim. Biophys. Acta **972:** 95–109.
40. URUSHIDANI, T., D. K. HANZEL & J. G. FORTE. 1987. Biochim. Biophys. Acta **930:** 209–219.
41. HALM, D. R., G. R. RECHKEMMER, R. A. SCHOUMACHER & R. A. FRIZZEL. 1988. Am. J. Physiol. **254:** C505–C511.
42. SORSCHER, E. J., A. TOUSSON, R. J. BRIDGES, B. R. BRINKLEY, D. J. BENOS & R. A. FRIZZELL. 1988. J. Cell. Biol. **107(6pt3):** 355a.
43. GASSER, K. W., J. DOMENICO & U. HOPFER. 1988. Am. J. Physiol. **254:** G93–G99.

# Regulation of Intracellular pH in Renal Mesangial Cells[a]

WALTER F. BORON,[b] GREGORY BOYARSKY, AND
MICHAEL GANZ[c]

*Department of Cellular and Molecular Physiology
and
Department of Medicine, Division of Nephrology[c]
Yale University School of Medicine
New Haven, Connecticut 06510
and
West Haven Veterans Administration Hospital[c]
West Haven, Connecticut*

The mesangial cell is a smooth-muscle-like cell that lives in the vascular core or mesangium of the renal glomerulus. It has been suggested that contraction of the mesangial cell may play a role in modulating glomerular filtration. On the other hand, it is clear that the proliferation and secretion of matrix substance by the mesangial cell plays a major role in the destruction of normal glomerular architecture that underlies most cases of chronic renal failure. Inasmuch as cell proliferation is known to be sensitive to changes in intracellular pH ($pH_i$) in other cells, we decided to investigate the mechanism of $pH_i$ regulation in mesangial cells and how $pH_i$ regulation is modified by various mitogens.

## METHODOLOGY

### Cell Culture

We isolated the MCs from rat kidneys and maintained them in culture,[1] performing experiments on cells in the second through fifth passages. The cells were identified by morphologic examination and positive staining with an antibody against vascular-smooth-muscle myosin.[2] MCs were incubated at 37°C in a 5% $CO_2$/95% air atmosphere, employing DMEM supplemented with 10–20% fetal calf serum (FCS), 5 $\mu$g/ml insulin, 10 mM L-glutamine, 400 ng/ml penicillin, 500 ng/ml streptomycin, 10 mM N-hydroxyethylpiperazine-N'-2-ethanesulfonic acid (HEPES), and 20 mM $HCO_3^-$. Twenty-four hours before the experiments, when the cells were 70–90% confluent, we induced quiescence by reducing the FCS to 0.5%.

[a] The research we report here was supported by NIH grants R01-DK30344 and R01-NS18400. Dr. Boyarsky was supported by a fellowship from the Connecticut Affiliate of the American Heart Association. Dr. Ganz was supported by a Veterans Administration Career Development Award.

[b] Address correspondence to: Dr. Walter F. Boron, Department of Cellular and Molecular Physiology, Yale University School of Medicine, 333 Cedar Street, New Haven, CT 06510.

## $pH_i$ Measurements

We loaded MCs with the pH-sensitive dye 2,7-biscarboxyethyl-5(6)-carboxy-fluorescein (BCECF) by exposing them to the dye's relatively permeant acetoxy-methyl ester derivative, which is hydrolyzed inside the cells to yield BCECF.[3] Our $pH_i$ measurements were made in two ways. In the first,[2] we placed a glass coverslip on which the MCs were grown into the cuvette of a Perkin-Elmer LS-5B spectrofluorometer and excited the intracellular dye at 500 or 440 nm while monitoring the emission at 530 nm. In the second approach,[4] we fixed the coverslip to the bottom of a chamber placed on a fluorescence microscope and illuminated only a 10-$\mu$m-diameter portion of a single MC. We used a computer to control the alternate excitation of the dye at 440 and 490 nm and the emission at 530 nm. In both approaches, we continuously superfused the cells with solution at 37°C, so we could make rapid solution changes and avoid the extracellular build-up of dye leaking from the cells. However, in the first approach we monitored the average $pH_i$ of thousands of cells, whereas in the second we recorded the $pH_i$ time course in a single MC.

## $pH_i$ REGULATION IN THE ABSENCE OF $HCO_3^-$

The steady-state $pH_i$ is determined by the balance between processes that load the cell with acid (thereby tending to lower $pH_i$) and those that extrude acid (thereby tending to raise $pH_i$). In the absence of a $CO_2/HCO_3^-$ buffer system, the only substantial acid-extrusion mechanism is Na-H exchange. FIGURE 1A illustrates an experiment in which a cell was acid loaded by applying and then withdrawing a $CO_2/HCO_3^-$-free solution containing 20 mM $NH_4^+$ ($NH_4^+$ replacing $Na^+$). Withdrawal of the $NH_4^+$ caused a rapid fall in $pH_i$ (segment $cd$ in FIG. 1). The MC recovered from this acid load in about 5 minutes ($de$). In other vertebrate cells, similar $pH_i$ recoveries are mediated by a Na-H exchanger in the plasma membrane. Indeed, as shown in FIGURE 1B, we found that $pH_i$ recovery in MCs is blocked by ethylisopropyl amiloride (EIPA). Other experiments (not shown) indicated that the $pH_i$ recovery is blocked by the removal of external $Na^+$, with an apparent Km for external $Na^+$ of about 27 mM. Thus, these data indicate that in the absence of $CO_2/HCO_3^-$, MCs recover from acid loads by means of a Na-H exchanger.

In the past, others have used data such as the EIPA results just summarized to conclude that (1) the Na-H exchanger is responsible for the entire $pH_i$ recovery and (2) the Na-H exchanger shuts off (i.e., has a "threshold") when the $pH_i$ recovery halts (e.g., point $e$ in FIG. 1A). However, the foregoing experiments determined the EIPA-sensitive component of the $pH_i$ recovery (i.e., Na-H exchange rate) at only one rather low $pH_i$ (~6.6 in FIG. 1B) and did not address the issue of the exchanger's $pH_i$ dependence. To determine whether the Na-H exchanger is inactive in the "normal" $CO_2/HCO_3^-$-free steady state, we added 50 $\mu$M EIPA to cells in the normal $CO_2/HCO_3^-$-free steady state. We found that application of EIPA caused $pH_i$ to fall rapidly by about 0.25.[4] If we assume that the only acute effect of EIPA is to block Na-H exchange, then the aforementioned result implies that the Na-H exchanger is normally active, and that blocking this acid extruder unmasks a background acid-loading process that normally counteracts the alkalinizing effects of the exchanger. Thus, if there is a threshold for the Na-H exchanger, this threshold must be substantially higher than the steady-state $pH_i$ prevailing in the absence of $CO_2/HCO_3^-$.

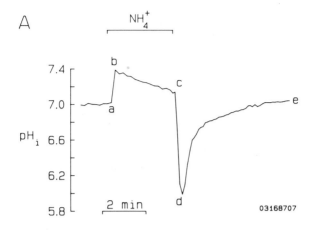

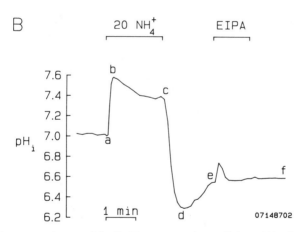

**FIGURE 1.** Response of mesangial cell pH to an acute intracellular acid load. **(A)** Intracellular pH ($pH_i$) recovery from an acid load under control conditions. A single mesangial cell (MC), grown on a glass coverslip and visualized on the stage of an inverted microscope, was acid loaded by applying and then withdrawing a solution in which 20 mM $Na^+$ was replaced by 20 mM $NH_4^+$. The application of the $NH_4^+$ caused a rapid $pH_i$ increase (*ab*) due to entry of the weak base $NH_3$, which is present in small amounts, and its subsequent combination with $H^+$ to produce $NH_4^+$. The slower entry of the weak acid $NH_4^+$ leads to the plateau phase acidification (*bc*) as a small fraction of the entering $NH_4^+$ dissociates into $NH_3$ and $H^+$. The removal of extracellular $NH_4^+$ causes the remaining intracellular $NH_4^+$ to dissociate into $NH_3$, which leaves the cell, and $H^+$, which remains trapped inside. Thus, $pH_i$ falls rapidly (*cd*). However, the cell recovers from this acid load as $pH_i$ returns to its initial value (*de*). **(B)** Effect of ethylisopropylamiloride (EIPA) on the $pH_i$ recovery. Application of EIPA, which is a potent inhibitor of Na-H exchange in other cells, blocks the recovery of $pH_i$ from the acid load (*ef*). (From ref. 4; reprinted by permission of the American Physiological Society.)

## EFFECT OF MITOGENS IN THE ABSENCE OF HCO$_3^-$

In 1983 Moolenaar and coworkers,[5] working on fibroblasts, and then Glaser and colleagues, working on NR6[6] and A431 cells,[7] showed that application of growth factors to quiescent cells caused a sustained pH$_i$ increase. In the latter two studies, the sustained alkalinization was preceded by a small transient acidification. Because the sustained, growth-factor-induced alkalinizations were dependent upon Na$^+$ and blocked by amiloride, they were ascribed to stimulation of the Na-H exchanger. Later work demonstrated that the growth-factor-dependent alkalinization produced by platelet-derived growth factor could be mimicked by 12-O-tetradecanoyl-phorbol-13-acetate (TPA), implicating the involvement of protein kinase C.[8] In experiments conducted in the nominal absence of $CO_2/HCO_3^-$, we[2] demonstrated that arginine vasopressin (AVP) elicits in MCs the same sort of biphasic pH$_i$ changes produced by platelet-derived growth factor in NR6 cells[6] and by EGF in A431 cells.[7] Arginine vasopressin is the most potent single agent in promoting the proliferation of previously quiescent MCs. As illustrated in FIGURE 2, arginine vasopressin produces a sustained alkalinization of ~0.1 pH units that reaches its maximal level in about 5 minutes, and is fully reversed upon removal of the peptide. In agreement with earlier data obtained by others, we found that the presence of EIPA or the absence of Na$^+$ accentuated the peptide-induced acidification, but blocked the alkalinization.

The data just summarized are certainly consistent with the hypothesis[9] that in the absence of $CO_2/HCO_3^-$, growth factors produce an increase in pH$_i$ by stimulating a Na-H exchanger. However, another possibility cannot be ruled out by these data: the predominant effect of the growth factor may not be to stimulate Na-H exchange, but to reduce intracellular acid loading. Indeed, we know that in the steady state the Na-H exchanger balances a substantial amount of background acid loading. Evidence for this is that application of EIPA (or removal of Na$^+$) in the steady state produces a fall in pH$_i$.[4] As pH$_i$ falls, this background acid loading gradually disappears. If the growth factor is now applied after EIPA (or removal of Na$^+$), pH$_i$ cannot rise, because both Na-H exchange and the previously existing background acid loading are inactive. The application of a growth factor after EIPA (or removal of Na$^+$) could produce a further fall in pH$_i$, as actually ob-

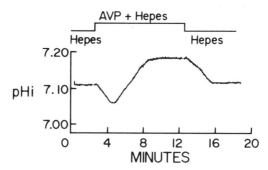

**FIGURE 2.** Effect of arginine vasopressin (AVP) on pH$_i$ in the nominal absence of $CO_2/HCO_3^-$. Mesangial cells grown on a glass coverslip were placed in a flow-through cuvette in a commercial spectrofluorometer. The application of $10^{-7}M$AVP in a HEPES-buffered medium caused a transient pH$_i$ decline, followed by a sustained but reversible alkalinization. (From ref. 3; reprinted by permission of the American Physiological Society.)

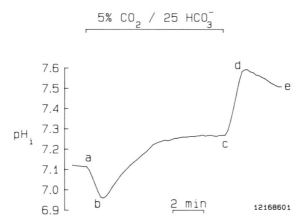

**FIGURE 3.** Effect on $pH_i$ of applying $CO_2/HCO_3^-$. A single mesangial cell, initially exposed to a pH-7.4 solution buffered with HEPES, was exposed to a solution of the same pH, but buffered with 5% $CO_2$/25 mM $HCO_3^-$. The influx of $CO_2$ into the cell caused a transient acidification (*ab*) that was followed by a sustained alkalinization (*bc*). $CO_2/HCO_3^-$ presumably caused a net increase in $pH_i$ because the activity of the $HCO_3^-$-dependent acid-extrusion mechanism (i.e., the $Na^+$-dependent $Cl$-$HCO_3$ exchanger) outstripped that of the $HCO_3^-$-dependent acid-loading mechanism (i.e., the $Na^+$-independent $Cl$-$HCO_3$ exchanger. Removal of the $CO_2/HCO_3^-$ caused the opposite sequence of $pH_i$ changes (*cde*). (From ref. 4; reprinted by permission of the American Physiological Society.)

served,[2] if the growth factor introduces a new component of intracellular acid loading. Additional experiments will be necessary to distinguish between the conventional hypothesis that growth factors raise $pH_i$ in the absence of $CO_2$/$HCO_3^-$ by stimulating Na-H exchange, and the alternative hypothesis, that growth factors actually inhibit acid loading.

## $pH_i$ REGULATION IN THE PRESENCE OF $HCO_3^-$

Although the experiments just summarized provide information about the effects of growth factors on $pH_i$ regulation, they are of uncertain physiologic relevance. After all, the cells were incubated in an unphysiologic $CO_2/HCO_3^-$-free environment, and it is well established that many cells possess potent $HCO_3^-$ transport systems. We therefore decided to explore the possibility that MCs also possess one or more $HCO_3^-$ transport systems. The first hint that $HCO_3^-$ transport is important for MCs was that exposing MCs to $CO_2/HCO_3^-$ caused a transient $pH_i$ decline that was followed by a rise to a value more than 0.1 higher than the initial one.[4] This $HCO_3^-$-dependent alkalinization, illustrated in FIGURE 3, could not be accounted for by either a decrease in acid-loading rate or a $CO_2/HCO_3^-$-induced stimulation of Na-H exchange. Rather, we[10] found that the $HCO_3^-$-induced increase in $pH_i$ was inhibited 90% by the removal of $Na^+$, and reduced about two thirds by preincubating the cells in SITS,[4] an inhibitor of several $HCO_3^-$ transport processes. Furthermore, the $HCO_3^-$-induced alkalinization was greatly inhibited by preincubating the cells in a $Cl^-$-free medium. These results suggest that in addition to a Na-H exchanger, MCs possess a $Na^+$-dependent $Cl$-$HCO_3$ exchanger that is approximately twice as potent as the Na-H exchanger. It is the newly introduced activity of this acid-extruding mechanism that is responsible for the increase in $pH_i$ brought about by the addition of $CO_2/HCO_3^-$.

Both the Na-H and the $Na^+$-dependent $Cl-HCO_3$ exchangers are acid-extrusion mechanisms. That is, they tend to raise $pH_i$. Work on certain other cells has identified a $Na^+$-independent $Cl-HCO_3$ exchanger (hereafter referred to simply as a $Cl-HCO_3$ exchanger) that functions as an acid loader.[11,12] We investigated the possibility that MCs possess a $Cl-HCO_3$ exchanger in two ways. First, we removed external $Cl^-$, finding that this evoked the rapid and reversible $pH_i$ increase that would be expected of a $Cl-HCO_3$ exchanger. Furthermore, this $pH_i$ increase was largely blocked by preincubating the cells with SITS, but it was unaffected by removing $Na^+$. Interestingly, we found the $Cl-HCO_3$ exchanger to have a $pH_i$ dependency opposite to that of the Na-H exchanger. That is, the $Cl-HCO_3$ exchanger was only minimally active at $pH_i$ values below ~7, gradually increased in activity between $pH_i$ values of 7.0 and 7.4, and then greatly increased in activity at $pH_i$ values in excess of 7.4. Similar conclusions regarding the $pH_i$ dependence of the $Cl-HCO_3$ exchanger of Vero cells were reached by Olsnes and colleagues,[13] using a different approach.

In summary, the MC possesses three acid-base transport systems. Of the two acid-extruders, only the Na-H exchanger is active in the absence of $HCO_3^-$. The $Na^+$-dependent $Cl-HCO_3$ exchanger becomes active only in the presence of $CO_2/HCO_3^-$, but it is approximately twice as potent as the Na-H exchanger at a $pH_i$ of ~6.8. The third acid-base transporter is the $Cl-HCO_3$ exchanger, which functions as an acid loader, thereby counteracting the effects of the two acid extruders. The sustained increase in $pH_i$ that occurs when MCs are transferred from a $HCO_3^-$-free to a $HCO_3^-$-containing environment presumably occurs because the acid-extruding effects of the $Na^+$-dependent $Cl-HCO_3$ exchanger exceed the acid-loading influence of the $Cl-HCO_3$ exchanger.

## EFFECT OF MITOGENS IN THE PRESENCE OF $HCO_3^-$

Although a variety of growth factors produces a sustained $pH_i$ increase in the absence of $CO_2/HCO_3^-$, this has not proven to be the case in the presence of $CO_2/HCO_3^-$. For example, Cassel et al.[14] found that EGF plus FCS causes A431 cells to alkalinize in the absence of $CO_2/HCO_3^-$, but to acidify in the presence of $CO_2/HCO_3^-$. We[2,15] made similar observations in MCs. In work on the fibroblastic MES-1 cell line, Bierman et al.[16] found that EGF causes a rise in $pH_i$ in the absence of $CO_2/HCO_3^-$, but no change in the presence of this physiologic buffer. In the last case, the authors presented evidence that $CO_2/HCO_3^-$ causes $pH_i$ to rise to such a level that the Na-H exchanger is inactivated and thus unable to respond to the growth factor.

To determine why AVP causes a small decline in $pH_i$ when MCs are incubated in $CO_2/HCO_3^-$, we examined the $pH_i$ recovery from acid and alkali loads in both control/quiescent MCs and MCs acutely exposed to AVP.[17] FIGURE 4 summarizes the results of four experiments in which we used the $NH_4^+$ prepulse technique to acid load MCs exposed to no inhibitors (FIG. 4A), exposed to the Na-H exchange inhibitor EIPA (FIG. 4B), pretreated with the anion transport inhibitor SITS (FIG. 4C), or treated with both EIPA and SITS (FIG. 4D). In each of the four experiments, we monitored the recovery of $pH_i$ from an acid load both before and after the application of $10^{-7}$ M AVP. In the absence of inhibitors, AVP caused a substantial increase in the rate at which $pH_i$ recovered from the $NH_4^+$-induced

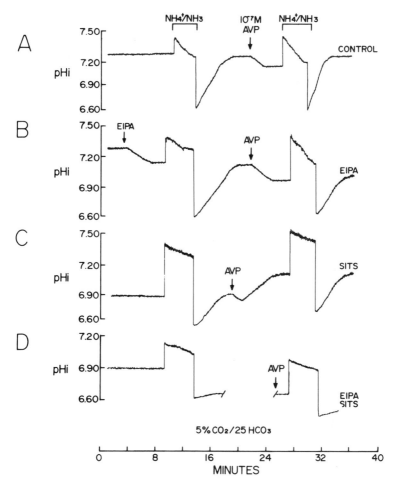

**FIGURE 4.** The effect of arginine vasopressin (AVP) on the recovery of $pH_i$ from acid loads in the presence of $CO_2/HCO_3^-$. The experiments were performed by measuring the average $pH_i$ of many mesangial cells in a commercial spectrofluorometer. **(A)** Effect of AVP in the absence of inhibitors. The $pH_i$ recovery from an acid load (imposed by applying and then withdrawing 20 mM $NH_4^+$) was much faster after the application of $10^{-7}$ AVP. In control experiments, we showed that the rates of $pH_i$ recovery were the same in two consecutive $NH_4^+$ pulses. **(B)** Effect of AVP in the presence of EIPA. By comparing the rates of $pH_i$ recovery in **(A)** with those in **(B)**, we were able to determine the EIPA-sensitive component of the $pH_i$ recovery (presumably due to Na-H exchange) both before and after application of EIPA. **(C)** Effect of AVP in the presence of SITS. Cells were pretreated with 0.5 mM SITS for 1 hour before the start of the experiment, and the drug was then washed away. By comparing the rates of $pH_i$ recovery in **(A)** with those in **(C)**, we were able to determine the SITS-sensitive component of the $pH_i$ recovery (presumably due to $Na^+$-dependent Cl-HCO$_3$ exchange) both before and after application of EIPA. **(D)** Effect of AVP in the presence of both EIPA and SITS. Simultaneous blockade of both Na-H exchange and $Na^+$-dependent Cl-HCO$_3$ exchange blocked nearly all $pH_i$ recovery both before and after application of EIPA. Thus, at least at the relatively acid $pH_i$ values at which the transporters were assayed, AVP stimulates both Na-H and $Na^+$-dependent Cl-HCO$_3$ exchange. (From ref. 18; reprinted by permission of Macmillan Magazines Ltd.)

acid load (FIG. 4A). EIPA slowed the $pH_i$ recoveries, both before and after addition of the AVP. By comparing the rates of $pH_i$ recovery at a $pH_i$ of 6.6 before AVP addition in FIGURE 4A and B, we determined that the EIPA-sensitive component of the $pH_i$ recovery (i.e., Na-H exchange) accounts for about one third of the total, corresponding to an average acid extrusion rate of 62 $\mu$M s$^{-1}$. By comparing the rates of $pH_i$ recovery after AVP addition in FIGURE 4A and B, we determined that AVP caused the EIPA-sensitive component of acid extrusion to increase to an average of 121 $\mu$M s$^{-1}$. Thus, in the presence of $CO_2/HCO_3^-$, AVP stimulated Na-H exchange by ~95%. Note that the application of EIPA in FIGURE 4B caused a fall in $pH_i$, indicating that the Mc Na-H exchanger is active in MCs in the normal steady state.

To determine the effects of AVP on the Na$^+$-dependent Cl-HCO₃ exchanger, we compared rates of $pH_i$ recovery in the absence of inhibitors (FIG. 4A) with those obtained in SITS-pretreated MCs (FIG. 4C). Analyzing the data as just summarized for the EIPA experiments, we found that before addition of AVP, the SITS-sensitive component of the $pH_i$ recovery corresponded to a mean acid extrusion rate of 121 $\mu$M s$^{-1}$. Thus, in the absence of a growth factor, the SITS-sensitive component (presumably Na$^+$-dependent Cl-HCO₃ exchanger) accounted for about two thirds of the cells' total acid-extruding capacity. After the addition of AVP, the SITS-sensitive component averaged 236 $\mu$M s$^{-1}$. Thus, AVP must have stimulated the exchanger by ~95%. These estimates of Na$^+$-dependent Cl-HCO₃ exchange activity are, if anything, likely to be low, inasmuch as SITS is also expected to block Cl-HCO₃ exchange, an effect that would speed the $pH_i$ recovery.

We approached the problem of examining the effect of AVP on the Cl-HCO₃ exchanger from the opposite direction.[17] We could acutely alkali load MCs by pretreating them with a solution equilibrated with twice the normal $CO_2$ level (i.e., 10% instead of 5%) and containing twice the normal amount of $HCO_3^-$ (i.e., 50 mM rather than 25 mM), but with $HCO_3^-$ replacing Cl$^-$. When exposed to this solution, which had a pH of 7.4, MCs slowly alkalinized, presumably due either to stimulation of Na$^+$-dependent Cl-HCO₃ exchange or to inhibition of Cl-HCO₃ exchange (FIG. 5, left panel). The rise in $[CO_2]_o$ would have been expected to cause a transient decline in $pH_i$. Although this acidification was observed in some experiments, it was generally overwhelmed by the alkalinization. When the MCs were then returned to a normal $CO_2/HCO_3^-$ solution (i.e., with a lower $[CO_2]_o$), they rapidly alkalinized because of the efflux of $CO_2$ from the cells. At this point the cells were alkali loaded. Note that this newly introduced $CO_2/HCO_3^-$ prepulse technique for alkali loading is just the inverse of the $NH_3/NH_4^+$ prepulse technique for acid loading. The important part of this experiment physiologically is that the cells spontaneously recovered from the alkali load, with $pH_i$ falling to its initial value within ~6 minutes. When similar experiments were performed in the absence of Cl$^-$, the $pH_i$ recovery was inhibited an average of 92%. DIDS, another blocker of anion transport, was similarly effective at blocking the recovery. Not illustrated are other experiments demonstrating that Na$^+$ was not required for the recovery of $pH_i$ from the alkali load. Thus, the $pH_i$ recovery observed in the left half of FIGURE 5 was most likely due to Cl-HCO₃ exchange. The right panel of FIGURE 5 illustrates the results of similar experiments performed on cells treated with AVP. As can be seen, AVP substantially increases the rate of $pH_i$ recovery from the alkali load. Analysis of the data shows that the DIDS-sensitive component of the $HCO_3^-$ influx at a $pH_i$ of 7.7 was 568 $\mu$M s$^{-1}$ in the absence of AVP, but 1,385 $\mu$M s$^{-1}$ in the presence of the growth factor. Thus, AVP stimulated Cl-HCO₃ exchange by 143%.

   The experiments just summarized showed that the growth factor AVP, which is thought to activate phospholipase C and thereby stimulate protein kinase C, activates all three MC acid-base transporters. However, we examined only the acute effects of AVP. In more recent studies, we have examined over a more prolonged period the effects of AVP as well as a second growth factor, epidermal growth factor (EGF), which is thought to act through a receptor-tyrosine kinase in

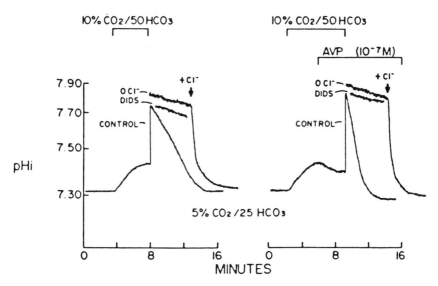

**FIGURE 5.** The effect of arginine vasopressin (AVP) on the recovery of $pH_i$ from alkali loads. The experiments were performed by measuring the average $pH_i$ of many mesangial cells (MCs) in a commercial spectrofluorometer. **(Left panel)** Recovery of $pH_i$ from an alkali load in the absence of AVP. The MCs were initially exposed to a solution buffered to pH 7.4 with 5% $CO_2$/25 mM $HCO_3^-$. Switching to a solution buffered to the same pH with 10% $CO_2$/ 50 mM $HCO_3^-$ caused a sustained $pH_i$ increase. When the cells were finally returned to the original 5% $CO_2$/25 mM $HCO_3^-$ solution, $pH_i$ rose rapidly, due to the efflux of $CO_2$. The cells then recovered from the alkali load, returning to the initial $pH_i$. The $pH_i$ recovery was largely blocked by 50 $\mu$M DIDS or by removal of external Cl⁻. Returning the Cl⁻ allowed the recovery to proceed. Not illustrated are experiments showing that the $pH_i$ recovery is not Na⁺ dependent. Thus, the $pH_i$ recovery was due to Cl-$HCO_3$ exchange. **(Right panel)** Recovery of $pH_i$ from an alkali load in the presence of AVP. The protocol was the same as in the right panel, except that $10^{-7}$M AVP was applied during the 10% $CO_2$/25 mM $HCO_3^-$ pulse. The $pH_i$ recovery from the alkali load was much faster in the presence than in the absence of AVP, although the recovery was still largely blocked by either DIDS or Cl⁻ removal. Thus, AVP stimulates Cl-$HCO_3$ exchange, at least at relatively alkaline $pH_i$ values. (From Figure 18; reprinted by permission of Macmillan Magazines Ltd.)

MCs. Our protocol was to render MCs quiescent, and then to expose them to either one of the two growth factors, maintaining the presence of the growth factor over the entire course of the experiment. Both growth factors produced their maximal stimulation of all three transporters immediately after application of the agonist. In both cases, however, transporter activity slowly decreased over a period of 4–6 hours, leveling off at activities that were still substantially above prestimulated levels. An interesting result in the EGF experiments was that the activity of all three transporters temporarily decreased to slightly above control levels between 10 and 20 hours after application of EGF. This is approximately the time of maximal cell division. In the AVP experiments, there was a similar temporary dip in the activities of all three transporters, but between 40 and 48 hours after addition of the agent. This is also the period in which cell division is most pronounced after stimulation with AVP. Thus, both AVP and EGF activate all three MC acid-base transporters in a time-dependent manner, the activation being most pronounced immediately after addition of the mitogen and transiently decreasing during a period corresponding to that of maximal proliferation. A novel twist to these data was that although time-dependent changes in transporter activity occurred, there were no significant changes in steady-state $pH_i$ in either the long-term AVP or long-term EGF experiments.

## CONCLUSIONS

Our data demonstrate that mesangial cells have three acid-base transporters that contribute to $pH_i$ regulation. As schematized in FIGURE 6, these are the Na-H

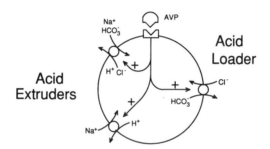

**FIGURE 6.** The three acid-base transporters in the renal mesangial cell and their activation by AVP.

exchanger and the $Na^+$-dependent $Cl$-$HCO_3$ exchanger, both of which are acid extruders, and the $Cl$-$HCO_3$ exchanger, which is an acid loader. Although the growth factor AVP or EGF both produce significant alkalinizations in the absence of $CO_2/HCO_3^-$, they produce slight acidifications (0.03–0.05 pH units) in the presence of this physiologic buffer. We believe that reversal of the direction of the growth-factor-induced $pH_i$ shift is a direct result of the activity of the two $HCO_3^-$ transporters. Support for this comes first from the observation that, even in $CO_2/HCO_3^-$, AVP can produce a $CO_2/HCO_3^-$-free-like alkalinization, provided $HCO_3^-$ transport is blocked by SITS. Therefore, it is not the presence of $CO_2/HCO_3^-$ *per se* that is required to elicit the mitogen-induced acidification, but rather a functioning pair of $HCO_3^-$ transporters.

Although we had not expected that $pH_i$ would be more or less stable in the presence of a mitogen-induced activation of the three acid-base transporters, this result is reasonable when interpreted within the scope of $pH_i$ regulation. One might imagine that it is in the cell's best interest to maintain a stable $pH_i$, one that on balance is optimal for cell function. Thus, it may be important for the growth factor to improve $pH_i$ regulation without producing deleterious changes in $pH_i$. Indeed, neither AVP nor EGF produces substantial $pH_i$ changes, but both substantially increase the cell's ability to recover from acid and alkali loads. If the growth factor were to stimulate only acid-extruding mechanisms, then steady-state $pH_i$ would increase. The cell would have to pay a price if it were to improve $pH_i$ regulation only in the acid direction: steady-state $pH_i$ would have to rise in proportion to the stimulation of acid extrusion. By stimulating both acid-loading and acid-extruding systems in parallel, the cell has the best of both worlds, a stable $pH_i$ and greatly improved $pH_i$ regulation in both the acid and alkaline directions. In the mesangial cell, the activities of the three acid-base transporters apparently can change without affecting $pH_i$, provided that the activities of the three transporters change in parallel. This independence also applies to the long-term changes in acid-base transport that are produced by AVP and EGF. Both growth factors produce degrees of stimulation of the transporters that vary substantially over the course of 2–3 days. However, because the activities of all three transporters vary more or less in parallel, there are no significant changes in $pH_i$.

Significant work remains to be done in the area of $pH_i$-growth factor interactions. One area that needs to be examined is the $pH_i$ dependence of each acid-base transporter in the absence and presence of $CO_2/HCO_3^-$, and how this $pH_i$ dependence is influenced by various growth factors as a function of time. A second area that we are actively investigating is cell shrinkage as a stimulus for activating the acid-base transporters. We have already determined that shrinking MCs affects $pH_i$ substantially more than does either AVP or EGF. Studying the interactions of shrinkage activation and growth factor activation may also be important in elucidating the control of these acid-base transporters. Finally, it will be important to learn more about the exchangers on a molecular level. A human Na-H exchanger has been cloned and sequenced.[18] It will be interesting to know whether some of the subtle differences in response of various cells to growth factors can be accounted for, at least in part, by differences in the primary structure of the acid-base transporters.

## REFERENCES

1. LOVETT, D. H., J. L. RYAN & R. B. STERZEL. 1983. Stimulation of rat mesangial cell proliferation by macrophage interleukin-1. J. Immunol. **131:** 2830–2836.
2. GANZ, M. B., G. BOYARSKY, W. F. BORON & R. B. STERZEL. 1988. Effects of angiotensin II and vasopressin on intracellular pH of glomerular mesangial cells. Am. J. Physiol. **254:** F787–F794.
3. RINK, T. J., R. Y. TSIEN & T. POZZAN. 1982. Cytoplasmic pH and free $Mg^{+2}$ in lymphocytes. J. Cell Biol. **95:** 189–196.
4. BOYARSKY, G., M. B. GANZ, B. STERZEL & W. F. BORON. 1988. pH regulation in single glomerular mesangial cells. I. Acid extrusion in absence and presence of $HCO_3$. Am. J. Physiol. **255:** C844–C856.
5. MOOLENAAR, W. H., R. Y. TSIEN, P. T. VAN DER SAAG & S. W. DE LAAT. 1983. $Na^+/H^+$ exchange and cytoplasmic pH in the action of growth factors in human fibroblasts. Nature **304:** 645–648.
6. CASSEL, D., P. ROTHENBERG, Y.-X. ZHUANG, T. F. DEUEL & L. GLASER. 1983. Platelet-derived growth factor stimulates $Na^+/H^+$ exchange and induces cytoplasmic alkalinization in NR6 cells. Proc. Natl. Acad. Sci. USA **80:** 6224–6228.
7. ROTHENBERG, P., L. GLASER, P. SCHLESINGER & D. CASSEL. 1983. Activation of $Na^+/H^+$ exchange by epidermal growth factor elevates intracellular pH in A431 cells. J. Biol. Chem. **258:** 12644–12653.
8. MOOLENAAR, W. H., L. G. J. TERTOOLEN & S. W. DE LAAT. 1984. Phorbol ester and diacylglycerol mimic growth factors in raising cytoplasmic pH. Nature **312:** 371–374.

9. MOOLENAAR, W. H. 1986. Effects of growth factors on intracellular pH regulation. Ann. Rev. Physiol. **48:** 363–376.

10. BOYARSKY, G., M. B. GANZ, B. STERZEL & W. F. BORON. 1988. pH regulation in single glomerular mesangial cells. II. Na-dependent and -independent Cl-HCO₃ exchangers. Am. J. Physiol. **255:** C857–C869.

11. CHAILLET, J. R., K. AMSLER & W. F. BORON. 1986. Optical measurement of intracellular pH in single LLC-PK1 cells: Demonstration of Cl⁻/HCO₃⁻ exchange. Proc. Natl. Acad. Sci. USA **83:** 522–526.

12. VAUGHAN-JONES, R. D. Chloride-bicarbonate exchange in the sheep cardiac purkinje fibre. *In* Intracellular pH: Its Measurement, Regulation and Utilization in Cellular. R. Nuccitelli & D. W. Deamer, eds.: 239–252. Alan R Liss, Inc. New York, NY.

13. OLSNES, S., J. LUDT, T. I. TONNESSEN & K. SANDVIG. 1987. Bicarbonate/chloride antiport in Vero cells. II. Mechanisms for bicarbonate-dependent regulation of intracellular pH. J. Cell Physiol. **132:** 192–202.

14. CASSEL, D., B. WHITELEY, Y. X. ZHUANG & L. GLASER. 1985. Mitogen-independent activation of Na⁺/H⁺ exchange in human epidermoid carcinoma A431 cells: Regulation by medium osmolarity. J. Cell Physiol. **122:** 178–186.

15. BOYARSKY, G., M. B. GANZ, R. B. STERZEL & W. F. BORON. 1987. Optical study of pH regulation in single cultured glomerular mesangial cells. Kidney Int. **31:** 161.

16. BIERMAN, A. J., E. J. CRAGOE JR., S. W. DE LATT & W. H. MOOLENAAR. 1988. Bicarbonate determines cytoplasmic pH and suppresses mitogen-induced alkalinization in fibroblastic cells. J. Biol. Chem. **263:** 15253–15256.

17. GANZ, M. B., G. BOYARSKY, R. B. STERZEL & W. F. BORON. 1989. Arginine vasopressin enhances pH_i regulation in the presence of HCO₃⁻ by stimulating three acid-based transport systems. Nature **337:** 648–651.

18. SARDET, C., A. FRANCHI & J. POUYSSEGUR. 1989. Molecular cloning, primary structure, and expression of the human growth factor-activatable Na⁺/H⁺ antiporter. Cell **56:** 271–280.

# Elementary Aspects of Acid-Base Permeation and pH Regulation[a]

OLAF SPARRE ANDERSEN

*Department of Physiology and Biophysics*
*Cornell University Medical College*
*New York, New York 10021*

This article presents an overview of the physicochemical principles underlying acid-base permeation through cell membranes and cellular pH regulation: first, the mechanisms of acid-base permeation; next, general aspects of pH regulation, which will be illustrated with a specific example; finally, how cell pH and volume changes are coupled. The central theme is the importance of nonionic diffusion through the plasma membrane, and the constraints this imposes on cells' ability to regulate their pH and volume.

Physicochemically, a central issue in acid-base physiology is that the aqueous concentrations of $H^+$ (actually $H_3O^+$ [e.g., ref. 1]) and $OH^-$ usually are low, $\sim 100$ M,[2] and that $H^+$ and $OH^-$ participate in numerous reactions of the types:

$$H_2O \rightleftharpoons H^+ + OH^-. \tag{I}$$

$$HA \rightleftharpoons H^+ + A^-, \tag{IIa}$$

$$HA + OH^- \rightleftharpoons H_2O + A^-. \tag{IIb}$$

$$B + H^+ \rightleftharpoons HB^+, \tag{IIIa}$$

$$BH^+ + OH^- \rightleftharpoons H_2O + B. \tag{IIIb}$$

Apart from their role in buffering $[H^+]$, these aqueous protonation/deprotonation reactions are important for three reasons.

First, because $H^+$ and $OH^-$ participate in the foregoing reactions, it is difficult to identify the actual molecular species involved in the movement of acid-base equivalents across cell membranes. A net efflux of acid-base equivalents can be brought about by an efflux of $H^+$ or by an influx of $OH^-$, which means, for example, that a $H^+$-driven cotransport (or symport) is difficult to distinguish from an $OH^-$-driven countertransport (or antiport). These two alternatives are difficult to distinguish when considering only changes in the intracellular or extracellular pH ($pH_i$ and $pH_o$, respectively) or in the concentration of the co- or counter-transported solute. These alternative mechanisms are not equivalent, however, because they do not induce equivalent changes in the cell's solute content. They would be equivalent only if the influx of $OH^-$ occurred in conjunction with an efflux of $H_2O$ (cf. Reaction I).

Second, cell membranes are much more permeable to the unionized forms of low-molecular acids and bases than to their charged counterparts (or to $H^+$ or $OH^-$), so that a net flux of acid-base equivalents often will occur in the guise of a flux of the unionized partner in an acid/base pair, which is the basis for the importance of nonionic diffusion in acid-base physiology.

[a] This work was supported in part by NIH grants GM21342 and GM40062.

333

Third, the aqueous concentrations of the partners in an acid/base pair (HA/A⁻ or BH⁺/B) will be much higher than the $H^+$ or ($OH^-$) concentration. Small (absolute) changes in [$H^+$] will thus be associated with large changes in the concentrations of any involved acid/base partners, so that $pH_i$ changes and $pH_i$ regulation become inextricably coupled to volume changes and volume regulation.

## PERMEATION MECHANISMS

$H^+$ appears in many disguises, and many permeation mechanisms mediate the movement of acid-base equivalents across plasma membranes (e.g., ref. 3). Some of these mechanisms are listed in TABLE 1, in which is also denoted whether under physiologic conditions they would be acid or base extruding (whether they would tend to increase or decrease $pH_i$). This table is not comprehensive, and many

TABLE 1. Mechanisms of Plasma Membrane Proton Movement[a]

| Mechanism | $pH_i$ |
|---|---|
| Movement of $H^+$/$OH^-$ Equivalents through the Lipid Bilayer | |
| Nonionic diffusion ($CO_2$, $NH_3$) | |
| Electrodiffusion of $H^+$/$OH^-$ | ↓ |
| Channel-Mediated Movement of $H^+$/$OH^-$ Equivalents | |
| $H^+$, $NH_4^+$ (cation channels) | |
| $HCO_3^-$, $OH^-$ (anion channels) | |
| $H^+$/$OH^-$ (water channels) | ↓ |
| Antiport-Mediated Movement of $H^+$/$OH^-$ Equivalents | |
| $H^+$/$Na^+$ exchange | ↑ |
| $HCO_3^-$/$Cl^-$ exchange | ↓ |
| $A^-$/$Cl^-$ exchange (e.g., formate/chloride exchange) | ↓ |
| $Na^+$-dependent $HCO_3^-$/$Cl^-$ exchange | ↑ |
| Symport-Mediated Movement of $H^+$/$OH^-$ Equivalents | |
| $A^-$,$Na^+$ cotransport | ↑ |
| $A^-$,$H^+$ cotransport | |
| $NaCO_3^-$,$HCO_3^-$ cotransport | ↓ |
| ATP-Driven Movement of $H^+$/$OH^-$ Equivalents | |
| $H^+$-ATPase | ↑ |
| $H^+$/$K+$-ATPase | ↑ |

[a] Modified after Table 1 in ref. 3.

classes of transport systems, which currently are identified in generic terms, are likely to be subtyped into isoforms, based on functional or protein chemical information or primary structures deduced from cDNA sequencing, as has been done for the $HCO_3^-$/$Cl^-$ exchangers.[4–6]

Mechanistically, acid-base equivalents permeate cell membranes through the lipid bilayer moiety *per se,* or they cross the membrane by a protein-mediated (or protein-catalyzed) process.

### Solute Movement Through Lipid Bilayers

The most elementary permeation mode is solubility-diffusion, in which a solute partitions into the nonpolar core of the lipid bilayer of biologic membranes and moves through the hydrophobic core, from one membrane/solution interface to the other by diffusion. The membrane's permeability coefficient ($P_m$) is given by:

$$P_m = \alpha \cdot D_m/\delta_m, \tag{1}$$

where $\alpha$ is the solute's partition coefficient ($= C_m/C_a$, where $C_m$ and $C_a$ denote the solute concentrations on the membrane and aqueous phase, respectively), $D_m$ is the solute's diffusion coefficient in the membrane, and $\delta_m$ is the thickness of the membrane's hydrophobic core. (Lipid bilayers are anisotropic, and both $C_m$ and $D_m$ can vary as a function of distance through the membrane [e.g., ref. 7]. The $\alpha \cdot D_m$ product is thus a membrane-averaged parameter.)

*Partition Coefficients*

Lipid bilayers are effective barriers for the nonselective movement of small ions (e.g., $NH_4^+$ and $HCO_3^-$), because these solutes have extremely small partition coefficients.[8] $\alpha$ can be estimated using the Boltzmann equation:

$$\alpha = \exp\{-\Delta E/RT\}, \tag{2}$$

where $\Delta E$ denotes the free energy difference for moving a mole of ions from the aqueous phase into the membrane interior, $R$ the gas constant, and $T$ the temperature in kelvins. The dominant contribution to $\Delta E$ is the electrostatic charging, or Born energy ($\Delta E_B$), for transferring the ion from the aqueous medium of dielectric constant $\varepsilon_{H_2O}$ (~80) into a hydrocarbon medium of dielectric constant $\varepsilon_m$ (~2) (e.g., ref. 9): and $\Delta E_B = N \cdot [(z \cdot e)^2/(8 \cdot \pi \cdot r \cdot \varepsilon_0)] \cdot (1/\varepsilon_m - 1/\varepsilon_{H_2O})$, where $N$ is Avogadro's number, $z$ the ion valency, $e$ the elementary charge, $r$ the ion radius, and $\varepsilon_0$ the permittivity of vacuum. $\Delta E_B$ is large, ~170 kJ/mol (or ~66 $kT$ at 37°C) for a monovalent ion with $r = 0.2$ nm, and $\alpha$, to a first approximation, is estimated to be ~$10^{-30}$! ($\Delta E$ is less than $\Delta E_B$ because ion-water interactions are overestimated in the Born model[9] and because ions in the membrane interior will polarize the membrane-solution interfaces, which will lower the ions' electrostatic energy in the membrane.[8] $\alpha$ will thus be larger than indicated by the foregoing estimate, but the qualitative result, that $\alpha \simeq 0$, is not affected by more refined calculations.)

For unionized substances, $\alpha$ is estimated from oil/water or hydrocarbon/water partition coefficients. (Because a solute's partition coefficient may vary as a function of distance through the membrane, the choice of organic solvent used to estimate $\alpha$ is somewhat arbitrary.[10] This will not affect the order-of-magnitude arguments presented here.) $\alpha$ is thus estimated to be ~1 (for $CO_2$), ~$10^{-3}$ (for $NH_3$), and ~$10^{-4}$ (for $H_2O$).[11,12] These unionized substances thus partition with comparative ease into the hydrophobic core of lipid bilayers, which explains why the unionized forms of many weak acids and bases are many orders of magnitude more permeant than are their charged counterparts.

*Nonionic Diffusion*

The quantitative importance of nonionic diffusion through the lipid bilayer moiety of biologic membranes is further highlighted by considering the last two terms in Eq. 1. Based on diffusion coefficients for small molecules in tetradecane,[13,14] $D_m$ is expected to be ~$10^{-5}$ cm²/s at 37°C; $\delta_m$ is ~4 nm.[15] $D_m/\delta_m$ is thus expected to be ~10 cm/s, and the membrane's permeability coefficient for $H_2O$ ($P_m^{H_2O}$) is predicted to be ~$10^{-3}$ cm/s; this is in good agreement with estimates of the diffusional water permeability through planar lipid bilayers[10] or the lipid bilayer moiety of red blood cells.[16] The permeability coefficient for $CO_2$ ($P_m^{CO_2}$) is predicted to be ~10 cm/s, in reasonable agreement with experimental determinations in planar lipid bilayers[17] and red blood cell membranes.[18]

A cautionary note: a membrane's water permeability varies as a function of its lipid composition,[10] and some cell membranes, notably the luminal membrane of the medullary thick ascending loop of Henle,[19] are essentially impermeable to $H_2O$. The molecular basis of this $H_2O$ impermeability is unknown, but this membrane is also impermeable to $NH_3$.[20]

## Acid-Base Electrodiffusion

$H^+$ permeation by nonionic diffusion does not involve a net charge transfer, and one would not expect a priori any significant electrodiffusion of acid-base equivalents across lipid bilayers. Nevertheless, there is a substantial electrodiffusive flux of $H^+$ equivalents across unmodified bilayers[21,22] and biologic membranes.[23]

The membrane conductance to acid-base equivalents ($G_{H/OH}$) is almost pH independent, $2 \leq pH \leq 12$,[21,22] which implies that the number of current carriers in the membrane is pH independent and that the underlying permeation process

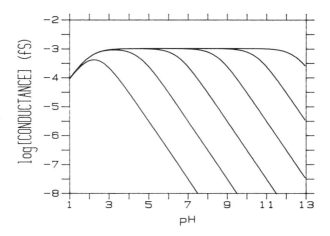

**FIGURE 1.** Conductance relation for a simple, symmetrical carrier. The curves were calculated based on the equation:

$$g = [(zF)^2/RT]\, k_{|S}^{S|}\, k_{|O}^{S}\, [S]/\{2(K_{S|} + [S])(2k_{|S}^{S|} + k_{|O}^{S} + (k_{S|}^{O}\, k_{|S}^{S|}/k_{O|}^{|O})[S]\},$$

where the $ks$ ($k_{\text{final state}}^{\text{initial state}}$) denote the transition rate constants among the different carrier states (see also FIG. 3): $O|$ and $|O$, where the empty binding site is exposed towards the left and right solution, respectively; $S|$ and $|S$, where the occupied binding site is exposed towards the left and right solution, respectively. $K_{S|}$ is the equilibrium constant for binding to the site $= k_{|O}^{S|}/k_{S|}^{O}$. For further details, see Andersen.[25] The curves were calculated for an $OH^-$ carrier; $K_{S|} = 10^{-12}$, $k_{|S}^{S|} = k_{|O}^{S} = 1$ s$^{-1}$, and (left to right) $k_{O|}^{|O} = 1$, $10^2$, $10^4$, $10^6$, and $10^8$ s$^{-1}$, respectively.

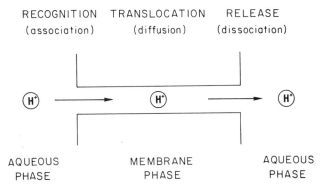

RECOGNITION     TRANSLOCATION      RELEASE
(association)        (diffusion)      (dissociation)

AQUEOUS          MEMBRANE          AQUEOUS
PHASE             PHASE             PHASE

**FIGURE 2.** Schematic representation of channel-mediated ion movement. Each of the three steps denoted in the figure involves multiple elementary steps. The association step involves, for example, a diffusive approach followed by a "solvation" of the incoming solute by the channel.

cannot be of the solubility-diffusion type. This pH independence of $G_{H/OH}$ is an enigma (cf. ref. 24), and it is generally believed that the high $G_{H/OH}$ cannot result from a (mobile) carrier-mediated process.[21,23,24]

The *a priori* exclusion of carrier-mediated processes is premature, however, because according to the simple (mobile) carrier model $G_{H/OH}$ can be independent of substrate concentration over a large concentration range (Fig. 1), similar to what is observed experimentally.

### Protein-Mediated Acid-Base Permeation

Integral membrane proteins mediate the transmembrane movement of acid-base equivalents by several mechanisms: channels, co- and countertransporting carriers, and pumps.

### Channel-Mediated Acid-Base Permeation

The mechanistically simplest (specific) acid-base permeation mechanism is the channel-mediated process, where $H^+$ (or some equivalent) traverses the plasma membrane through a water-filled pore that connects the two aqueous phases with each other (Fig. 2). If the permeating species is $H^+$ (or $OH^-$), the solute movement is presumably by a Grotthus-type hopping motion,[26] that is, a sequence of steps of the type: $H_3O^+ + H_2O \rightarrow H_2O + H_3O^+$, where amino acid side chains can substitute for $H_2O$ (e.g., ref. 27).

All water-filled channels will likely be permeable to acid-base equivalents. The vasopressin-induced increase in luminal acid-base permeability in the toad urinary bladder apparently is mediated by water channels.[28] Voltage-dependent sodium channels from myelinated nerve have a $H^+/Na^+$ permeability ratio ~200.[29] Not surprisingly, therefore, in neurons the $pH_i$ as well as the extracellular cell surface pH varies as a function of voltage.[30,31] The voltage dependence of the pH changes is consistent with the notion that they result from a flux of acid-base equivalents through voltage-dependent channels. These channels may be specific for acid-base equivalents, but they result most likely from the sizable $H^+$ (or $OH^-$) permeability that is expected for any voltage-dependent channel.

In any case, acid-base permeable channels promote the passive (electrodiffusive) movement of acid-base equivalents across cell membranes, and $pH_i$ will tend towards a value where $H^+$ would be in electrochemical equilibrium across the cell membrane:

$$pH_i = pH_o + F \cdot \Delta V/(2.3 \cdot RT), \qquad (3)$$

where F is Faraday's constant. Acid-base permeable channels can thus be acidifying or alkalinizing, depending on their activation voltage range.

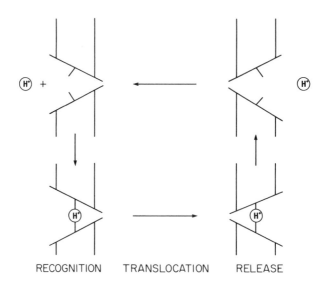

**FIGURE 3.** Schematic representation of carrier-mediated ion movement. Note that carrier- and channel-mediated solute translocation depends on similar steps.

*Carrier- and Pump-Mediated Acid-Base Permeation*

From a regulatory point of view, the important mechanisms of protein-mediated movement of acid-base equivalents are the carrier-mediated co- and countertransport systems (sym- and antiports) and ATP-driven $H^+$ pumps. The molecular basis for the function of any of these transport systems is not known, but all of them are probably of the conformational carrier type,[32] where substrate binding triggers conformational changes that mediate the substrate translocation through the membrane (FIG. 3). This basic scheme is easily extended to formally depict how the movement of a desired substrate (e.g., $H^+$) can be coupled to the transmembrane movement of other solutes or ATP hydrolysis.[33]

*Carrier vs Channel*

In cell membranes, channels, carriers, and pumps are integral membrane-spanning proteins, and the solute movement mediated by carriers and pumps most likely occurs through fixed "pores" in the protein.[34-37] The distinction between channels and carriers has thus become blurred, because they both exhibit saturating flux-concentration relations and are likely to encompass aqueous pores through which the transported solutes traverse the protein. Nevertheless, there is a fundamental difference between carriers and channels, as can be seen by comparing FIGURES 2 and 3; the unoccupied channel can be accessed from either side of the membrane, whereas the unoccupied carrier can be accessed from only one side.[38,39] Carriers (and pumps) thus have a (transient) "memory" of the last transport event. The difference between the curves in FIGURE 1, for example, is based on a different "memory" in the different simulations: when $k_o^o$ is small (relative to $k_{||}$), the carrier has a fairly good "memory" and exhibits the classical biphasic flux-concentration relation; when $k_o^o$ is much larger than $k_{||}$, the carrier has a poor "memory" and exhibits an apparently monotonic flux-concentration relation, similar to what would be expected for a channel.

The restricted accessibility to the carriers' substrate binding site(s) forms the basis for both co- and countertransport. This is particularly important for antiports, where the coupling between opposing ion fluxes arises because the unoccupied states interconvert only poorly, or not at all (when the molecular "memory" is perfect) in which case perfect coupling can occur between the ions.

### Energetics of Transport

For sym- and antiports, the maximal $\Delta pH$ that can be achieved when the carrier is at equilibrium is determined by the membrane potential difference $(\Delta V = V_i - V_o)$ and the difference in electrochemical potential for the driving solute $(\Delta \bar{\mu}_S = RT \cdot \ln\{[S]_i/[S]_o\} + z \cdot F \cdot \Delta V$, where S denotes the solute and $z$ its valence). For $H^+$ (or $NH_4^+$) transporters:

$$\Delta pH_{max} = +F \cdot \Delta V/(2.3 \cdot RT) \pm \Delta \bar{\mu}_S/n, \qquad (4)$$

where $n$ denotes the $H^+/S$ stoichiometry, and the upper signs (+) pertain to symports and the lower (−) to antiports. For $OH^-$ (or $HCO_3^-$ or lactate) transporters, all signs in Eq. 4 are inverted. (In the expression for the chemical potential, one should properly use ion activities rather than concentrations. To a first approximation, however, the activity coefficient for a given ion depends on the total ionic strength and not on the detailed composition of the electrolyte composition (e.g., ref. 39). The activity *ratio* can thus be replaced by the concentration *ratio*.)

*Symports*

For $H^+$, lactate-symports, for example, the equilibrium $pH_i$ will be given by:

$$pH_i = pH_o + \log\{[lact^-]_i/[lact^-]_o\}. \qquad (5a)$$

In the case of $Na^+$, lactate-symports, the corresponding expression is:

$$pH_i = pH_o - \log\{[Na^+]_i/[Na^+]_o\}, \tag{5b}$$

where it is assumed that lactic acid is in equilibrium across the cell membrane (see below).

### Antiports

For $H^+/Na^+$-antiports, for example, the equilibrium $pH_i$ will be given by:

$$pH_i = pH_o - \log\{[Na^+]_i/[Na^+]_o\}. \tag{6a}$$

Formally, Eqs. 5b and 6a are identical, even though they reflect the operation of different molecular mechanisms: $Na^+$-coupled symport of an $OH^-$ equivalent and $Na^+$-coupled antiport of $H^+$. This formal equivalence is related to a functional equivalence (see below). For $HCO_3^-/Cl^-$-antiports, the corresponding expression is:

$$pH_i = pH_o + \log\{[Cl^-]_i/[Cl^-]_o\}, \tag{6b}$$

where it is assumed that $CO_2$ is in equilibrium across the cell membrane (see below).

### ATP-Driven $H^+$ Translocation

For primary (ATP-driven) transport systems, such as $H^+$-ATPases, the equilibrium $pH_i$ will be given by:

$$pH_i = pH_o + (F \cdot \Delta V)/(2.3 \cdot RT) - \Delta G_{ATP}/(n \cdot 2.3 \cdot RT), \tag{7a}$$

where $n$ denotes the $H^+/ATP$ stoichiometry and $\Delta G_{ATP}$ the free energy for ATP hydrolysis. For $H^+,K^+$-ATPases, the corresponding expression is:

$$pH_i = pH_o - \log\{[K^+]_i/[K^+]_o\} - \Delta G_{ATP}/(2.3 \cdot RT). \tag{7b}$$

### $\Delta pH_{max}$

The maximal $\Delta pHs$ that can be established by sym- and antiports are large. The extracellular $Na^+$ activity ($a_i^{Na}$) is ~110 mM, whereas the intracellular activity ($a_i^{Na}$) seems to vary between ~6 (in heart muscle[40]) and ~20 mM (in glial cells [e.g., ref. 41]). The $Na^+$, lactate-symport and $Na^+/H^+$-antiport can thus establish $\Delta pHs > 1$ (cf. Eq. 6a).

$Cl^-$ is likewise not in electrochemical equilibrium across cell membranes,[42] and the $Cl^-/HCO_3^-$-antiport will tend to drive $pH_i$ below $pH_o$ ($\Delta pH < 0$), but to a value above that expected if $H^+$ were in electrochemical equilibrium (cf. Eq. 4).

Much larger pH differences are established by the $H^+$-translocating ATPases. $\Delta G_{ATP} \simeq -60$ kJ/mol (e.g., ref. 41), and $\Delta pH_{max} \simeq 4$ for simple $H^+$-ATPases with a stoichiometry of 2 $H^+/ATP$ (cf. ref. 36), and $\simeq 8$ for $H^+,K^+$-ATPases.

## Regulation of Transport Activity

Given the large and opposing $\Delta$pHs that can be established by these transport systems, it is necessary that they are tightly regulated (cf. Boron, *this volume*). If this were not the case, a cell that was endowed with both $H^+/Na^+$- and $HCO_3^-/Cl^-$-antiports, for example, would be loaded with $Na^+$ and $Cl^-$, whereas $H^+$ and $HCO_3^-$ would recirculate across the membrane in the form of $H_2O$ and $CO_2$. This transport regulation cannot only be based on kinetic factors (changes in substrate concentrations), because the transport rate in that case would change by at most 10-fold for a 10-fold change in substrate concentration, which is not sufficient to account for the steepness of the activation curves (e.g., refs. 43 and 44). This implies that the $\Delta$pH at which the transport system appears to stop may not be the thermodynamic reversal $\Delta$pH predicted from Eqs. 5 and 6. Changes in transport rate that occur during cellular pH regulation reflect both kinetic factors (changes in pH and solute concentrations) and regulatory factors (e.g., changes in protonation at a regulatory site at the intracellular surface of the transporter[45]).

With ATP-driven pumps, pump activity likewise ceases far from the equilibrium $\Delta$pH, because of the highly nonlinear relation between flux and $\Delta\bar{\mu}_H$ (cf. refs. 36 and 46).

## UNSTIRRED LAYER LIMITATIONS

The cell membrane constitutes only a very small fraction of the distance that a solute diffuses from one (well-mixed) aqueous solution to another (well-mixed) solution. It is therefore necessary to consider not only how a given solute is handled by the membrane, but also the solute's access to the membrane. Adjacent to the membrane, the aqueous solutions are stagnant, even with vigorous convective stirring of the bulk solutions, and the solute must traverse these "unstirred" layers by diffusion.[47] (See FIGURE 4.) The high $P_m$ of many unionized solutes is a consequence of the very small membrane thickness, but the unstirred layers will contribute to the measured permeability barrier. It is therefore necessary to compare $P_m$ to the solute's permeability coefficient in the unstirred layers ($P_{ul}$) in the extra- and intracellular solutions:

$$P_{ul} = D_a/\delta_{ul}, \tag{8}$$

where $D_a$ is the aqueous solute diffusion coefficient ($\sim 10^{-5}$ cm$^2$/s[48]), and $\delta_{ul}$ the combined thickness of the extra- and intracellular unstirred layers ($= \delta^I + \delta^{II}$ in FIG. 1). *In vivo*, $\delta_{ul}$ can be approximated by one half the average capillary separation, or $\sim 20$ $\mu$m,[49] and $P_{ul}$ is $\sim 10^{-2}$ cm/s; *in vitro*, unstirred layers are likely to be much larger, $\geq 100$ $\mu$m,[50] and $P_{ul}$ is $\leq 10^{-3}$ cm/s, and could be much less! (Note, however, that the dimensions of most cells are such that one usually can disregard intracellular diffusion limitations. Cf. ref. 51.)

Thus, because $\delta_m$ is $\ll \delta_{ul}$, the permeation of even quite polar molecules with an $\alpha \leq 10^{-4}$ could be affected by unstirred layer limitations, because the *measured* solute permeability coefficient ($P$) is given by:

$$P = P_m \cdot P_{ul}/(P_m + P_{ul}). \tag{9}$$

That is, $P < P_m$ unless $P_m \ll P_{ul}$. This is an important restriction, because many neutral species have $P_m$s that are comparable to, or larger than, $P_{ul}$. TABLE 2

**TABLE 2.** Unstirred Layer and Membrane Permeability Coefficients

| Substance | $P$ (cm/s) | Refs. |
|---|---|---|
| Unstirred layer permeability | $10^{-4}$–$10^{-2}$ | 46, 49 |
| Membrane permeability coefficients | | |
| Unionized species | | |
| $CO_2$ | 1 | 17, 18 |
| $NH_3$ | $10^{-1}$ | 12 |
| Acetic acid | $10^{-2}$ | 12 |
| $H_2O$ | $10^{-4}$–$10^{-3}$ | 10 |
| Lactic acid | $10^{-4}$ | 12 |
| Charged species | | |
| $H^+/OH^-$ (pH 7) | $10^{-5}$ | 22 |
| $Na^+$, $NH_4^+$, $Cl^-$, $HCO_3^-$ | $10^{-11}$ | a |
| Red cell anion exchanger | | |
| net exchange permeability | $10^{-3}$ | 51 |

[a] Calculated using a specific membrane conductance $(G)$ of $\sim 10^{-9}$ S/cm² in 1 M salt and the relation: $P = G \cdot RT/(F^2 \cdot C)$, where $C$ denotes the aqueous salt concentration.

summarizes information on $P_{ul}$, $P_m$ of lipid bilayers for some physiologically important solutes, and the apparent permeability coefficient for a high-capacity transport system, the red cell anion exchanger.

When unstirred layer limitations are significant, solute concentrations at the two membrane/solution interfaces will differ from their bulk values (FIG. 4). This

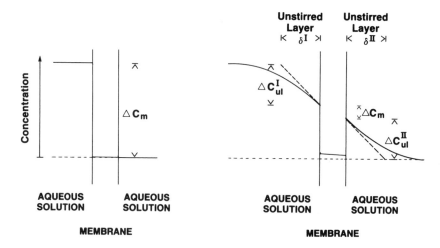

**FIGURE 4.** Unstirred layers and solute concentration profiles adjacent to the membrane. (*Left*) For a poorly permeant solute, with a low partition coefficient. The aqueous solutions do not impose a significant barrier for solute movement, and the solute concentrations adjacent to the membrane are equal to the bulk concentrations. (*Right*) For a highly permeant solute that partitions well into the membrane. The aqueous solutions impose a significant barrier for solute movement, and the solute concentrations adjacent to the membranes differ from the bulk concentrations. The "thickness" of the unstirred layers ($\delta^I$ and $\delta^{II}$, respectively) are related to the slope of the solute concentration profile at the membrane/solution interfaces by the relation: $\Delta C_{ul}/\delta = -D_a \cdot dC/dx$, evaluated at either interface.

concentration polarization is particularly important for weak acids and bases, because reactions II and III occur throughout the unstirred layer. Appreciable pH differences may thus develop across the unstirred layers as a result of concentration polarization of the unionized species.[53–55] The extent of this concentration polarization, and of the associated pH changes, diminishes as *impermeant* buffers are added. These impermeant buffers act as "carriers" of acid-base equivalents through the unstirred layers (cf. ref. 53). Importantly, the pH profile will be determined not only by the buffer concentration, but also by the magnitude of the deprotonation/protonation rate constants, because the protonation/deprotonation reactions will not be in equilibrium close to the membrane.[53,56] Generally, this will be of little consequence, because deprotonation/protonation rate constants usually are very high.[57] The $CO_2/HCO_3^-$ pair, however, in the absence of carbonic anhydrase will be a poor unstirred-layer buffer, because the very slow hydration reaction ($CO_2 + H_2O \rightleftarrows H_2CO_3$) will limit the rate at which $H^+$ can be buffered.

## CELL pH AND THE INTRACELLULAR CONCENTRATIONS OF WEAK ACIDS AND BASES

The high membrane permeability to the unionized form of many metabolically important weak acids and bases implies that one, to a first approximation, can assume that the unionized forms of these solutes occur in equal concentrations in the extra- and intracellular solutions. Cell pH regulation results primarily from changes in the intracellular concentration of the relevant anions or cations (e.g., ref. 58).

The deprotonation of a weak acid, for example, is described by Scheme IIa with a dissociation constant $(K)$ given by $K = [H^+] \cdot [A^-]/[HA]$. Because

$$[HA]_i = [HA]_o, \tag{10a}$$

it follows that $[H^+]_o \cdot [A^-]_o = [H^+]_i \cdot [A^-]_i$ (or $[H^+]_i/[H^+]_o = [A^-]_o/[A^-]_i$), and that (cf. Eq. 5a)

$$\Delta pH = \log\{[A^-]_i/[A^-]_o\}. \tag{11a}$$

To the extent that the deprotonation/protonation reactions of HA are fast, Eq. 11a will be valid whenever Eq. 10a is satisfied. Thus, at a constant $pH_o$ and $[A^-]_o$, changes in $pH_i$ can be brought about only by changing $[A^-]_i$. Regulatory changes in $pH_i$, consequent to an acidifying insult, are thus related to changes in the intracellular anion concentrations:

$$\Delta pH_{reg} = pH_i^{ss} - pH_i^p = \log\{[A^-]_i^{ss}/[A^-]_i^p\}, \tag{12a}$$

where $\Delta pH_{reg}$ is the intracellular pH change produced by the cell's pH regulatory systems, and $pH_i^p$, $pH_i^{ss}$, $[A^-]_i^p$, and $[A^-]_i^{ss}$ denote values of $pH_i$ and $[A^-]_i$ at the peak acidification and steady state, respectively. The corresponding expressions for weak bases (cf. Scheme IIIa) are:

$$[B]_i = [B]_o; \tag{10b}$$

$$\Delta pH = -\log\{[BH^+]_i/[BH^+]_o\}; \tag{11b}$$

and

$$\Delta pH_{reg} = -\log\{[BH^+]_i^{ss}/[BH^+]_i^p\}. \tag{12b}$$

For either situation, the absolute changes in $[H^+]_i$ are exceedingly small. But the absolute changes in $[A^-]_i$ or $[BH^+]_i$ can be very large, and regulatory changes in $[A^-]_i$ or $[BH^+]_i$ must be associated with changes in other intracellular electrolytes in order to maintain electroneutrality. Cell pH regulation must thus be coupled to cell volume regulation.

### $H^+/Na^+$-Antiport Functions As an $A^-/Na^+$-Symport

The intracellular anion concentrations could be changed by the operation of specific anion transport systems, such as the $HCO_3^-/Cl^-$ antiport. But a general mechanism for increasing the intracellular anion concentration is to couple the non-ionic influx of the protonated weak acid and the consequent release of $H^+$ to an extrusion of $H^+$ (FIG. 5, top). This extrusion can be mediated by the $H^+/Na^+$-antiport. Thus, although the $H^+/Na^+$-antiport at the molecular level exchanges $H^+$ for $Na^+$, at the cell physiological level its function becomes modified by nonionic diffusion: taken together, the two permeation mechanisms act as a general anion transporter, a $A^-,Na^+$,-symport. It can indeed be difficult to identify the transport systems that are involved in cellular pH regulation; see also FIGURE 5 (bottom) and FIGURE 6.

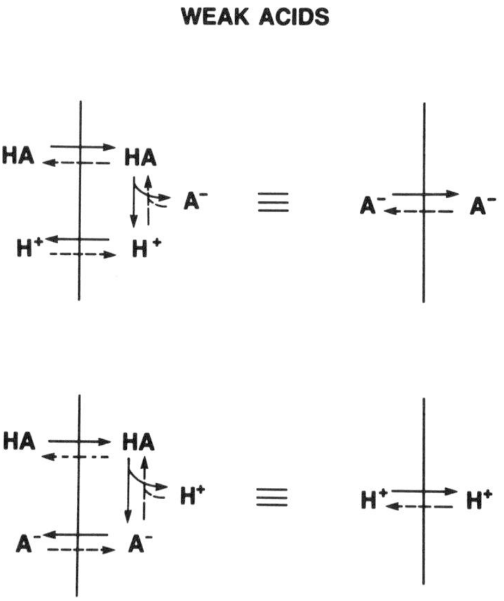

**WEAK ACIDS**

FIGURE 5. Weak acid permeation across membranes. (*Top*) Nonionic diffusion coupled to the extrusion of $H^+$ is equivalent to an influx of $A^-$. (*Bottom*) Nonionic diffusion coupled to the extrusion of $A^-$ is equivalent to an influx of $H^+$.

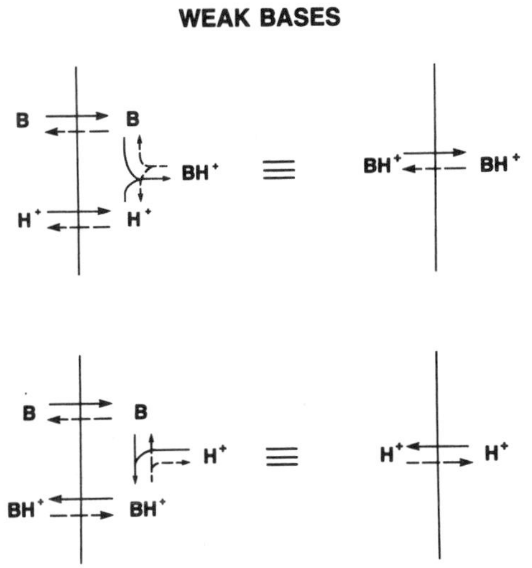

## WEAK BASES

FIGURE 6. Weak base permeation across membranes. (*Top*) Nonionic diffusion coupled to an uptake of $H^+$ is equivalent to an influx of $BH^+$. (*Bottom*) Nonionic diffusion coupled to an efflux of $BH^+$ is equivalent to an efflux of $H^+$.

Given the importance of non-ionic diffusion and the need to regulate the intracellular anion concentrations (as well as the concentration of protonated bases, such as $NH_4^+$), it is apparent that the conventional depiction of cellular acid-base regulation is incomplete (FIG. 7). It is not sufficient to consider only the membrane transport events (FIG. 7, top). At a minimum, the weak acid anions (FIG. 7, middle) and more generally the weak base cations must also be included. The steady-state pH can be described at this level. For the analysis of $pH_i$ transients, however, it becomes necessary to include the contributions of impermeant intracellular buffers as well (FIG. 7, bottom).

## pH REGULATION

Generally, alterations in cell pH ($pH_i$) result from changes in metabolic activity (primarily aerobic $CO_2$ or anaerobic lactic acid production), from changes in the transmembrane flux of acid-base equivalents, or from a combination of these mechanisms. These events will trigger the regulatory response. The multitude of mechanisms that mediate the transmembrane movement of acid-base equivalents (TABLE 1) complicates attempts to analyze pH regulation. Fortunately, however, the problem of cellular pH regulation can be simplified by noting that some of the mechanisms listed in TABLE 1 are involved not so much in cellular pH regulation *per se* as in the transepithelial movement of $Na^+$, $HCO_3^-$, organic anions, and the like (e.g., see Ref. 59). I will therefore limit myself to a discussion of the elementary properties of the pH regulatory response in nonepithelial cells, specifically in glial cells.

The basic phenomenon is illustrated in FIGURE 8, which shows the result of an experiment in which a rat astrocyte was maintained in a nominally $CO_2/HCO_3^-$-free solution containing 150 mM Na and 130 mM $Cl^-$, buffered to pH 7.3 with HEPES. At the first arrow the extracellular solution was changed and 20 mM $Cl^-$ was replaced by 20 mM lactate at a constant $pH_o$ of 7.3. Over the next 2 minutes, $pH_i$ decreased from ~7.0 to ~6.7. This decrease was followed by a regulatory phase, in which $pH_i$ returned to ~7.0 over the next 10 minutes. At that time

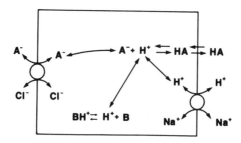

**FIGURE 7.** Cellular acid-base permeation and regulation. (*Top*) The membrane transport systems in an idealized cell. (*Middle*) Different transport systems are coupled through the intracellular concentrations of $A^-$ and $H^+$ (and $Na^+$ and $Cl^-$). (*Bottom*) For a complete analysis, it is necessary to incorporate the role of intracellular buffers as well.

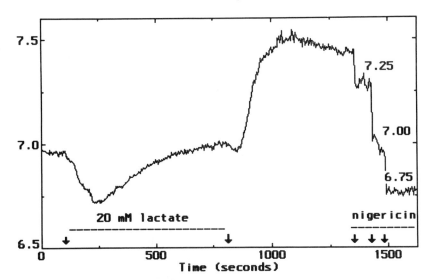

**FIGURE 8.** $pH_i$ changes in a rat astrocyte in primary culture that is exposed to 20 mM lactate at a constant $pH_o$ of 7.3. See text for details. $pH_i$ was measured by microspectrophotometry using BCECF as an intracellular pH indicator. The fluorescence signals are calibrated at the far right in the figure. Unpublished experiment of Dr. M. Nedergaard.

(second arrow) the lactate was replaced by $Cl^-$ at $pH_o = 7.3$, and $pH_i$ increased to ~7.5, with little change over the next 5 minutes.

The shape of the $pH_i$ transient is a classic response, even though the lactate-mediated influx of $H^+$ equivalents primarily occurs as a lactate-$H^+$ symport.[60][b] What is important for the present purpose is that cell acidification is a consequence of a net influx of lactic acid equivalents. Conceptually, therefore, one can analyze these pH transients by considering only two permeation mechanisms: non-ionic diffusion and a $H^+/Na^+$-antiport. (This is the simplest possible situa-

---

[b] In this experiment, $pH_i$ decreases at a rate of ~0.2/min (cf. FIG. 8). If the intracellular acidification is solely the result of non-ionic diffusion of lactic acid (HA) into the cell, the membrane permeability coefficient to lactic acid ($P_{HA}$) can be estimated from the relation:

$$dpH_i/dt = P_{HA} \cdot [HA]_o \cdot A/(V \cdot \beta),$$

where $A$ and $V$ denote the cell's surface area and volume, respectively, and $\beta$ is the intracellular buffer capacity (10–20 mM/pH). For a spherical cell with radius $r$, $A/V = 3/r$, and

$$P_{HA} = dpH_i/dt \cdot \beta \cdot r/(3 \cdot [HA]_o).$$

For astrocytes in primary culture, $r$ is ~5 $\mu$m, and $P_{HA}$ is estimated to be ~$10^{-3}$ cm/s, or ~10-fold higher than the lactic acid permeability coefficient in planar lipid bilayers, ~$10^{-4}$ cm/s.[12] It thus appears that the intracellular acidification results primarily from an influx of lactic acid equivalents via a specific transport system, most likely a $H^+$-lactate symport. This conclusion is strengthened by the finding that the intracellular acidification rate is a saturating function of the extracellular lactate concentration.[60]

tion. Many other pH regulatory systems may be involved, but they are not necessary for describing the general shape of the $pH_i$ transient.)

## The Regulatory Response

In terms of cellular pH homeostasis, the quantitatively most important mechanism is nonionic diffusion—of $CO_2$—from tissue cells into red blood cells in peripheral capillaries and from red cells in pulmonary capillaries into the alveoli. This is not, however, a regulatory mechanism but an equilibrative one. Nonionic diffusion serves only to equalize the extra- and intracellular concentrations of unionized substances.

Cellular pH regulation is achieved through the actions of the $H^+/Na^+$-antiport(s), which will tend to alkalinize the cytoplasm, and of the $HCO_3^-/Cl^-$-antiport(s), which will tend to acidify the cytoplasm. I will only consider the former. FIGURE 9 illustrates a sequence of events that pertains to lactate exposure in the experiment in FIGURE 8.

The lactate-containing solution contains 20 mM lactate and $\sim$6 $\mu$M lactic acid (at pH 7.3, pK = 3.8). As the cell is exposed to this solution, lactic acid equivalents enter the cell and $pH_i$ decreases. Because $pH_i$ is much greater than the pK for lactic acid, the incoming lactic acid equivalent will be fully dissociated and the $H^+$ is initially buffered by intracellular buffers (FIG. 9, lower left). Assuming that

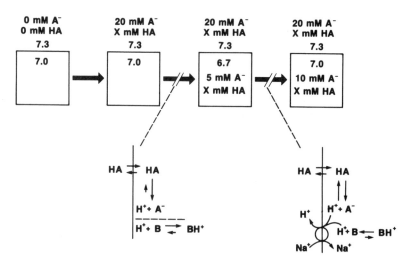

**FIGURE 9.** Schematic representation of the events underlying $pH_i$ transients in experiments similar to that illustrated in FIGURE 8. (*Top*) Solute concentrations and $pH_i$s before exposure to the weak acid anion $A^-$ (e.g., lactate); immediately after the insult; at the peak acidification (before regulation has begun); and at the end of the regulatory phase, when $pH_i$ has returned to its original value. Note that the cell now has gained 10 mM $A^-$ (and 10 mM $Na^+$, not illustrated). (*Bottom*) The membrane permeation events that underlie the initial acidification and the regulatory phase. See text for details.

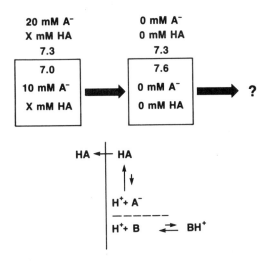

**FIGURE 10.** Schematic representation of the events that underlie the alkaline $pH_i$ transient that is seen when $A^-$ is removed. (*Top*) Solute concentrations and $pH_i$ immediately before $A^-$ removal, and after all $A^-$ has left the cell. Note that the cell is left with a net gain of 10 mM $Na^+$ (not illustrated). See text for details.

no other processes are involved, the "peak" acidification $pH_i$ will be determined by Eqs. 10a and 11a together with the condition:

$$\Delta pH_i = -[A^-]_i^p/\beta, \qquad (13)$$

where $[A^-]_i^p$ is the intracellular lactate concentration at the maximal acidification, and $\beta$ is the buffer capacity (= $d[SB]/dpH$, where SB denotes a strong base). Setting $\beta$ at 17 mM/pH (which simplifies the calculations), $pH_i$ drops from ~7.0 to ~6.7, and $[A^-]_i$ increases to 5 mM. This initial acidifying insult is thus associated with a net solute gain for the cell.[c]

With time, acid-base equivalents will be extruded from the cell by the $H^+/Na^+$-antiport that is activated by intracellular acidification. The extrusion of $H^+$ will reverse the initial buffer reactions, but it will also tend to make $[HA]_i$ less than $[HA]_o$ (or $[H^+]_i \cdot [A^-]_i$ less than $[H^+]_o \cdot [A^-]_o$). Consequently, there will be a further influx of lactic acid equivalents, and a new steady state is reached when $[A^-]_i = 10$ mM, $pH_i = 7.0$, and the pH regulatory machinery again is quiescent (Fig. 9, lower right). The regulation is incurred at a cost, however; the intracellular concentration of lactate, *and of $Na^+$*, is increased (both $[A^-]_i$ and $[Na^+]_i$ have increased by 10 mM). pH regulation thus imposes a further osmotic stress on the cell (see below).

At the end of the experiment, when the lactate-containing solution is replaced by the control solution, lactic acid equivalents will leave the cell and the cell becomes alkaline. This situation is schematized in FIGURE 10. The extent of the

---

[c] The increased cellular solute content will cause the cell to swell and thus alter the concentration of all intracellular solutes. I will not consider this complication further.

alkalinization is determined by the intracellular lactate concentration at the time of the solution change ($[A^-]_i^{ss}$) and the buffer capacity:

$$\Delta pH_i = [A^-]_i^{ss}/\beta, \tag{14}$$

and $pH_i$ increases to $\sim$7.6. $pH_i$ now decreases only slowly towards its initial value, the cell does not appear to possess mechanisms that enable it to recover from alkaline pH shifts (for such mechanisms depend on the presence of $CO_2$/$HCO_3^-$ and are inactive under the conditions used for this experiment). More interestingly, however, the cell is left with a net gain of 10 mM $Na^+$ and with the problem of normalizing the intracellular electrolyte composition. To the extent that this regulation is mediated by $Na^+,K^+$-ATPase, one may find that changes in the $pH_i$ induce changes in $[K^+]_i$ and so forth.

## pH REGULATION AND VOLUME CHANGES

In the foregoing example, all three $pH_i$ transients were associated with changes in intracellular electrolyte composition. The qualitative time courses of the changes in $[A^-]_o$, $[A^-]_i$, and $pH_i$ are illustrated in FIGURE 11.

$pH_i$ changes (whether imposed or regulatory) are associated with large changes in intracellular solute concentrations and thus with cell volume changes, just as cell volume regulation is associated with $pH_i$ changes (e.g., ref. 61). This

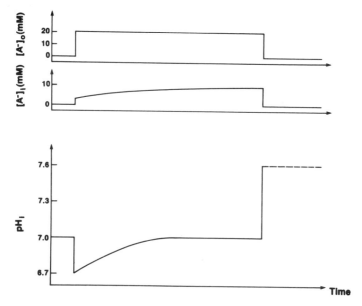

**FIGURE 11.** Schematic time course of the changes in $[A^-]_o$ and $[A^-]_i$ and in $pH_i$ that occur following exposure to, and removal of, a weak acid anion. Cellular acid-base regulation is associated with large solute concentration changes, which implies that cell volume will change as $pH_i$ changes.

**TABLE 3.** Cell Volume Changes during pH Regulation[a]

| | | Regulatory Changes | | |
|---|---|---|---|---|
| | Primary Maneuver | $[A^-]_i$ | $[Na^+]_i$ | Volume |
| Acidification | $[HA]_o$ ↑<br>$[Na^+]_o$ ↓ | ↑ | ↑ | ↑ |
| | | ↓ | ↓ | ↓ |
| Alkalinization | $[HA]_o$ ↓<br>$[Na^+]_o$ ↑ | ↓ | (↓) | ↓ |
| | | (↑) | (↑) | (↑) |

[a] Regulatory changes in parentheses may not occur because of the regulation of the $H^+/Na^+$-antiport.

coupling between $pH_i$ and cell volume is based on the thermodynamic constraints that relate $pH_i$ changes to changes in the concentration of the relevant intracellular solutes (Eqs. 11 and 12), in conjunction with the fact that $Na^+$, lactate, and the like are present at much higher concentrations than are $H^+$ (or $OH^-$). Intracellular buffering determines the magnitude of the initial acidifying and final alkalinizing $pH_i$ changes, but the changes in protonation of impermeant electrolytes and the associated Donnan effects are of secondary importance for the volume changes. This latter point is further emphasized by the absence of a direct relation between $pH_i$ and cell volume; dependent on the experimental maneuver, cell volume may increase *or* decrease when $pH_i$ is decreased and/or cell volume may increase when $pH_i$ is decreased *or* increased.

In the foregoing example (FIGS. 9 and 10), the initial $pH_i$ drop was imposed by increasing the extracellular concentration of lactate (and lactic acid). The solute influx caused the cell to swell. During the regulatory phase, additional solute was added to the cell through the operation of the $H^+/Na^+$-antiport and the associated influx of lactic acid equivalents. At the end of the regulatory phase, the cell volume had thus increased even further, but $pH_i$ had returned to its initial, resting value. Removing the extracellular lactate caused $pH_i$ to rise, and the cell lost solute, but the cell volume would nevertheless be larger than that preceding the experiment.

Other experimental maneuvers may also be used to acidify the cell. If, for example, the extracellular $Na^+$ is removed, the $Na^+/H^+$-antiport would tend to operate in reverse, that is, exchange intracellular $Na^+$ for extracellular $H^+$, and $pH_i$ would decrease. In this case, the acidification would be associated with a loss of intracellular solutes, and cell volume would decrease. Depending on the experimental conditions, even the initial acidifying insult may thus be associated with cell swelling or shrinkage. There is no simple relation between cell pH and volume. (See also TABLE 3.)

## ACKNOWLEDGMENTS

I thank Dr. M. Nedergaard for permission to use FIGURE 8, and E. Heinz, M. Nedergaard, D. B. Sawyer, R. Silver, and P. R. Steinmetz for helpful discussions.

## REFERENCES

1. CONWAY, B. E. 1981. Ionic Hydration in Chemistry and Biophysics.: 385–408. Elsevier. Amsterdam, The Netherlands.

2. Roos, A. & W. F. Boron. 1981. Physiol. Rev. **61:** 296–433.
3. Thomas, R. C. 1988. *In* Proton Passage Across Cell Membranes. Ciba Foundation Symposium 139.: 254–257. John Wiley. New York, NY.
4. Alper, S. L., R. R. Kopito, S. M. Libresco & H. F. Lodish. 1988. J. Biol. Chem. **263:** 17092–17099.
5. Kudrycki, K. E. & G. E. Shull. 1989. J. Biol. Chem. **264:** 8185–8192.
6. Brosius, F. C. III, S. L. Alper, A. M. Garcia & H. F. Lodish. 1989. J. Biol. Chem. **264:** 7784–7787.
7. Andersen, O. S. 1978. *In* Membrane Transport in Biology. G. Giebisch, D. C. Tosteson & H. H. Ussing, eds. Vol. 1: 369–446. Springer, Berlin.
8. Parsegian, V. A. 1969. Nature **221:** 844–846.
9. Bockriss, J. O'M. & A. K. N. Reddy. 1970. Modern Electrochemistry. Vol. 1: 45–174. Plenum Press. New York, NY.
10. Finkelstein, A. 1976. J. Gen. Physiol. **68:** 127–135.
11. Simon, S. A. & J. Gutknecht. 1980. Biochim. Biophys. Acta **596:** 352–358.
12. Walter, A. & J. Gutknecht. 1986. J. Membr. Biol. **90:** 207–217.
13. Schatzberg, P. 1965. J. Polymer Sci. C **10:** 87–92.
14. Evans, D. F., T. Tominaga & C. Chan. 1979. J. Solution Chem. **8:** 461–478.
15. Levine, Y. K. & M. H. F. Wilkins. 1971. Nature **230:** 69–72.
16. Moura, T. F., R. I. Macey, D. Y. Chien, D. Karan & H. Santos. 1984. J. Membr. Biol. **81:** 105–111.
17. Gutknecht, J., M. A. Bisson & D. C. Tosteson. 1977. J. Gen. Physiol. **69:** 779–794.
18. Gros, G. & W. Moll. 1971. Pflügers Arch. **324:** 249–266.
19. Rocha, A. S. & J. P. Kokko. 1973. J. Clin. Invest. **52:** 612–623.
20. Kikeri, D., A. Sun, M. L. Zeidel & S. C. Hebert. 1989. Nature **339:** 478–480.
21. Nichols, J. W. & D. W. Deamer. 1980. Proc. Natl. Acad. Sci. USA **77:** 2038–2042.
22. Gutknecht, J. 1987. Biochim. Biophys. Acta **898:** 97–108.
23. Verkman, A. S. & H. E. Ives. 1986. Biochemistry **25:** 2876–2882.
24. Nagle, J. F. 1987. J. Bioenerg. Biomembr. **19:** 413–426.
25. Andersen, O. S. 1989. Methods Enzymol. **171:** 62–112.
26. Eigen, M. & L. De Maeyer. 1958. Proc. Roy. Soc. A **247:** 505–533.
27. Schulten, Z. & K. Schulten. 1986. Methods Enzymol. **127:** 419–438.
28. Gluck, S. & Q. Al-Awqati. 1980. Nature **284:** 631–632.
29. Mozhaeva, G. N. & A. P. Naumov. 1983. Pflügers Arch. **396:** 163–173.
30. Thomas, R. C. & R. W. Meech. 1982. Nature **299:** 826–828.
31. Thomas, R. C. 1988. *In* Proton Passage Across Cell Membranes. Ciba Foundation Symposium 139.: 168–183. John Wiley. New York, NY.
32. Patlak, C. S. 1957. Bull. Math. Biophys. **19:** 209–235.
33. Mitchell, P. 1967. Adv. Enzymol. **29:** 33–87.
34. Rothstein, A., Z. I. Cabantchik & P. Knauf. 1976. Fed. Proc. **35:** 3–10.
35. Laüger, P. 1979. Biochim. Biophys. Acta **552:** 143–161.
36. Steinmetz, P. R. & O. S. Andersen. 1982. J. Membr. Biol. **65:** 155–174.
37. Wieth, J. O., P. J. Bjerrum, J. Brahm & O. S. Andersen. 1982. Tokai J. Exp. Clin. Med. **7**(Suppl.): 91–101.
38. Laüger, P. 1980. J. Membr. Biol. **57:** 163–178.
39. Kielland, J. 1937. J. Am. Chem. Soc. **59:** 1675–1678.
40. Lee, C. O. & H. A. Fozzard. 1975. J. Gen. Physiol. **65:** 695–708.
41. Erecińska, M. & I. A. Silver. 1989. J. Cereb. Blood Flow Metab. **9:** 2–19.
42. Mauro, A. 1954. *Fed. Proc.* **13:** 96 (Abstr.).
43. Vaughan-Jones, R. D. 1988. *In* Proton Passage Across Cell Membranes. Ciba Foundation Symposium 139.: 23–46. John Wiley. New York, N.Y.
44. Grinstein, S., J. Garcia-Soto & M. J. Mason. 1988. *In* Proton Passage Across Cell Membranes. Ciba Foundation Symposium 139.: 70–86. John Wiley. New York, NY.
45. Aronson, P. S. 1985. Ann N. Y. Acad. Sci. **456:** 220–228.
46. Andersen, O. S., J. E. N. Silveira & P. R. Steinmetz. 1985. J. Gen. Physiol. **86:** 215–234.
47. Dainty, J. 1963. Adv. Bot. Res. **1:** 279–326.
48. Weast, R. C. (ed.) 1972. Handbook of Chemistry and Physics. CRC Press, Boca Raton. P. F-47.

49. FOLKOW, B. & E. NEIL. 1971. Circulation.: 39. Oxford University Press. New York, NY.
50. BARRY, P. H. & J. M. DIAMOND. 1981. Physiol. Rev. **64:** 763–872.
51. PIETRZYK, C. & E. HEINZ. 1972. *In* Na linked Transport of Organic Solutes, E. Heinz, ed.: 84–90. Springer. Berlin.
52. WIETH, J. O., O. S. ANDERSEN, J. BRAHM, P. J. BJERRUM & C. L. BORDERS, JR. 1982. Philos. Trans. Roy. Soc. Lond. B **299:** 383–399.
53. LEBLANC, O. H., JR. 1971. J. Membr. Biol. **4:** 227–251.
54. STHELE, R. G. & W. I. HIGUCHI. 1972. J. Pharm. Sci. **61:** 1922–1930.
55. GUTKNECHT, J. & D. C. TOSTESON. 1973. Science **182:** 1258–1261.
56. KORYTA, J. & J. DVORAK. 1987. Principles of Electrochemistry. John Wiley. New York, NY.
57. EIGEN, M., W. KRUSE, G. MAASS & L. DE MAEYER. 1964. *In* Progress in Reaction Kinetics, G. Porter, ed. Vol 2: 285–318. Pergamon Press, New York.
58. BORON, W. F. 1986. *In* Physiology of Membrane Disorder. 2nd Ed. T. E. Andreoli, J. F. Hoffman, D. D. Fanestil & S. G. Schultz, eds.: 423–435. Plenum. New York, NY.
59. MALNIC, G. 1987. Kidney Int. **32:** 136–150.
60. NEDERGAARD, M., S. A. GOLDMAN, S. DESAI & W. A. PULSINELLI. 1989. Soc. Neurosci. Abstr. **15:** 804.
61. HOFFMANN, E. K. & L. O. SIMONSEN. 1989. Physiol. Rev. **69:** 315–382.

# Down-Regulation of pH-Regulating Transport Systems in BC3H-1 Cells[a]

ROBERT W. PUTNAM

*Department of Physiology and Biophysics*
*Wright State University*
*School of Medicine*
*Dayton, Ohio 45401–0927*

Changes in intracellular pH ($pH_i$) can affect a variety of functions in muscle cells[1-3] including contraction,[4,5] metabolism,[6] and the response to external agents such as hormones, neurotransmitters, and growth factors.[7-11] Regulation of a constant $pH_i$ in virtually all cells is mediated by transmembrane transport systems.[3] These pH-regulating transport systems usually fall into one of two general types: Na/H exchangers and $HCO_3$-dependent transporters. In many cells these two types of transporters appear to coexist.

A great deal of attention has been focused recently on the regulation of $pH_i$ in smooth muscle cells. A number of features of pH regulation appear to be unusual in these cells. Many smooth muscle cells[b] have shown a low intrinsic buffering power compared to other muscle cells.[12,13] Furthermore, a variety of transport systems are involved in pH regulation in various smooth muscle cells: (1) alkalinizing Na/H exchange;[7-9,13-17] (2) acidifying $Cl/HCO_3$ exchange,[14,15,17,18] and (3) alkalinizing (Na + $HCO_3$)/Cl exchange.[13,14] A fourth transport system, which alkalinizes cells in a SITS-inhibitable, Na- and $HCO_3$-dependent, and Cl-independent fashion has also been proposed in smooth muscle and may be unique to these cells.[14,17,18] Finally, many smooth muscle cells maintain a more alkaline steady-state $pH_i$ in the presence of $CO_2$ than in its absence,[12,13-15,17] a response that is markedly different from skeletal and cardiac muscle.[2,3]

The present review summarizes our studies on a smooth muscle-like cell line, BC3H-1 cells. These cells will be shown to exhibit all of the foregoing features of pH regulation reported in other smooth muscle cells. Furthermore, evidence will be presented which suggests that (Na + $HCO_3$)/Cl and Na/H exchangers down-regulate in response to exposure to $CO_2/HCO_3$ in BC3H-1 cells.

## MEASUREMENT OF $pH_i$

BC3H-1 cells were seeded into 35-mm diameter polystyrene dishes containing a 10 × 13 mm Aclar plastic chip. Cells were grown at 37°C in a Dulbecco's minimal essential medium supplemented with 20% calf serum in an environment

[a] This work was supported by a grant from the American Heart Association, Ohio Affiliate, Inc., Columbus, Ohio, by a State of Ohio Research Challenge Fund Grant, and by a Biomedical Sciences Research Grant from Wright State University.

[b] The term "smooth muscle cell" will be used throughout to denote both true smooth muscle cells and smooth muscle-like cells (either cell lines with smooth muscle-like properties or primary cultures of smooth muscle-like cells such as mesangial cells).

of 10% $CO_2$. Within 4–6 days cells became confluent. During this period, growth-promoting serum components are either degraded or exhausted. Within another 4 days cells differentiated, as determined by assaying creatine kinase activity, into muscle-like cells. All pH measurements were done on cells within 10–15 days of plating.

Cells were loaded with dye by placing them in 2.5 ml of a Na HEPES-buffered medium (NHB) containing 5 $\mu$M of the acetoxymethyl ester (AM) of 2',7'-bis (2'-carboxyethyl)-5'(and 6) carboxyfluorescein (BCECF). Cells were washed in dye-free NHB after loading for 1 hour, placed in a special chip holder that allowed for perfusion of the cells, and fit into a standard polystyrene cuvette (FIG. 1A). The cuvette was placed in the thermostatted (37°C) cuvette holder of an SLM 8000C Spectrofluorometer.

Emitted fluorescence (at 530 nm) was determined at excitation wavelengths of 507 nm (where fluorescence is maximally pH sensitive) and 440 nm (where fluorescence is essentially pH insensitive) and a fluorescence ratio ($R_{fl}$) determined ($Fl_{507}/Fl_{440}$). Although $Fl_{507}$ gave a larger and less noisy signal, several transient decreases in the signal appeared which made analysis of the data difficult (FIG. 2A). These same transient decreases appeared in the $Fl_{440}$ record (FIG. 2B) including a rapid decrease in the signal upon solution injection (arrow in FIG. 2) and a slow downward drift. These transient decreases are interpreted as reflecting (1)

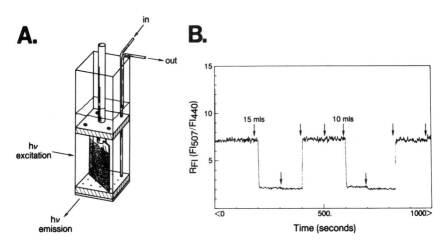

**FIGURE 1.** (A) The chamber used to measure $pH_i$ in cultured BC3H-1 cells. Cells were placed in a teflon holder which allowed perfusion through stainless steel tubes. Solution flowed up from the bottom across the cells and was removed by suction from the top of the cuvette. Emitted fluorescence was measured at a 90° angle to excitation light. (B) Control experiment to determine the conditions for complete exchange of the fluid in the cuvette. When 15 ml of a solution containing BCECF at a more acid pH were injected through the cuvette, $R_{fl}$ changed rapidly to a lower value. A second injection of 15 ml caused no additional change in $R_{fl}$. The change in $R_{fl}$ was completely reversible upon injection of 15 ml of BCECF-containing solution at the original pH. A second injection of this solution caused no further change in $R_{fl}$. Repeating this experiment, but injecting 10 ml instead of 15 ml, showed that the change in $R_{fl}$ amounted to about 95% of its previous change. A second injection of 10 ml resulted in a small additional change in $R_{fl}$. Two injections of 10 ml were also required to return $R_{fl}$ to its original value.

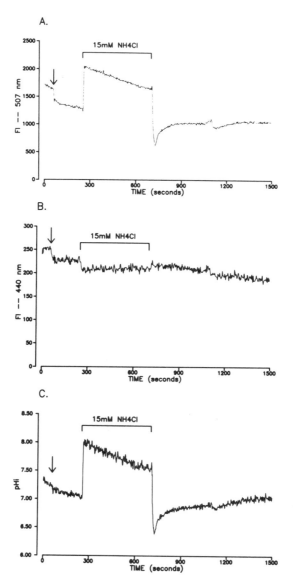

**FIGURE 2.** **(A)** The response of maximum fluorescence at an excitation wavelength of 507 nm to an $NH_4Cl$ pulse. At the *arrow*, a second 15 ml of NHB were perfused through the cuvette. Note the large drop in fluorescence upon this injection. **(B)** The response of the fluorescence at an excitation wavelength of 440 nm to an $NH_4Cl$ pulse. Note the rapid drop in fluorescence upon the injection of NHB (*arrow*) and upon injection of $NH_4Cl$. A slow downward drift in fluorescence after removal of $NH_4Cl$ is also evident. **(C)** The ratio of fluorescence ($Fl_{507}/Fl_{440}$), converted to $pH_i$ by the calibration equation given in the text, in response to an $NH_4Cl$ pulse. Note the rapid recovery of $pH_i$ from acidification induced upon $NH_4Cl$ removal. Also note the lack of the rapid downward deflection in the trace upon injection of NHB.

the loss of dye-loaded cells from the chip upon solution injection (especially during the first few injections), (2) the leakage of dye from the cells, and (3) photobleaching of the dye. Because these transient decreases were largely eliminated by taking a ratio of $Fl_{507}$ versus $Fl_{440}$ (FIG. 2C), $R_{fl}$ values were used in these studies despite the fact that such records are inherently noisy. During an experiment, an $R_{fl}$ value was collected every 1.5 seconds. $R_{fl}$ was calibrated using the high K/nigericin technique.[19] Such a calibration revealed a linear relation between pH and $R_{fl}$ over a pH range of 6.2 to 7.8:

$$pH = 0.227(R_{fl}) + 5.76 \qquad (r^2 = 0.98).$$

$R_{fl}$ became markedly nonlinear with respect to pH outside this range. At the end of each experiment, dye fluorescence was calibrated at a $pH_o$ of 7.4 (e.g., FIGS. 5 and 7) and the calibration curve adjusted assuming the slope remained constant.

During the course of an experiment, cuvettes were perfused with various solutions injected from 20-ml syringes through exteriorized PE tubing. Solution entered the bottom of the cuvette, flowed up across the cells, and was removed by suction after pooling above the teflon holder (FIG. 1A). Preliminary experiments using dye-containing solutions at different pH values showed that 15 ml of injectate are required to completely turn over the solution in the cuvette (FIG. 1B). The injection of 15 ml of solution required about 15–20 seconds.

The standard Na HEPES-buffered solution contained 127 mM NaCl, 4 mM KCl, 1 mM $CaCl_2$, 0.5 mM $MgCl_2$, 20 mM NaHEPES, and 5.6 mM glucose, $pH_o = 7.4$ at 37°C, equilibrated with 100% $O_2$. The standard $HCO_3$-buffered solution was identical except that 24 mM $NaHCO_3$ replaced the NaHEPES and the solution was equilibrated with 5% $CO_2$ (balance air). The Na replacement was N-methyl-D-glucammonium (NMDG) and the Cl replacement was gluconate. (Amiloride was the generous gift of Dr. Clement Stone [Merck, Sharp and Dohme, Rahway, NJ] and 4-acetamino-4'-isothiocyano stilbene-2,2'-disulfonic acid [SITS] was purchased from ICN Biomedicals, Inc. [Plainview, NY]). The calibration solution contained 131 mM KCl, 1 mM $CaCl_2$, 0.5 mM $MgCl_2$, 20 mM NMDG HEPES, 5.6 mM glucose, and 5 $\mu$g/ml nigericin.

All values are reported as the mean ± one standard error of the mean (SE).

## PRESENCE OF Na/H EXCHANGE

In the nominal absence of $CO_2$, the intracellular pH ($pH_i$) of BC3H-1 cells is 6.89 ± 0.01 ($n = 178$). This steady-state $pH_i$ is unaffected by the anion exchange inhibitor SITS, but acidifies on exposure to amiloride (0.07 ± 0.01 pH unit, $n = 4$) or Na-free solutions (0.21 ± 0.04 pH unit, $n = 8$). The rate of acidification on exposure to amiloride is $2.0 ± 0.7 × 10^{-4}$ pH/s, which is a very slow rate of acidification. These data suggest that the steady-state $pH_i$ of BC3H-1 cells in NHB is largely maintained by a Na/H exchanger that counterbalances a slow rate of acid loading (due presumably to H influx and metabolic acid production).

The presence of a Na/H exchanger in BC3H-1 cells is also shown by the response of these cells to an acid load induced after an $NH_4Cl$ pulse.[20] Upon removal of $NH_4Cl$, $pH_i$ acidifies and exhibits a rapid recovery back to the initial $pH_i$ (FIG. 2C). The initial rate of this recovery (as estimated by the slope of a least squares regression line fit to the first several seconds of recovery) amounts to $40.2 ± 2.6 × 10^{-4}$ pH/s ($n = 6$). This rate of recovery can be converted into an equivalent H flux (recognizing that an acid flux in one direction is equivalent to a

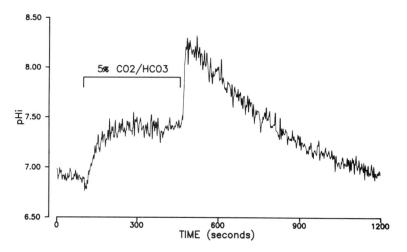

**FIGURE 3.** The response of BC3H-1 cells to $CO_2/HCO_3$ solution. Note the small initial acidification followed by an alkalinization to a new steady-state $pH_i$. Upon removal of $CO_2$, $pH_i$ rapidly alkalinized and then returned to a steady-state $pH_i$ close to the original value in NHB.

base flux in the opposite direction) by multiplying the recovery rate by the small intrinsic cellular buffering power[3,13,14] (10 mM in NHB; data not shown). The rate of this recovery is equivalent to an H efflux ($J_H$) of about $400 \times 10^{-4}$ mM/s. This recovery is reversibly inhibited 82% by 1 mM amiloride and 88% by the removal of Na from the external medium. Thus, virtually all of the recovery from an acid load (in the nominal absence of $CO_2$) is mediated by a Na/H exchanger in BC3H-1 cells.

In summary, a Na/H exchanger is present in BC3H-1 cells and is the major transport system responsible for the regulation of $pH_i$ in the absence of $CO_2/HCO_3$.

## PRESENCE OF HCO₃-DEPENDENT TRANSPORTERS

Most cells acidify upon exposure to $CO_2$ due to formation of protons by the entry, hydration, and dissociation of $CO_2$.[2,3,20] Intracellular pH may recover to its initial steady-state value in these cells or to a value below the initial $pH_i$. In an increasing number of cells, including smooth muscle cells[12-15] and fibroblasts,[21] a different $pH_i$ response is observed. Upon exposure to $CO_2$, these cells briefly acidify and then alkalinize to a new steady-state value *above* the initial $pH_i$. Such a pattern is also seen in BC3H-1 cells (FIG. 3). In the presence of $CO_2$, BC3H-1 cells achieve a new steady-state $pH_i$ about 0.4 pH unit more alkaline than in the presence of NHB, $7.27 \pm 0.01$ ($n = 65$). That this alkalinization is mediated by a transport system is suggested by the marked alkaline overshoot of $pH_i$ (to above pH 8.0) upon the removal of $CO_2$ (FIG. 3). Such an alkaline overshoot is indicative of the removal of acid equivalents (or the addition of base equivalents) to the cell during $CO_2$ exposure.[3,20] $pH_i$ recovers from this alkaline overshoot to the initial steady-state $pH_i$ in NHB, well below the steady-state $pH_i$ observed in $CO_2$ (FIG. 3). The most likely explanation for this response of $pH_i$ to $CO_2$ exposure is that BC3H-1 cells have $HCO_3$-dependent transport systems that are responsible for the maintenance of a higher steady-state $pH_i$ in the presence of $CO_2$ than in its absence.

## $(Na + HCO_3)/Cl$ EXCHANGE

The membrane transport systems responsible for the $CO_2$-induced alkalinization of BC3H-1 cells have been characterized. The initial rate of this alkalinization is inhibited 95% in the absence of external Na and 84% in the presence of 0.5 mM SITS (FIG 4A). In contrast, 1 mM amiloride causes no reduction in the rate of alkalinization. To eliminate contributions from intracellular Cl, BC3H-1 cells can be maintained in Cl-free NHB for 25 minutes before $CO_2$ exposure. Under such conditions, the initial rate of alkalinization is decreased by 85% (FIG. 4B). Finally, when BC3H-1 cells are depolarized by exposure to elevated external K (55 mM K with 10 $\mu$M of the K ionophore valinomycin), there is no reduction in the rate of $CO_2$-induced alkalinization (FIG. 4A). Taken together, these data suggest that a major portion of $CO_2$-induced alkalinization is mediated in BC3H-1 cells by a $HCO_3$-, Na-, and Cl-dependent transporter that is inhibitable by SITS. This trans-

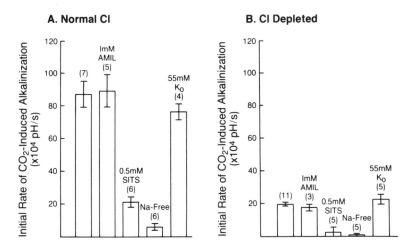

**FIGURE 4. (A)** Summary of the initial linear rate of alkalinization upon exposure to $CO_2$, derived from records such as shown in FIGURE 3. Note that this recovery is unaffected by amiloride or cellular depolarization in elevated external K, but is nearly abolished by SITS or the removal of external Na. **(B)** Summary of the initial linear rate of alkalinization upon exposure to $CO_2$ in cells that have been exposed for 25 minutes to Cl-free NHB. Note that the control rate of recovery is reduced by about 85% compared to the rate in normal Cl. However, as in normal Cl, the recovery that remains in Cl-depleted cells is unaffected by amiloride or cellular depolarization, and abolished by SITS or by the removal of external Na. The height of a bar represents the mean $\pm$ 1 SE (number of experiments given in parentheses).

porter is most likely the electroneutral $(Na + HCO_3)/Cl$ exchanger, previously described in a number of cells including smooth muscle cells.[3,14]

Several lines of evidence suggest that an additional $HCO_3$-dependent, alkalinizing transport system may be present in smooth muscle cells. After prolonged exposure to Cl-free NHB (25–55 minutes), BC3H-1 cells are still able to alkalinize in a SITS-inhibitable fashion in response to $CO_2$ exposure (FIG. 4B). This remaining recovery is abolished by SITS or by Na-free solutions, but is not affected by amiloride (FIG. 4B). Furthermore, this recovery is not affected by depolarization in high K solutions (FIG. 4B). Presuming that intracellular Cl is indeed depleted by prolonged exposure to Cl-free NHB, these data suggest an electroneutral Na-$HCO_3$ cotransporter that is independent of Cl. A similar type of transporter has been suggested for other smooth muscle cells. A portion of $CO_2$-induced alkalinization remains in Cl-depleted vas deferens smooth muscle cells[18] and in mesangial cells,[14] suggesting the presence of a Cl-independent alkalinizing transport system in these cells. A Na-$HCO_3$ cotransporter has been demonstrated in kidney cells,[22] although this transporter is electrogenic. An electroneutral Na-$HCO_3$ exchanger, as suggested in BC3H-1 cells, has not previously been described. Whatever its properties, an additional $HCO_3$-dependent transporter, besides the $(Na + HCO_3)/Cl$ exchanger, appears to contribute to $CO_2$-induced alkalinization in smooth muscle cells.

## $Cl/HCO_3$ EXCHANGE

In addition to alkalinizing transport systems (Na/H exchange, $(Na + HCO_3)/Cl$ exchange, and putative Na-$HCO_3$ cotransport), BC3H-1 cells contain an acidifying $Cl/HCO_3$ exchanger. Upon exposure to Cl-free $CO_2/HCO_3$ solution, BC3H-1 cells rapidly alkalinize by about 0.8 pH unit. This alkalinization occurs at an initial rate of $135 \pm 9 \times 10^{-4}$ pH/s ($n = 12$). The extent and rate of this alkalinization is unaffected by 1 mM amiloride. The addition of 0.5 mM SITS abolishes the alkalinization induced by Cl-free $CO_2$ solutions and reduces the rate of this alkalinization by 87% to $18 \pm 2 \times 10^{-4}$ pH/s ($n = 6$). Na-free solutions do not markedly affect the extent of alkalinization, but slow its rate by about 50% to $65 \pm 4 \times 10^{-4}$ pH/s ($n = 4$). These data suggest that a portion of the alkalinization induced in Cl-free $CO_2$-buffered solutions is due to activation of the $(Na + HCO_3)/Cl$ exchanger, but that a sizable fraction is due to reversal of the Na-independent $Cl/HCO_3$ exchanger.

The presence of $Cl/HCO_3$ exchange in BC3H-1 cells is also indicated by the rate of acidification seen in response to $NH_4Cl$-induced alkalinization (referred to as plateau phase acidification;[20] see, for example, FIG. 2C). The rate of this acidification, expressed as an equivalent H influx, is increased over fourfold in the presence of $CO_2/HCO_3$, suggesting the mediation of a $HCO_3$-dependent acidifying transport system in response to cellular alkalinization. The rate of plateau phase acidification in the presence of $CO_2/HCO_3$ is reduced 50% by 0.5 mM SITS (from $10.8 \pm 0.6 \times 10^{-4}$ pH/s, $n = 11$, to $5.3 \pm 0.6 \times 10^{-4}$ pH/s, $n = 11$). Furthermore, the acidification seen in response to alkalinization induced upon removal of $CO_2/HCO_3$ (see, for example, FIG. 3) is reduced in the absence of external Cl. Taken together, these data clearly indicate a $Cl/HCO_3$ exchanger in BC3H-1 cells that results in cellular acidification in response to an alkaline load. The ability of this exchanger to acidify the cell even in the nominal absence of $CO_2/HCO_3$ indicates that it either has a very high affinity for $HCO_3$ or can transport OH ions.

## MODEL OF pH REGULATION IN BC3H-1 CELLS

The foregoing data indicate that the regulation of $pH_i$ in BC3H-1 cells is mediated by at least three and possibly four transport systems. In the nominal absence of $CO_2/HCO_3$, the regulation of $pH_i$ is largely mediated by Na/H exchange, which is activated by cellular acidification and returns $pH_i$ to about 6.9. Steady-state $pH_i$ under these conditions results from the balance between the alkalinizing influence of Na/H exchange and the acidifying processes of cellular metabolism, passive H influx, and perhaps a small contribution from $Cl/HCO_3$ exchange. Upon addition of $CO_2/HCO_3$, powerful $HCO_3$-dependent transport systems contribute to the regulation of $pH_i$ and result in the maintenance of an elevated steady-state $pH_i$ of about 7.3. This new $pH_i$ is also a balance between acidifying processes (cellular metabolism, H influx, $Cl/HCO_3$ exchange, and possibly $HCO_3$ efflux) and alkalinizing transport systems, predominantly (Na + $HCO_3$)/Cl exchange and perhaps Na-$HCO_3$ cotransport. Thus, under physiologic conditions, $pH_i$ regulation is largely due to the mediation of $HCO_3$-dependent transport systems which maintain a steady-state pH of 7.3. Such a pattern of pH regulation has previously been reported in other smooth muscle cells, including nonvascular smooth muscle[12,18] and mesangial cells.[13,14] However, primary cultures of vascular smooth muscle cells have been shown to have a high activity of Na/H exchange and very large intrinsic buffering power,[7] suggesting that $pH_i$ regulation may differ between vascular and nonvascular smooth muscle cells.

## EFFECT OF AN NH₄Cl PREPULSE ON SUBSEQUENT pH_i REGULATION

The effect of an initial $NH_4Cl$ pulse on subsequent $pH_i$ regulation was assessed. If BC3H-1 cells are allowed to recover fully from acidification induced after an $NH_4Cl$ pulse and then exposed to another $NH_4Cl$ pulse, a virtually identical pattern of pH response is seen (FIG. 5). The minimum $pH_i$ obtained upon removal of $NH_4Cl$ is clearly the same after the first $NH_4Cl$ pulse ($6.19 \pm 0.05$, $n = 10$) as after the second ($6.11 \pm 0.05$). The initial rate ($58.3 \pm 9.6 \times 10^{-4}$ vs $53.6 \pm 11.1 \times 10^{-4}$ pH/s) and extent of $pH_i$ recovery are also the same for the first and second $NH_4Cl$ pulses (FIG. 6). A paired comparison, expressing the second rate of recovery as a fraction of the initial rate of recovery for the same cells, indicates that $pH_i$ recovery is not reduced significantly after the second $NH_4Cl$ exposure (the second rate of recovery was $89.5 \pm 7.8\%$ of the first). Because this recovery is nearly entirely due to Na/H exchange, these data indicate that Na/H exchange activity remains constant despite previous activation by $NH_4Cl$ prepulse-induced acidification.

The response of BC3H-1 cells to $CO_2$ exposure after an $NH_4Cl$ pulse was also studied. Cells are allowed to recover from acidification induced after an $NH_4Cl$ pulse back to the initial $pH_i$. Exposure to $CO_2/HCO_3$ at that point results in a normal $pH_i$ response consisting of a small, transient acidification followed by alkalinization to a $pH_i$ some 0.4 pH unit more alkaline than steady-state $pH_i$ in the absence of $CO_2$ (FIG. 7). Upon removal of $CO_2$, a marked alkaline overshoot is seen with $pH_i$ recovering back towards its initial level. The initial rate of alkalinization upon $CO_2$ exposure after an $NH_4Cl$ pulse is the same as the initial rate of alkalinization upon $CO_2$ exposure without a previous $NH_4Cl$ pulse (FIG. 8),

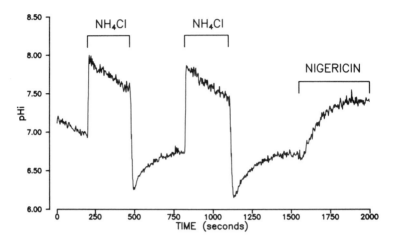

**FIGURE 5.** The response of $pH_i$ to two pulses of 15 mM $NH_4Cl$. The response of $pH_i$ to the second $NH_4Cl$ pulse is indistinguishable from the first. Exposure of the cells to a high K solution containing nigericin at $pH_o$ of 7.4 served to calibrate the fluorescence.

whether the $CO_2$ pulse came early in the experiment (early pulse) or after about 15 minutes in NHB (late pulse). These data indicate that previous exposure to $NH_4Cl$ does not affect the $pH_i$ response of BC3H-1 cells to $CO_2$ exposure.

In summary, an initial exposure to $NH_4Cl$, with the associated alkaline and acid shifts in $pH_i$ and the activation of Na/H exchange, does not affect the $pH_i$ response of BC3H-1 cells to either another $NH_4Cl$ pulse or subsequent $CO_2$ exposure.

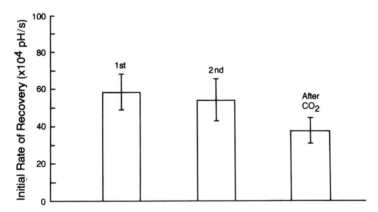

**FIGURE 6.** A summary of the initial rate of recovery from acidification induced after removal of $NH_4Cl$. The rates for the first and second pulses were derived from records as in FIGURE 5. The last bar is the rate of $pH_i$ recovery after an $NH_4Cl$ pulse that had been preceded by a 5-minute exposure to $CO_2$ (as in FIG. 9). The height of a bar represents mean ± 1 SE (the bars represent the mean of 10, 10, and 7 experiments, respectively).

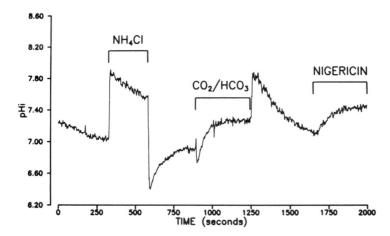

**FIGURE 7.** The response of $pH_i$ to a 5-minute exposure to $CO_2/HCO_3$ after an $NH_4Cl$ pulse. The response to $CO_2$ exposure is very similar to that without a previous $NH_4Cl$ exposure (FIG. 3).

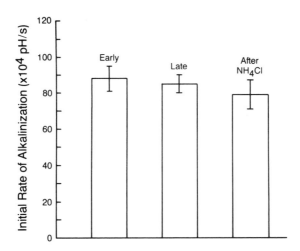

**FIGURE 8.** The summary of data on the initial rate of cellular alkalinization upon exposure to $CO_2$ (from experiments as shown in FIGS. 3 and 7). For the early pulses, $CO_2$ exposure occurred at about 100–200 seconds, for the late pulses and the pulses after an $NH_4Cl$ pulse, at about 1,000 seconds. Note that the recovery was the same in all cases. The height of a bar represents the mean ± 1 SE (the bars represent the mean of 6, 6, and 13 experiments, respectively).

## EFFECT OF $CO_2$ EXPOSURE ON SUBSEQUENT $pH_i$ REGULATION

An initial exposure to $CO_2$ appears to reduce $pH_i$ recovery from a subsequent exposure to $NH_4Cl$ (FIG. 9). The minimum $pH_i$ obtained upon removal of $NH_4Cl$ under these conditions is $6.24 \pm 0.04$ ($n = 7$). The rate of $pH_i$ recovery from this acidification is $38.3 \pm 6.6 \times 10^{-4}$ pH/s. Thus, the rate of $pH_i$ recovery from $NH_4Cl$-induced acidification after a previous exposure to $CO_2$ is reduced about 30% compared to normal $pH_i$ recovery (FIG. 6). These data indicate that a previous exposure to $CO_2$ (for about 4–5 minutes) results in a small reduction in the observed rate of $pH_i$ recovery from $NH_4Cl$-induced acidification and thus presumably a reduction in the rate of Na/H exchange activity.

The inhibitory effect of an initial $CO_2$ exposure is even more striking upon a subsequent exposure to $CO_2$ than it is upon a subsequent $NH_4Cl$ pulse. If $pH_i$ is

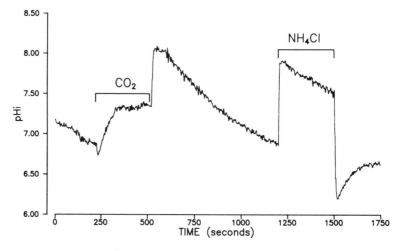

**FIGURE 9.** The response of $pH_i$ to an $NH_4Cl$ pulse after an initial exposure to $CO_2$. Note that the response of $pH_i$ to $NH_4Cl$ appears normal except for a slight reduction in the initial rate of $pH_i$ recovery from acidification induced upon removal of $NH_4Cl$ (compare to FIG. 2 and see FIG. 6).

allowed to return to its initial value in NHB after a 4-minute exposure to $CO_2$ and then exposed to a second pulse of $CO_2$, BC3H-1 cells acidify more dramatically and $pH_i$ recovery is markedly reduced (FIG. 10). Reduced recovery during $CO_2$ exposure is also clearly indicated by the smaller alkaline overshoot upon removal of the second $CO_2$ pulse (FIG. 10). This effect of $CO_2$ is dependent on the duration of the initial exposure to $CO_2$. If the initial $CO_2$ pulse is quite short (less than 1 minute), the response to the second $CO_2$ exposure is fairly normal, with $pH_i$ rapidly alkalinizing in the presence of $CO_2$ (FIGS. 11 and 12). When the initial $CO_2$ pulse is longer (2–4 minutes), the rate of alkalinization in response to the second $CO_2$ pulse is markedly reduced (FIG. 12). No further reduction of this rate of alkalinization is seen even if the duration of the first $CO_2$ pulse is increased to 10 minutes (FIG. 12). The initial rate of $CO_2$-induced alkalinization is about $100 \times 10^{-4}$ pH/s. In FIGURE 12, the difference between the rate of $CO_2$-induced alkalinization during the second versus the first pulse (done as paired data) is plotted as a

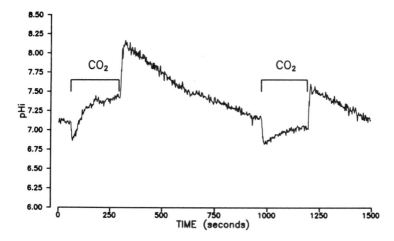

**FIGURE 10.** The response of $pH_i$ to repeated 5-minute exposures of $CO_2$. Note the normal $pH_i$ response to the initial exposure of $CO_2$ (compare to FIG. 3). After $pH_i$ is allowed to return to its initial steady-state value in NHB, a second exposure to $CO_2$ yields a markedly different $pH_i$ response. The initial acidification is more pronounced and maintained, and the alkaline overshoot upon the removal of $CO_2$ is reduced.

function of the duration of the first pulse. The maximum reduction in the rate of alkalinization during the second pulse amounts to about a 60% decrease as compared to the rate during the initial pulse. This inhibition requires several minutes of previous $CO_2$ exposure to be fully expressed (FIG. 12).

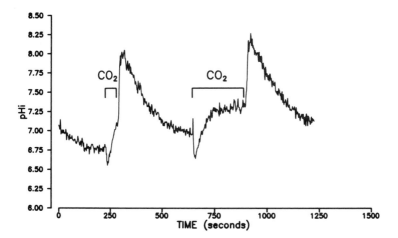

**FIGURE 11.** The response of $pH_i$ to $CO_2$ exposure after an initial brief (30-second) $CO_2$ exposure. During the initial $CO_2$ pulse, $pH_i$ alkalinized but did not achieve a new steady state. Upon the second $CO_2$ exposure, $pH_i$ acidified somewhat more than initially and the rate of alkalinization was somewhat slowed, but a steady-state $pH_i$ alkaline to that in NHB was achieved and a marked alkaline overshoot upon the removal of $CO_2$ is evident.

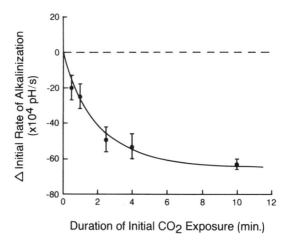

Duration of Initial $CO_2$ Exposure (min.)

**FIGURE 12.** The difference between the initial rate of alkalinization upon $CO_2$ exposure during a second pulse minus the first pulse as a function of the duration of the first $CO_2$ pulse (derived from experiments as shown in FIGS. 10 and 11). If the initial $CO_2$ exposure is of short duration, a nearly normal $pH_i$ response to the second $CO_2$ exposure is observed. However, if the initial $CO_2$ exposure is at least a few minutes long, the recovery rate during the second exposure is reduced by about 60%. The average rate of alkalinization during the first $CO_2$ exposure is about $100 \times 10^{-4}$ pH/s. Each point represents the mean difference in initial rate (paired data) $\pm$ 1 SE of five experiments. The curve was fit by eye.

In summary, unlike $NH_4Cl$ pulses, prior exposure to $CO_2$ causes a marked reduction in pH recovery in response to acidification induced by either an $NH_4Cl$ prepulse or a second exposure to $CO_2$. This reduction in pH recovery requires several minutes of exposure to $CO_2$ to fully develop. It appears that the activity of $HCO_3$-dependent transport systems is more markedly affected by prior $CO_2$ exposure than is the activity of the Na/H exchanger.

## MECHANISM OF $CO_2$-INDUCED INHIBITION OF $pH_i$ RECOVERY

There are three possible mechanisms by which previous $CO_2$ exposure could result in reduction of subsequent $pH_i$ recovery: (1) $CO_2$-induced stimulation of acid-loading processes in the cell; (2) $CO_2$-induced reduction in the activity of alkalinizing membrane transport systems; or (3) $CO_2$-induced endocytosis.

It is possible that prior $CO_2$ exposure does not alter the activity of the alkalinizing membrane transport systems. If acid-loading processes are increased by exposure to $CO_2$, then both the rate of $CO_2$-induced alkalinization and the steady-state $pH_i$ would be reduced.[3,14] The main acid-loading processes in the absence of $CO_2$ are passive H influx and metabolic acid production. It could be that $CO_2$ exposure induces a sustained increase in H permeability, but it is more likely that metabolic acid production would be stimulated. It is known that increasing $pH_i$ can activate key glycolytic enzymes,[23] and cellular acidification would thus inhibit glycolysis.[24] In most cells, $CO_2$ exposure results in cellular acidification and thus would be expected to reduce the metabolic acid loading, but in BC3H-1 cells, glycolysis could be activated due to $CO_2$-induced alkalinization. The problem with this explanation is that after removal of $CO_2$, $pH_i$ returns to a lower steady-state value, and thus metabolic acid production should once again be low. This is inconsistent with the reduced recovery rate seen in response to $NH_4Cl$-induced acidification after a $CO_2$ pulse (FIG. 9). However, it is still possible that some other modification of the intracellular environment induced upon $CO_2$ exposure could activate metabolic acid production for a prolonged period.

The alternative explanation for the effect of $CO_2$ on $pH_i$ recovery systems is that an initial pulse of $CO_2$ results in the down-regulation of these transporters. This could reflect a decrease in activity of the transport systems due to some modifying agent generated upon the initial exposure to $CO_2$. Candidate intracellular agents that have been shown to affect the activity of pH-regulating transporters include calcium,[25] protein kinase C,[26] and cyclic AMP.[27] A role of calcium in the down-regulation of these transport systems can be ruled out, because exposure to $NH_4Cl$ causes even greater increases in intracellular Ca than does exposure to $CO_2$ (data not shown), and yet preexposure to $NH_4Cl$ does not result in down-regulation of the transporters (FIGS. 5 and 7). A detailed knowledge of the factors that confer kinetic control over the pH-regulating transport systems in BC3H-1 cells will be needed before this possibility can be fully assessed.

The second possible way by which these transport systems could be down-regulated is by a $CO_2$-induced endocytosis. Such a process would actually remove transporters from the surface. A $CO_2$-induced increase in the trafficking of surface membrane has previously been observed in turtle bladder cells[28] and some mammalian kidney cells.[29] In these cells, acidification upon $CO_2$ exposure results in an increase in intracellular Ca which induces exocytosis of intracellular vesicles. In this way, transporters that reside in the vesicular membrane are incorporated into the surface membrane. Such a model, however, is not very appropriate for BC3H-1 cells. Firstly, $CO_2$ induces alkalinization in BC3H-1 cells, although an increase in intracellular Ca still seems to occur (data not shown). Secondly, a $CO_2$-induced *endocytosis* would have to be hypothesized to account for the reduction in the rate of alkalinization after $CO_2$ exposure. Cellular acidification, including that generated by $CO_2$ exposure, has been shown to *inhibit* endocytosis in other cells.[30,31] It is clear that if $CO_2$ exposure promotes endocytosis in BC3H-1 cells, it must do so in a way that is independent of changes in $pH_i$ or Ca, because an $NH_4Cl$ pulse causes marked changes in both $pH_i$ and Ca but does not result in down-regulation of alkalinizing transport systems.

In summary, the basis for $CO_2$-induced down-regulation of pH-regulating transporters is not known but could involve increased cellular metabolism, reduction in the rate of transport of these systems, or increased endocytosis which removes transport proteins from the surface membrane. The mechanism for this down-regulation is currently under investigation.

## CONCLUSIONS

The regulation of $pH_i$ in smooth muscle cells appears to be mediated by at least three and as many as four transport systems: (1) Na/H exchange; (2) $Cl/HCO_3$ exchange; (3) $(Na + HCO_3)/Cl$ exchange; and (4) a putative $Na-HCO_3$ cotransport. Under typical experimental conditions (nominal absence of $CO_2$), the Na/H exchange is predominantly responsible for the regulation of $pH_i$. Under more physiologic conditions (5% $CO_2$/24 mM $HCO_3$) the $HCO_3$-dependent transport systems regulate $pH_i$ at a value more alkaline than that in the absence of $CO_2$. In addition to this variety of transport systems, the activity of any one transporter may be markedly affected by its previous history. After an exposure to $CO_2$, lasting for at least 4 minutes, the alkalinizing transport systems appear to be down-regulated. This loss of activity is more marked for the $HCO_3$-dependent transporters than for the Na/H exchanger. It thus appears that the activity of the $HCO_3$-dependent transport systems is somewhat more labile than is the activity of the Na/H exchange. These data point out the importance of studying pH regulation in smooth muscle cells in the presence of $CO_2/HCO_3$. In such studies, however, $CO_2$-induced down-regulation of pH-regulating transport systems may be a complicating factor.

## ACKNOWLEDGMENTS

The superb technical assistance of Phyllis Douglas in all aspects of this work is gratefully acknowledged. Invaluable assistance with tissue culturing was given by Patricia Beltz and Dr. Robert Grubbs. Some of the later experiments were done with the technical assistance of Desiree Renaux Kosan. The expert advice of Dr. Anne Walter on fluorescence measurements is also gratefully acknowledged.

## REFERENCES

1. WRAY, S. 1988. Am. J. Physiol. **254:** C213–C225.
2. AICKIN, C. C. 1986. Ann. Rev. Physiol. **48:** 349–361.
3. ROOS, A. & W. F. BORON. 1981. Physiol. Rev. **61:** 296–434.
4. AALKJAER, E. & E. J. CRAGOE, JR. 1988. J. Physiol. **402:** 391–410.
5. GARDNER, J. P. & F. P. J. DIECKE. 1988. Pflügers Arch. **412:** 231–239.
6. KNEHR, H. E. & A. M. LINKE. 1980. Resp. Physiol. **42:** 155–169.
7. BERK, B. C., M. S. ARONOW, T. A. BROCK, E. CRAGOE, JR., M. A. GIMBRONE, JR. & R. W. ALEXANDER. 1987. J. Biol. Chem. **262:** 5057–5064.
8. HATORI, N., B. P. FINE, A. NAKAMURA, E. CRAGOE, JR. & A. AVIV. 1987. J. Biol. Chem. **262:** 5073–5078.
9. HUANG, C. -L., M. G. COGAN, E. J. CRAGOE, JR. & H. E. IVES. 1987. J. Biol. Chem. **262:** 14134–14140.
10. OWEN, N. E. 1984. Am. J. Physiol. **247:** C501–C505.
11. OWEN, N. E. 1986. J. Cell Biol. **103:** 2053–2060.
12. AICKIN, C. C. 1984. J. Physiol. **349:** 571–585.
13. BOYARSKY, G., M. B. GANZ, R. B. STERZEL & W. F. BORON. 1988. Am. J. Physiol. **255:** C844–C856.
14. BOYARSKY, G., M. B. GANZ, R. B. STERZEL & W. F. BORON. 1988. Am. J. Physiol. **255:** C857–C869.
15. KORBMACHER, C., H. HELBIG, F. STAHL & M. WIEDERHOLT. 1988. Pflügers Arch. **412:** 29–36.
16. WEISSBERG, P. L., P. J. LITTLE, E. J. CRAGOE, JR. & A. BOBIK. 1987. Am. J. Physiol. **253:** C193–C198.
17. PUTNAM, R. W. 1988. J. Gen Physiol. **92:** 52a.
18. AICKIN, C. C. & A. F. BRADING. 1984. J. Physiol. **349:** 587–606.
19. THOMAS, J., R. BUCHSBAUM, A. ZIMNIAK & E. RACKER. 1979. Biochemistry **18:** 2210–2218.
20. BORON, W. F. & P. DE WEER. 1976. J. Gen. Physiol. **67:** 91–112.
21. BIERMAN, A. J., E. J. CRAGOE, JR., S. W. DE LAAT & W. H. MOOLENAAR. 1988. J. Biol. Chem. **263:** 15253–15256.
22. BORON, W. F. & E. L. BOULPAEP. 1983. J. Gen. Physiol. **81:** 53–94.

23. TRIVEDI, B. & W. H. DANFORTH. 1966. J. Biol. Chem. **241:** 4110–4111.
24. FIDELMAN, M. L., S. H. SEEHOLZER, K. B. WALSH & R. D. MOORE. 1982. Am. J. Physiol. **242:** C87–C93.
25. HESKETH, T. R., J. P. MOORE, J. D. MORRIS, M. V. TAYLOR, J. ROGERS, G. A. SMITH & J. C. METCALFE. 1985. Nature **313:** 481–484.
26. MOOLENAAR, W. H. 1986. Ann. Rev. Physiol. **48:** 363–376.
27. REUSS, L. & K. -U. PETERSEN. 1985. J. Gen. Physiol. **85:** 409–429.
28. GLUCK, S., C. CANNON & Q. AL-AWQATI. 1982. Proc. Natl. Acad. Sci. USA **79:** 4327–4331.
29. SCHWARTZ, G. J. & Q. AL-AWQATI. 1985. J. Clin. Invest. **75:** 1638–1644.
30. REEVES, W., S. GLUCK & Q. AL-AWQATI. 1983. Kidney Int. **23:** 237.
31. SANDVIG, K., S. OLSNES, O. W. PETERSON & B. VAN DEURS. 1987. J. Cell Biol. **105:** 679–689.

# Regulation of Transepithelial Chloride Transport by Amphibian Gallbladder Epithelium[a]

LUIS REUSS

*Department of Physiology and Biophysics*
*University of Texas Medical Branch*
*Galveston, Texas 77550*

Mammalian and amphibian gallbladder epithelia have been used for over two decades as model systems to study the mechanisms of transepithelial isosmotic fluid transport. In *Necturus* gallbladder epithelium, quantitative intracellular-microelectrode techniques, including the use of ion-sensitive microelectrodes, have permitted detailed characterization of the mechanisms of ion transport across the apical[1-4] and basolateral membrane domains,[5-7] quantitation of the hydraulic water permeabilities of both cell membranes,[8] and assessment of some of the mechanisms of regulation of fluid transport. Among the latter, significant progress has been made in understanding the inhibitory effect of cyclic AMP[9-11] and the stimulatory effect of $HCO_3^-/CO_2$.[3,6,7]

## MECHANISMS OF ION TRANSPORT AT APICAL AND BASOLATERAL MEMBRANES

The major transepithelial transport function of the gallbladder is the absorption of NaCl and water in isosmotic proportions.[12-15] Transepithelial salt transport involves entry of both $Na^+$ and $Cl^-$ across the apical membrane and extrusion of both ions across the basolateral membrane. In *Necturus* gallbladder under steady-state conditions, intracellular-microelectrode measurements of the electrochemical driving forces across the individual cell borders indicate that the passive forces favor $Na^+$ entry and $Cl^-$ exit. Therefore, because vectorial NaCl transport is from the apical solution to the cell and from the cell to the basolateral solution, at the apical membrane $Na^+$ entry is a downhill and $Cl^-$ entry is an uphill process; at the basolateral membrane, $Na^+$ extrusion is uphill, whereas $Cl^-$ exit is downhill. The data supporting these conclusions are summarized in FIGURE 1.

Abundant experimental results indicate that the $Na^+$ and $Cl^-$ fluxes across the apical membrane are electrically silent, that is, they must be carrier mediated and consist of ion cotransport and/or countertransport, with the net result of zero net charge transfer across the apical membrane. Although experimental results from some laboratories support the possibility that NaCl cotransport plays a role in salt entry,[16-17] the overall evidence favors the view that the main entry mechanism is the parallel, independent operation of $Na^+/H^+$ and $Cl^-/HCO_3^-$ exchangers.[18-21] The main arguments for this conclusion have recently been reviewed in detail.[12,18]

[a] This work was supported by grants DK38588 and DK38734 from the National Institute of Diabetes, Digestive and Kidney Diseases.

370

They include demonstrations that the rate of apical membrane anion exchange under control conditions suffice to account for the rates of transepithelial $Na^+$ and $Cl^-$ transport.

Controversy remains, however, as to whether there is NaCl or NaKCl cotransport across the apical membrane of gallbladder epithelium, on the basis of kinetic analysis in rabbit[16] and of pharmacologic studies, that is, the use of bumetanide, in *Necturus*.[22,23] The differences in results from diverse laboratories have been discussed,[12,18] but they are not fully understood. Among several lines of evidence supporting the hypothesis of double-ion exchange, perhaps the most convincing is the demonstration that the initial rates of fall of intracellular $Na^+$ or $Cl^-$ activities upon removal of either $Na^+$ or $Cl^-$ from the mucosal solution are radically different, although direct coupling between $Na^+$ and $Cl^-$ transport across the membranes would predict under these conditions similar rates of fall in the intracellu-

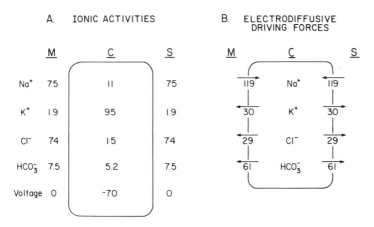

**FIGURE 1. (A)** Membrane voltages, extracellular and intracellular ionic activities in *Necturus* gallbladder under control conditions (1% $CO_2$, 99% air). **(B)** Net driving forces for electrodiffusive ion movements across the cell membranes, expressed as $\Delta\bar{\mu}_i/F$, in millivolts. Note that the passive driving forces favor outward transport of $K^+$, $Cl^-$, and $HCO_3^-$, and inward transport of $Na^+$, at both membranes. Reproduced from Reuss and Stoddard,[12] with permission of *Annual Reviews*.

lar ionic activities of both ions. A typical intracellular activity trace illustrating this point is shown in FIGURE 2; average rates of change in intracellular activities upon $Na^+$ or $Cl^-$ removal are shown in FIGURE 3.

$Na^+$ transport across the basolateral membrane is an active process, namely, extrusion via the $Na^+$ pump, that is, the $(Na^+,K^+)$-activated ATPase, as shown in a variety of epithelial tissues. The mechanism of $Cl^-$ extrusion is less clear. There is agreement in that the driving force favors passive basolateral efflux of this anion (FIG. 1), but estimations of the apparent $Cl^-$ conductance of the basolateral membrane, with intracellular microelectrode techniques, suggest that under control conditions the electrodiffusive $Cl^-$ efflux is small[5] or nil.[24] Recent studies from our laboratory have demonstrated that the electrodiffusive $Cl^-$ permeability of the basolateral membrane is significantly increased when the tissue is bathed in $HCO_3^-/CO_2$-buffered solutions[6] (see below). However, even in these conditions the estimated electrodiffusive $Cl^-$ flux is insufficient to account for the steady-

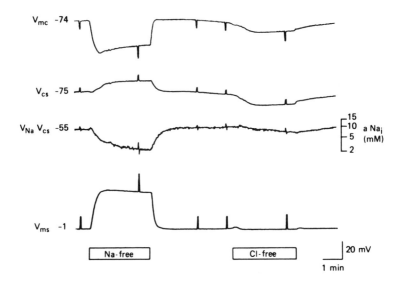

**FIGURE 2.** Effects of mucosal $Na^+$ removal (*left*) and mucosal $Cl^-$ removal (*right*) on membrane potentials and $aNa_i$. Note the large difference in the rates of fall of $aNa_i$ in response to these perturbations. Reproduced from Reuss,[3] with permission of *The Journal of General Physiology*.

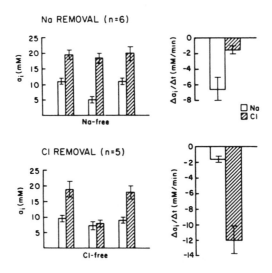

**FIGURE 3.** Summary of changes in intracellular $Na^+$ and $Cl^-$ activities on $Na^+$ or $Cl^-$ removal from the mucosal side. *Top bar graphs:* $Na^+$ removal; *bottom bar graphs:* $Cl^-$ removal. Values in $Na^+$- or $Cl^-$-free media were measured at 3 minutes. Bar graphs on the right-hand side are the initial rates of fall of activities. Note that there are large differences in the effects of removal of one ion on the two intracellular activities and on the initial rate of fall. Reproduced from Reuss,[18] with permission of *Physiological Reviews*.

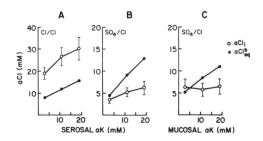

**FIGURE 4.** Effects of changes in serosal or mucosal solution [K$^+$] on intracellular Cl$^-$ activity (aCl$_i$). The control serosal bathing solution was Na-Ringer's (NaCl main salt) with 3.5 mM K$^+$. [K$^+$] was increased by isomolar Na$^+$ substitution. Main anions in the mucosal and serosal side, respectively, are indicated at the top of each panel. The measured intracellular Cl$^-$ activity (aCl$_i$, *open symbols*) and the calculated Cl$^-$ equilibrium activity across basolateral membrane (aCl$_{eq}^b$, *filled symbols*) are plotted as functions of K$^+$ activity (aK) in the serosal (**A,B**) or mucosal side (**C**). Means or means ± SEM are shown. (**A**) During bilateral exposure to Cl$^-$ media, increasing serosal aK from 2.6 to either 11 or 19 mM significantly increased aCl$_i$ ($p < 0.01$ and $p < 0.025$, respectively, $n = 5$). (**B**) In the absence of mucosal Cl$^-$ (SO$_4$ replacement), increasing serosal aK resulted in significant, but smaller increases of aCl$_i$ ($p < 0.05$, $n = 6$). (**C**) Increasing mucosal aK in the absence of Cl$^-$ on that side had no effect on aCl$_i$. Reproduced from Reuss,[26] with permission of *Nature*.

state rate of transepithelial Cl$^-$ transport, indicating that an electrically silent Cl$^-$ transport pathway must exist in parallel with the Cl$^-$ conductance. Studies in *Necturus* gallbladder have demonstrated that this parallel pathway is a KCl cotransporter.[25,26] Experimental observations supporting this conclusion are illustrated in FIGURE 4.

Under control conditions, both cell membranes exhibit predominant K$^+$ conductive pathways. Recent single-channel studies with the patch-clamp technique

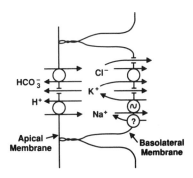

**FIGURE 5.** Transepithelial ion transport model for *Necturus* gallbladder. Na$^+$ and Cl$^-$ enter at the same steady-state rate by apical membrane exchange with H$^+$ and HCO$_3^-$, respectively. Basolateral Na$^+$ exit is via the Na$^+$-K$^+$ pump; Cl$^-$ exit is principally by KCl cotransport, but also via a channel. There are K$^+$-selective channels in both membranes; K$^+$ recycles predominantly at the basolateral membrane, which has a much higher K$^+$ conductance. Basolateral Na$^+$ recycling is likely. Not depicted, there is an apical Na$^+$ conductive pathway whose role in Na$^+$ entry under control conditions is small.

indicate that at the apical membrane the dominant channel is a voltage and $Ca^{2+}$-gated maxi-$K^+$ channel.[27] The basolateral $K^+$-conductive pathway has not been characterized yet at the single-channel level. Both membranes have other ionic permeabilities in parallel with the $K^+$ channels, which account for the difference between the cell membrane voltage and the $K^+$ equilibrium potential. At the basolateral membrane a $Cl^-$ conductive pathway exists; at the apical membrane, ion substitution experiments suggest a $Na^+$ conductance that is amiloride insensitive. Patch-clamp experiments have not revealed $Na^+$ channels.

The major ion transport mechanisms across the cell membranes of *Necturus* gallbladder epithelium are summarized in FIGURE 5.

## MECHANISM OF TRANSEPITHELIAL WATER TRANSPORT

Optical and electrophysiologic techniques have permitted direct estimates of the hydraulic permeability coefficients of apical and basolateral membranes of *Necturus* gallbladder epithelium. A major aim of these studies has been to assess in the transporting tissue the magnitude of the steady-state differences in osmolality required to account for the measured rates of transepithelial water transport, according to the relationship

$$J_v = L_p \sigma \Delta\pi \tag{1}$$

where $J_v$ is the fluid transport rate ($cm \cdot sec^{-1}$), $L_p$ is the hydraulic permeability coefficient ($cm \cdot sec^{-1} [osmol/kg]^{-1}$), $\sigma$ is the mean solute reflection coefficient, and $\Delta\pi$ is the difference in osmolality ([osmol/kg]).

The experimental approach consists in exposing rapidly one of the cell surfaces to a known change in external osmolality ($\Delta\pi$), to measure the change in cell volume (from which the fluid flow, $J_v$, can be estimated), and to calculate $L_p$ from Eq. 1.

Spring and Hope[28] used a computerized optical-sectioning technique to estimate changes in cell volume. Cell cross-sections were obtained at fixed distances across the epithelium and the images were digitized, stored, and later reconstructed to yield the time course of the change in cell volume. Acquisition of data required to calculate the volume of the whole cell takes several seconds. The time resolution of this method can be improved by calculating the change in volume of a single cell slice,[29] but validation of the latter technique is difficult.[8]

An alternative technique is to use an intracellular volume marker whose concentration can be measured with an intracellular microelectrode. We have used the impermeant ion tetramethylammonium ($TMA^+$), which is not transported by the native cell membranes of *Necturus* gallbladder epithelium. The cells can be loaded with the cation by transient exposure to the pore-forming antibiotic nystatin, which increases reversibly the cation permeability of the apical membrane.[30] The intracellular $TMA^+$ concentration can be rapidly and accurately monitored with $K^+$-sensitive intracellular microelectrodes, inasmuch as their selectivity for $TMA^+$ over $K^+$ is almost three orders of magnitude.[30]

This technique has been used to estimate changes in cell volume during transient exposure of the apical surface of the epithelium to anisosmotic solutions.[8] To assess the time course of the change in osmolality at the cell surface, which is slowed down by mixing in the chamber and diffusion of the osmotic probe in the unstirred layer, we have used the tracer tetrabutylammonium (TBA⁺), which can be measured with an extracellular K⁺-sensitive microelectrode. This cation was chosen because its diffusion coefficient is similar to that of sucrose, the osmotic probe in these experiments.[31] Typical traces of surface osmolality and simultaneous time courses of changes in cell volume upon an increase or a reduction in bathing solution osmolality are shown in FIGURE 6. Appropriate modeling of the system allows for calculation of the $L_p$ of the individual cell membranes.

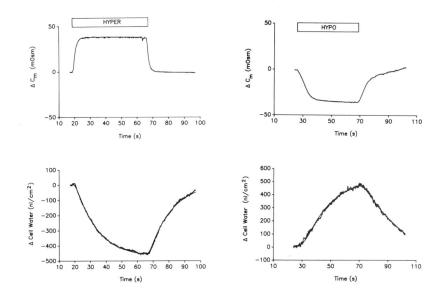

**FIGURE 6.** Changes in mucosal solution osmolality, cell water volume, and the fitted change in cell water volume in response to exposure to a hyperosmotic mucosal solution (*left*) or a hyposmotic mucosal solution (*right*). Changes in osmolality at the cell surface were measured using TBA⁺ as a "marker" for sucrose concentration changes. Cell volume changes were calculated from the changes in intracellular [TMA⁺] measured with a microelectrode. In the hyperosmotic experiment, the mucosal solution osmolality was increased from 206 to 244 mOsmol/kg water and the [TBA⁺] from 1 to 2 mM. In the hyposmotic experiment, the mucosal solution osmolality was decreased from 206 to 163 mOsmol/kg water and the [TBA⁺] was reduced from 2 to 1 mM. The cell volume data in both experiments were fit to a water transport model,[8] to calculate the apparent $L_p$ values of the cell membranes. Reproduced in part from Cotton *et al.*,[8] with permission of *The Journal of General Physiology.*

Our results, and those obtained with the microscopic technique, indicate that both cell membranes have high $L_p$ values of about $1 \cdot 10^{-3}$ cm·sec⁻¹ [osmol/kg]⁻¹. From the average spontaneous fluid transport rate we calculate that transcellular fluid transport requires very small steady-state differences in osmolality of 1 to 3 mosmol/kg, across the cell membranes. These results suggest strongly that water absorption by gallbladder epithelium is transcellular and driven by small osmotic forces generated by salt transport.

## REGULATION OF ION TRANSPORT: ROLE OF $HCO_3^-/CO_2$

In gallbladders from both mammals and amphibians exposure to $HCO_3^-/CO_2$ media, at constant extracellular pH, causes an increase in the rate of transepithelial fluid absorption which is in part attributable to absorption of $NaHCO_3^-$, but which involves also stimulation of NaCl absorption.[3,32,33] Under these conditions, the catalytic action of carbonic anhydrase clearly plays a role in enhancing apical membrane NaCl entry via $Na^+/H^+$ and $Cl^-/HCO_3^-$ exchangers.[34] The increased availability of intracellular substrates for cation and anion exchange, namely, $H^+$ and $HCO_3^-$, would in principle suffice to account for the stimulation of entry of both $Na^+$ and $Cl^-$. In agreement with this view, the rate of intracellular alkalinization upon removal of apical solution $Cl^-$ is higher in the presence than in the nominal absence of $HCO_3^-$, although the intracellular buffering power is much higher when the tissue is incubated in media buffered with $HCO_3^-/CO_2$.[2]

In the steady state, however, elevated rates of transepithelial salt transport must involve not only increased entry of salt across the apical membrane, but also increased effluxes of both $Na^+$ and $Cl^-$ across the basolateral membrane.[35] The mechanisms of these processes have not been fully established, but several lines

TABLE 1. Equivalent Circuit Parameters in Gallbladders Incubated in HEPES and $HCO_3^-$-Buffered Solutions

| Solution | $V_{mc}$ | $V_{cs}$ | $fR_a$ | $R_t$ |
|---|---|---|---|---|
| 1 mM HEPES | $-67 \pm 4$ | $-67 \pm 4$ | $0.48 \pm 0.08$ | $210 \pm 20$ |
| 10 mM $HCO_3^-$/1% $CO_2$ | $-68 \pm 1$ | $-68 \pm 1$ | $0.87 \pm 0.06^a$ | $160 \pm 10$ |

[a] Significantly different from value in 1 mM HEPES.

$V_{mc}$ and $V_{cs}$ are the apical and basolateral membrane voltages (in mV). $fR_a$ is the fractional resistance of the apical membrane. (See text.) $R_t$ is the transepithelial electrical resistance, in $\Omega \cdot cm^2$.

Values are mean $\pm$ SEM from seven experiments in each group.

Modified from Stoddard and Reuss.[6]

of evidence suggest that elevations of $HCO_3^-/CO_2$ concentrations in the bathing media cause stimulation of the $Na^+$ pump and increases in basolateral membrane $Cl^-$ and $K^+$ conductances that would suffice to account for stimulation of basolateral NaCl transport and $K^+$ recycling.

TABLE 1 summarizes recent results from equivalent circuit analysis in *Necturus* gallbladders studied under steady-state conditions in media buffered with either 1 mM HEPES or 10 mM $HCO_3^-$/1% $CO_2$. The fractional apical membrane resistance $(fR_a)$, defined as $R_a/(R_a + R_b)$, where $R_a$ and $R_b$ are the apical and basolateral resistances, respectively, was significantly increased in the $HCO_3^-$/$CO_2$ medium, although the steady-state membrane voltages were not different. Cable analysis experiments showed that the elevation of $fR_a$ was due to both an increase in $R_a$ and a decrease in $R_b$, compared with values obtained in several previous studies in low-$HCO_3^-$ media.[6] The fall in basolateral membrane resistance is due to increases in both $K^+$ and $Cl^-$ electrodiffusive permeabilities. FIGURES 7 and 8 show the changes in cell membrane voltages elicited by elevating basolateral solution $[K^+]$ from 2.5 to 25 mM, and by reducing basolateral solution $[Cl^-]$ from 98 to 8 mM, respectively. Although the basolateral membrane voltage is more dependent on external $[K^+]$ than on external $[Cl^-]$, it is clear that under these conditions the electrodiffusive $Cl^-$ permeability is sizable. Calculations based on the driving forces for passive basolateral $Cl^-$ transport and the estimated

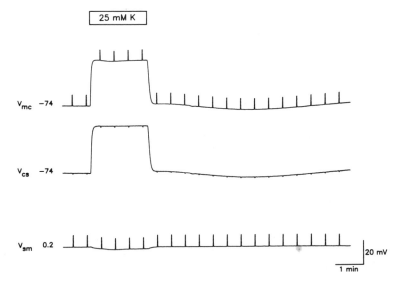

**FIGURE 7.** Effect of serosal $K^+$ Ringer's solution (25 mM) on transepithelial ($V_{sm}$) and cell membrane voltages ($V_{mc}$ and $V_{cs}$) in tissues incubated in 10 mM $HCO_3^-/1\%$ $CO_2$ Ringer's solution. Initial voltages are given to the left of each trace. Vertical deflections are the result of 1-second transepithelial current pulses (50 $\mu$A/cm²) applied at 30-second intervals to obtain resistance estimates. Note the long time course of hyperpolarization of cell membrane voltages following return to control after the exposure to $K^+$ Ringer's solution. Reproduced from Stoddard and Reuss,[6] with permission of *The Journal of Membrane Biology*.

change in $P_{Cl}$ suggest that the change in permeability can account completely for the stimulation of basolateral $Cl^-$ efflux by $HCO_3^-/CO_2$.[6,7] Because of the simultaneous changes in basolateral membrane $P_K$ and $P_{Cl}$, during exposure to $HCO_3^-/CO_2$-buffered solutions there is little change in resting basolateral membrane volt-

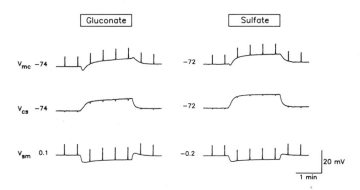

**FIGURE 8.** Effects of serosal solution $Cl^-$ reduction (98.1 to 8.1 mM) on $V_{mc}$, $V_{cs}$, and $V_{sm}$ using gluconate and sulfate as $Cl^-$ substitutes. The voltage records were obtained from the same cell impalement as in FIGURE 7. Format as in FIGURE 7. Reproduced from Stoddard and Reuss,[6] with permission of *The Journal of Membrane Biology*.

age. This fact facilitates conductive efflux, via the $K^+$ channels, of the excess $K^+$ transported into the cell by the stimulated $Na^+$ pump. If the sole effect of exposure to $HCO_3^-/CO_2$ media were to increase $P_K$, the basolateral membrane would hyperpolarize and the driving force for conductive $K^+$ efflux would be reduced.[6]

## REGULATION OF ION TRANSPORT: ROLE OF CYCLIC AMP

It is well known that elevating intracellular cAMP levels in salt-absorbing epithelia causes a decrease in NaCl uptake across the apical membrane, and in some cases a reversal of the direction of net fluid transport from absorption to secretion. On the basis of measurements of transepithelial and apical membrane $Na^+$ and $Cl^-$ unidirectional tracer fluxes in mammalian gallbladder, this effect was attributed to inhibition of apical membrane NaCl cotransport.[36] In agreement with this notion, intracellular microelectrode studies in *Necturus* gallbladder showed cAMP-induced falls in the intracellular activities of both $Na^+$ and $Cl^-$, with no appreciable changes in membrane voltages.[37] These results have not been confirmed. Instead, other investigators observed that cAMP treatment elicited a moderate depolarization of both cell membranes, and a large decrease of the apparent

**TABLE 2.** Evidence for cAMP-Induced Increase in Apical Membrane Electrodiffusive $P_{Cl}$

| Observation | References |
| --- | --- |
| Depolarization of both cell membranes to voltages near $E_{Cl}$ | 9, 10, 11, 37 |
| Large reduction in the apparent ratio of cell membrane resistances (apical: basolateral) | 9, 10, 11, 37 |
| Fall in intracellular $Cl^-$ activity | 9, 11 |
| Rapid, transient apical membrane depolarization upon lowering apical solution $[Cl^-]$ | 9, 11, 21 |
| Patch-clamp demonstration of apical membrane $Cl^-$ channels | 39 |

ratio of cell membrane resistances (apical:basolateral).[38] We confirmed these results and showed that the mechanism is a large increase in apical membrane electrodiffusive $Cl^-$ permeability.[9] The arguments supporting this conclusion are summarized in TABLE 2. The most convincing one is the comparison of the magnitude and time course of the changes in cell membrane voltages elicited by transient reductions in apical solution $[Cl^-]$ under control conditions and during elevations of cAMP levels (FIG. 9). Similar effects are obtained during treatment with 8-Br-cAMP, or upon exposure to the adenylate cyclase activator forskolin or the phosphodiesterase inhibitor theophylline.[9,10] We also showed that cAMP causes a fall of intracellular $Cl^-$ activity to electrochemical-equilibrium values (FIG. 10). The simplest interpretation of these results is that the $Cl^-$ entering the cells via anion exchange leaks from cytosol to apical solution via the cAMP-induced conductive pathway. Clearly, the fall in cell $Cl^-$ activity will reduce basolateral $Cl^-$ efflux, both electrodiffusive and by cotransport.

Somewhat surprisingly, however, we found that reductions in apical solution $[Cl^-]$ under control conditions and in cAMP-stimulated cells resulted in falls of intracellular $Cl^-$ activities at nearly identical rates, although in cAMP a faster efflux would be expected from the sum of electrodiffusive and exchange fluxes. Typical records of the time courses of intracellular $Cl^-$ activities in such experi-

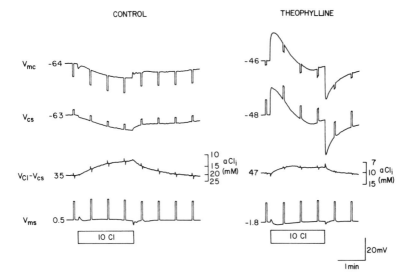

**FIGURE 9.** Effects of reducing mucosal solution [Cl⁻] to 10 mM on membrane voltages, resistances, and aClᵢ, under control conditions and in theophylline. Both sets of traces were obtained in the same preparation before and ~20 minutes after exposure to theophylline. Under control conditions, the low-Cl⁻ medium causes hyperpolarization of both cell membranes, little change in $V_{ms}$, $R_t$, or $R_a/R_b$, and a rapid fall of aClᵢ. In theophylline, the same maneuver causes rapid depolarization of both cell membranes, followed by repolarization during the period of exposure to the low-Cl⁻ medium, and a spiking, transient hyperpolarization upon returning to control solution. In contrast with the control traces, $R_a/R_b$ is very low in theophylline and rises when luminal [Cl⁻] is lowered. The initial rates of fall of aClᵢ are similar in the two conditions. Reproduced from Reuss,[11] with permission of *The Journal of General Physiology.*

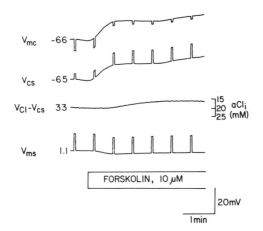

**FIGURE 10.** Effects of forskolin ($10^{-5}$ M, mucosal side) on membrane voltages and resistances and intracellular Cl⁻ activity (aClᵢ) in a tissue incubated in control Ringer's solution. Theophylline causes a slow depolarization of both cell membranes, a mucosal side-negative change in transepithelial voltage, an increase in $R_t$, a decrease in $R_a/R_b$, and a fall in aClᵢ toward equilibrium levels. Reproduced from Reuss,[11] with permission of *The Journal of General Physiology.*

379

**TABLE 3.** Effects of cAMP on Apical Membrane $Cl^-/HCO_3^-$ Exchange[a]

| Observation | Control | cAMP |
|---|---|---|
| $Cl^-$-dependent mucosal alkalinization rate[b] | $-2.3 \pm 0.6$ | $-0.8 \pm 0.5$[c] |
| Initial rate of change of $aCl_i$ via anion exchange[d] | $-11.6$ | $-5.1$ |
| Initial rate of change of $aHCO_{3i}$ upon reducing $[Cl^-]_o$ to 10 mM | $14.3 \pm 3.4$ | $8.7 \pm 1.9$[c] |
| Initial rate of change of $aCl_i$ upon reducing $[HCO_3]_o$ to 1 mM | $4.5 \pm 1.2$ | $1.7 \pm 0.2$[c] |
| Initial rate of change of $aHCO_{3i}$ upon reducing $[HCOl_3^-]_o$ to 1 mM | $6.2 \pm 0.8$ | $-4.0 \pm 0.8$[c] |

[a] Results are summarized from ref. 11. Values shown are mean or mean $\pm$ SEM. Rates of change in intracellular activities ($aCl_i$, $aHCO_{3i}$) are in mM/min.

[b] Fluxes normalized to control acidification rate in NaCl-Ringer's solution.

[c] Significantly different from control, paired comparison.

[d] Calculated as the difference between $daCl_i/dt$ upon lowering $[Cl]_o$ to 10 mM and upon lowering both $[Cl]_o$ and $[HCO_3^-]_o$, to 10 mM and 1 mM, respectively.

mental conditions are shown at the bottom of FIGURE 9. Additional studies revealed that in cAMP-treated tissues the $HCO_3^-$ efflux from cell to mucosal solution is decreased. The cell alkalinization elicited by lowering external $Cl^-$ (i.e., caused by reversal of the fluxes via the anion exchanger) was also decreased. Finally, the effects of lowering apical solution $HCO_3^-$ concentration on both intra-

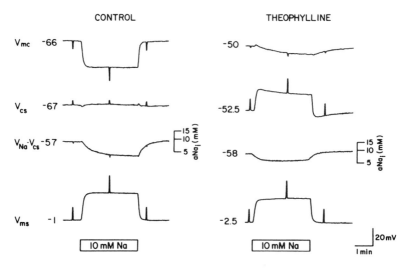

**FIGURE 11.** Effect of theophylline on the change in $aNa_i$ produced by lowering mucosal $[Na^+]$. Both sets of traces were obtained from the same tissue under control conditions and ~30 minutes after serosal addition of theophylline, respectively. During the periods indicated by the lower bars, the mucosal solution $[Na^+]$ was transiently decreased from 100 to 10 mM. Note the differences in the effects of lowering $[Na^+]$ on $V_{mc}$, $V_{cs}$, and $V_{ms}$ (control $vs$ theophylline), which are caused by the high apical membrane $P_{Cl}$ in theophylline. The change in $aNa_i$ was smaller and slower in theophylline as compared with control. Reproduced from Reuss and Petersen,[10] with permission of *The Journal of General Physiology*.

cellular pH and intracellular $Cl^-$ activity were also decreased by cAMP. These results are summarized in TABLE 3. Taken together, they indicate that cAMP inhibits anion exchange by a mechanism that cannot be ascribed to the small steady-state change in intracellular pH. Kinetic analysis of unidirectional $Cl^-$ entry from the apical solution suggests strongly that the effect of cAMP is due to a decrease in $V_{max}$ to about 50% of the control value.[11] In conclusion, then, inhibition of net apical $Cl^-$ influx in cAMP-treated tissues is the result of two processes: decrease in unidirectional influx via anion exchange, and increase in conductive $Cl^-$ efflux. The pathway of the latter flux is a channel, as evidenced by patch-clamp experiments.[39] The $Cl^-$ channel is not found in unstimulated cells.

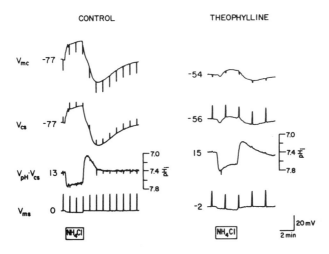

**FIGURE 12.** Recovery of $pH_i$ from an acid load under control conditions and in theophylline. During exposure to 25 mM $NH_4Cl$ (3 minutes, indicated by the bars) the cells alkalinize rapidly ($NH_3$ entry) and then $pH_i$ recovers somewhat ($NH_4^+$ entry), while the cell membranes depolarize. Upon $NH_4Cl$ removal, $pH_i$ falls below its control value (exit of both intracellular $NH_3$ and $NH_4^+$ in the form of $HN_3$) and then recovers. Simultaneously, both cell membranes transiently hyperpolarize. Note the differences between the control traces and those in theophylline, that is, the smaller changes in cell membrane voltages and the slower $pH_i$ recovery after $NH_4Cl$ removal. Reproduced from Reuss and Petersen,[10] with permission of *The Journal of General Physiology.*

These results do not explain why cAMP reduces transepithelial $Na^+$ transport. To address this question, we measured the rate of apical membrane $Na^+/H^+$ exchange under a variety of conditions.[10] These studies showed that cAMP causes a decrease in $Na^+$-dependent, amiloride-sensitive $H^+$ transport from the cells to the apical solution, inhibits the rate of $Na^+$ entry via $Na^+/H^+$ exchange (FIG. 11), and decreases the rate of $Na^+$-dependent $pH_i$ recovery upon intracellular acidification (FIG. 12). These results are summarized in TABLE 4. As observed for the inhibition of anion exchange, the main effect of cAMP on $Na^+/H^+$ exchange was a reduction in $V_{max}$ to about half of control.[10] The inhibition of $Na^+/H^+$ exchange by cAMP cannot be attributed to the change in intracellular pH, because the observed small intracellular acidification would be expected to stimulate this process. It is tempting to speculate that the similarity of the effects of cAMP on $Na^+/H^+$ exchange and on $Cl^-/HCO_3^-$ exchange (about 50% reduction in $V_{max}$ in both cases) could be due to removal of transporters from the apical membrane by endocytosis. We are currently testing this possibility.

Our results appear to provide a complete picture of the effects of cAMP on the apical membrane of this epithelium, but there is less experimental evidence pertaining to the necessary effects on basolateral $Na^+$ and $Cl^-$ fluxes. It is possible to account for the reduction in $Cl^-$ efflux on the basis of the fall in intracellular $Cl^-$ activity, but it is difficult to explain the inhibition of $Na^+$ transport on the same grounds, inasmuch as the steady-state reduction in cell $Na^+$ activity is modest, typically from about 10 to about 8 mM.[10] It is conceivable that, directly or indirectly, cAMP causes reductions in $Na^+$ pump activity and in $K^+$ permeability of the basolateral membrane. In fact, there is some evidence supporting a fall in $P_K$.[40] A decrease in cell volume brought about by the apical membrane events is an attractive hypothesis to account for the pari-passu reduction in basolateral transport, but further experiments will be necessary to test this possibility.

**TABLE 4.** Effects of cAMP on Apical Membrane $Na^+/H^+$ Exchange[a]

| Observation | Control | cAMP |
|---|---|---|
| Apparent $H^+$ flux from cell to apical medium ($nmol \cdot cm^{-2} \cdot min^{-1}$)[b] | $1.8 \pm 0.3$ | $0.9 \pm 0.2^c$ |
| Steady-state intracellular $Na^+$ activity (mM) | $10.0 \pm 0.7$ | $8.7 \pm 0.7^c$ |
| Steady-state intracellular pH | $7.40 \pm 0.04$ | $7.30 \pm 0.05^c$ |
| Initial rate of change of $aNa_i$ upon reducing $[Na^+]_o$ to 10 mM | $-5.7 \pm 1.1$ | $-2.8 \pm 0.7^c$ |
| Initial rate of change of $aNa_i$ upon exposure to 1 mM amiloride | $-2.4 \pm 0.4$ | $-1.3 \pm 0.4^c$ |
| Initial rate of $pH_i$ recovery in 100 mM $Na^+$ | $0.16 \pm 0.03$ | $0.08 \pm 0.01^c$ |
| Initial rate of $pH_i$ recovery in 10 mM $Na^+$ | $0.09 \pm 0.01$ | $0.05 \pm 0.01^c$ |

[a] Results summarized from ref. 10. Values shown are mean $\pm$ SEM. Rates of change in intracellular $Na^+$ ($aNa_i$) are in mM/min, and rates of change of $pH_i$ in $min^{-1}$.

[b] Apparent fluxes estimated from the volume and rate of change in $pH_o$ of a mucosal solution buffered with 1 mM HEPES.

[c] Significantly different from control, paired comparison.

## ACKNOWLEDGMENTS

I wish to acknowledge the major contributions of C. U. Cotton, K.-U. Petersen. Y. Segal, J. S. Stoddard, and S. A. Weinman to the work reviewed in this article. I thank also Margaret Jost and Jennifer Chilton for software development, Jozianne Bazile, James L. Costantin, and Margaret Simpson for technical assistance, and Lynette Morgan for secretarial help.

## REFERENCES

1. WEINMAN, S. A. & L. REUSS. 1984. $Na^+$-$H^+$ exchange and $Na^+$ entry across the apical membrane of *Necturus* gallbladder. J. Gen. Physiol. **83:** 57–74.
2. REUSS, L. & J. L. CONSTANTIN. 1984. $Cl^-/HCO_3^-$ exchange at the apical membrane of *Necturus* gallbladder. J. Gen. Physiol. **83:** 801–818.
3. REUSS, L. 1984. Independence of apical membrane $Na^+$ and $Cl^-$ entry in *Necturus* gallbladder epithelium. J. Gen. Physiol. **84:** 423–445.

4. BAERENTSEN, H., F. GIRALDEZ & T. ZEUTHEN. 1983. Influx mechanisms for $Na^+$ and $Cl^-$ across the brush border membrane of leaky epithelia: A model and microelectrode study. J. Membr. Biol. **75:** 205–218.

5. REUSS, L. 1979. Electrical properties of the cellular transepithelial pathway in *Necturus* gallbladder. III. Ionic permeability of the basolateral cell membrane. J. Membr. Biol. **47:** 239–259.

6. STODDARD, J. S. & L. REUSS. 1988. Dependence of cell membrane conductances on bathing solution $HCO_3^-/CO_2$ in *Necturus* gallbladder. J. Membr. Biol. **102:** 163–174.

7. STODDARD, J. S. & L. REUSS. 1989. Electrophysiological effects of mucosal $Cl^-$ removal in *Necturus* gallbladder epithelium. Am J. Physiol.:Cell Physiol. **257:** C568–C578.

8. COTTON, C. U., A. M. WEINSTEIN & L. REUSS. 1989. Osmotic water permeability of *Necturus* gallbladder epithelium. J. Gen. Physiol. **93:** 649–679.

9. PETERSEN, K.-U. & L. REUSS. 1983. Cyclic AMP-induced chloride permeability in the apical membrane of *Necturus* gallbladder epithelium. J. Gen. Physiol. **81:** 705–709.

10. REUSS, L. & K.-U. PETERSEN. 1985. Cyclic AMP inhibits $Na^+/H^+$ exchange at the apical membrane of *Necturus* gallbladder epithelium. J. Gen. Physiol. **85:** 409–425.

11. REUSS, L. 1987. Cyclic AMP inhibits $Cl^-/HCO_3^-$ exchange at the apical membrane of *Necturus* gallbladder epithelium. J. Gen. Physiol. **90:** 172–196.

12. REUSS, L. & J. S. STODDARD. 1987. Role of $H^+$ and $HCO_3^-$ in salt transport in gallbladder epithelium. Ann. Rev. Physiol. **49:** 35–49.

13. DIAMOND, J. M. 1968. Transport mechanisms in the gallbladder. *In* Handbook of Physiology: Section 6, Alimentary Canal, Bile, Digestion, Ruminal Physiology. W. Heidel & C. F. Cole, eds. Vol. V: 2451–2482. Am. Physiol. Soc., Washington, DC.

14. FRIZZELL, R. A. & K. HEINTZE. 1980. Transport functions of the gallbladder. *In* International Review of Physiology; Liver and Biliary Tract Physiology I. N. B. Javitt, ed. :221–247. University Park Press, Baltimore.

15. REUSS, L. 1979. Transport in gallbladder. *In* Membrane Transport in Biology. Transport Organs. G. Giebisch, D. C. Tosteson & H. H. Ussing, eds. Vol IVB:853–898. Springer-Verlag, Berlin.

16. CREMASCHI, D., G. MEYER, C. ROSSETTI, G. BOTTA & P. PALESTINI. 1987. The nature of the neutral $Na^+$-$Cl^-$ coupled entry at the apical membrane of rabbit gallbladder epithelium. I. $Na^+/H^+$, $Cl^-/HCO_3^-$ double exchange and $Na^+$-$Cl^-$ symport. J. Membr. Biol. **95:** 209–218.

17. ERICSON, A.-C. & K. R. SPRING. 1982. Coupled NaCl entry into *Necturus* gallbladder epithelial cells. Am. J. Physiol. **243:** C140–C145.

18. REUSS, L. 1989. Ion transport across gallbladder epithelium. Physiol. Rev. **69:** 503–545.

19. WEINMAN, S. A. & L. REUSS. 1982. $Na^+$-$H^+$ exchange at the apical membrane of *Necturus* gallbladder. Extracellular and intracellular pH studies. J. Gen. Physiol. **80:** 299–321.

20. ZEUTHEN, T. & T. MACHEN. 1984. $HCO_3^-/CO_2$ stimulates $Na^+/H^+$ and $Cl^-/HCO_3^-$ exchange in *Necturus* gallbladder. *In* Hydrogen Ion Transport in Epithelia. J. G. Forte, D. G. Warnock & F. C. Rector, Jr., Eds. :97–108. Wiley, New York.

21. REUSS, L., J. L. CONSTANTIN & J. E. BAZILE. 1987. Diphenylamine-2-carboxylate blocks $Cl^-$-$HCO_3^-$ exchange in *Necturus* gallbladder epithelium. Am. J. Physiol. **253:** C79–C89.

22. DAVIS, C. W. & A. L. FINN. 1985. Effects of mucosal sodium removal on cell volume in *Necturus* gallbladder epithelium. Am. J. Physiol. **249:** C304–C312.

23. LARSON, M. & K. R. SPRING. 1983. Bumetanide inhibition of NaCl transport by *Necturus* gallbladder. J. Membr. Biol. **74:** 123–129.

24. FISHER, R. S. 1984. Chloride movement across basolateral membrane of *Necturus* gallbladder epithelium. Am. J. Physiol. **247:** 495–500.

25. CORCIA, A. & W. McD. ARMSTRONG. 1983. KCl cotransport: A mechanism for basolateral chloride exit in *Necturus* gallbladder. J. Membr. Biol. **76:** 173–182.

26. REUSS, L. 1983. Basolateral KCl co-transport in a NaCl-absorbing epithelium. Nature (Lond.) **305:** 723–726.

27. SEGAL, Y. & L. REUSS. 1988. Large conductance $K^+$ channels in *Necturus* gallbladder epithelium. J. Gen. Physiol. **92:** 34a.
28. SPRING, K. R. & A. HOPE. 1979. Fluid transport and the dimensions of cells and interspaces of living *Necturus* gallbladder. J. Gen. Physiol. **73:** 287–305.
29. PERSSON, B.-E. & K. R. SPRING. 1982. Gallbladder epithelial cell hydraulic water permeability and volume regulation. J. Gen. Physiol. **79:** 481–505.
30. REUSS, L. 1985. Changes in cell volume measured with an electrophysiologic technique. Proc. Natl. Acad. Sci. USA **82:** 6014–6018.
31. COTTON, C. U. & L. REUSS. 1989. Measurement of the effective thickness of the mucosal unstirred layer in *Necturus* gallbladder epithelium. J. Gen. Physiol. **93:** 631–647.
32. DIAMOND, J. M. 1964. Transport of salt and water in rabbit and guinea pig gallbladder. J. Gen. Physiol. **48:** 1–14.
33. HEINTZE, K., K.-U. PETERSEN, P. OLLES, S. H. SAVERYMUTTU & J. R. WOOD. 1979. Effects of bicarbonate on fluid and electrolyte transport by the guinea pig gallbladder: A bicarbonate-chloride exchange. J. Membr. Biol. **45:** 43–59.
34. PERSSON, B. E. & M. LARSON. 1986. Carbonic anhydrase inhibition and cell volume regulation in *Necturus* gallbladder. Acta Physiol. Scand. **128:** 501–507.
35. SCHULTZ, S. G. 1981. Homocellular regulatory mechanisms in sodium-transporting epithelia: Avoidance of extinction by "flush-through." Am. J. Physiol. **241:** F579–F590.
36. FRIZZELL, R. A., M. C. DUGAS & S. G. SCHULTZ. 1975. Sodium chloride transport by rabbit gallbladder. Direct evidence for a coupled NaCl influx process. J. Gen. Physiol. **65:** 769–795.
37. DIEZ DE LOS RIOS, A., N. E. DeROSE & W. McD. ARMSTRONG. 1981. Cyclic AMP and intracellular ionic activities in *Necturus* gallbladder. J. Membr. Biol. **63:** 25–30.
38. DUFFEY, M. E., B. HAINAU, S. HO & C. J. BENTZEL. Regulation of epithelial tight junction permeability by cyclic AMP. Nature (Lond.) **294:** 451–453.
39. SEGAL, Y. & L. REUSS. 1989. $Cl^-$ channels in cyclic AMP-stimulated gallbladder epithelium. FASEB J. **3:** A862.
40. ZELDIN, D. C., A. CORCIA & W. McD. ARMSTRONG. 1985. Cyclic AMP-induced changes in membrane conductance of *Necturus* gallbladder epithelial cells. J. Membr Biol. **84:** 193–206.

# Ionic Mechanisms of Cell Volume Regulation in Isolated Rat Hepatocytes

JAMES G. CORASANTI, DERMOT GLEESON,
AND JAMES L. BOYER

*Department of Medicine and Liver Center*
*Yale University School of Medicine*
*New Haven, Connecticut 06510*

When exposed to anisotonic conditions, most cells behave initially as osmometers, but subsequently they regulate their volumes back toward the resting (isotonic) level by activating a variety of ion transport processes.[1] Regulatory volume decrease (RVD) and regulatory volume increase (RVI) are the terms used to describe recovery from hypotonic swelling and hypertonic shrinkage, respectively. Although RVD has been demonstrated in hepatocytes,[2-4] its mechanisms are far from clear, and a regulatory response to hypertonic stress (RVI) has not yet been demonstrated in liver cells.

## METHODS AND RESULTS

In the present study we examine the effects of hypo- and hypertonic stresses on cell volumes of isolated hepatocyte suspensions using a Coulter counter method modified from that of Grinstein *et al.*[5] We also examine changes in intracellular pH (pH$_i$) in subconfluent hepatocyte monolayers using the fluorescent dye BCECF. In isotonic (300 mOsm) buffer, median cell volume was 5,720 ± 460 $\mu$m$^3$ (mean ± SD) ($n = 28$). Following exposure to hypotonic medium (160 mOsm), cells swelled to 9,614 ± 310 $\mu$m$^3$ and within 20 minutes underwent RVD to 7,710 ± 846 $\mu$m$^3$ ($n = 16$) (FIG. 1); RVD was bicarbonate independent but was inhibited by barium (1 mM), quinine (0.5 mM), DIDS (0.05 and 0.5 mM), and high extracellular K$^+$ (65 mM). Ouabain and sodium removal had little effect on RVD. In addition, RVD was associated with a decrease in pH$_i$ from 6.98 ± 0.12 to 6.85 ± 0.06 (TABLE 1, $n = 10$) which was sodium dependent, bicarbonate independent, and amiloride sensitive.

After RVD, on reexposure to isotonic medium, cells shrunk further to 5,121 ± 535 $\mu$m$^3$ and then underwent RVI to 5,958 ± 650 $\mu$m$^3$ (FIG. 1). This RVI was bicarbonate, sodium, and (partially) chloride dependent and amiloride sensitive and was associated with a sodium-dependent, amiloride-sensitive, but bicarbonate- and chloride-independent increase in pH$_i$ from 6.84 ± 0.06 to 7.12 ± 0.15 (TABLE 1, $n = 11$). In contrast, cells transferred directly from iso- to hypertonic medium (550 mOsm) shrunk to 3,825 ± 112 $\mu$m$^3$ ($n = 6$) but did not undergo RVI; however, they exhibited a similar rise in pH$_i$.

**TABLE 1.** Effects of Osmotic Stress on Intracellular pH (pH$_i$)[a]

| | | pH$_i$ (Mean ± SD) | | | | | |
| | | Hypotonic Stress | | | Hypertonic Stress (relative) | | |
| Expt. | Buffer | Isotonic | Hypotonic[b] | (n) | Hypotonic | Isotonic | (n) |
|---|---|---|---|---|---|---|---|
| 1 | HEPES[c] | 6.98 ± 0.12 | 6.85 ± 0.06* | (10) | 6.84 ± 0.08 | 7.12 ± 0.15* | (11) |
| 2 | KRB[d] | 7.24 ± 0.11 | 7.17 ± 0.12** | (9) | 7.17 ± 0.11 | 7.34 ± 0.12* | (9) |
| 3 | Na-free HEPES | 6.57 ± 0.19 | 6.69 ± 0.22 | (4) | 6.62 ± 0.20 | 6.61 ± 0.21 | (3) |
| 4 | Na-free KRB | 6.66 ± 0.14 | 6.73 ± 0.15 | (2) | 6.73 ± 0.14 | 6.76 ± 0.12 | (2) |
| 5 | HEPES + amiloride | 6.68 ± 0.07 | 6.84 ± 0.05 | (3) | 6.78 ± 0.07 | 6.69 ± 0.11 | (5) |

[a] Intracellular pH (pH$_i$) was measured in subconfluent hepatocyte monolayers using the pH-sensitive dye BCECF. For hypotonic stress cells were perifused for 20 minutes with isotonic buffer (HEPES or KRB) to attain a stable pH$_i$ and then the perfusate was changed to hypotonic buffer and pH$_i$ measured for 20 minutes at 30-second intervals. For (relative) hypertonic stress, cells that had been exposed to hypotonic buffer for 20 minutes were then reexposed to isotonic buffer and pH$_i$ measured for an additional 20 minutes. The values given in the table are those recorded at t = 20 minutes because changes in pH$_i$ paralleled changes in volume (FIG. 1) and were maximal at that time point. Experiments 1 and 2 demonstrate that hypotonic stress results in intracellular acidification in both bicarbonate-free (HEPES) and bicarbonate-containing (KRB) buffers; in contrast, reexposure to isotonic buffer following hypotonically induced RVD resulted in intracellular alkalinization. Experiments 3, 4, and 5 demonstrate that both these osmotically induced changes in pH$_i$ were sodium dependent and amiloride sensitive, although these interventions result in a lower basal pH$_i$. These data suggest that Na/H exchange in rat hepatocytes is down- and up-regulated by hypo- and hypertonic stresses, respectively.

[b] Hypotonic buffer = isotonic HEPES or KRB minus 65 mM choline chloride.

[c] HEPES = (mM) NaCl 65, KCl 4.5, CaCl$_2$ 1.25, MgSO$_4$ 1.2, KH$_2$PO$_4$ 1.2, choline chloride 65, HEPES 20, pH 7.40 @ 37°C.

[d] KRB = (mM) NaCl 40, KCl 4.5, CaCl$_2$ 1.25, MgSO$_4$ 1.2, KH$_2$PO$_4$ 1.2, choline chloride 65, NaHCO$_3$ 25, gassed with 95% O$_2$/5% CO$_2$ to pH 7.40.

* $p < 0.005$ (hypotonic vs. isotonic); ** $p < 0.05$ (hypotonic vs. isotonic).

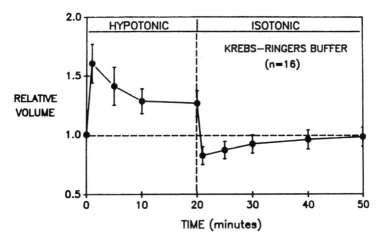

**FIGURE 1.** At t = 0 minutes isolated hepatocyte suspensions, preincubated for 20 minutes in warm isotonic Krebs-Ringer's buffer (KRB), were exposed to hypotonic KRB and median cell volumes measured at t = 1, 5, 10, and 20 minutes. At t = 20 minutes cells were reexposed to isotonic KRB and volume measurements made at t = 21, 25, 30, 40, and 50 minutes. Cell volumes are expressed as relative volume (measured volume/volume in isotonic medium) and represent mean ± SD of 16 pairs.

## CONCLUSIONS

These data suggest that RVD in rat hepatocytes depends on barium- and quinine-sensitive $K^+$ and DIDS-sensitive chloride channels, and is associated with down-regulation of the Na/H exchanger. RVI, seen only on reexposure to isotonic medium following hypotonically induced RVD, depends on activation of Na/H exchange coupled with a $Cl^-$ and $HCO_3^-$-dependent process.

### REFERENCES

1. EVELOFF, J. L. & D. G. WARNOCK. 1987. Activation of ion transport systems during cell volume regulation. Am. J. Physiol. **252:** F1–F10.
2. VAN ROSSUM, G. D. V. & M. A. RUSSO. 1981. Ouabain-resistant mechanism of volume control and the ultrastructural organization of liver slices recovering from swelling in vitro. J. Membr. Biol. **59:** 191–209.
3. BERTHON, B., M. CLARET, J. L. MAZET & J. POGGIOLI. 1980. Volume- and temperature-dependent permeabilities in isolated rat liver cells. J. Physiol. **305:** 267–277.
4. BAKKER-GRUNWALD, T. 1983. Potassium permeability and volume control in isolated rat hepatocytes. Biochim. Biophys. Acta **731:** 239–242.
5. GRINSTEIN, S., J. D. GOETZ, W. FURUYA, A. ROTHSTEIN & E. W. GELFAND. 1984. Amiloride-sensitive $Na^+$-$H^+$ exchange in platelets and leukocytes: Detection by electronic cell sizing. Am. J. Physiol. **247:** C293–C298.

# Characterization of Na$^+$/H$^+$ Exchange in the Murine Macrophage Cell Line J774.1

L. C. McKINNEY AND A. MORAN

*Armed Forces Radiobiology Research Institute*
*Bethesda, Maryland 20814*
*and*
*Ben-Gurion University of the Negev*
*Beer Sheva, Israel*

Characterization of the Na$^+$/H$^+$ exchanger has been carried out in several types of leukocytes and leukocyte cell lines, where it was shown to be activated by substances that modulate white cell function, including fmet-leu-phe (FMLP),[1] lipopolysaccharide (LPS),[2] concanavalin A,[3] phorbol ester,[4] and interleukin-2.[5] The purpose of this study was to demonstrate the existence of the Na$^+$/H$^+$ exchanger in a macrophage cell line and to determine the effect of several modulators of macrophage function on its activity. We used the murine cell line J774.1, whose functional similarities to macrophages have been extensively characterized.

pHi of suspended or adherent cells was measured using BCECF. We expressed fluorescence as the ratio of emission at 529 nm for excitation at 497 versus 437 nm. Calibration was carried out in 145 mM K Hanks' solution containing 3 $\mu$M nigericin + 3 $\mu$M valinomycin. Na Hanks' solution consisted of (in mM) 135 NaCl, 4.5 KCl, 1.6 CaCl$_2$, and 1.2 MgCl$_2$, and was buffered either with 20 HEPES (NaOH) or 21 NaHCO$_3$ (5% CO$_2$), pH 7.4. K Hanks' solution replaced NaCl with KCl; Na-free Hanks' solution contained N-methylglucamine (NMG).

Resting pHi values of suspended and adherent J774.1 cells in HEPES buffered Na Hanks' solution at 37°C were 7.53 ± 0.02 ($n$ = 86) and 7.59 ± 0.02 ($n$ = 97). Resting pHi values in the presence of bicarbonate/CO$_2$ were significantly lower (7.41 ± 0.02 [$n$ = 12] and 7.40 ± 0.01 [$n$ = 28], respectively, $p$ < 0.001). 100 $\mu$M amiloride did not affect resting pHi; average change in pHi for a 1–10-minute exposure was −0.02 ± 0.01 ($n$ = 4).

J774.1 cells recover from intracellular acidification in a Na- and amiloride-dependent fashion, indicating the presence of a Na$^+$/H$^+$ exchanger in the membrane of these cells. Cells were acidified by exposing them to (isotonic) Na Hanks' solution containing 30 mM NH$_4$Cl for 15 minutes, then to Na-free Hanks' solution for 2 minutes. Cells were then allowed to recover in Na Hanks' Na Hanks' + amiloride, or Na-free Hanks' solution. The initial rate of recovery was dependent on external [Na]o (apparent Km = 79 mM, FIG. 1A) and was inhibited by amiloride (apparent Ki = 2 $\mu$M in 135 mM Na, FIG. 1B). Inhibition of Na$^+$/H$^+$ exchange reduced the initial rate of recovery by approximately 85%.

We measured the effect of two substances known to modulate macrophage function on intracellular pHi. Lipopolysaccharide (LPS) at a concentration of 1 $\mu$g/ml induces spreading and vesiculation in J774.1 cells after 24 hours. It primes the cells to produce an oxidative burst in response to phorbol ester stimulation, and it has been reported to stimulate Na$^+$/H$^+$ exchange in the pre-B cell line 70Z/3.[2] J774.1 cells showed no change in pHi after short exposures to 1, 10, or 50 $\mu$g/ml LPS. After longer exposures (up to 24 hours) acidification was observed for 10 $\mu$g/ml LPS (TABLE 1).

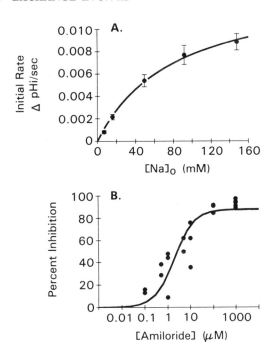

**FIGURE 1. (A)** Dependence of initial rate of recovery from intracellular acidification on [Na]o. Data were fit by the Michaelis-Menten equation. **(B)** Initial rate of recovery is inhibited by amiloride. Data were fit by a modified Michaelis-Menten equation.

Platelet activating factor (PAF), at a concentration of 5–100 ng/ml (approximately 10–200 nM), is known to increase intracellular calcium in murine macrophages[6] and J774 cells.[7] At high concentrations (1–10 mM), it has been reported to stimulate spreading, secretion, and the oxidative burst.[8] We tested the effect of PAF on pHi in J774.1 cells at low doses (0.5–100 ng/ml) over a time course of 0–7 minutes. No stimulation of Na⁺/H⁺ exchange was observed.

**TABLE 1.** Effect of LPS on pHi

| LPS ($\mu$g/ml) | $\Delta$pHi (0–10 min) | $\Delta$pHi (5 min-1 h) | $\Delta$pHi (> 24 h) |
|---|---|---|---|
| 1 | $-0.06 \pm 0.03$ ($n = 4$) | $-0.01 \pm 0.06$ ($n = 6$) | $-0.02 \pm 0.01$ ($n = 4$) |
| 10 | $+0.04 \pm 0.03$ ($n = 4$) | $-0.05 \pm 0.01$ ($n = 4$) | $-0.17 \pm 0.05$ ($n = 10$) |
| 50 | $0.0 \pm 0.04$ ($n = 4$) | | |

## REFERENCES

1. SIMCHOWITZ, L. 1985. Chemotactic factor-induced activation of Na⁺/H⁺ exchange in human neutrophils. II. Intracellular pH changes. J. Biol. Chem. **260:** 13248–13255.
2. ROSOFF, P. M., L. F. STEIN & L. C. CANTLEY. 1984. Phorbol esters induce differentiation in a pre-B-lymphocyte cell line by enchancing Na⁺/H⁺ exchange. J. Biol. Chem. **259:** 7056–7060.

3. GRINSTEIN, S., J. D. SMITH, C. ROWATT & S. J. DIXON. 1987. Mechanism of activation of lymphocyte Na$^+$/H$^+$ exchange by concanavalin A. J. Biol. Chem. **262:** 15277–15284.
4. BESTERMAN, J. M. & P. CUATRECASAS. 1984. Phorbal esters rapidly stimulate amiloride-sensitive Na$^+$/H$^+$ exchange in a human leukemic cell line. J. Cell Biol. **99:** 340–343.
5. MILLS, G. B., E. J. CRAGOE, E. W. GELFAND & S. GRINSTEIN. 1985. Interleukin-2 induces a rapid increase in intracellular pH through activation of a Na$^+$/H$^+$ antiport. J. Biol. Chem. **260:** 12500–12507.
6. CONRAD, G. W. & T. J. RINK. 1986. Platelet activating factor raises intracellular calcium ion concentration in macrophages. J. Cell Biol. **103:** 439–450.
7. DIVIRGILIO, F., T. H. STEINBERG, J. A. SWANSON & S. C. SILVERSTEIN. 1988. Fura-2 secretion and sequestration in macrophages. J. Immunol. **140:** 915–920.
8. HARTUNG, H. P. 1983. Acetyl glyceryl ether phosphorylcholine (platelet activating factor) mediates heightened metabolic activity in macrophages. FEBS Lett. **160:** 209–212.

# Participation of Ion Transport Systems in the Regulation of the Cytolytic Function of Interleukin-2 Activated Killer Cells

KUO-CHIEH WANG

*Pittsburgh Cancer Institute*
*3343 Forbes Avenue*
*Pittsburgh, Pennsylvania 15213*

LAK cells are effective in the treatment of murine and human cancer metastases.[1] Therefore, it is important to understand the physiology of LAK cells, such as proliferation and cytolytic function. It has been shown that interleukin-2 (IL-2)-induced lymphocyte proliferation is proceeded by activation of an amiloride-sensitive $Na^+/H^+$ antiport.[2] Because IL-2-induced lymphocyte proliferation could also occur in the presence of amiloride in $HCO_3^-$-free medium, it seemed plausible that other alternative ion transport systems might be involved in proliferation. However, the role of neither $Na^+/H^+$ antiport nor any other ion transport system in the cytolytic function of LAK cells is known.

Chemical agents such as amiloride, nigericin, monensin, 4-acetamide-4'-isothiocyanostilbene-2,2'-disulfonic, and DIDS have been used to study ion transport systems in lymphocytes, neurons, renal cells, muscle fibers, red blood cells, and fibroblasts.[3] These chemical agents were applied to the study of the cytolytic function of LAK cells.

Firstly, cytoplasms of LAK cells were acidified in $NH_4$ Cl-defined medium for 30 minutes. The cells were washed with various RPMI-1640 media containing either amiloride, nigericin, monensin, SITS, DIDS, or combinations of these agents, such as amiloride/nigericin or amiloride/nigericin/monensin, and then resuspended in the same. In TABLE 1, the results of a 3-hour $^{51}Cr$-release assay in complete culture media containing these agents indicated that: (1) amiloride, nigericin, and monensin strongly inhibited LAK cytolytic activity, that is, 90%, 84%, and 93% inhibition, respectively; (2) DIDS inhibited cytolytic activity by only 22%, whereas SITS had no effect under this condition; and (3) more interestingly, combinations of either amiloride/nigericin or amiloride/nigericin/monensin nearly abolished cytolytic acitivty at higher effect or to target (E/T) ratios, such as 8 : 1 and 4 : 1, whereas they were less inhibitory at an E/T ratio of 1 : 1.

Secondly, when cytolytic function was assessed in three defined media, that is, $HCO_3^-$-free medium, $HCO_3^-$-defined medium, and $NH_4$ Cl-defined medium, the following inhibitory effects of amiloride, nigericin, and SITS were observed: (1) only nigericin was inhibitory in $HCO_3^-$-free medium; (2) nigericin and amiloride were inhibitory in $HCO_3^-$-defined medium, but amiloride was less potent; and (3) nigericin and SITS were both strong inhibitors in $NH_4$ Cl-defined medium (TABLE 2).

In summary, depending on the assay condition used, $Na^+$, $K^+$, and $Cl^-$ transport systems are apparently all involved to different degrees in the cytolytic function of LAK cells.

**TABLE 1.** Cytolytic Function of Acid-Loaded LAK[a] Cells in the Presence of Ion Transport Blockers and Ionophores

| Blocker or Ionophore (20 $\mu$M) | % Cytotoxicity[b] at E:T Ratio | | | | |
|---|---|---|---|---|---|
| | 8:1 | 4:1 | 2:1 | 1:1 | LU$_{25}$/10$^7$ LAK |
| None | 71.4 | 61.8 | 45.8 | 35.2 | 2,800 |
| Amiloride[d] | 26.2 | 17.0 | 8.5 | 3.0 | 260 |
| Nigericin | 44.3 | 22.2 | 25.0 | 20.5 | 440 |
| Amiloride + nigericin | 2.9 | 3.4 | 9.4 | 15.6 | 30–1,200[e] |
| Monesin | 20.2 | 10.7 | 4.0 | 6.2 | 200 |
| Amiloride + nigericin + monesin | 14.0 | 2.7 | 20.0 | 45.6 | 50–3,640[e] |
| SITS | 72.6 | 61.8 | 48.2 | 32.3 | 2,600 |
| DIDS | 67.3 | 51.0 | 39.0 | 27.1 | 2,200 |

[a] Fourteen- to 16-day-old LAK cells were generated from C57BL/6 splenocytes in complete culture medium (CM) containing 1,000 U Human recombinant interleukin-2 (a gift of Cetus Corp., Emeryville, CA). LAK cells were acid loaded with NH$_4$ Cl-defined medium (115 mM NH$_4$ Cl, 4 mM KCl, 1.8 mM CaCl$_2$, 11 mM glucose, and 10 mM HEPES, pH 7.4) for 30 minutes. Acid-loaded LAK cells were washed and resuspended with CM containing the listed ion transport blockers or ionophores.

[b] Cytolytic function of LAK cells was assessed in CM containing ion transport blockers and/or ionophores in a 3-hour $^{51}$Cr-release assay. Five thousand $^{51}$Cr-labeled YAC-1 cells (target) in 0.1 ml volume were mixed with LAK cells (v/v) at effector to target ratios (E:T ratio) of 8:1, 4:1, 2:1, or 1:1. Tests and controls, that is, maximal $^{51}$Cr-release in the presence of 1.0% triton-100 containing H$_2$O and spontaneous $^{51}$Cr-release, were set up triplicately in a round-bottom 96-well microtiter plate. The cells were centrifuged in the microtiter plate at 100 × g, incubated at 37°C for 3 hours, then centrifuged at 300 × g for 10 minutes. The cell-free supernates at 0.1 ml volume were removed and counted by an LKB gamma counter. The percentage cytolysis was calculated by the following formula:

$$\frac{\text{test release} - \text{spontaneous release}}{\text{maximal release} - \text{spontaneous release}} \times 100.$$

[c] One LU$_{25}$ was defined as the number of LAK cells required to kill 25% $^{51}$Cr-labeled target cells.

[d] Amiloride was used at 200 $\mu$M instead of 20 $\mu$M.

[e] Cytolysis did not follow the common and typical ratio-dependent relationship. It was not possible to determine LU$_{25}$ correctly.

**TABLE 2.** Cytolytic Function of LAK Cells Assessed in Three Defined Media Containing Amiloride, Nigericin, or SITS

| Blocker or Ionophore (20 $\mu$M) | % Cytotoxicity[a] in: | | | | | | | | |
|---|---|---|---|---|---|---|---|---|---|
| | HCO$_3^-$-Free Medium[b] at E:T Ratio | | | HCO$_3^-$-Defined Medium[b] at E:T Ratio | | | NH$_4$ Cl-Defined Medium[b] at E:T Ratio | | |
| | 16:1 | 8:1 | 4:1 | 16:1 | 8:1 | 4:1 | 16:1 | 8:1 | 4:1 |
| None | 16.5 | 6.9 | 3.3 | 46.8 | 31.4 | 23.7 | 6.3 | 8.1 | 4.3 |
| Amiloride[c] | 11.7 | 5.6 | 1.5 | 16.7 | 14.1 | 8.8 | 11.2 | 8.0 | 9.2 |
| Nigericin | 0 | 0 | 0 | 0 | 0 | 0 | 0 | 0 | 0 |
| SITS | 22.8 | 15.0 | 6.6 | 42.6 | 34.9 | 27.3 | 0 | 0 | 0 |

[a] Cytotoxicity studies were performed as described in footnote of TABLE 1. Percentage of cytotoxicity was calculated in a 3-hour $^{51}$Cr-release assay.

[b] Basal medium was supplemented with 11 mM NaCl and 10 mM HEPES to obtain HCO$_3^-$-free medium; 115 mM NaCl and the HCO$_3^-$-defined medium; or 115 mM NH$_4$Cl and 10 mM HEPES, the NH$_4$Cl-defined medium. The basal medium contained 4 CaCl$_2$ and 11 mM glucose with pH adjusted to 7.4.

[c] Amiloride was used at 200 $\mu$M.

## REFERENCES

1. ROSENBERG, S. A. & M. T. LOTZE. 1986. Cancer immunotherapy using interleukin-2 and interleukin-2 activated lymphocytes. Ann. Rev. Immunol. **4:** 681–709.
2. MILLS, G. B., E. T. GRAGSE, E. W. GELFAND & S. GRINSTEIN. 1985. IL-2 induces a rapid increase in intracellular pH through activation of a $Na^+/H^+$ antiport. J. Biol. Chem. **260:** 12500–12507.
3. ARONSON, P. S., W. F. BORON, J. F. HOFFMAN & G. GIEBISCH, EDS. 1986. Current Topics in Membranes and Transport. Academic Press, New York.

# Intracellular pH (pH$_i$) Regulation in Rabbit Inner Medullary Collecting Duct Cells

D. KIKERI AND M. L. ZEIDEL

*Harvard Medical School*
*Boston, Massachusetts 02115*

To evaluate mechanisms involved in pH$_i$ regulation in inner medullary collecting duct cells (IMCD), pH$_i$ and membrane potential (PD) were assessed using fluorescent dyes BCECF and DiSC3(5), respectively, in fresh suspensions of rabbit IMCD cells prepared by collagenase digestion and density gradient centrifugation.

The resting pH$_i$ of IMCD cells in nonbicarbonate Ringer's solution, pH 7.4, was 7.21 ± 0.03. Cells were acidified by incubation with 20 mM NH$_4$Cl for 20 minutes and subsequent dilution in an ammonium-free medium, pH 7.4. The rate of pH$_i$ recovery after acidification was more rapid in the presence of extracellular Na$^+$ than in its absence (0.33 ± 0.02 *vs* 0.23 ± 0.01 pH unit/min, 130 mM Na$^+$ *vs* 130 mM K$^+$); amiloride (amil, 0.1 mM) inhibited Na-dependent pH$_i$ recovery (0.20 ± 0.02 *vs* 0.33 ± 0.02 pH unit/min, +amil *vs* −amil). Similar results were obtained in cells acid loaded by incubation with acidic Ringer's solution. Na$^+$-dependent pH$_i$ recovery rates were independent of the PD, which was varied by setting K$^+$ gradients. The pH$_i$ recovery rate in the absence of extracellular Na$^+$ (0.24 ± 0.02 pH unit/min) was inhibited by cellular ATP depletion (0.13 ± 0.02 pH unit/min), sensitive to 1 mM NEM (0.16 ± 0.01), and insensitive to 0.01 mM oligomycin in the presence of glucose (0.27 ± 0.01 pH unit/min). Graded hyperpolarization of PD (achieved by reducing extracellular K$^+$ from 130 to 1 mM in the presence of valinomycin) reduced and then abolished Na$^+$-independent pH$_i$ recovery. Thus, pH$_i$ recovery from acidification was mediated by the activities of both an electroneutral Na$^+$:H$^+$ antiporter and a H$^+$-ATPase that was rheogenic and sensitive to NEM.

When nonacidified cells were diluted into alkaline Ringer's solution (pH 8.3), the rate at which the pH$_i$ increased was independent of the presence of extracellular Na$^+$. In addition, membrane hyperpolarization led to rapid acidification of pH$_i$. The data indicate that the activity of the Na$^+$:H$^+$ antiporter is markedly reduced at or above the resting pH$_i$, and that proton extrusion at resting pH$_i$ is probably mediated by the H$^+$-ATPase. These data suggest that H$^+$-ATPase participates in transepithelial proton transport in this segment under basal conditions, and that Na$^+$:H$^+$ exchange plays a significant role in homeostatic mechanisms such as defense of cell pH and volume regulation.

Cells exposed to HCO$_3^-$/CO$_2$ and then diluted into HCO$_3^-$/CO$_2$-free media alkalinized and then rapidly acidified to resting pH$_i$. The acidification was inhibited by removal of extracellular Cl$^-$ and was more rapid in the absence of extracellular Na$^+$, indicating that Cl$^-$- and Na$^+$-dependent pathway(s) mediate HCO$_3^-$ exit.

We conclude that (a) Na$^+$ : H$^+$ exchange, H$^+$-ATPase activity, and Cl$^-$- and Na$^+$-dependent HCO$_3^-$ transport pathway(s) mediate pH$_i$ regulation in the IMCD, and (b) in nonacidified cells, H$^+$-ATPase but not Na$^+$ : H$^+$ exchange mediates the bulk of H$^+$ extrusion.

# Basolateral Na/H Exchange in the Hamster Inner Medullary Collecting Duct

YOHKAZU MATSUSHIMA, KOJI YOSHITOMI, CHIZUKO
KOSEKI, MASASHI IMAI, SATOSHI AKABANE,
MASAHITO IMANISHI, AND MINORU KAWAMURA

*Departments of Cardiovascular Dynamics*
*Pharmacology and Internal Medicine*
*National Cardiovascular Center Research Institute*
*Suita, Osaka 565, Japan*

Although the inner medullary collecting duct (IMCD) has been reported to participate in the acidification of urine,[1] the underlying cellular mechanisms are unknown. The present study was designed to examine the mechanisms of $H^+$ transport in the *in vitro* perfused hamster IMCD by measuring the intracellular pH (pHi) by microscopic fluorometry.

## MATERIALS AND METHODS

Segments of mid-IMCD isolated from male golden hamsters were perfused *in vitro*. The tubules were then loaded with a pH-sensitive fluorescent probe, BCECF, by adding 2 $\mu$M of BCECF-AM to the bath for 15 minutes at 37°C. Then, fluorescence of epithelia was measured with a microscopic fluorometer (Olympus, model OSP3) at the excitation wavelength 490 nm (pH sensitive) and 440 nm (pH insensitive) and at the emission wavelength 530 nm. At the end of the experiments, the pHi was calibrated on perfused tubules by the method of Thomas *et al.*[2] using nigericin.

## RESULTS AND DISCUSSION

In the basal condition, pHi of the hamster IMCD was 6.74 ± 0.04 ($n = 45$) in the $HCO_3^-$-free Ringer's solution. Because luminal as well as basolateral membrane voltage was about $-80$ mV (Imai *et al.*, unpublished data), some active process for $H^+$ transport must exist in cell membranes. In $NH_4^+$-loaded tubules, acidic pHi recovered with an initial rate of 0.096 ± 0.012 unit/min ($n = 23$), indicating the existence of $H^+$ transport systems. The process of pHi recovery was critically dependent on the presence of $Na^+$ in the bath but not in the lumen (FIG. 1A) and was inhibited by 1 mM amiloride added to the bath (FIG. 1B). Elimination of $Na^+$ from the bath, but not from the lumen, decreased pHi from 6.62 ± 0.05 to 6.02 ± 0.05 ($n = 8$) in $HCO_3^-$-free Ringer's solution or from 6.97 ± 0.06 to 6.44 ± 0.05 ($n = 12$) in $HCO_3^-$-Ringer's solution. These observations are consistent with the view that the $Na^+/H^+$ exchanger exists in the basolateral but

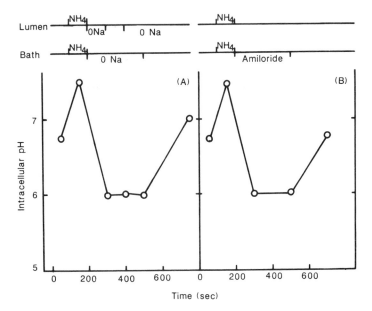

**FIGURE 1.** Characteristics of pHi recovery from $NH_4^+$-induced acid loading in $HCO_3^-$-free Ringer's solution. **(A)** Dependence of pHi recovery on basolateral $Na^+$. **(B)** Effect of basolateral amiloride (1 mM) on pHi recovery. The IMCD was acid loaded by exposing to Ringer's solution containing 25 mM $NH_4Cl$ for 2 minutes and then withdrawing the $NH_4^+$ in the absence of ambient $Na^+$ **(A)** or in the presence of amiloride in the bath **(B)**.

not in the luminal membranes and that it plays the major role in the regulation of pHi. In the absence of ambient $Na^+$, the acid pHi gradually tended to recover in $HCO_3^-$-Ringer's solution, but it was unchanged in $HCO_3^-$-free Ringer's solution (FIG. 2). These observations suggest that a process of $Na^+$-independent $H^+$ transport ($H^+$-pump) may also be operative in the presence of $CO_2/HCO_3^-$.

**FIGURE 2.** Effect of elimination of $Na^+$ from the bath on the pHi in the presence (*open circles*) or absence (*closed circles*) of $CO_2/HCO_3^-$. The pHi was monitored continuously, but only those data at the selected time scale are shown.

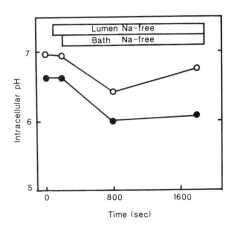

## REFERENCES

1. GRABER, M. L., H. H. BENGELE, J. H. SCHWARTZ & E. A. ALEXANDER. 1981. pH and pco₂ profiles of the rat inner medullary collecting duct. Am. J. Physiol. **241:** F659–F668.
2. THOMAS, J. A., R. N. BUCHSHUM, A. ZIMNIAK & E. RACKER. 1979. Intracellular pH measurements in Ehrlich ascites tumor cells utilizing spectroscopic probes generated in situ. Biochemistry **81:** 2210–2218.

# Cellular Mechanisms in Parathyroid Hormone Control of Apical $Na^+/H^+$ Exchange in Cultured Kidney (OK) Cells

C. HELMLE, M. H. MONTROSE,[a] AND H. MURER

*Physiology Institute*
*University of Zurich*
*Zurich, Switzerland*

In the kidney, parathyroid hormone (PTH) regulates phosphate and bicarbonate reabsorption by the renal proximal tubule.[1] OK cells, an established cell line from opossum kidney, contain proximal tubular transport functions (e.g., $Na^+/H^+$ exchange and $Na^+$/phosphate cotransport) regulated by PTH.[2] The current work examines the polarized $Na^+/H^+$ exchanger expressed by OK cells and examines regulatory cascades controlling its activity.

To study epithelial polarity, cells loaded with BCECF (a fluorescent dye with a pH-sensitive spectrum) were examined as single cells within a confluent monolayer on a filter support.[3] As shown in FIGURE 1, cells acid loaded by transient exposure to $NH_4$ have a $pH_i$-recovery mechanism that is $Na^+$ dependent (the white bars denote isosmotic medium containing $TMA^+$ as a $Na^+$ replacement). The $pH_i$ recovery from this acid load could be attributed to $Na^+/H^+$ exchange (apparent $K_t$ [Na] = 35 mM) localized in the apical membrane. As shown in the first and third $pH_i$ recovery in FIGURE 1, the majority of $pH_i$ recovery is due to apical $Na^+$ addition, although the addition of basolateral $Na^+$ does initiate a limited $pH_i$ recovery. In addition, apical $Na^+/H^+$ exchange was shown to be important to maintenance of steady-state $pH_i$. Removal of basolateral $Na^+$ (mol : mol replacement with $TMA^+$) caused no acidification, whereas removal of apical $Na^+$ caused prompt acidification, consistent with inhibition (or reversal) of $Na^+/H^+$ exchange in the apical membrane. We conclude that OK cells express apical $Na^+/H^+$ exchange which is important for $pH_i$ regulation.

To study regulatory cascades controlling $Na^+/H^+$ exchange activity, cells were placed into suspension by mild trypsinization (3–5-minute exposure to 0.1% trypsin and 0.5 mM EDTA in $Ca^{2+}$, $Mg^{2+}$-free saline solution) and allowed to reequilibrate for 2 hours in $Na^+$ medium (see legend to FIG. 1). Cells loaded with BCECF were then examined in a standard fluorometer cuvette following a 15 minute prepulse with 5-15 mM $NH_4Cl$ (to vary the magnitude of an imposed acid load).[4] At any given acid load, PTH caused inhibition of $pH_i$ recovery in a dose-dependent manner, as did forskolin and tumor-promoting phorbol esters (4 $\beta$ phorbol 12,13 dibutyrate, phorbol 12-myristate 13-acetate; PMA). In suspended OK cells, net $Na^+/H^+$ exchange ceased at $pH_i$ values above pH 7.1 ("set point"). PTH, forskolin, and phorbol ester (PMA) inhibited $Na^+/H^+$ exchange over a wide range of $pH_i$, but had no effect on the set point value. It is concluded that activation of kinase A and/or kinase C could mediate the PTH response. No synergistic effects were produced using submaximal doses of phorbol esters with

[a] Address for correspondence: M. H. Montrose, Johns Hopkins Department of Medicine, Hunterian 515, 725 N. Wolfe St., Baltimore, MD 21205.

either forskolin or PTH. Therefore, the two regulatory cascades of kinase A and C appear to operate independently (with no detectable interaction) to inhibit $Na^+/H^+$ exchange. Potentially, these cascades could all act at the same final target site (phosphorylation site?), because the maximal inhibition due to one agent (e.g., PMA) is not further increased due to the addition of a second agent (e.g., forskolin).

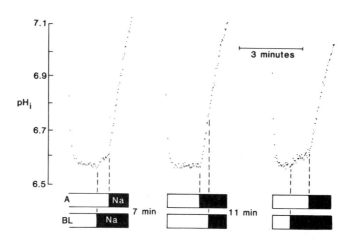

**FIGURE 1.** $Na^+$-dependent recovery of $pH_i$ from sequential acid loads. Cells were grown to confluency on collagen-coated filters (Biopore, Millipore). For experiments, cells were dye loaded (60-minute exposure to 1.2 $\mu$M acetoxymethyl ester of BCECF) in "$Na^+$ medium" (containing [in mM] 130 NaCl, 4 KCl, 1 $CaCl_2$, 1 $MgSO_4$, 18 glucose, 20 HEPES, 1 $(Na)PO_4$; pH 7.4) and mounted in a miniature chamber on the stage of a Zeiss IM35 microscope. This chamber allowed separate apical (A) and basolateral (BL) perfusion and either $Na^+$ medium (*black bars*) or medium with tetramethylammonium (TMA) replacing $Na^+$ (*white bars*) were used. Single cells within the monolayer were examined by excitation ratioing microfluorometry (photomultiplier tube measurements of fluorescent emission in response to alternating excitation wavelengths) and dye response was calibrated as described previously.[3] Using this preparation, cells were acid loaded by exposure to 20 mM $NH_4Cl$ ($NH_4$ prepulse). Data are presented for three separate acidifications, with data presentation starting immediately after removal of $NH_4Cl$ (i.e., during initiation of the acid load).

## REFERENCES

1. Iino, Y. & M. B. Burg. 1979. Am. J. Physiol. **236:** F378–F391.
2. Pollock, A. S., D. G. Warnock & G. J. Strewler. 1986. Am. J. Physiol. **250:** F217–F225.
3. Montrose, M. H., T. Friedrich & H. Murer. 1987. J. Membr. Biol. **97:** 63–78.
4. Montrose, M. H. & H. Murer. 1986. J. Membr. Biol. **93:** 33–42.

# Characteristics of Two Different $Na^+$-$H^+$ Antiport Systems in Several Epithelial and Nonepithelial Cell Lines

CARLOS A. RABITO, EDWARD J. CRAGOE, JR., AND
SALVADOR VINIEGRA

*Nuclear Medicine Division*
*Department of Radiology*
*Massachusetts General Hospital and*
*Harvard Medical School*
*Boston, Massachusetts 02114*

To study the differential characteristics of the $Na^+$-$H^+$ antiport systems involved in multiple cellular functions, we analyzed the changes in activity of these systems induced by cell proliferation, mitogenic factors, extracellular $Na^+$, and different amiloride analogs in several epithelial and nonepithelial cell lines.

Confluent monolayers of LLC-PK$_{1A}$ cells, an epithelial cell line with the differentiated characteristics of renal proximal cells, have a $Na^+$-$H^+$ antiport system located in the apical membrane of the cell. This system, however, is not expressed during cell proliferation or after incubation in the presence of different mitogenic agents. In contrast, confluent monolayers of MDCK$_4$ and LLC-PK$_{1B4}$, two epithelial cell lines with differentiated characteristics of renal distal tubular cells, do not express $Na^+$-$H^+$ antiport activity in the confluent monolayer state. Expression of this system does occur, however, during cell proliferation or after activation of the cells by different mitogenic agents.

Similar results were obtained with the renal fibroblastic cell line BHK. In LLC-PK$_{1A}$ cells, an increase in the extracellular $Na^+$ concentration produces a hyperbolic increase in the activity of the $Na^+$-$H^+$ antiporter. In MDCK$_4$ and BHK cells, however, an increase in external $Na^+$ produces a sigmoid activation of the system. An increase in intracellular $H^+$ concentration activates, whereas an increase in extracellular $H^+$ inhibits the $Na^+$-$H^+$ antiport system of both LLC-PK$_{1A}$ and MDCK$_4$ cells. In LLC-PK$_{1A}$ cells, however, the maximal activation in MDCK$_4$ cells occurs at the same internal and external pH (that is, pH 7.0). The $Na^+$-$H^+$ antiporter of LLC-PK$_{1A}$ cells is more sensitive to the inhibitory effect of amiloride ($K_i$: $1.8 \times 10^{-7}$ M) than is the antiporter of MDCK$_4$ cells ($K_i$: $7.0 \times 10^{-6}$). Moreover, the 5-amino substitute analog of amiloride, methylisobutyl-amiloride, is the most effective inhibitor of $Na^+$-$H^+$ exchange in LLC-PK$_{1A}$ cells, but the least effective inhibitor in MDCK$_4$ cells. Conversely, the analog methyl-methylamiloride is the most effective inhibitor of $Na^+$-$H^+$ exchange in MDCK$_4$ cells, but the least effective inhibitor in LLC-PK$_{1A}$ cells. These results suggest that the $Na^+$-$H^+$ exchange system observed in LLC-PK$_{1A}$ and other cell lines represents the activity of different $Na^+$-$H^+$ antiporters. The system present in MDCK$_4$ cells is localized in the basolateral membrane of the epithelial cell. The location of the system on the opposite side as compared with the apical location in LLC-PK$_{1A}$ cells suggests that the antiporter in MDCK$_4$ cells is not involved in transepithelial transport of $H^+$, but it may serve other cellular functions such as regulation of cytoplasmic pH, cell volume regulation, or, when required, mitogenic activation.

To differentiate between these systems, we have designated them as follows: (a) *regulatory* $Na^+$-$H^+$ antiport system, the system(s) involved in intracellular pH regulation, cell volume regulation, mitogenic activation, and so forth, and (b) *nonregulatory* $Na^+$-$H^+$ antiport system, the system localized in the apical membrane of certain epithelial cells and involved in the transepithelial transport of $H^+$.

# Mechanisms of Chloride Transport in Secretory Epithelia[a]

RAINER GREGER, KARL KUNZELMANN,
AND LARS GERLACH

*Physiologisches Institut*
*Albert-Ludwigs-Universität*
*D7800 Freiburg, FRG*

The role of chloride channels in NaCl secretion by a variety of epithelia has been well recognized over the last few years.[1–5] It has become apparent that the mechanism of chloride secretion involves, in its simplest constellation, the functional components summarized in FIGURE 1.

Chloride uptake on the blood side is conducted by a cotransport system that couples the chloride ion to sodium.[1] This ingenious concept has made it possible to understand how chloride secretion is coupled to $(Na^+ + K^+)$-ATPase. Sodium enters the cell down its electrochemical gradient and enables chloride to enter the cell and achieve a cytosolic concentration far above its electrochemical equilibrium.[3,6,7] Even though the direct coupling of $Na^+$ and $Cl^-$ fulfills all theoretic requirements, a NaCl cotransporter has not yet been unequivocally identified. In fact, it is far more likely that the uptake of chloride occurs via a $Na^+2Cl^-K^+$-cotransporter.[3,8] The general model in FIGURE 1 explains why the uptake of $Na^+$ and $2Cl^-$ should be coupled to that of $K^+$. This $Na^+2Cl^-K^+$ cotransporter works electroneutrally like a $Na^+Cl^-$ cotransporter, but it can couple the passive movement of *two* $Cl^-$ to the active transport of only *one* $Na^+$. It is apparent from this figure that $Na^+$ taken up by the cotransporter is recycled entirely by $(Na^+ + K^+)$-ATPase. Hence, the net secretion of $Na^+$ is accounted for entirely by paracellular $Na^+$ movement, and the latter is driven by transepithelial voltage. This voltage is generated by the conductive properties of the epithelium, that is, the apical membrane is exclusively $Cl^-$ conductive and the basolateral membrane is entirely $K^+$ conductive. This general principle was first recognized for the NaCl reabsorbing epithelium of the thick ascending limb of the loop of Henle, and it was deduced that this basic concept may theoretically generate a stoichiometry of 6 NaCl transported per 1 ATP consumed.[9] Along these lines, oxygen consumption in the rectal gland of *Squalus acanthias* has been compared to the rate of $Cl^-$ transport, and the rate of transport was found to be almost twice as economical as it could possibly be on the basis of a 3 $Na^+$ per 1 ATP stoichiometry.[10] It is highly likely that the universal distribution of the $Na^+$-$K^+$-$2Cl^-$ cotransporter is related to this energetic advantage.

In many of the chloride-secreting epithelia the rate of transport is regulated by hormones and autakoids. The concept in FIGURE 1 makes it appear equally likely that hormonal upregulation of transport is geared by upregulation of any of the transport systems involved. One might envisage an increase in the rate of turn-

[a] The work from the authors' laboratory was supported by Deutsche Forschungsgemeinschaft Gr 480/9 and by Common Market ST 2J-0095-2-D(CD).

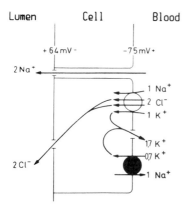

**FIGURE 1.** Model of a NaCl secreting cell. $\dotplus$ = ion channel; $\sigma^{\rightarrow}$ = carrier; $\bullet^{\rightarrow}$ = active pump.

over or the number of (1) ($Na^+ + K^+$)-pumps, (2) $Na^+2Cl^-K^+$ cotransporters, (3) $Cl^-$-channels, (4) $K^+$-channels, and even (5) the paracellular shunt pathway. Most of these alternatives have some experimental proof. Initially, regulation of the $Na^+2Cl^-K^+$ cotransporter was favored as the key step.[11] Then it was recognized that primary regulation occurred at the $K^+$-channel[2] in pancreatic acini and other salivary glands. Shortly thereafter, it was found that in other glands like the rectal gland of *Squalus acanthias* the first step involved upregulation of $Cl^-$ conductance.[12] Recently, simultaneous upregulation of the $Cl^-$ channels and $K^+$ channels has been postulated for other excretory glands such as the lacrimal and salivary glands.[13-15] Further studies are needed to elucidate how the different functional components are turned on synchronously to avoid undue disequilibria in these heavily transporting cells which turn over the entire NaCl content in less than a few seconds. The present short report will specifically address only the properties and the regulation of chloride channels in a variety of epithelia in which current evidence favors the view that $Cl^-$-channel regulation is the key event.

## $Cl^-$ CHANNELS, CONDUCTANCE PROPERTIES, AND VOLTAGE DEPENDENCE

TABLE 1 summarizes the conductance properties of $Cl^-$ channels in several $Cl^-$-secreting epithelia. In these epithelia, namely, the rectal gland of the dogfish *Squalus acanthias,* colonic carcinoma cells, and respiratory epithelial cells, $Cl^-$ channels of intermediate conductance have been found.[4] It is not certain, however, if these epithelia also contain $Cl^-$ channels of such small conductance that they are undetected in single channel recordings. A review of the current literature[5] reveals that intermediate conductance chloride channels share many properties. They all are of some 20–80-pS conductance when the $Cl^-$ concentration on both sides of the channel is around 100–150 mmol/l. Like most channels, the chloride channels are also voltage dependent. This dependence has two facets: (1) the slope conductance is higher for positive clamp voltages (the sign of the voltage refers to the cytosolic side) as compared to negative voltages; and (2) the open state probability is strongly voltage dependent. It is high for positive voltages and low for negative voltages. Both factors taken together result in about a fivefold increase in macroscopic conductance when the clamp voltage is depolarized from

**TABLE 1.** Properties of Epithelial Chloride Channels of Intermediate Conductance

| Tissue | Conductance (pS) | Function | Selectivity | Regulation | Inhibition | References |
|---|---|---|---|---|---|---|
| TAL | 20–50 | Cl-reabs. | Cl $\gg$ Na | cAMP | DPC, NPPB | Greger and Kunzelmann,[4] 1989 |
| RGT | 45 | Cl-secr. | Cl > Br > J > Gluc $\gg$ Na | cAMP, voltage | DPC, NPPB | Greger *et al.*,[17] 1987 |
| Colon rat | 50 | Cl-secr. | I > Br > Cl > F = HCO$_3$ > SO$_4$ | Voltage | DIDS | Summarized by Gögelein,[5] 1988 |
| T$_{84}$ | 30–50 | Cl-secr. | ClO$_4$ > I > NO$_3$ Cl $\gg$ F | Voltage | NPPB | Summarized by Gögelein,[5] 1988 and unpublished work from the authors' laboratory |
| HT$_{29}$ | 50 | Cl-secr. | Cl = Br $\gg$ Na Cl = Br = I > HCO$_3$ > Gluc = Na = 0 | Voltage Voltage pH no effect Ca$^{2+}$ no effect | NPPB | Hayslett *et al.*,[19] 1987 and unpublished work from the authors' laboratory |
| Trachea | 20–50 | Cl-secr. | Cl > Na | Voltage | DPC | Welsh,[16] 1987 |
| Respiratory epithelial cells | 30–80 | Cl-secr. | Cl = Br = I $\gg$ Gluc = Na = 0 | Voltage pH no effect Ca$^{2+}$ no effect | NPPB | Greger and Kunzelmann,[4] 1989 |

*Abbreviations*: DPC = diphenylamine-2-carboxylate; NPPB = 5-nitro-2-(3-phenylpropylamino)-benzoate; reabs. = reabsorption; RGT = rectal gland tubule of *Squalus acanthias*; secr. = secretion; T$_{84}$ and HT$_{29}$ = colonic carcinoma cell lines; TAL = thick ascending limb of the loop of Henle.

$-50$ to $+50$ mV.[4] It is not possible at present to translate this voltage dependence into a valuable prediction for an intact cell. The variation in voltage would be comparably small under physiologic conditions with hormonally induced depolarizations of not more than 20 mV.[3,12,16] Whole cell current recordings will be required to answer this question quantitatively.

The voltage dependence of open state probability has been studied in some detail in the rectal gland of the dogfish *Squalus acanthias*[17] and in respiratory epithelial cells.[4,18] It was found that these chloride channels can be characterized by two open state and two closed state time constants. Depolarization reduced the incidence of the long-lasting closed state and increased that of the long-lasting open state. Hyperpolarization had the opposite effect.

## SELECTIVITY OF CHLORIDE CHANNELS

Selectivity sequences of chloride channels in several $Cl^-$-secreting epithelia are summarized in TABLE 1. Determination of these sequences is no trivial issue and encounters two problems. First, large ions replacing $Cl^-$ not only may not permeate, but also may inhibit the channel. This claim has been made for the rectal gland of the dogfish *Squalus acanthias*,[17] for the colonic carcinoma $HT_{29}$ cell line,[19] and more recently for the renal collecting tubule.[20] Another problem arises from the curvilinear current voltage relationship of these chloride channels. This curve cannot be described adequately by the Goldman-Hodgkin-Katz (GHK) formalism. Consequently, the zero current potentials obtained with replacing ions cannot be determined by GHK-curve fitting. An example is shown in FIGURE 2. These data were obtained in respiratory epithelial cells. It becomes evident that all substituting halides and also nitrate (not shown in this figure) can permeate this channel to some extent. Being aware of the dilemma to determine accurate zero current potentials, we decided to utilize the different slope conductances rather than permeability sequences to define conductance. Large anions such as gluconate, sulfate, and cyclamate do not permeate these chloride channels. Recently, we (unpublished results from the authors' laboratory) examined the bicarbonate conductance of these chloride channels and found that bicarbonate permeates

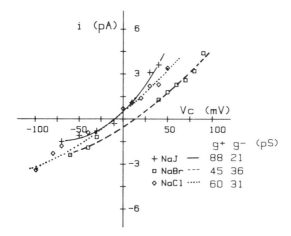

**FIGURE 2.** Current voltage relation of an excised chloride channel from a respiratory epithelial cell. $Cl^-$ was replaced by $Br^-$ and $I^-$. Note the curvilinear shape of the hand-drawn regressions. $g^+$ refers to the slope conductance of the positive voltage range and $g^-$ to the negative voltage range.

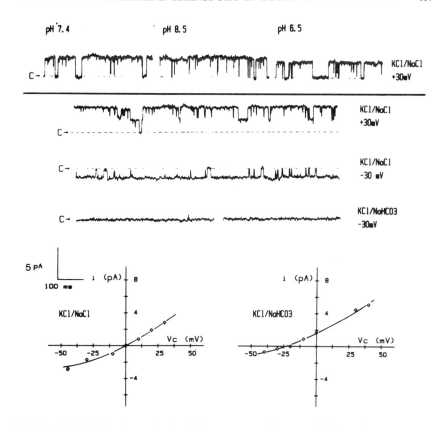

**FIGURE 3.** Chloride channel excised from an HT$_{29}$ cell. *Arrows* indicate the zero current level. The pipette and bath contained 150 mmol/l KCl and NaCl or NaHCO$_3$, respectively. *Upper panel:* pH dependence. The bath solution pH (cytosolic side) was varied between 8.5 and 6.5 with no effect on open state probability or conductance. *Middle panel:* Typical current traces with NaCl and NaHCO$_3$. Note that no measurable current is obtained with negative clamp voltage where HCO$_3^-$ would have to carry the current. *Lower panel:* Current voltage relations with NaCl (*left*) and NaHCO$_3^-$ (*right*) on the cytosolic side.

these channels poorly. An example in a colonic carcinoma cell (HT$_{29}$) chloride channel is shown in FIGURE 3. It is apparent from this figure that the channel is not influenced by alkaline or acid pH, and that the complete replacement of Cl$^-$ by bicarbonate leads to a substantial zero current potential shift and also to a marked reduction in the slope conductance for the bicarbonate current. These preliminary data suggest that bicarbonate is conducted less than half as well as Cl$^-$.

## CHLORIDE CHANNEL BLOCKERS

Apparently all the chloride channels listed in TABLE 1 can be blocked reversibly by the chloride channel blocker 5-nitro-2-(3-phenylpropylamino)-benzoate (NPPB). An example of the inhibitory effect of NPPB on a chloride channel in a respiratory epithelial cell is shown in FIGURE 4. Only fairly low concentrations of NPPB apparently are required to inhibit these chloride channels and to induce flickering. The kinetic effects of NPPB have been examined in $HT_{29}$ cells,[21] and the flickering effect of NPPB was analyzed. At higher concentrations of NPPB, chloride channels appear entirely closed. The high concentration of the blocker may also induce such rapid flickering, however, that it will not be resolved with our methodology, in which filtering occurs at approximately 1–2 kHz.

Current evidence favors the view that NPPB acts from the outside. (1) The effect of NPPB is delayed in inside-out patches, and the washout of the block is also slow. These findings are compatible with the assumption that NPPB has to permeate the membrane before it reaches the binding site. (2) NPPB acts instantaneously in outside-out membrane patches. The molecular requirements for NPPB-like compounds to inhibit chloride channels have been examined with a large number of analogs by Wangemann *et al.*,[22] who found that the interaction of NPPB with the chloride channel was complex and required both polar and nonpolar sites. It has been claimed that a variety of other well-known blockers of anion transport systems such as the stilbene derivatives SITS or DIDS also block chloride channels.[23] (For reviews see also refs. 5 and 24.) We cannot confirm this possibility for the channels studied thus far in our laboratory. Also, unlike previous reports,[25] we are unable to document any effect of loop diuretics such as piretanide or furosemide. This is shown for an $HT_{29}$ cell chloride channel in FIGURE 5. It was found that for up to 0.1 mmol/l, neither compound had any effect on chloride channels in respiratory epithelial cells or in the rectal gland of the dogfish *Squalus acanthias* (unpublished from the authors' laboratory and ref. 24). It is important to note that we do not use concentrations higher than 0.1 mmol/l,

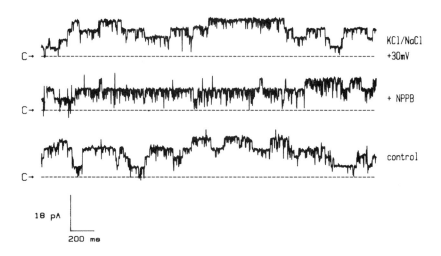

**FIGURE 4.** Typical current traces of chloride channels in an excised inside-out patch of a respiratory epithelial cell. Effect of a low dose ($10^{-7}$ mol/l) of 5-nitro-2-(3-phenylpropylamino)-benzoate (NPPB). The patch pipette was filled with 150 mmol/l KCl, and 150 mmol/l NaCl was present in the bath. The clamp voltage was 30 mV. *Arrows* indicate the zero level. Note the channel flickering induced by the low concentration of the blocker. In addition, NPPB reduces markedly the open state probability of the four channels present in this patch.

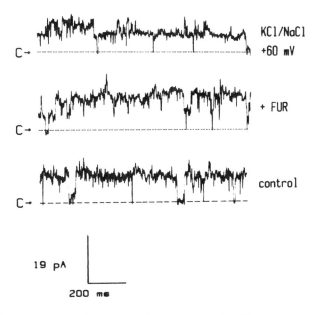

**FIGURE 5.** No effect of the loop diuretic furosemide ($10^{-4}$ mol/l) on chloride channels was seen in an excised inside-out patch of an $HT_{29}$ cell. The pipette and the bath contained 150 mmol/l of KCl and NaCl, respectively. *Arrows* indicate the zero current level.

because we believe that inhibitory effects at high concentrations may not be very specific.

Several groups[26-28] have attempted to use the inhibitor approach for the purification of chloride channels. The molecular weight of the purified chloride channel is still a matter of debate. Until now all attempts to purify chloride channels have been carried out with reversibly binding probes, and some of the probes used were of limited specificity.

## REGULATION OF CHLORIDE CHANNELS

As mentioned in the introduction the rate of $Cl^-$ net transport is controlled by the magnitude of $Cl^-$ conductance in a variety of epithelia. The increases in hormonally or toxin-induced $Cl^-$ conductance can be one order of magnitude.[12] The consequence of this increase in $Cl^-$ conductance is a strong depolarization which may persist as long as the stimulus is present[28] or may only be transient.[16] Pathophysiologically, disturbances in the regulation of chloride channels are highly relevant, as in toxin-induced intestinal oversecretion or in cystic fibrosis-induced undersecretion in a variety of epithelia. The mechanisms involved in the toxin- or hormone-induced increase in chloride conductance are not yet fully understood.

General agreement exists that cAMP is an important second messenger in this respect. An example of the rectal gland of the dogfish *Squalus acanthias* is given

**TABLE 2.** Incidence of Chloride Channels in Apical Membrane Patches of Rectal Gland Tubules of *Squalus acanthias*[a]

| Data | Nonstimulated Tissue | Pretreatment (in mol/l) with $10^{-4}$ dbcAMP, $10^{-4}$ Adenosine, and $10^{-6}$ Forskolin |
|---|---|---|
| Number of seals | 114 | 87 |
| Number of chloride channels | 1 | 37 |

[a] Data taken from Greger *et al.*,[29] 1985.

in TABLE 2. Chloride channels were absent in resting tissue but were present in glands pretreated with the membrane-permeable form of cAMP (dbcAMP).[29,30] Furthermore, the sudden appearance of chloride channels in previously silent membrane patches could be demonstrated when dbcAMP was added to bathe the medium in cell-attached recordings. Similar data have meanwhile been reported for respiratory epithelial cells.[16] These data suggest that chloride channels may be activated by cAMP-dependent phosphorylation. This hypothesis has been examined in excised inside-out membrane patches of rectal glands of the dogfish *Squalus acanthias*. An example[30] is shown in FIGURE 6. The addition of the catalytic subunit of proteinkinase A and ATP generates channel activity in a previously silent patch. Comparable data have also been reported for respiratory epithelial cells.[31,32] In the three studies just cited, the data appear entirely convincing. Still it should be noted that spontaneous channel activation may occur in excised membrane patches. Therefore, it would be preferable to repeat these studies with a postexperimental control in which phosphatase abolishes the kinase-induced activation. Furthermore, the same hypothesis should also be tested with whole cell recording in a strictly paired fashion, that is, the catalytic subunit and the phosphatase should be dialyzed into the cell using a patch pipette perfusion system.

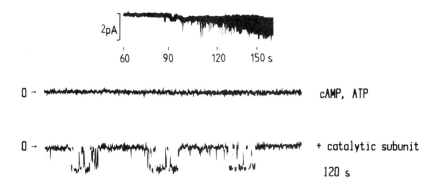

**FIGURE 6.** Effect of the catalytic subunit of proteinkinase A (kindly provided by Prof. Dr. F. Hofmann, Homburg Saar, FRG) on an excised luminal membrane patch of a rectal gland tubule cell of *Squalus acanthias*. *Upper trace:* Condensed current trace. At zero time catalytic subunit of proteinkinase A was added to the cytosolic side. Note the increasing channel activity between 90 and 150 seconds. *Middle trace:* Current trace with high time resolution (1 second per trace) at zero time. *Arrows* indicate the zero current level. *Lower trace:* Current trace 120 seconds after the addition of the catalytic subunit. Data taken from Greger *et al.*[30] (1988).

cAMP-dependent chloride channel regulation may be sufficient to explain hormonal regulation in some but certainly not all $Cl^-$-secreting epithelia. For example, a stimulatory effect of cGMP has been noted in intestinal cells;[33] phorbol esters have a complex effect in cultures of respiratory epithelial cells;[34,35] increases in cytosolic $Ca^{2+}$ stimulate $Cl^-$ secretion in colonic[36] and respiratory epithelial cells.[37] The detailed mechanisms of these second messengers, and even more so, of their interplay, are largely unknown. It has been claimed that $Ca^{2+}$ on the cytosolic side of the channel increases the open probability very much like it does in several $K^+$ channels.[38] We and others have reexamined this issue and find that $Ca^{2+}$ activities between $<10^{-9}$ and $10^{-3}$ mol/l do not alter chloride channel properties in excised membrane patches. Therefore, modulation of $Cl^-$ conductance by cytosolic $Ca^{2+}$ requires other cytosolic components to become effective.

Another candidate in chloride channel regulation is cytosolic pH. As shown above, pH variations on a large scale have little if any influence on $Cl^-$ conductance. Changes in voltage alter the properties of chloride channels (as described above), but the impact of voltage is probably moderate because (1) the encoun-

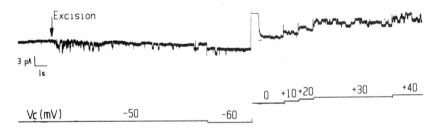

**FIGURE 7.** Excision activation of chloride channels in a cystic fibrosis respiratory epithelial cell. The pipette and the bath contained modified Ringer's solution with 150 mmol/l NaCl. *Upper panel:* Current; *lower panel:* Voltage trace. Note the absence of channels before excision (cell attached). Excision, here performed at a clamp voltage of $-50$ mV, leads to instantaneous channel activation. Note also that this channel, like may other chloride channels, shows a marked increase in open state probability with positive clamp voltages.

tered changes in voltage are rather small and also (2) the voltage dependence of these channels is not very strong.

Much attention has been paid lately to the possible defect of chloride channel function in cystic fibrosis respiratory epithelial cells. Several groups[31,32,39] found that no chloride channels are present in cell-attached membrane patches. On the other hand, it was a surprise to find that chloride channels were present in excised inside-out patches of cystic fibrosis respiratory epithelial cells. This indicates that in cystic fibrosis not the channel itself but its regulation is defective. It was then claimed that the defect in regulation was at the site of kinase A phosphorylation, because long-lasting exposure to depolarized voltages generated channel activity in cystic fibrosis respiratory epithelial cells, but the addition of kinase A and ATP had no effect.[31,32] Also, both studies emphasize the importance of the polarity of the voltage and the delay in activation of cystic fibrosis chloride channels. We have reexamined this issue and arrive at somewhat different conclusions.

We confirm that excision leads to chloride channel activation in cystic fibrosis respiratory epithelial cells, but we did not observe any delay in the activation process. FIGURE 7 shows that activation occurs immediately after the membrane patch is ripped off the cystic fibrosis cell. Another important difference from previous studies relates to voltage dependence. Previously, voltage activation was only seen if the excised membrane patches were clamped to positive, that is, depolarized, voltages. No activation was seen with negative, that is, hyperpolarized, voltages.[31,32] We observed excision activation of chloride channels equally frequently whether the patch was excised into positive or negative clamp voltages. The latter is also shown in FIGURE 7. We do not have a ready explanation for why the results obtained in our laboratory are different from those reported from other laboratories. One obvious methodologic difference is the temperature at which the studies were carried out. The other groups worked at room tempera-

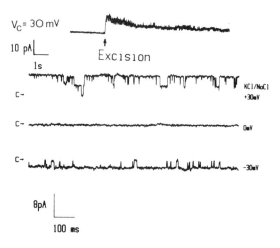

**FIGURE 8.** Excision activation of chloride channels in an $HT_{29}$ cell. The pipette and the bath contained modified Ringer's solution with 150 mmol/l NaCl. The $HT_{29}$ cell was not pretreated with any secretagogue. *Upper panel:* Condensed current trace. Note the instantaneous channel activation when the patch was ripped off the cell at a clamp voltage of +30 mV. *Lower panel:* Typical current traces of this channel at high time resolution and different clamp voltages. *Arrows* indicate the zero current level. Note that two current levels are apparent at positive, but only one at negative clamp voltages. Note also that, like that for most of these chloride channels, the current amplitude is smaller for hyperpolarized than for depolarized voltages.

ture, which is customary in patch clamp studies. We study all tissues of mammalian origin at 37°C.

Data like those shown in FIGURE 7 enlighten a new means of regulation for chloride channels. It is likely that the chloride channels are regulated not only by activating processes but also by inhibitory pathways, and that excision of the patch removes the cytosolic inhibitor and hence leads to channel activation. This interpretation explains why cystic fibrosis membrane patches, which are devoid of channel activity, when they are exposed to the cytosol, show chloride channel activity after excision. Cystic fibrosis is thus viewed as a regulatory defect that leads to an enhanced inhibitory signal.

Obviously a normal cell should also possess some of this inhibitory factor. Therefore, we examined hormonally untreated $HT_{29}$ cells to specifically determine if these cells also exhibit excision activation of chloride channels. A typical experiment is shown in FIGURE 8. This phenomenon apparently is easily demonstrable in these cells, which otherwise respond normally to hormonal activation. Thus, these preliminary data suggest that the cytosolic inhibitor pathway is probably part of the normal regulatory machinery controlling chloride channel activity.

## CONCLUSION

The present report summarizes properties of chloride channels in a variety of NaCl-transporting cells. The data suggest that these chloride channels have many properties in common. The mechanisms of regulation are complex and diversified. It is highly likely that hormones acting via the second messenger cAMP do so by an A-kinase-dependent phosphorylation at the channel itself or at some regulatory site. Clearly other possible signals such as cytosolic pH and $Ca^{2+}$ do not act directly at the channel. Proteinkinase C-dependent regulatory mechanisms have not yet been fully elucidated. Recent observations suggest strongly that chloride channels are also controlled by inhibitory pathways. The cytosolic factor responsible for this tonic inhibition of chloride channels[40] has not been identified. Future research in this field will largely depend on the successful use of whole cell recordings with the option to modify the composition of the cytosol by perfusion techniques.[41]

## ACKNOWLEDGMENTS

We wish to thank Dr. G. Gard for his help with the manuscript. The technical assistance of R. Braitsch, A. Hauser, and J. Siebert is gratefully acknowledged.

## REFERENCES

1. FRIZZELL, R. A., M. FIELD & S. G. SCHULTZ. 1979. Sodium-coupled chloride transport by epithelial tissue. Am. J. Physiol. **236:** F1–F8.
2. PETERSEN, O. H. & Y. MARUYAMA. 1984. Calcium-activated potassium channels and their role in secretion. Nature **307:** 693–696.
3. GREGER, R. & E. SCHLATTER. 1984. Mechanism of NaCl secretion in rectal gland tubules of spiny dogfish (squalus acanthias). I. Experiments in isolated in vitro perfused rectal gland tubules. Pflügers Arch. **402:** 63–75.
4. GREGER, R. & K. KUNZELMANN. 1989. Epithelial chloride channels. *In* Epithelial Secretion of Water and Electrolytes. J. A. Young & P. Y. D. Wong, eds. Springer, New York (in press).
5. GOGELEIN, H. 1988. Chloride channels in epithelia. Biochim. Biophys. Acta **947:** 521–547.
6. WELSH, M. J. 1983. Intracellular chloride activities in canine tracheal epithelium. J. Clin. Invest. **71:** 1392–1401.
7. SAITO, Y., T. OZAWA, H. HAYASHI & A. NISHIYAMA. 1985. Acetylcholine-induced change in intracellular Cl- activity of the mouse lacrimal acinar cells. Pflügers Arch. **405:** 108–111.

8. HANNAFIN, J., E. KINNE-SAFFRAN, D. FRIEDMAN & R. KINNE. 1983. Presence of a sodium-potassium chloride cotransport system in the rectal gland of *Squalus acanthias*. J. Membr. Biol. **75:** 73–83.
9. GREGER, R. 1985. Ion transport mechanisms in thick ascending limb of Henle's loop of mammalian nephron. Physiol. Rev. **65:** 760–797.
10. SILVA, P., J. S. STOFF, R. J. SOLOMON, R. ROSA, A. STEVENS & J. EPSTEIN. 1980. Oxygen cost of chloride transport in perfused rectal gland of *Squalus acanthias*. J. Membr. Biol. **53:** 215–221.
11. PALFREY, H. C., S. L. ALPER & P. GREENGARD. 1980. cAMP-stimulated cation cotransport in avian erythrocytes: Inhibition by "loop" diuretics. Am. J. Physiol. **238:** 103–115.
12. GREGER, R., E. SCHLATTER, F. WANG & J. FORREST. 1984. Mechanism of NaCl secretion in rectal gland tubules of spiny dogfish (*Squalus acanthias*). III. Effects of stimulation of secretion by cyclic AMP. Pflügers Arch. **402:** 376–384.
13. MARTY, A., Y. P. TAN & A. TRAUTMANN. 1984. Three types of calcium-dependent channel in rat lacrimal glands. J. Physiol. **357:** 293–325.
14. COOK, D. J., M. P. CHAMPION & J. A. YOUNG. 1988. Patch-clamp studies on β-adrenergic stimulus-secretion coupling in rat mandibular gland endpiece cells. *In* Exocrine Secretion. P. Y. D. Wong & J. A. Young, eds.: 51–54. Hong Kong University Press, Hong Kong.
15. FINDLAY, I. & O. H. PETERSEN. 1985. Acetylcholine stimulates a $Ca^{2+}$-dependent Cl-conductance in mouse lacrimal acinar cells. Pflügers Arch. **403:** 328–330.
16. WELSH, M. J. 1987. Electrolyte transport by airway epithelia. Physiol. Rev. **67:** 1143–1184.
17. GREGER, R., E. SCHLATTER & H. GOGELEIN. 1987. Chloride channels in the luminal membrane of the rectal gland of the dogfish (*Squalus acanthias*). Properties of the "larger" conductance channel. Pflügers Arch. **409:** 114–121.
18. KUNZELMANN, K., Ö. ÜNAL, C. BECK, P. EMMRICH, H. J. ARNDT & R. GREGER. 1988. Ion channels of cultured respiratory epithelial cells of patients with cystic fibrosis (Abstr.). Pflügers Arch. **411:** R68.
19. HAYSLETT, J. P., H. GOGELEIN, K. KUNZELMANN & R. GREGER. 1987. Characteristics of apical chloride channels in human colon cells ($HT_{29}$). Pflügers Arch. **410:** 487–494.
20. MATSUZAKI, K., J. B. STOKES & V. L. SCHUSTER. 1988. Inhibition of cortical collecting tubule chloride transport by organic acids. J. Clin. Invest. **82:** 57–64.
21. DREINHOFER, J., H. GOGELEIN & R. GREGER. 1988. Blocking kinetics of Cl⁻ channels in colonic carcinoma cells ($HT_{29}$) as revealed by 5-nitro-2-(3-phenylpropylamino)-benzoic acid (NPPB). Biochim. Biophys. Acta **956:** 135–142.
22. WANGEMANN, P., A. DI STEFANO, M. WITTNER, H. C. ENGLERT, H. J. LANG, E. SCHLATTER & R. GREGER. 1986. Cl⁻-channel blockers in the thick ascending limb of the loop of Henle. Structure activity relationship (Abstr.). Pflügers Arch. **407S:** 128–141.
23. FRIZZELL, R. A., R. A. SCHOUMACHER, R. L. SHOEMAKER, R. L. BRIDGES & D. R. HALM. 1988. Disorders of chloride channel regulation in cystic fibrosis (CF) and secretory diarrhea (SD). Disorders of Chloride Channel Regulation: Cystic Fibrosis and Secretory Diarrhea. Amsterdam.
24. GREGER, R. 1990. Chloride channel blockers. *In* Methods in Enzymology. B. Fleischer & S. Fleischer, eds. Academic Press, Orlando (in press).
25. EVANS, M. G., A. MARTY, Y. P. TAN & A. TRAUTMANN. 1986. Blockage of Ca-activated Cl conductance by furosemide in rat lacrimal glands. Pflügers Arch. **406:** 65–68.
26. AL-AWQATI, Q., D. LANDRY, M. H. AKABAS, C. REDHEAD, A. EDELMAN & E. J. CRAGOE. 1988. Purification of an epithelial chloride channel. (Abstr.). Disorders of Chloride Channel Regulation: Cystic Fibrosis and Secretory Diarrhea. Koninklijke Nederlandse Academie van Wetenschappen, Amsterdam.
27. DUBINSKY, W. P. & L. B. MONTI. 1986. Solubilization and reconstitution of a chloride transporter from tracheal apical membrane. Am. J. Physiol. **251:** C713–C720.

28. ZEUTHEN, T., O. CHRISTENSEN & B. CHERKSEY. 1987. Electrodiffusion of Cl⁻ and K⁺ in epithelial membranes reconstituted into planar lipid bilayers. Pflügers Arch. **408:** 275–281.

29. GREGER, R., E. SCHLATTER & H. GOGELEIN. 1985. Cl⁻ channels in the apical cell membrane of the rectal gland "induced" by cAMP. Pflügers Arch. **403:** 446–448.

30. GREGER, R., H. GÖGELEIN & E. SCHLATTER. 1988. Stimulation of NaCl secretion in the rectal gland of the dogfish *Squalus acanthias*. Comp. Biochem. Physiol. **90A:** 733–737.

31. SCHOUMACHER, R. A., R. L. SHOEMAKER, D. R. HALM, E. A. TALLANT, R. W. WALLACE & R. A. FRIZZELL. 1987. Phosphorylation fails to activate chloride channels from cystic fibrosis airway cells. Nature **330:** 752–754.

32. LI, M., J. D. McCANN, C. M. LIEDTKE, A. C. NAIRN, P. GREENGARD & M. J. WELSH. 1988. Cyclic AMP-dependent protein kinase opens chloride channels in normal but not cystic fibrosis airway epithelium. Nature **331:** 358–360.

33. DE JONGE, H. R. 1984. The mechanism of action of *Escherichia coli* heat-stable toxin. Biochemical Soc. Trans. **12:** 180–184.

34. BARTHELSON, R. A., D. B. JACOBY & J. H. WIDDICOMBE. 1987. Regulation of chloride secretion in dog tracheal epithelium by protein kinase C. Am. J. Physiol. **253:** C802–C808.

35. WELSH, M. J. 1987. Effect of phorbol ester and calcium ionophore on chloride secretion in canine tracheal epithelium. Am. J. Physiol. **253:** C828–C834.

36. CARTWRIGHT, C. A., J. A. McROBERTS, K. G. MANDEL & K. DHARMSATHAPHORN. 1985. Synergistic action of cyclic adenosine monophosphate- and calcium-mediated chloride secretion in a colonic epithelial cell line. J. Clin. Invest. **76:** 1837–1842.

37. WELSH, M. J. & J. D. McCANN. 1985. Intracellular calcium regulates basolateral potassium channels in chloride-secreting epithelium. Proc. Natl. Acad. Sci. **82:** 8823–8826.

38. FRIZZELL, R. A., D. R. HALM, G. R. RECHKEMMER & R. L. SHOEMAKER. 1986. Chloride channel regulation in secretory epithelia. Fed. Proc. **45:** 2727–2731.

39. KUNZELMANN, K., Ö. ÜNAL, C. BECK, P. EMMRICH, H. J. ARNDT & R. GREGER. 1988. Regulation of ion channels in respiratory cells of cystic fibrosis patients and normal individuals (Abstr.). Pflügers Arch. **412:** R10.

40. WELSH, M. J. 1986. An apical-membrane chloride channel in human tracheal epithelium. Science **232:** 1648–1649.

41. SOEJIMA, M. & A. NOMA. 1984. Mode of regulation of ACh-sensitive K-channel by the muscarinic receptor in rabbit atrial cells. Pflügers Arch. **400:** 424–431.

# Regulatory Mechanisms of Chloride Transport in Corneal Epithelium[a]

OSCAR A. CANDIA

*Department of Ophthalmology*
*and*
*Department of Physiology and Biophysics*
*Mount Sinai School of Medicine*
*New York, New York 10029*

Since the initial description of a $Na^+$-dependent active $Cl^-$ transport in the frog corneal epithelium,[1] this tissue has been extensively studied for its electrophysiologic and pharmacologic characteristics,[2–16] and it is now included within a group collectively known as $Cl^-$-secreting epithelia. In these epithelia $Cl^-$ is transported transcellularly in the basolateral to apical direction.[17–20] $Cl^-$ enters the cell, coupled with $Na^+$, across the basolateral membrane via a symport driven by the $Na^+$ gradient. $Cl^-$ then leaves the cell compartment down its electrochemical gradient through $Cl^-$ channels in the apical membrane. Also common among these epithelia is that cAMP and/or $Ca^{2+}$ are the intracellular messengers involved in the regulation of $Cl^-$ channel permeability.[21,22] The rate of transepithelial $Cl^-$ transport can be substantially increased by the action of secretagogues that increase apical $Cl^-$ and basolateral $K^+$ permeabilities.[16,23] Thus it follows that under basal conditions apical $Cl^-$ permeability may be the rate-limiting step for $Cl^-$ secretion. In fact, in some epithelia, $Cl^-$ secretion is minimal under basal conditions and must be stimulated for it to be studied.[24,25] It is not clear, however, if $Cl^-$ permeability can be increased to such a degree that the basolateral membrane symport becomes the rate-limiting process. Whether or not ion gradients across the basolateral membrane can be manipulated to increase the rate of transport and shift the burden of rate limitation back to the apical membrane is also unknown. Similar questions relating to the kinetics and regulation of $Cl^-$ transport remain unanswered.

The frog corneal epithelium offers several practical advantages over other tissues for the study of $Cl^-$ transport. This epithelium is easy to obtain, isolate, and maintain viable *in vitro* for many hours. $Cl^-$ transport proceeds under basal conditions at an easily measurable rate (0.5 $\mu$eq/hr $\cdot$ cm$^2$) without the need of exogenous stimulation. $Cl^-$ is the prevalent ion that is transported transcellularly under most experimental conditions, and its movement represents about 95% of the short-circuit current ($I_{SC}$). The cell membrane includes $\alpha$- and $\beta$-adrenergic receptors which can be stimulated with selective agonists to modify the rate of $Cl^-$ transport.[3,6] The basolateral membrane appears relatively simple with the $Na^+$-$K^+$ pump, the $Cl^-$ cotransporter, and high conductance $K^+$ channels as the main components. The apical membrane is highly and almost exclusively permeable to $Cl^-$.[14] A simplified electrical model of the frog corneal epithelium is shown in FIGURE 1. Despite the enviable potency with which the patch-clamp technique

[a] This work was supported by grants EY00160, EY01867, and EY04810 from the National Eye Institute, and a Research to Prevent Blindness Senior Investigator Award to O.A.C.

can characterize single channels, the method has not yet been applied to the apical membrane of the frog corneal epithelium. Gögelein[26] erroneously cites that a 66-pS Cl⁻ channel was described in the rabbit corneal epithelium. The actual study by Koniarek *et al.*[27] (although improperly entitled) was on the rabbit corneal endothelium, a cell layer that is neither Cl⁻ absorbing nor secreting.

We and others[10,14,24,28] suggested that the Cl⁻ entry step mechanism at the basolateral membrane of the corneal epithelium is a Na-Cl symport. We also previously indicated that although there are several agents that increase apical

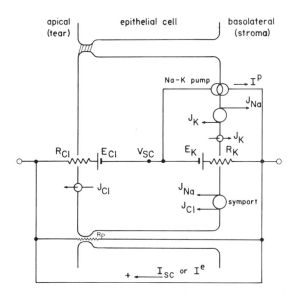

**FIGURE 1.** Schematic electrical model of the frog corneal epithelium. At the apical membrane $J_{Cl}$ is the net Cl⁻ flux across Cl⁻ permeable channels. $E_{Cl}$ represents the Cl⁻ gradient across the apical membrane. $V_{SC}$ is the intracellular electrical potential difference under short-circuit conditions. $R_{Cl}$ is the resistance to the movement of Cl⁻. At the basolateral membrane the Na⁺-K⁺ pump is represented by a current source which produces the pump current $I^P$, the difference between the fluxes of Na⁺ and K⁺ ($J_{Na}$ and $J_K$). $J_K$ recirculates across K channels (represented as $R_K$) driven by the concentration gradient $E_K$ minus $V_{SC}$. The symport is shown as a neutral element without electrical equivalent, producing equal fluxes of Na⁺ and Cl⁻. $R_P$ is the resistance of the paracellular pathway. The current in the external clamping circuit is represented as $I_{SC}$ (when $PD_t = 0$) or $I^e$ (when $PD_t \neq 0$).

Cl⁻ permeability via the cAMP pathway, an exogenous agent that decreases Cl⁻ permeability by acting on this physiologic pathway had not been found.[29] Very recently, however, we determined that $\alpha_2$-adrenergic stimulation of the frog corneal epithelium can indeed reduce apical Cl⁻ permeability and transepithelial Cl⁻ transport.[30]

After originally reporting that furosemide inhibits Cl⁻ transport in a Cl⁻-secreting epithelium,[4] we suggested that the "loop diuretic" acts by decreasing apical Cl⁻ permeability.[13] However, it is presently considered that an effect by furosemide or bumetanide is indicative of the existence of a $Cl^-$ cotransport system. Additional work, particularly by Greger et al.,[31] indicated that what was initially thought to be a $1Cl^-$-$1Na^+$ neutral carrier was in reality a $2Cl^-$-$1Na^+$-$1K^+$ neutral carrier. Most authors have now adopted this model for their $Cl^-$-absorbing or $Cl^-$-secreting epithelia.

In this paper we examine the following: (1) Is the $Cl^-$ cotransporter of the frog corneal epithelium a $1Cl^-$-$1Na^+$ or a $2Cl^-$-$1Na^+$-$1K^+$ system? (2) What are some of the mechanisms that regulate and synchronize $Cl^-$ exit at the apical membrane with $Cl^-$ entry at the basolateral side?

## MATERIALS AND METHODS

Corneas were isolated from bullfrogs (*Rana catesbeiana*) that were obtained from biologic suppliers and double pithed before the experiments. Excised corneas were then mounted as a membrane in a modified Ussing-type chamber following a procedure previously described.[2] The bathing solutions were bubbled with air and the pH was 8.6. When decreases in ionic concentration were made, the osmolality was compensated for with sucrose.

The transepithelial potential difference ($PD_t$) was monitored through agar-NaCl-filled polyethylene bridges that were connected to calomel half-cells. External current was passed through the cornea via agar-NaCl-filled polyethylene bridges. Except when offset voltages were applied, the corneas were kept short-circuited. Transepithelial conductance was determined by measuring the amount of current necessary to offset the short-circuited condition by $\pm 10$ mV. Pharmacologic agents were applied as previously described.[4,6]

## RESULTS AND DISCUSSION

### The $Cl^-$ Cotransporter

The main requirement when considering a coupled system is whether or not the process is energetically feasible, that is, the free energy dissipation ($\Delta G/\Delta t$) of the driving ion(s) must be larger than the free energy gain of driven ion(s). For the case of a $Na^+$-$Cl^-$ symport:

$$[RT/Fz \ln (a_{Na}^b/a_{Na}^c) + \Delta\psi]J_{Na} + [RT/Fz \ln (a_{Cl}^b/a_{Cl}^c) + \Delta\psi]J_{Cl} > 0$$

where "a" denotes ionic activities in the basolateral bathing solution (b) or cell compartment (c), J is flux, $\Delta\psi$ is the electrical potential difference, and the symbols R, T, F, and z have their usual meanings. For a 1-to-1 neutral process such as the $1Cl^-$-$1Na^+$ symport (RT/Fz), $\Delta\psi$ and the fluxes can be factored out so that as long as:

$$(a_{Na}^b/a_{Na}^c) > (a_{Cl}^c/a_{Cl}^b)$$

the process will be energetically feasible. Because cellular activities of $Na^+$ and $Cl^-$ are in the 10–20 mM range, bath activities need to be reduced to similar values for the process to stop. It is important that even though cell $Cl^-$ is above its electrochemical potential with respect to the bathing solution, this parameter is only relevant to its downhill movement across the apical membrane. Basolateral

$Cl^-$-$Na^+$ uptake will be dictated by the combined concentration gradients of the two ions. In fact, in the frog corneal epithelium, the negativity of the cell compartment combined with the high $Cl^-$ permeability of the apical membrane and the operation of the $Na^+$-$K^+$ pump keep the cellular $Cl^-$ and $Na^+$ activities low and render the $Cl^-$-$Na^+$ symport a highly efficient mechanism mutually driven by the $Na^+$ and $Cl^-$ concentration gradients.

Similar energetic considerations apply for the $2Cl^-$-$1Na^+$-$1K^+$ system:

$$RT/Fz \ln (a_{Na}^b/a_{Na}^c) J_{Na} + \Delta\psi J_{Na} +$$

$$RT/Fz \ln (a_K^b/a_K^c) J_K + \Delta\psi J_K +$$

$$RT/Fz \ln (a_{Cl}^b/a_{Cl}^c) 2J_{Cl} + \Delta\psi 2J_{Cl} > 0$$

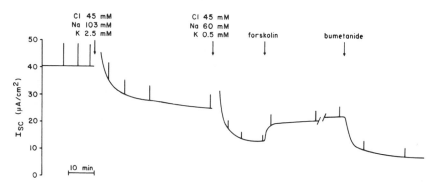

**FIGURE 2.** Effect of reduced $Na^+$, $Cl^-$, and $K^+$ concentration on $I_{SC}$. The $I_{SC}$ declined from 40 to 25 $\mu A/cm^2$ when the $Cl^-$ concentration was reduced to 45 mM. When $Na^+$ and $K^+$ concentrations were also reduced as indicated, $I_{SC}$ further declined to 14 $\mu A/cm^2$. This current was stable and could be stimulated by forskolin ($10^{-5}$ M) and inhibited by bumetanide ($10^{-4}$ M).

Again, if the three ions move in the same direction, the $\Delta\psi \cdot J$ terms cancel out and the direction in which the ions move is dictated by the prevailing gradients. Once the electrical field is eliminated, the movement of $K^+$, because of its high cellular activity, is an impediment to the operation of the carrier, and one must wonder what is the object of this system. Of course, dogmatic considerations are scientifically irrelevant and the point must be made experimentally.

Frog corneal epithelia were mounted in Ussing-type chambers. Identical solutions bathed each side of the tissue. Under control conditions extracellular $Na^+$, $Cl^-$, and $K^+$ concentrations were 104, 75, and 2.5 mM, respectively. Under similar conditions cellular $Na^+$, $Cl^-$, and $K^+$ activities were reported to be about 14, 18, and 106 mM.[13,14] After recording a stable $I_{SC}$ for about 15 minutes, the concentrations of $Na^+$, $Cl^-$, and $K^+$ in the bathing solutions were lowered individually for each ion in several steps or simultaneously in one single change to obtain a final concentration of 60 mM $Na^+$, 45 mM $Cl^-$, and 0.5 mM $K^+$. Records of two such experiments are shown in FIGURES 2 and 3. Changes in medium composition were made on each side of the cornea to avoid transepithelial gradients, so that the $I_{SC}$ always represented transepithelial movement, namely, that of Cl⁻. Average results from seven experiments are given in TABLE 1.

Rearranging the latter equation and replacing its terms with both control experimental values and appropriate ionic activities:

$$RT/Fz[ln\ (a_{Na}^b/a_{Na}^c)J_{Na}\ +\ ln(a_K^b/a_K^c)J_K\ +\ ln(a_{Cl}^b/a_{Cl}^c)2J_{Cl}]$$

25.3 mV[1.76 · 12.4 μA/cm² − 3.96 · 12.4 μA/cm² + 1.19 · 24.8 μA/cm²] =
56 nJoules/s

shows that the process is energetically feasible for control conditions. However, if the values in TABLE 1 are used after the concentrations of Na⁺, Cl⁻, and K⁺ have been reduced and assume proportional changes in cellular Na⁺ and Cl⁻, then:

25.3 mV[1.76 · 3.05 μA/cm² − 5.58 · 3.05 μA/cm² + 1.19 · 6.1 μA/cm²] =
−111 nJoules/s.

It can be seen that the energy dissipated by the movement of Na⁺ and Cl⁻ is insufficient to transfer K⁺ from the basolateral solution into the cell compartment. However, under these conditions Cl⁻ transport (represented by the $I_{SC}$), although only 25% of control, proceeded at a steady rate. This Cl⁻ transport could be stimulated by forskolin, and a relatively large transport rate was maintained for about 1 hour. Bumetanide inhibited this $I_{SC}$, further confirming that it represented Cl⁻ transport. Samples of the basolateral bathing solution were taken at the end of the experiment to measure Na⁺, Cl⁻, and K⁺ concentrations. They were exactly the same as those of the last solution used in the experiment. This corroborated that there was no appreciable gain or loss of these ions by the tissue. For K⁺ to enter the cell compartment coupled to 1Na⁺ and 2Cl⁻ under these experimental conditions, cellular K⁺ activity must be less than 25 mM.

These results demonstrate that it is energetically unlikely for K⁺ to be cotransported with Na⁺ and Cl⁻ at the basolateral membrane of the frog corneal epithelium, suggesting that the uptake mechanism is a Na⁺-Cl⁻ symport.

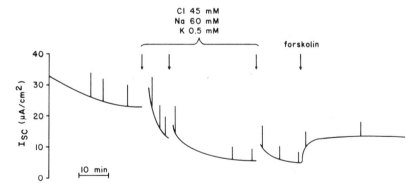

**FIGURE 3.** Effect of reduced Na⁺, Cl⁻, and K⁺ concentrations on $I_{SC}$. After several rinses with the modified solution, the $I_{SC}$ stabilized at 5 μA/cm² and was stimulated by forskolin.

**TABLE 1.** Effects of Reduced $[Cl^-]$, $[K^+]$, and $[Na^+]$ in the Bathing Medium and of Forskolin Addition on the $Cl^-$-Dependent $I_{SC}$ Across the Isolated Frog Corneal Epithelium

| Concentration (mM) | | | |
|---|---|---|---|
| $Na^+$ | $K^+$ | $Cl^-$ | $I_{SC}$ $(\mu A/cm^2)^a$ |
| 104 | 2.5 | 75 | $24.8 \pm 9.3$ |
| 60 | 0.5 | 45 | $6.1 \pm 3.6$ |
| 60 | 0.5 | 45 | $13.2 \pm 4.5^b$ |

[a] Mean $\pm$ SD ($n = 7$).
[b] Plus forskolin.

## Synchronization of $Cl^-$ Entry with $Cl^-$ Exit

At steady state, regardless of the mechanism, $Cl^-$ entry at the basolateral membrane must equal $Cl^-$ exit at the apical membrane. To understand the regulation of transepithelial $Cl^-$ transport it is important to determine which of the two steps is rate limiting under both control conditions and upon stimulation by secretagogues. Because secretagogues apparently increase apical $Cl^-$ and basolateral $K^+$ permeabilities without a direct effect on the carrier, it has been assumed that the permeability to $Cl^-$ of the apical membrane limits the rate of transepithelial transport. $Cl^-$ entry via the $Na^+$-$Cl^-$ symport would simply react to the demand imposed by the rate of $Cl^-$ exit across the apical membrane. However, which of the two barriers is indeed rate limiting when $Cl^-$ transport is stimulated by secretagogues is an open question. It is possible that despite a large increase in apical $Cl^-$ permeability, $Cl^-$ exit is limited by the inability of the symport to increase its rate of $Cl^-$ transport. To investigate this, steps of depolarizing current were applied across the isolated short-circuited cornea to force an additional exit of $Cl^-$ in control and secretagogue-stimulated conditions.

Under short-circuited conditions, $I_{SC}$ across the apical and basolateral membrane is given, respectively, by:

$$I_{SC}^a = \frac{E_{Cl} - V_{SC}}{R_{Cl}} = J_{Cl} \qquad (PD_t = 0)$$

$$I_{SC}^b = \frac{E_K + V_{SC}}{R_K} + I^P \qquad (PD_t = 0)$$

with symbols defined in the legend to FIGURE 1.

By clamping the tissue at a depolarizing voltage, $V_t$ (apical side positive), the current across the apical membrane is given by:

$$I^a = \frac{E_{Cl} - V_{SC}}{R_{Cl}} + \frac{V_t}{R_{Cl} + R_K} \qquad (PD_t = V_t)$$

The current across the basolateral membrane is:

$$I^b = \frac{E_K + V_{SC}}{R_K} + I^P + \frac{V_t}{R_{Cl} + R_K} \qquad (PD_t = V_t)$$

And the current across the paracellular pathway is:

$$I^{Pa} = \frac{V_t}{R_p}$$

The current measured in the external clamping circuit is:

$$I^e = I^a + I^{Pa}$$

When changing the clamping voltage from zero to $V_t$, $I_{SC}$ will increase by an amount equal to $V_t/(R_a + R_K) + V_t/R_P$. Across the cellular pathway the additional current will be carried by $Cl^-$ at the apical membrane and by $K^+$ at the basolateral membrane. Thus, there will be a tendency for the cell to lose KCl. If the symport and the $Na^+-K^+$ pump immediately react to maintain cellular KCl, the step in current produced by $V_t$ will remain constant. Alternatively, if the symport and pump do not provide the additional KCl, their cellular concentration will decline, resulting in an increase in $E_{Cl}$, a decrease in $E_K$, and increases in both $R_{Cl}$ and $R_K$ with a consequent reduction in $I^a$ (and $I^b$). The expected results for both situations are exemplified in FIGURE 4.

Three types of experiments were performed to examine whether the $Na^+-Cl^-$ symport can adapt its rate to that imposed at the apical surface or if it becomes rate limiting. For these, corneas were mounted in Ussing-type chambers and bathed in normal NaCl Ringer's solution. At least four experiments of each kind

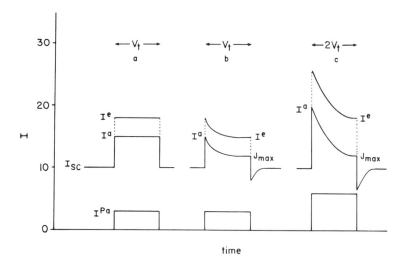

time

**FIGURE 4.** Theoretic changes in transcorneal current upon voltage clamping at a depolarizing voltage. When an additional depolarizing voltage ($V_t$) is applied across the cornea, $I_{SC}$ increases to $I^e$ which is comprised of a cellular component, $I^a$, and a paracellular component, $I^{Pa}$. If the symport and $Na^+-K^+$ pump can supply the additional demand, $I^a$ will remain constant (*panel a*). If the symport (and $Na^+-K^+$ pump) can only supply a maximum flux indicated as $J_{max}$ (*panel b*), $I^a$ will decline to that value. Doubling the value of $V_t$ (*panel c*) will initially double $I^a$ which will then decline to the same $J_{max}$ value. $I^{Pa}$ responds linearly to $V_t$, so that the externally measured current ($I^e$) will be larger in panel c than b.

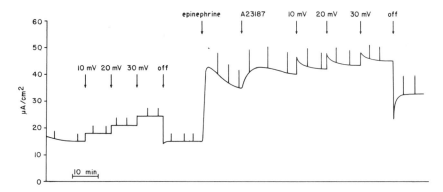

**FIGURE 5.** Responses of the frog corneal epithelial $I_{SC}$ to step increases in depolarizing voltage ($V_t$) in control conditions and after stimulation by epinephrine ($10^{-4}$ M) and $Ca^{2+}$ ionophore A23187 ($10^{-5}$ M).

were performed. An experiment of the first type is shown in FIGURE 5. Spikes represent conductance (g) measurements. After stabilization of $I_{SC}$ the tissue was clamped at 10 mV (tear side positive). As a result, the current increased from 15 to 18 $\mu A/cm^2$ and remained stable. When $V_t$ was increased to 20 and 30 mV, $I^e$ increased, respectively, to 21 and 24.5 $\mu A/cm^2$ in a stepwise manner and remained stable for at least 10 minutes at each step. When $V_t$ was turned off, $I_{SC}$ returned to 15 $\mu A/cm^2$ without an appreciable overshoot. There was no change in conductance between controls and any of the voltage steps, suggesting that cellular permeability and ion concentrations remained stable. The addition of $10^{-4}$ M epinephrine produced a large increase in $I_{SC}$ up to 43 $\mu A/cm^2$, which then slowly decreased. The addition of another $Cl^-$ secretagogue (the $Ca^{2+}$ ionophore, A23187, at $10^{-5}$ M) produced an increase in $I_{SC}$ which began to stabilize at about 40 $\mu A/cm^2$. The conductance of the epithelium increased from 0.3 to 0.8 mS, presumably due to a 0.5-mS increase in transcellular conductance. At this point, consecutive steps of 10, 20, and 30 mV were again applied. At 10 mV, the current increased proportionally to the conductance of the tissue, but it immediately declined and stabilized 2.0 $\mu A$ above the stimulated control value. Also, the conductance decreased from 0.8 to 0.6 mS. Similar patterns were obtained when the clamping voltage was increased to 20 and 30 mV. The current stabilized at 3.5 and 5.0 $\mu A$ above the stimulated control $I_{SC}$, while the conductance decreased to 0.5 and 0.4 mS, respectively. When the clamping voltage was instantly returned to zero, there was a large overshoot with a quick return of $I_{SC}$ and conductance towards control values. At the new steady state, $I_{SC}$ and conductance were about 80% of the previous values, probably because of attenuation of the secretagogue action.

In another experimental series (FIG. 6), application of a single 20-mV offset produced a step in current that then slightly declined. When $V_t$ was turned off, a small overshoot preceded the return of $I_{SC}$ and conductance to their control values. After stimulation by epinephrine and A23187, a single 40-mV offset was applied. The current increased by 47 $\mu A/cm^2$ in proportion to the tissue conductance, but it immediately declined and stabilized at 60% of the initial increase while the conductance diminished to 70% of the control. When $V_t$ was turned off

18 minutes later, a large overshoot occurred before $I_{SC}$ and g returned to the stimulated control values.

For the third type of experiment (FIG. 7), the short-circuit condition was changed to one in which the tissue was hyperpolarized from the open circuit PD of 37 mV to 63 mV. This required a reversed current of about 35 $\mu$A/cm$^2$ which was maintained for periods of 4 to 15 minutes. Upon return to short-circuit, $I_{SC}$ overshot by 10 $\mu$A, but then rapidly decreased to its previous control value. The conductance was 33% higher under the hyperpolarized condition.

The general conclusion from these experiments is that under control short-circuited conditions the permeability of the apical membrane to Cl$^-$ and, to a lesser degree, the basolateral K$^+$ permeability are the rate-limiting factors in the transepithelial transport of Cl$^-$. When the tissue was clamped at 10–30 mV, there were square, linear, and stable increases in current amounting to 9–12 $\mu$A/cm$^2$ (range from four experiments). Because cellular pathway conductance is 40–60% of the total, an increase in Cl$^-$ exit of 5 $\mu$A/cm$^2$ is estimated. The linearity of the responses to the imposed $V_t$ indicates that the carrier quickly adapted to the additional 5-$\mu$A demand which, when added to the mean $I_{SC}$ of 18 $\mu$A/cm$^2$, represents 23 $\mu$A/cm$^2$. After stimulation by epinephrine, $I_{SC}$ within 2 minutes reached a mean ($n = 6$) peak value of 45 $\mu$A, but declined to a stable value (with the addition of A23187) of 32 $\mu$A/cm$^2$. At this level, voltage-driven increases in current rapidly declined to a third. This amount probably represents current across the paracellular pathway, suggesting that 32 $\mu$A/cm$^2$ was the limit the carrier could provide. The decrease in conductance observed when $V_t$ was applied over the secretagogue-stimulated current was probably caused by a decrease in cellular K$^+$ and Cl$^-$. This salt loss may have also caused cell shrinkage and K$^+$ channel closure.

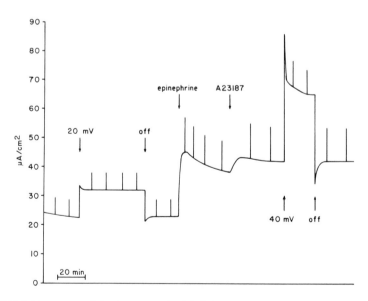

**FIGURE 6.** Responses of the frog corneal epithelial $I_{SC}$ to a large depolarizing voltage ($V_t$) before and after stimulation by epinephrine (10$^{-4}$ M) and A23187 (10$^{-5}$ M).

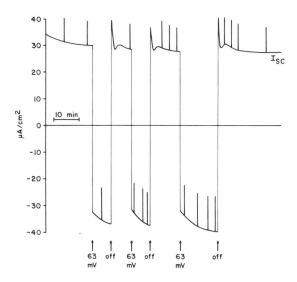

**FIGURE 7.** Responses of the frog corneal epithelial $I_{SC}$ to a large hyperpolarizing voltage. Conductance was consistently larger during the hyperpolarizing periods than during short circuit.

When $V_t$ is turned off, the force driving $Cl^-$ and $K^+$ out of the cell is suddenly reduced, and thus $Cl^-$ and $K^+$ entries exceed their exit until their concentrations are restored to the control values.

When the epithelium is hyperpolarized, the exit of $Cl^-$ and $K^+$ is sharply reduced, and entry possibly occurs. In any case, $Cl^-$ and $K^+$ will accumulate until the carrier adjusts to the new condition. If the tissue is then short-circuited, $Cl^-$ and $K^+$ will leave the cell as an excess $I_{SC}$ which rapidly declines until it equals $Cl^-$ entry via the symport.

These results support the idea that changes in the rate of $Cl^-$ transport across the symport are slow and limited. Conversely, the rates of apical $Cl^-$ and basolateral $K^+$ exit quickly react to changes in permeability and driving forces. It seems that after secretagogue stimulation the synchronism of the rate of $Cl^-$ transport between apical and basolateral membranes is mainly achieved by changes in $Cl^-$ and $K^+$ concentrations which modify the fluxes of these ions across permselective channels.

It is recognized that the $Na^+$-$K^+$ pump must participate in this regulation. Because $Na^+$ and $K^+$ recirculate across the basolateral membrane, changes in the pump rate must parallel those in the symport. A more detailed analysis of this interaction is beyond the scope of this work.

## ACKNOWLEDGMENTS

Phil Cook and Larry Alvarez assisted with the experiments and the preparation of the manuscript.

## REFERENCES

1. ZADUNAISKY, J. A. 1966. Active transport of chloride in frog cornea. Am. J. Physiol. **211:** 506–512.
2. CANDIA, O. A. 1972. Ouabain and sodium effects on chloride fluxes across the isolated bullfrog cornea. Am. J. Physiol. **223:** 1053–1057.
3. CHALFIE, M., A. H. NEUFELD, & J. A. ZADUNAISKY. 1972. Action of epinephrine and other cyclic AMP-mediated agents on the chloride transport of the frog cornea. Invest. Ophthalmol. **11:** 644–650.
4. CANDIA, O. A. 1973. Short-circuit current related to active transport of chloride in frog cornea: Effects of furosemide and ethacrynic acid. Biochim. Biophys. Acta **298:** 1011–1014.
5. BEITCH, B. R., I. BEITCH, & J. A. ZADUNAISKY. 1974. The stimulation of chloride transport by prostaglandins and their interaction with epinephrine, theophylline, and cyclic AMP in the corneal epithelium. J. Membrane Biol. **19:** 381–396.
6. MONTOREANO, R., O. A. CANDIA, & P. COOK. 1976. α- and β-adrenergic receptors in regulation of ionic transport in frog cornea. Am. J. Physiol. **230:** 1487–1493.
7. CANDIA, O. A., R. MONTOREANO, & S. M. PODOS. 1977. Effect of ionophore A23187 on chloride transport across isolated frog cornea. Am J. Physiol. **233**(Renal Fluid Electrolyte Physiol. 2): F94–F101.
8. CANDIA, O. A. & H. F. SCHOEN. 1978. Selective effects of bumetanide on chloride transport in bullfrog cornea. Am. J. Physiol. **234**(Renal Fluid Electrolyte Physiol. 3): F297–F301.
9. SPINOWITZ, B. S. & J. A. ZADUNAISKY. 1979. Action of adenosine on chloride active transport of isolated frog cornea. Am. J. Physiol. **237**(2): F121–127.
10. NAGEL, W. & P. REINACH. 1980. Mechanism of stimulation by epinephrine of active transepithelial Cl transport in isolated frog cornea. J. Membr. Biol. **56:** 73–79.
11. CANDIA, O. A. & S. M. PODOS. 1981. Inhibition of active transport of chloride and sodium by vanadate in the cornea. Invest. Ophthalmol. Visual Sci. **20:** 733–737.
12. CANDIA, O. A., H. F. SCHOEN, L. LOW & S. M. PODOS. 1981. Chloride transport inhibition by piretanide and MK-196 in bullfrog corneal epithelium. Am. J. Physiol. **240**(Renal Fluid Electrolyte Physiol. 9): F25–F29.
13. PATARCA, R., O. A. CANDIA & P. S. REINACH. 1983. Mode of inhibition of active chloride transport in the frog cornea by furosemide. Am. J. Physiol. **245**(Renal Fluid Electrolyte Physiol. 14): F660–F669.
14. REUSS, L., P. REINACH, S. A. WEINMAN & T. P. BRADY. 1983. Intracellular ion activities and Cl⁻ transport mechanisms in bullfrog corneal epithelium. Am. J. Physiol. **244** (Cell Physiol. 13): C336–C347.
15. CANDIA, O. A., L. R. GRILLONE & T. C. CHU. 1986. Forskolin effects on frog and rabbit corneal epithelium ion transport. Am. J. Physiol. **251**(Cell Physiol. 20): C448–C454.
16. WOLOSIN, J. M. & O. A. CANDIA. 1987. Cl⁻ secretagogues increase basolateral K⁺ conductance of frog corneal epithelium. Am. J. Physiol. **253**(Cell Physiol. 22): C555–C560.
17. AL-BAZZAZ, F. J. & Q. AL-AWQATI. 1979. Interaction between sodium and chloride transport in canine tracheal mucosa. J. Appl. Physiol. **46:** 111–119.
18. DEGNAN, K. J., K. J. KARNAKY & J. A. ZADUNAISKY. 1977. Active chloride transport in the in vitro opercular skin of a teleost (Fundulus heteroclitus), a gill-like epithelium rich in chloride cells. J. Physiol. **271:** 155–191.
19. SILVA, P., J. STOFF, M. FIELD, L. FINE, J. N. FORREST & F. H. EPSTEIN. 1977. Mechanism of active chloride secretion by shark rectal gland: Role of the Na-K-ATPase in chloride transport. Am. J. Physiol. **233:** F298–F306.
20. FRIZZELL, R. A., M. FIELD & S. G. SCHULTZ. 1979. Sodium-coupled chloride transport by epithelial tissues. Am. J. Physiol. **236:** F1–F8.
21. LEVITAN, I. B. 1985. Phosphorylation of ion channels. J. Membr. Biol. **87:** 177–190.
22. DE LISLE, R. C. & U. HOPFER. 1986. Electrolyte permeabilities of pancreatic zymogen granules: Implications for pancreatic secretion. Am. J. Physiol. **250:** G489–G496.

23. SMITH, P. L. & R. A. FRIZZELL. 1984. Chloride secretion by canine tracheal epithelium. IV. Basolateral membrane $K^+$ permeability parallels secretion rate. J. Membr. Biol. **77:** 187–199.
24. KLYCE, S. D. & R. K. S. WONG. 1977. Site and mode of adrenaline action on chloride transport across the rabbit corneal epithelium. J. Physiol. Lond. **266:** 777–799.
25. WELSH, M. J., P. L. SMITH & R. A. FRIZZELL. 1983. Intracellular chloride activities in the isolated perfused shark rectal gland. Am. J. Physiol. **245**(Renal Fluid Electrolyte Physiol. **14**): F640–F644.
26. GOGELEIN, H. 1988. Chloride channels in epithelia. Biochim. Biophys. Acta **947:** 521–547.
27. KONIAREK, J. P., L. S. LEIBOVITCH & J. FISCHBARG. 1985. Patch-clamp studies of in vitro rabbit corneal epithelial cells (abstr.). J. Gen. Physiol. **86:** 20a–21a.
28. CANDIA, O. A. 1982. The active translocation of Cl and Na by the frog corneal epithelium: Cotransport or separate pumps? *In* Chloride Transport in Biological Membranes. J. A. Zadunaisky, ed.: 223–242. Academic Press, New York, NY.
29. CANDIA, O. A. 1984. Secretion of chloride by the frog cornea. *In* Chloride Transport Coupling in Biological Membranes and Epithelia. G. A. Gerencser, ed.: 393–414. Elsevier Science Publishers B.V., Amsterdam.
30. CHU, T. C. & O. A. CANDIA. 1988. The role of $\alpha_1$- and $\alpha_2$-adrenergic receptors in $Cl^-$ transport across the frog corneal epithelium. Am. J. Physiol. **255:** C724–C730.
31. GREGER, R. & E. SCHLATTER. 1984. Mechanism of NaCl secretion in the rectal gland of spiny dogfish (*Squalus acanthias*). I. Experiments in isolated in vitro perfused rectal gland tubules. Pflügers Arch. **402:** 63–75.

# Characterization of the Proton-Secreting Cell of the Rabbit Medullary Collecting Duct

HARRY R. JACOBSON, VICTOR L. SCHUSTER, AND
MATTHEW D. BREYER

*Division of Nephrology*
*Vanderbilt University Medical Center*
*Nashville, Tennessee 37232*
*and*
*Division of Nephrology*
*Albert Einstein College of Medicine*
*Bronx, New York 10461*

In the mammalian kidney, reabsorption of a significant fraction of the filtered load of bicarbonate (5–10%) and the formation of ammonium and titratable acid are the responsibility of the collecting duct system. Over the last decade, much has been learned about the mechanism and regulation of acidification of the urine by the mammalian collecting duct. In reviewing the experimental progress made over the recent past, two central themes have emerged. The first is that of axial heterogeneity. Numerous morphologic and functional studies have characterized important segmental differences in the cellular composition and transport properties of cortical, outer medullary, and inner medullary collecting duct segments.[1–7] Indeed, differences in transepithelial voltage, sodium and potassium transport rates, and net rates of acidification have been described for the different anatomic regions of the collecting duct. Also, many attempts have been made to ascribe these different transport processes to the different numbers of the two major cell types present in collecting ducts, the principal cell and the intercalated cell. What has evolved from these studies is a generally accepted model that ascribes sodium and potassium transport to the principal cell and acidification to the intercalated cell.

The second major important theme of the research on collecting duct function is devoted to structure/function correlation at the individual cell level. Specifically, this relates to a whole body of work that followed the surprising observation that mammalian cortical collecting tubules may be either net acid secreters or net bicarbonate secreters depending on the experimental circumstances.[3,8] Subsequent important studies utilizing fluorescence measurements of intracellular pH, electron microscopy, and immunocytochemistry have characterized the intercalated cells of collecting ducts as belonging to two general populations, the alpha cell and the beta cell[9–20] (FIG. 1).

With peanut lectin binding, and localization of antibodies to the erythrocyte anion exchanger band 3, as well as antibodies to bovine medullary proton ATPase, the intercalated cells of the rabbit collecting duct can be divided into two major groups.[13,18,19] The beta cell is apical membrane lectin positive and basolateral membrane band 3 negative with diffuse distribution of proton ATPase. The alpha cell demonstrates apical proton ATPase and is negative for lectin staining at the apical membrane. However, the alpha cell exhibits positive staining for band 3 at the basolateral membrane. This cytochemical characterization supports the notion that the beta cell is involved in bicarbonate secretion and the alpha cell in proton secretion. Although some differences exist among mammalian species (rat, rabbit, and human), this general division of intercalated cells into two types is a universal observation. An additional consistent observation is that the beta cell type is present only in the cortical region of the collecting duct and not in the outer or inner medullary regions.[13,18,19] This observation is consistent with functional studies examining net acidification by collecting duct segments perfused *in vitro* which failed to find any evidence for bicarbonate secretion in medullary collecting ducts.[3]

In studying the mechanism and regulation of acidification by the collecting duct with the *in vitro* perfused tubule, the presence of both beta and alpha cells in the cortical segment makes interpretation of data somewhat difficult. We reasoned that characterization of the alpha cell type of intercalated cells would be most appropriate utilizing collecting duct segments from the outer medulla and, specifically, the inner stripe of the outer medulla. The cytochemical and electron

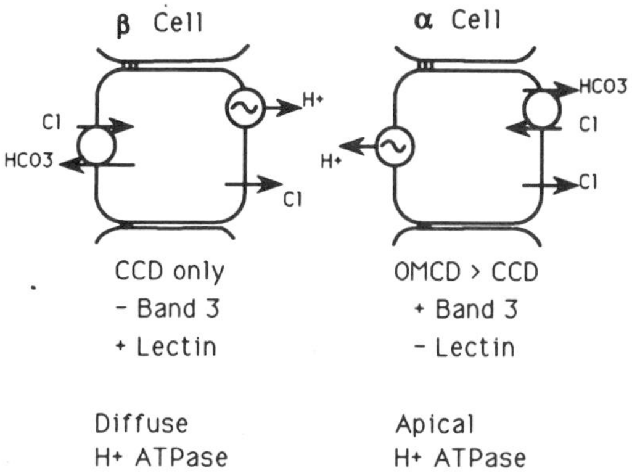

**FIGURE 1.** Characterization of the $\beta$ (HCO$_3^-$-secreting) and $\alpha$ (H$^+$-secreting) intercalated cells of the collecting duct.

microscopic characterization of this segment in the rabbit has confirmed that only the alpha cell type is present and, indeed, it suggests that the innermost region of the inner stripe of outer medulla as well as the outer portion of the inner medullary collecting duct is made up predominantly of alpha cells.[7] Thus, we initiated a series of experiments designed to further characterize the alpha cell, with specific aims being: (1) to provide functional proof for the presence of chloride bicarbonate exchange on the basolateral membrane; (2) to confirm the presence of sodium-independent proton extrusion from the cell (presumptive evidence for proton ATPase); and (3) to determine if a sodium-dependent proton transport process is present in these cells.

## GENERAL METHODS

Standard *in vitro* microperfusion techniques were used with tubules microdissected from the kidneys of New Zealand White rabbits. Experiments were performed only on tubules originating from the most distal part of the outer medullary collecting duct inner stripe and close to its junction with the inner medullary collecting duct. Tubules were perfused in a low volume laminar flow perfusion chamber which allowed for very rapid bath changes. All tubules were perfused with a sodium-free, chloride-containing solution buffered with HEPES (pH 7.40). This perfusate contained (in mM) tetramethyl ammonium (TMA$^+$) 125, Cl 110, H$^+$ HEPES 30, and Ca$^{2+}$ 2.4. Control bath solution contained (in mM) Na 135, Cl 120, bicarbonate 25, and Ca$^{2+}$ 2.4. A number of experimental baths were used in which either sodium or chloride was partially or totally replaced with either TMA$^+$ or N-methyl-D-glucamine (NMDG$^+$) in the case of sodium, and gluconate in the case of chloride. In certain experiments, bicarbonate- and CO$_2$-free solutions were used in the bath, in which case HEPES served as the buffer. Also, all solutions contained (in mM) K$^+$ 5.0, PO$_4$ 2.4, Mg$^{2+}$ 1.0, SO$_4^{2-}$ 1.0, alanine 5, and glucose 8.

Measurement of intracellular pH of perfused tubules was determined with the acetoxymethyl ester of the pH-sensitive dye 2,7-bis (carboxy ethyl)-5,6-carboxy fluorescein (BCECF-AM). Two micromoles BCECF-AM were placed in the bath, and tubules were loaded for 10 minutes at 37°C. Subsequently, tubules were alternately excited at 24 Hz with 495 and 440 nM light from two monochromaters (Delta Scan, Photon Technology, Inc., New Brunswick, NJ). The emitted fluorescent light intensity was quantitated using a photomultiplier tube linked to a per-

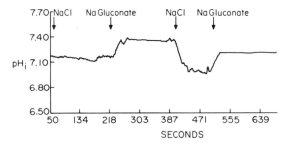

**FIGURE 2.** Effect of bath chloride replacement with gluconate on intracellular pH of medullary collecting duct cells.

sonal computer. The emission was monitored at 520–570 nM. The ratio of fluorescent intensity using 495 versus 440 nM excitation was measured in real-time and used as an index of intracellular pH. At the end of each experiment, the fluorescent ratio was calibrated to cell pH by the addition of nigericin (10 $\mu$M) and 135 mM KCl HEPES buffers of four different pHs (6.3, 6.8, 7.3, and 7.8). In our experience, the 495/440 ratio was a linear function of cell pH over this pH range. In the studies reported, the mean slope was 1.95 ratio units per pH unit.

## CHARACTERIZATION OF BASOLATERAL CHLORIDE BASE EXCHANGE

### Resting Cell pH

Resting cell pHi was 7.15 ± 0.04 ($n = 14$) in tubules bathed with bicarbonate CO$_2$ solution of pH 7.4. In tubules bathed with the HEPES buffered solution of pH 7.4, pHi was 7.19 ± 0.05, a value not significantly different from that seen during exposure to a CO$_2$- and bicarbonate-containing bath.

Our first approach to characterizing the anion exchanger at the basolateral membrane was to determine the effects of complete removal of peritubular chlo-

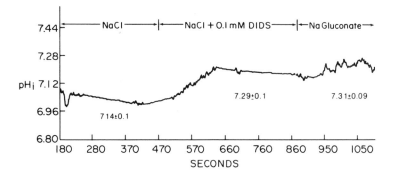

**FIGURE 3.** Effect of peritubular DIDS on cell pH in medullary collecting duct cells. Superimposed peritubular chloride removal does not result in further alkalinization.

ride. Chloride was replaced with gluconate, while peritubular pH was maintained constant with a bicarbonate concentration of 25 mM and a pH of 7.37. FIGURE 2 illustrates a representative experiment showing reversible alkalinization of intracellular pH. The mean increase in cell pH in nine tubules was $0.386 \pm 0.86$ pH units. To confirm further the presence of basolateral cell membrane chloride base exchange in additional experiments, 0.1 mM DIDS was added to the bath. The mean resting pHi increased from $7.14 \pm 0.09$ to $7.34 \pm 0.09$, as illustrated in FIGURE 3. The superimposition of complete peritubular chloride removal on peritubular DIDS did not significantly further alkalinize the cell (mean pH with chloride and DIDS, $7.29 \pm 0.10$; pH without chloride and DIDS, $7.31 \pm 0.09$). Thus, DIDS alkalinizes the cell through a mechanism not additive to that seen with peritubular chloride replacement.

We next pursued experiments designed to elucidate the chloride concentration dependence of this basolateral exchanger. In these studies, tubules were bicar-

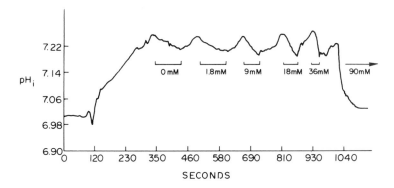

**FIGURE 4.** Chloride concentration dependence of intracellular pH recovery in alkaliloaded medullary collecting duct cells. Chloride concentrations from 0–90 mM were tested, and the slope of the pH decline was measured. Bath chloride concentration was returned to 0 between each chloride concentration tested.

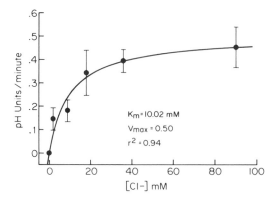

**FIGURE 5.** Nonlinear regression analysis of the data presented in FIGURE 4, illustrating a $K_m$ for the basolateral chloride base exchanger for chloride of 10 mM and a $V_{max}$ of intracellular pH change of 0.50 pH units/min.

bonate loaded by first exposing them to a 0 peritubular chloride bath containing 50 mM bicarbonate (120 mM of chloride was replaced by 95 mM of gluconate and 25 mM of bicarbonate). After cell pH rose to a maximum steady-state level, each tubule was exposed sequentially to a number of different peritubular chloride concentrations ranging from 1.8 to 19 mM. Between each chloride concentration tested, tubules were reexposed to the 0 chloride, 50 mM bicarbonate bath. FIGURE 4 is a representative tracing of data from a single tubule, showing the pH recovery rate at each peritubular chloride concentration. This recovery rate was determined from the first 15 seconds of intracellular pH change after exposure to chloride. The results from seven tubules were analyzed by nonlinear regression, as illustrated in FIGURE 5. The half-maximal recovery rate was observed at a mean peritubular chloride concentration of 10 mM, and the maximal rate of pH recovery ($V_{max}$) was 0.50 pH units/min.

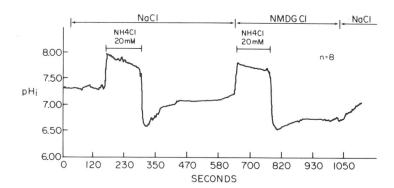

**FIGURE 6.** Comparison of intracellular pH recovery from an acid load (ammonium chloride pulse) in the presence (*left side of the figure*) and absence (*right side of the figure*) of bath sodium. At the end of the experiment (*far right of the figure*), sodium was added to the bath and intracellular pH rose.

## SODIUM-INDEPENDENT AND SODIUM-DEPENDENT PROTON
## EXTRUSION

To test the ability of cells from the OMCDi to extrude protons, we examined their response to acute reduction of pHi. pHi reduction was accomplished by the technique of ammonium chloride pulse with monitoring of the pH recovery rate after maximal acidification. Experiments were performed in the presence and absence of peritubular sodium. (Recall that all experiments were performed with 0 luminal sodium.) FIGURE 6 illustrates a representative experiment. When sodium chloride was present in the bath and ammonium chloride pulsed (accomplished by acutely adding and then removing 20 mM ammonium chloride from the bath), the cell interior was acidified to a nadir of approximately 6.5 pHi. However, pHi recovered rapidly with a mean recovery rate of 0.96 ± 0.25 pH units/min. In the absence of peritubular and luminal sodium, pHi recovery rate from the same acid

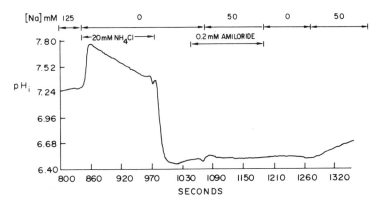

**FIGURE 7.** Intracellular pH recovery after an acid load is inhibited by amiloride even in the presence of 50 mM bath sodium. With amiloride removal intracellular pH does not recover in the absence of sodium, but when 50 mM bath sodium is present intracellular pH rises.

intracellular pH was significantly slower with a mean rate of 0.23 ± 0.05 pH units/ min. These studies clearly demonstrate the presence of both sodium-independent and sodium-dependent intracellular alkalinization. Although the relative rates of these two acid extrusion mechanisms are different, and because the studies were done in the complete absence of $CO_2$ and bicarbonate, we are reluctant at this time to claim that the sodium-dependent process predominates. Indeed, because that the sodium-independent process likely represents the apical membrane proton ATPase, it is possible that the conditions under which these intracellular pH recovery experiments were performed favored a minimal presence of these proton pumps in the apical membrane.[21-23] (In the absence of $CO_2$ and bicarbonate apical proton pumps may have been endocytosed.) As illustrated in FIGURE 6, the readdition of sodium to the peritubular fluid after pHi recovery in the absence of sodium produced a brisk further pHi recovery.

We wished to characterize further the sodium-dependent process, and therefore we repeated the pHi recovery experiment in the absence of sodium, and then exposed the tubule to 0.2 mM bath amiloride to inhibit sodium proton exchange. Then, superimposed on the bath amiloride, bath sodium was restored. As shown in FIGURE 7, there is a slow rate of pHi recovery in the absence of sodium and no further recovery during peritubular amiloride exposure. When sodium was restored in the presence of amiloride, there was no further pHi recovery. However, when sodium was restored subsequently in the absence of amiloride, pHi recovery occurred. These experiments document the presence of an intracellular alkalinizing process that is sodium dependent and amiloride inhibitable, likely a sodium proton antiporter.

Just as we had done with chloride, we determined the sodium concentration dependence of pHi recovery. Tubules were acid loaded and then exposed to varying concentrations of peritubular sodium. Between sodium exposures, bath sodium was returned to 0. As shown in FIGURE 8, pHi recovery depended on peritubular sodium concentration. Indeed, when the data for six tubules was analyzed by nonlinear regression (FIG. 9), this basolateral sodium proton antiporter was shown to have a $K_m$ for sodium of 6.93 mM and a $V_{max}$ of 0.605 pH units/min.

## DISCUSSION AND MODEL OF THE OMCDi CELL

Our studies provide strong functional evidence for the existence of a chloride-dependent base exit mechanism across the antiluminal membrane of the OMCDi cell. This observation is consistent with the recent demonstration of band-3–like immunoreactivity on this membrane.[13,19] The kinetics of this exchanger deserve some comment. Similar studies have been performed on suspensions of rabbit OMCDi cells, yielding a $K_m$ for chloride of approximately 30 mM and a $V_{max}$ of only 0.118 pH units/min.[24] However, significant methodologic differences between our study and previous studies exist, including the manner of harvesting the tissue for study and, importantly, the procedure for exposing tissue to the varying chloride concentrations. Also, we determined the initial rate of acidification by

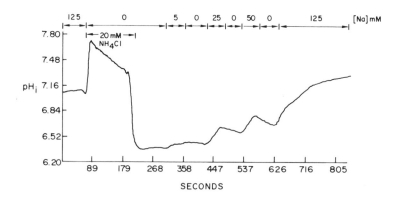

**FIGURE 8.** Sodium concentration dependence of intracellular pH recovery after acid load. The peritubular sodium concentrations tested are listed at the top of the figure. The initial 15 seconds of the pH recovery rate were analyzed.

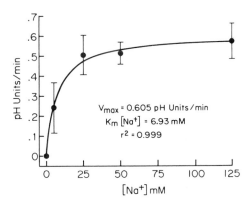

**FIGURE 9.** Nonlinear regression analysis of the data presented in FIGURE 8. The sodium-dependent proton exchange on basolateral cell membranes of medullary collecting duct cells exhibit a $K_m$ for sodium of 6.9 mM and a $V_{max}$ of 0.605 pH units/min.

taking measurements in the first 15 seconds. In the previous study, the first time point was examined at 30 seconds after exposure to chloride and thus could have underestimated the rate of pHi recovery.

In comparing the OMCDi basolateral anion base exchanger to that of band 3 from the standpoint of chloride concentration dependence, we find that our $K_m$ for chloride is not much different from that observed previously. Thus, in the red cell band 3 a $K_m$ of approximately 3 mM has been determined.[25,26] In addition, anion exchangers in other nucleated mammalian cells reveal a $K_m$ of similar magnitude, that is, 6–7 mM. It should be pointed out, however, that the external anion binding site of this exchanger is highly dependent on the experimental conditions, especially intracellular chloride concentration, temperature, and external pH. While the $K_m$ of approximately 10 mM observed by us would suggest that this exchanger might not be sensitive to physiologic changes in peritubular chloride concentration *in vivo*, we do not wish to make this extrapolation at present.

In a number of cells, chloride base exchange has been shown to be sodium dependent. Generally, the inwardly directed sodium gradient energizes bicarbonate entry in exchange for cell chloride. This exchanger is usually active when intracellular pH is low and functions as an alkali loader. Amiloride does not block this exchanger. Because our sodium-dependent regulation of intracellular pH was amiloride inhibitable and separate from the DIDS-sensitive chloride regulation of intracellular pH, we believe it is unlikely that sodium-dependent chloride bicarbonate exchange is present on the basolateral membrane of the OMCDi cells. Further support for this stems from the fact that in our study the sodium-dependent recovery of pHi from an acid load occurred in the absence of extracellular bicarbonate and $CO_2$.

Our observation of basolateral sodium proton exchange was somewhat surprising but not totally unexpected, considering the ubiquitous nature of the transporter. The $K_m$ for sodium of 7 mM is certainly within the range of those reported for other exchangers, 3–50 mM. A sodium proton exchanger has been described on the basolateral membrane of rabbit cortical collecting tubules, but it is unclear if this exchanger is present in intercalated or principal cells.[27] Previous studies on suspensions of inner medullary collecting duct cells from the rabbit fail to demonstrate sodium-dependent pHi regulation.[28] However, in these studies, the pHi recovery rate in the absence of sodium was 0.177 pH units/min, a value somewhat lower than our sodium-independent rate and markedly lower than our sodium-dependent rate. The reasons for this discrepancy are unclear, given the markedly different tissue preparation protocols and the possibility that intracellular sodium was significantly elevated in the previously published studies. It is premature to comment further.

What the function of basolateral sodium proton exchange is in the OMCDi is not clear. This segment normally does not demonstrate any net sodium transport in the absence of sodium concentration gradients. Thus, it is likely that the sodium proton antiporter is involved in cell volume regulation. This may be especially pertinent, because the OMCDi is in a region of the kidney that undergoes significant changes in tonicity depending on salt and water balance.

As stated earlier, we concentrated on the OMCDi because of the likelihood that the proton-secreting cell is the predominant, if not the only, cell type in this region of the rabbit collecting duct. It is important to point out that our experiments measuring intracellular pH involve analysis of fluorescent data from a region of the tubule approximately 15 $\mu$ in length and 40 $\mu$ in width (i.e., the entire wall of the tubule for a length of 2–3 cells). It is possible that the chloride base exchanger we examined and the sodium proton antiporter we demonstrated do not reside in the same cell. We will test this further by performing single cell measurements of intracellular pH.

In summary, we have further characterized the acid-secreting or alpha cell in the collecting duct from the inner stripe of outer medulla. We have confirmed the functional presence of a basolateral chloride base exchanger and have demonstrated that its kinetics are similar to those observed with the classical anion exchange protein band 3. This is consistent with the immunohistochemical demonstration of band-3–like protein on the basolateral cell membrane of these cells. In addition, we demonstrated that two other processes are involved in regulating cell pH in the OMCDi cell. These include a sodium-independent acid extrusion process which, we feel, represents the apical cell membrane proton translocating ATPase. The second acid extrusion mechanism is sodium dependent, resides on the basolateral membrane, and has properties consistent with those of a sodium proton antiporter. Having functionally characterized these transporters, we are now in a position to determine their regulation and their integrated contribution to net proton secretion by this nephron segment.

## REFERENCES

1. STOKES, J. B., C. C. TISHER & J. H. KOKKO. 1978. Structural-functional heterogeneity along the rabbit collecting tubule. Kidney Int. **14:** 585–593.
2. STOKES, J. B., M. J. INGRAM, A. D. WILLIAMS & D. INGRANE. 1981. Heterogeneity of the rabbit collecting tubule: localization of mineralocorticoid hormone action to the cortical portion. Kidney Int. **20:** 340–347.
3. LOMBARD W. E., J. H. KOKKO & H. R. JACOBSON. 1983. Bicarbonate transport in cortical and outer medullary collecting tubules. Am. J. Physiol. **244:** F289–F296.
4. LASKI, M. E. & N. A. KURTZMAN. 1983. Characterization of acidification in the cortical and medullary collecting tubule of the rabbit. J. Clin. Invest. **72:** 2050–2059.
5. MCKINNEY, T. D. & K. K. DAVIDSON 1987. Bicarbonate transport in collecting tubules from outer stripe of outer medulla of rabbit kidneys. Am. J. Physiol. **253:** F816–F822.
6. MADSEN, K. M. & C. C. TISHER. 1986. Structural-functional relationships along the distal nephron. Am. J. Physiol. **250:** F1–F15.
7. RIDDERSTRALE, Y., M. KASHGARIAN, B. KOEPPEN, G. GIEBISCH, D. STETSON, T. ARDITO & B. STANTON. 1988. Morphological heterogeneity of the rabbit collecting duct. Kidney Int. **34:** 655–670.

8. McKINNEY, T. D. & M. B. BURG. 1977. Bicarbonate transport by rabbit cortical collecting tubules. Effect of acid and alkali loads *in vivo* on transport *in vitro*. J. Clin. Invest. **60:** 766–768.

9. SCHWARTZ, G. J., J. BARASCH & Q. AL-AWQATI. 1985. Plasticity of functional epithelial polarity. Nature **318 28:** 368–371.

10. BROWN, D., J. ROTH & L. ORCI. 1985. Lectin-gold cytochemistry reveals intercalated cell heterogeneity along rat kidney collecting ducts. Am. J. Physiol. **248:** C348–C356.

11. LeHIR, M. B., B. KAISSLING, B. M. KOEPPEN & J. B. WADE. 1982. Binding of peanut lectin to specific epithelial cell types in kidney. Am. J. Physiol. **242:** C117–C120.

12. BROWN, D., S. GLUCK & J. HARTWIG. 1987. Structure of the novel membrane-coating material in proton-secreting epithelial cells and identification as an $H^+ATPase$. J. Cell Biol. **105:** 1637–1648.

13. SCHUSTER, V. L., S. M. BONSIB, & M. L. JENNINGS. 1986. Two types of collecting duct mitochrondria-rich (intercalated) cells: Lectin and band 3 cytochemistry. Am. J. Physiol. **251:** C347–C355.

14. BROWN, D., S. GLUCK & J. HARTWIG. 1987. Structure of the novel membrane-coating material in proton-secreting epithelial cells and identification as an $H^+ATPase$. J. Cell Biol. **105:** 1637–1648.

15. WAGNER, S., R. VOGEL, R. LIETZKE, KOOB R. & D. DRENCKHAHN. 1987. Immunochemical characterization of a band 3-like anion exchanger in collecting duct of human kidney. Am. J. Physiol. **253:** F213–F221.

16. HOLTHOFER, H., B. A. SCHULTE, G. PASTERNACK, G. J. SIEGEL & S. S. SPICER. 1987. Three distinct cell populations in rat kidney collecting duct. Am. J. Physiol. **253:** C323–C328.

17. VERLANDER, J. W., K. M. MADSEN & C. C. TISHER. 1987. Effect of acute respiratory acidosis on two populations of intercalated cells in rat cortical collecting duct. Am. J. Physiol. **253:** F1142–F1156.

18. BROWN, D., S. HIRSCH, & S. GLUCK. 1988. An $H^+$-ATPase in opposite plasma membrane domains in kidney epithelial cell subpopulations. Nature **331 18:** 622–624.

19. VERLANDER, J. W., K. M. MADSEN, P. S. LOW, D. P. ALLEN, & C. C. TISHER. 1988. Immunocytochemical localization of band 3 protein in the rat collecting duct. Am. J. Physiol. **255:** F115–F125.

20. SCHWARTZ, G. J., L. M. SATLIN, & J. E. BERGMANN. 1988. Fluorescent characterization of collecting duct cells: A second $H^+$-secreting type. Am. J. Physiol. **255:** F1003–F1014.

21. GLUCK, S., C. CANNON & Q. AL-AWQATI. 1982. Exocytosis regulates urinary acidification in turtle bladder by rapid insertion of $H^+$ pumps into the luminal membrane. Proc. Natl. Acad. Sci. USA **14:** 4327–4331.

22. CANNON, C., J. VAN ADELSBERG, S. KELLY & Q. AL-AWQATI. 1985. Carbon-dioxide-induced exocytotic insertion of $H^+$ pumps in turtle-bladder luminal membrane: Role of cell pH and calcium. Nature **324 4:** 443–446.

23. SCHWARTZ, G. J. & Q. AL-AWQATI. 1985. Carbon dioxide causes exocytosis of vesicles containing $H^+$ pumps in isolated perfused proximal and collecting tubules. J. Clin. Invest. **75:** 1638–1644.

24. ZEIDEL, M. L., P. SILVA & J. L. SEIFTER. 1986. Intracellular pH regulation in rabbit renal medullary collecting duct cells: Role of chloride-bicarbonate exchange. J. Clin. Invest. **77:** 1682–1688.

25. DIX, J. A., A. S. VERKMAN & A. K. SOLOMAN. 1986. Binding of chloride and a disulfonic stilbene transport inhibitor to red cell band 3. J. Membr. Biol. **89:** 211–223.

26. KNAUF, P. A. 1979. Erythrocyte anion exchange and the band 3 protein: Transport kinetics and molecular structure. Curr. Top. Membr. Transp. **12:** 251–363.

27. CHAILLET, J. R., A. G. LOPES & W. F. BORON. 1985. Basolateral Na-H exchange in the rabbit cortical collecting tubule. J. Gen. Physiol. **86:** 795–812.

28. ZEIDEL, M. L., P. SILVA & J. L. SEIFTER. 1986. Intracellular pH regulation and proton transport by rabbit renal medullary collecting duct cells. J. Clin. Invest. **77:** 113–120.

# Cl⁻ Conductance and Acid Secretion in the Human Sweat Duct

PAUL M. QUINTON AND M. M. REDDY

*Division of Biomedical Sciences*
*University of California*
*Riverside, California 92521*
*and*
*Department of Physiology*
*University of California*
*Los Angeles, California 92521*

Humans are uniquely endowed with an extraordinary ability to thermoregulate in the face of large heat loads. This efficiency to dissipate heat derives from eccrine sweat glands that are present over most of the body surface and that are capable of secreting liters of fluid per hour to support evaporative heat loss. Large quantities of $Na^+$ and $Cl^-$ could potentially be lost in these excreted fluid volumes and could thereby threaten the circulating volume. Primates, especially humans, which do not pant, have evolved a unique defense against such volume threats in that the eccrine sweat gland contains a ductal region that is highly efficient in reabsorbing hypertonic concentrations of NaCl before fluid is excreted onto the surface of the skin. That is, as in apparently all mammalian exocrine secretory systems, the sweat begins as an isotonically secreted fluid in the secretory coil region of the gland and then passes through a ductal region where the fluid is modified by active electrolyte transport processes to render a final hypotonic fluid for evaporation.[1] FIGURE 1 shows the concentration of $Na^+$ and $Cl^-$ in the final sweat emerging from single sweat glands of normal subjects and those with cystic fibrosis as a function of secretory rate. By inspection, it is clear that the electrolyte composition of fluid in the duct becomes more concentrated as the secretory load (sweat rate) increases.[2] Likewise, the $HCO_3^-$ concentration also increases dramatically with sweating so that virtually no $HCO_3^-$ is present in the final sweat at secretory rates that are below about half maximum, but at maximal rates, $HCO_3^-$ is almost isotonic. Likewise, the pH of sweat increases from acidic values of about pH 5 at low secretory rates to approximately neutral values at the highest secretory rates. Parenthetically, we have observed individual pH values for sweat of less than 5.0 at very low sweat rates (FIG. 2).[3]

These observations from single sweat glands *in vivo* indicate that the cells of the eccrine sweat duct face large changes in the fluid environment at their luminal surface as secretory rates vary. These changes in the fluid environment suggest that different components of electrolyte transport mechanisms may be necessary to cope with the different electrochemical gradients that result. This paper deals with several observations, both direct and indirect, which suggest that $Cl^-$ conductance and activation of proton secretion may be important components needed to support efficient electrolyte absorption from the duct lumen during sweating.

The sweat duct is characterized by an extremely high $Cl^-$ conductance. TABLE 1 shows the general conductance properties of the isolated microperfused eccrine sweat duct epithelium from normal subjects and those with cystic fibrosis. It now seems clear that the diagnostically higher concentrations of NaCl in the sweat of cystic fibrosis subjects (TABLE 1)[4] are due to the inherently poor permeability of their sweat ducts to $Cl^-$.[5,6] $Cl^-$ conductance represents almost 90% of the total epithelial conductance of normal sweat ducts, that is, about 110 ms/cm[2], but less than 10 ms/cm[2] of cystic fibrosis sweat duct conductance is contributed by $Cl^-$.[7] $Na^+$ conductance, however, seems to contribute only about 5–10 ms/cm[2] and to be about the same for both groups. Results of studies from voltage divider ratios, which give the relative resistances of the apical membrane (Ra) to the basal membrane (Rb), indicate that both the apical and basal membrane of the normal sweat duct cell is significantly permeable to $Cl^-$. That is, removing $Cl^-$ from the

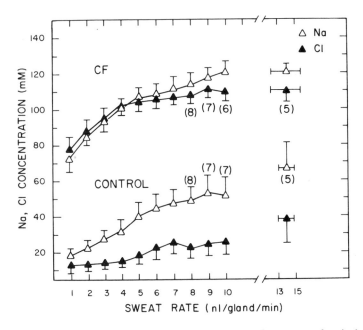

**FIGURE 1.** Concentrations of $Na^+$ and $Cl^-$ as a function of sweat rate for single glands from persons with cystic fibrosis and control subjects *in vivo*. Glands were stimulated to secrete by iontophoresing a 2% solution of acetylcholine through the epidermis. Sweat composition and rates were measured as described previously, and the figure was modified from FIGURE 3 in ref. 2.

apical surface caused Ra/Rb to increase from approximately 4.8 to 7.0, whereas removing $Cl^-$ from the contraluminal bath caused the ratio to decrease from 3.2 to 1.9.[8]

On the surface it might appear that this large $Cl^-$ conductance simply facilitates $Na^+$ absorption from the lumen. However, closer inspection, and results from experiments in which the luminal concentration was decreased from 150 to 15 mM $NaCl^-$ to more closely mimic the luminal fluid composition that occurs *in vivo* suggest that such a simple supposition may not be compatible with completely electrodiffusive $Cl^-$ uptake (FIG. 3). Although we do not as yet have

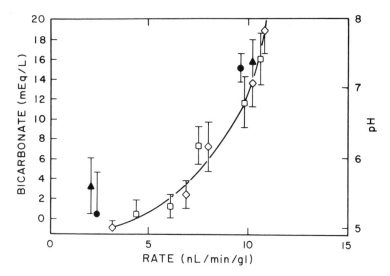

**FIGURE 2.** Changing $HCO_3^-$ concentration and the pH of sweat collected from single sweat glands from persons with cystic fibrosis and normal subjects. The $HCO_3^-$ concentration, pH, and rates were determined as described previously. Figure was taken from FIGURE 2 in ref. 3. At low secretory rates, $HCO_3^-$ was not detectable in final sweat.

measurements of intracellular $Cl^-$ concentration, the chemical gradient for $Cl^-$ entry into the cell should decrease when the luminal NaCl concentration falls. And it is clear from FIGURE 3 that the electrical gradient into the cell across the apical membrane also becomes much less favorable when the luminal NaCl concentration falls to low levels. The problem becomes very clear in view of the possible Nernst distributions for $Cl^-$. That is, for $Cl^-$ to move passively across the apical membrane when its concentration in the lumen is 15 mM and the apical membrane potential is $-40$ mV, the cell $Cl^-$ concentration must fall below about 3 mM. However, for $Cl^-$ to move passively out of the cell across the basal membrane into the bath (150 mM Cl) through a potential of $-65$ mV, the $Cl^-$ concentration in the cell must be greater than 12 mM. From a qualitative point of view, we must conclude that if $Cl^-$ enters the cell electroconductively, its concentration in the cell must be too low for it to exit electroconductively. Vice versa, if it exits the cell electroconductively, its concentration in the cell must be too high for it to enter electroconductively.

**TABLE 1.** $Cl^-$, $Na^+$, and Residual Ion Transepithelial Conductances ($G_{Cl}$, $G_{Na}$, and $G_x$, respectively) in Control and Cystic Fibrosis Microperfused Sweat Ducts[a]

|  | Control (ms/cm²) | Cystic Fibrosis (ms/cm²) |
|---|---|---|
| $G_{Cl}$ | 110 | 6 |
| $G_{Na}$ | 5 | 7 |
| $G_x$ | 10 | 8 |

[a] Values are taken from Quinton.[1]

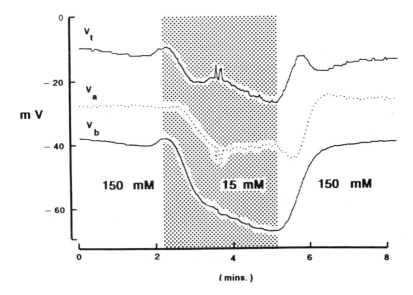

**FIGURE 3.** Effect of low luminal NaCl concentration on membrane potentials of sweat duct cells. The apical membrane potential ($V_a$) is calculated from the difference between the basal membrane potential ($V_b$) and the transepithelial potential ($V_t$) and is shown as the dotted line ($V_a$). The ordinate shows the potentials in millivolts (mV) as a function of time in minutes (mins.). As shown, exposure of the duct lumen to 15 mM NaCl with the osmotic difference made up with mannitol causes the transepithelial as well as both membrane potentials to hyperpolarize reversibly.

While investigating the electrical properties of cystic fibrosis duct cells, we were presented with a possible insight into a mechanism by which the duct may overcome this problem. As shown in TABLE 2, we noted that amiloride in the presence of $Cl^-$ essentially abolishes the transepithelial potential difference across normal ducts. However, when amiloride was applied to cystic fibrosis ducts which are almost $Cl^-$ impermeable, their transepithelial potential becomes positive (lumen with respect to bath). We also observed that when we completely replaced $Cl^-$ with gluconate in the lumen and bath, in the presence of luminal amiloride in normal ducts, the luminal potential of these normal ducts also be-

**TABLE 2.** Transepithelial Potential (in millivolts) of Microperfused Sweat Ducts from Cystic Fibrosis and Normal Subjects before and after Adding Amiloride ($10^{-4}$ M) to the Luminal Perfusate with and without $Cl^-$ in the Bathing Solutions ($Cl^-$ was replaced with gluconate)[a]

|  | No Amiloride | | Amiloride | |
|---|---|---|---|---|
|  | $Cl^-$ | $Cl^-$ Free | $Cl^-$ | $Cl^-$ Free |
| Cystic fibrosis | −88 | −62 | +20 | +19 |
| Normal | −10 | −61 | −0.6 | +14 |

[a] Data are taken from Bijman and Quinton.[9]

came positive. Thus, in the absence of a Cl⁻ permeability, an underlying lumen positive electromotive force (EMF) appears to be unmasked when the apical $Na^+$ conductance is blocked. For the luminal potential to become positive, the apical membrane must hyperpolarize more than the basal membrane. A logical source of a hyperpolarizing apical EMF would be an apical $K^+$ conductance because $K^+$ in the sweat is low (5–15 mM).[2] We have previously shown by ion substitution experiments across the entire duct that it is unlikely that the apical membrane is

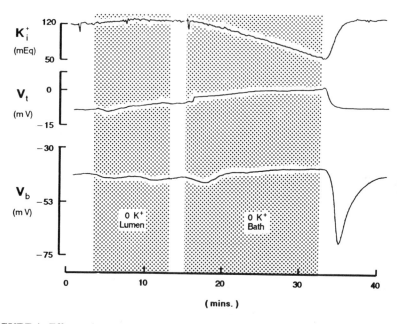

**FIGURE 4.** Effects of $K^+$ removal from bath and lumen on intracellular $K^+$ concentration and membrane potentials. Normally the perfusing solutions in the lumen and bath contain 5 mM $K^+$. When $K^+$ is removed from the perfusing solution, there is virtually no effect on the cytoplasmic $K^+$ concentration or the membrane potentials. However, when $K^+$ is removed from the bath, intracellular $K^+$ decreases from about 120 to 50 mM over a period of about 15 minutes, while both the transepithelial and basal membrane depolarize. These responses, as well as the rebound responses in $K^+$ accumulation and hyperpolarization (probably due in part to stimulation of the Na-$K^+$ ATPase), demonstrate the absence of $K^+$ permeability in the apical membrane and the presence of $K^+$ permeability in the basal membrane.

permeable to $K^+$.[9] Intracellular $K^+$-sensitive electrodes (FIG. 4) show more definitely that when $K^+$ is removed from and replaced in the lumen, there is no change in either the luminal membrane potential or the intracellular $K^+$ concentration; that is, luminal $K^+$ changes apparently have no effect on the electrical properties of the ductal cells. In contrast, removal of $K^+$ from the bath caused both a decrease in cytoplasmic $K^+$ concentration and a decrease in the basal membrane and transepithelial potentials. Restitution of the serosal bath $K^+$

**TABLE 3.** Transepithelial Conductances of Normal and Cystic Fibrosis Sweat Ducts in the Presence and Absence of $Cl^-$ and Amiloride[a]

|  | 150 mM Cl | 0 mM Cl | 150 mM Cl + Amiloride | 0 mM Cl + Amiloride |
|---|---|---|---|---|
| Control | 125 | 15 | 55 | 10 |
| Cystic fibrosis | 21 | 15 | 16 | 8 |

[a] Amiloride ($10^{-4}$ M) was added only to the luminal perfusate. Values are given in ms/cm². Conductances were determined by cable analysis as previously described and data are taken from Quinton.[7]

causes immediate hyperpolarization of both membranes and restoration of intracellular $K^+$. From these observations we conclude that the basal membrane is $K^+$ permeable, but the apical membrane is $K^+$ impermeable. Hence, we exclude the possibility that $K^+$ is the source of the hyperpolarizing EMF on the apical membrane.

Because $Na^+$ conductance was blocked with amiloride (its EMF is in the wrong direction, in any case), because $Cl^-$ conductance either was inherently absent as in cystic fibrosis or was removed by replacement with an impermeable anion gluconate, and because apical $K^+$ conductance was absent, we conclude that none of these major ionic constituents could be the source of positive EMF in the apical membrane. Thus, the capacity to excrete acidic sweat suggests that the most likely source of this EMF may be active proton secretion across the apical membrane into the duct lumen.

The presence of an apical membrane electrogenic proton secreting pump seems highly compatible with our observations of duct function. It provides a mechanism by which $HCO_3^-$ can be removed as $CO_2$ from the luminal fluid by nonionic diffusion. It explains the acidity of final sweat, and if coupled to a $Cl^-/HCO_3^-$ exchange, it could be an energy source for $Cl^-$ absorption. However, for hydrogen secretion to support $Cl^-$ absorption via a $Cl^-/HCO_3^-$ exchange when luminal $Cl^-$ is low, it seems likely that the $Cl^-$ conductance of the apical membrane must be decreased. Otherwise, higher cytoplasmic concentrations of $Cl^-$ would leak back across the apical membrane into the lumen.

Although direct demonstration of varying $Cl^-$ conductance in the apical membrane has not been possible thus far, we have obtained indirect evidence from the

**TABLE 4.** Anion Selectivity of the Microperfused Sweat Duct from Normal Subjects[a]

|  | Gluconate = < | $HCO_3^-$ < | I < | Cl < | Br < | $NO_3^-$ |
|---|---|---|---|---|---|---|
| $\Delta V_t$ | −44 | −37 | −17 | 0 | +2.9 | +6.0 |
| $\dfrac{P_x}{P_{Cl}}$ | — | .014 | 0.39 | 1 | 1.16 | 1.34 |

[a] The lumen of microperfused sweat ducts was perfused with 25 mM $Cl^-$ plus 125 mM gluconate in the lumen with 150 mM $Cl^-$ in the serosal bath. The luminal chloride concentration was substituted with the ions indicated on an equimolar basis, and the resulting transepithelial potentials of lumen referenced to bath are given in millivolts as the mean for seven ducts. The permeability of the anions relative to $Cl^-$ permeability ($P_x/P_{Cl}$) was estimated from the Goldmann-Hodgkin-Katz relationship after assigning $Na^+$ a permeability equal to 5% $Cl^-$ permeability as suggested from the data in TABLES 1 and 3.

**TABLE 5.** Effect of Amiloride ($10^{-4}$ M) on Anion Selectivity of the Microperfused Normal Sweat Duct[a]

| | Gluconate | = | HCO$_3^-$ | < | Cl | = | I | = < | Br | = | NO$_3^-$ |
|---|---|---|---|---|---|---|---|---|---|---|---|
| $\Delta V_t$ | −11.4 | | −11.1 | | 0 | | +0.7 | | −2.1 | | −2.2 |
| $\dfrac{P_x}{P_{Cl}}$ | 0.65 | | 0.65 | | 1 | | 1.03 | | 1.09 | | 1.09 |

[a] Data are recorded and permeability ratios calculated as described in TABLE 4, except that in the presence of amiloride the permeability of Na$^+$ was assumed to be 0. The apical membrane was also assumed to be impermeable to other ionic species. (See text.)

effects of amiloride which suggest that Cl$^-$ conductance may change substantially. First, in the presence of Cl$^-$, amiloride reduces the total epithelial conductance by about 70 ms/cm$^2$ in the normal duct (TABLE 3). In the absence of Cl$^-$, amiloride reduces the transepithelial conductance by only about 5 ms/cm$^2$. These unequal changes in conductance as a function of amiloride and Cl$^-$ removal can be explained as being due to either a decrease in Cl$^-$ conductance that follows the blocking of Na$^+$ conductance or a decrease in Na$^+$ conductance that follows the loss of Cl$^-$. Amiloride also seems to influence Cl$^-$ conductance, because it substantially reduces the anion selectivity of the epithelium. That is, in the absence of luminal amiloride, we find that relative to Cl$^-$, gluconate appears to be impermeable across the epithelium. HCO$_3^-$ is <2% of P$_{Cl}$ and the epithelium shows good discrimination among the halides and nitrate (TABLE 4). But when amiloride is applied to the lumen, gluconate and HCO$_3^-$ appear to be almost two thirds P$_{Cl}$ and the halides and nitrate appear to have almost equal permeabilities (TABLE 5). The apparent loss of anion selectivity is consistent with the loss of significant anionic conductance across the epithelium, which would occur if Cl$^-$ conductance were reduced by amiloride. We do not know how this effect is exerted, but one possibility might be via changes in intracellular pH. That is, amiloride might leave the cell acidic, which may depress Cl$^-$ conductance, or hyperpolarization of the apical membrane may shut off Cl$^-$ conductance in a voltage-dependent manner.

We use these components and features to formulate a model of the absorbing sweat duct (FIG. 5). In this model, Na$^+$ diffuses across the apical membrane electrodiffusively under conditions of both high and low Na$^+$ concentrations in the duct lumen, because low cytoplasmic Na$^+$ concentrations and a hyperpolarized apical membrane continuously support a favorable electrochemical gradient for Na$^+$ entry into the cell. Na$^+$ absorption across the cell is supported by Na$^+$/K$^+$ ATPase located in the basolateral membrane.[10] At relatively high Cl$^-$ concentrations, Cl$^-$ may diffuse passively and electroconductively across the apical membrane "down" a chemical gradient but up a smaller electrical gradient. Loss of Cl$^-$ from the cell across the basal membrane could also occur electroconductively down a combined electrochemical gradient. However, as the apical electrochemical gradient for Cl$^-$ transport becomes unfavorable when luminal Cl$^-$ falls, Cl$^-$ conductance may decrease so that Cl$^-$ absorption could now occur predominantly via a Cl$^-$/HCO$_3^-$ exchange in the apical membrane. A favorable gradient for HCO$_3^-$ entry into the luminal compartment may be maintained by a proton pump which neutralizes luminal HCO$_3^-$ to form carbonic acid. CO$_2$ from this acid should be highly membrane permeable and, with the high concentration of carbonic anhydrase found in the duct cell cytoplasm,[11] should provide a continuous source of HCO$_3^-$ to be exchanged for Cl$^-$. In circumstances of high secretory rates where luminal $HCO_3^-$ is supplied from the primary secretory fluid, $HCO_3^-$ absorption in the duct would depend directly on ductal $H^+$ secretory activity.

In conclusion, we suggest that NaCl absorption across the eccrine sweat duct occurs principally via a large electroconductive transcellular pathway for $Cl^-$ under large salt loads (high luminal $Cl^-$ concentration). But when the salt load is low, large apical $Cl^-$ concentration gradients may be maintained and $Cl^-$ absorption may be supported by the concerted actions of a $Cl^-/HCO_3^-$ exchanger and an active proton pump in the apical membrane which provide the driving force for moving $Cl^-$ up an electrochemical gradient into the cell. The movement of $Cl^-$ out of the cell could then be passive and electroconductive across the basolateral membrane.

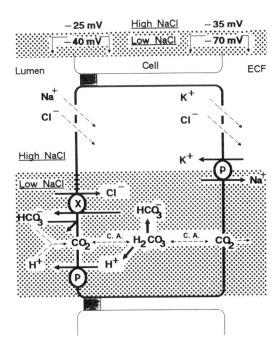

**FIGURE 5.** Model of NaCl absorption under the conditions of high and low NaCl concentrations in the lumen. When the concentration of NaCl is high in the lumen (e.g., 150 mM), the high $Cl^-$ conductance of the apical membrane may permit passive electrodiffusion of $Cl^-$ into the cell and, likewise, out of the cell through a basal membrane conductance. However, under conditions of low NaCl concentration in the lumen (e.g., 15 mM), the electrical and chemical gradient for electroconductive $Cl^-$ movement into the cell appear to become more unfavorable. The model suggests that under these conditions $Cl^-$ could be exchanged for intracellular $HCO_3^-$ which is immediately removed from the lumen by neutralization with protons actively secreted across the apical membrane by a $H^+$ pump. Presumably, under these conditions, the apical $Cl^-$ conductance would be substantially decreased. (See text.) The hyperpolarization of the apical membrane ensures that $Na^+$ enjoys a favorable electrochemical gradient under all luminal conditions. $Na^+$, of course, is actively extruded in exchange for $K^+$ which recycles electroconductively at the basal membrane. Because the ductal cells stain positively for the presence of carbonic anhydrase (Briggman et al.[11]), intracellular $HCO_3^-$ should be readily available from $CO_2$ which should be highly permeable through both membranes.

## REFERENCES

1. QUINTON, P. M. 1979. Water metabolism in animals: Protozoa to man. *In* Comparative Animal Nutrition. Miloslav Recighl, ed. Vol. III: 100–231. S. Karger, Basel.
2. BIJMAN, J. & P. M. QUINTON. 1984. Influence of abnormal Cl permeability on sweating in cystic fibrosis. Am. J. Physiol. **247:** C3–C9.
3. BIJMAN, J. & P. M. QUINTON. 1987a. Lactate and bicarbonate uptake in the sweat duct of cystic fibrosis and normal subjects. Ped. Res. **21:** 79–82.
4. DI SANT'AGNESE, P. A. & G. F. POWELL. 1962. The eccrine sweat defect in cystic fibrosis of the pancreas (mucoviscidosis). Ann. N.Y. Acad. Sci. **93:** 555–599.
5. QUINTON, P. M. 1983. Chloride impermeability in cystic fibrosis. Nature **301:** 421–422.
6. QUINTON, P. M. & J. BIJMAN. 1983. Higher bioelectric potentials in sweat glands due to decreased Cl absorption in patients with cystic fibrosis. N. Engl. J. Med. **308:** 185–189.
7. QUINTON, P. M. 1986. Missing Cl conductance in cystic fibrosis. Am. J. Physiol. **251:** C649–C652.
8. REDDY, M. M. & P. M. QUINTON. 1988. Chloride and bicarbonate transport in the human sweat duct. *In* Cellular and Molecular Basis of Cystic Fibrosis. G. Mastella and P. M. Quinton, eds.: 125–132. San Francisco Press, San Francisco.
9. BIJMAN, J. & P. M. QUINTON. 1987b. Permeability properties of cell membranes and tight junctions of normal and cystic fibrosis sweat ducts. Pflügers Arch. **408:** 505–510.
10. QUINTON, P. M. & J. McD. TORMEY. 1976. Localization of Na-K ATPase in the secretory and reabsorptive epithelia of perfused eccrine sweat glands. A question of the role of the enzyme in secretion. J. Membr. Biol. **29:** 383–399.
11. BRIGGMAN, J. V., R. E. TASHIAN & S. S. SPICER. 1983. Immunohistochemical localization of carbonic anhydrase I and II in eccrine sweat glands from control subjects and patients with cystic fibrosis. Am. J. Pathol. **112:** 250–257.

# H and $HCO_3$ Transport across the Basolateral Membrane of the Parietal Cell[a]

TERRY E. MACHEN, MALCOLM C. TOWNSLEY, ANTHONY M. PARADISO, ETIENNE WENZL, AND PAUL A. NEGULESCU

*Department of Physiology/Anatomy*
*University of California*
*Berkeley, California 94720*

The parietal cells (PC) of the mammalian stomach are responsible for secreting large amounts of highly concentrated HCl (pH <1.0) into the gastric lumen, usually three times a day. This creates special problems in terms of the regulation of intracellular pH (pHi), because the cells have to generate large transcellular fluxes of protons as well as guard against the normal tendencies of cells to acidify from passive forces and from the accumulation of metabolic waste products in a highly active cell.[1] To maintain a relatively constant pHi the PC must have effective membrane mechanisms for regulating pHi, and there must be very good coordination among the activities of the transporters at the apical and basolateral membranes.

H secretion across the apical membrane is generally accepted to occur by the operation of the well-known H/K-ATPase in parallel with K and Cl permeability pathways:[2,3] K and Cl leak out of the cell, K then recycles through the ATPase in exchange for H, and the net result is the secretion of HCl. During this process the cytoplasm produces one OH for every H secreted into the lumen, and this base eventually is transported across the basolateral membrane into the blood.

Both we[4,5] and Muallem, Sachs, and their colleagues[6] have shown that there are Na/H and Cl/$HCO_3$ exchangers in the basolateral membranes of PC. Recent experiments[7] have also shown a Na/$HCO_3$ cotransporter. This paper reviews the relevant experiments from our laboratory and shows how the activities of the various transporters change during activation of HCl secretion by the powerful stimulants histamine plus isobutylmethylxanthine (IBMX), a potent phosphodiesterase inhibitor.

We have used microspectrofluorimetry of the pH-sensitive dye 2′,7′-bis-(2-carboxyethyl)-5(and 6)-carboxyfluorescein, BCECF,[8] to assess the activities of different pHi regulatory mechanisms in the PC of isolated, but intact, rabbit gastric glands. These techniques have been described in detail.[5] Glands that are isolated from rabbit stomachs by high pressure perfusion and collagenase digestion are incubated with the (membrane-permeable) acetoxymethylester derivative of BCECF for 20 to 45 minutes. The dye enters all the cells of the glands, and intracellular esterases cleave the dye into its membrane-impermeable, pH-sensitive, fluorescent form. The glands are then stuck to a coverglass that is mounted into a perfusion chamber on the stage of an inverted microscope. A high numeri-

[a] Work in this laboratory is supported by NIH grants DK19520 and DK38664.

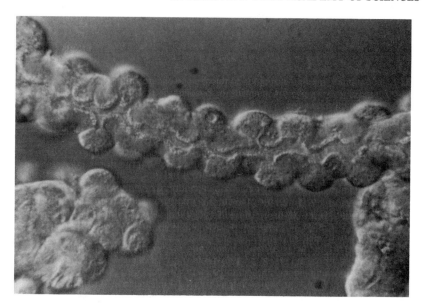

**FIGURE 1.** Nomarski micrograph of an isolated rabbit gastric gland that was mounted in the perfusion chamber and perfused with NaCl Ringer's solution. The lumen can be seen running down the center of the gland. Parietal cells are the large ones that bulge out from the long axis; the intracellular canaliculus can be seen at the entrance to a few of the PC. The smaller cells in between the PC are chief or mucus neck cells.

cal aperture lens allows easy identification of PC from neighboring chief and mucus neck cells (FIG. 1) and also allows efficient collection of the emitted fluorescence. Because one end of the gastric gland is blind ended, and the other end pinches off when the glands are isolated, the solutions that rapidly flow through the chamber perfuse only the basolateral surfaces of the cells.

Measurement of pHi is accomplished by exciting the dye alternately at its pH-sensitive (490 nm) and pH-insensitive (440 nm) wavelengths once every 0.5 seconds using either the commercially available SPEX spectrofluorimeter or a computer-controlled set of filters. The emitted fluorescent light (520–550 nm) is collected by a photomultiplier and computer, and the 490/440 ratio (1 point every second) then yields a concentration and pathlength-independent estimate of pHi. Calibration of the 490/440 signal occurs at the end of an experiment by treating the gland with high [K]-Ringer's solution containing nigericin (the artificial K/H ionophore) at three different (at least) pH values. This so-called *in situ* calibration is more accurate than are calibrations performed on the *in vitro* dye in a cuvette.[9]

To obtain information about only the PC (FIG. 1) an image-plane pinhole is inserted into the microscope. The pinhole effectively eliminates light emitted from adjacent chief cells. This is important because the PC and the chief cells have very different pHi regulatory mechanisms.[4] Our approach also offers the advantage of being able to look at the behavior of PC on a cell-by-cell basis, and this has yielded unique insights into their mechanisms of ionic homeostasis.

## CHANGES OF pHi DURING HISTAMINE TREATMENT

In the resting state PC have pHi 7.1, and this remains steady for long periods of measurement (up to 2 hours). Apparently the light exposure required to obtain reproducible measurements using the photomultiplier does not harm the PC. When PC are incubated in HEPES-buffered, nominally HCO$_3$- and CO$_2$-free solutions and then stimulated with 100 $\mu$M histamine plus 100 $\mu$M IBMX (maximally stimulating doses), pHi undergoes a variety of different responses.[10] In about 60% of the cells pHi increases (FIG. 2A), although the rate of increase was sometimes rapid (within 1 minute) and other times slow (5–10 minutes); the maximal increase was 0.2 pH units, but the average increase (in those PC that showed an increase) was only 0.09 pH units.[10] In the other 40% of the cells pHi either decreased (maximum decrease was 0.2 pH units) or remained constant. When pHi decreased, it usually did so rather rapidly (within 1 minute). An example in which pHi first decresaed then increased back to the baseline is shown in FIGURE 2B.

These data are slightly different from those of Muallem, Sachs, and their colleagues.[11] For isolated PC in suspension in a cuvette they found that 100 $\mu$M histamine always caused pHi to increase by an average of 0.13 pH units (largest increase was 0.22 pH units). Their approach shows the average behavior of millions of cells. Our technique indicates that not all PC behave the same, and an accurate model of how the cells operate must take account of the fact that pHi increases in most PC, but it either decreases or remains constant in a significant number of cells, at least in response to histamine+IBMX.

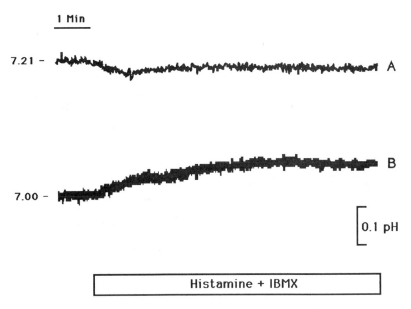

**FIGURE 2.** Effects of 100 $\mu$M histamine + 100 $\mu$M IBMX on pHi in single PC of intact rabbit gastric gland. The gland was incubated with a NaCl, HEPES-buffered Ringer's solution at 37°C, and the stimulants were added as shown. The pHi increased slowly in one cell **(A)**, whereas pHi first decreased then increased in the other **(B)**.

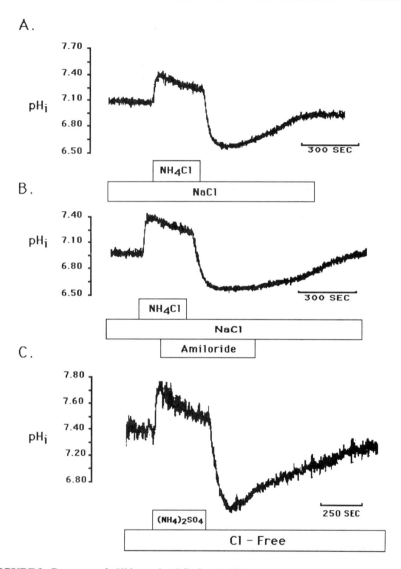

**FIGURE 3.** Recovery of pHi in resting PC after an $NH_4$ prepulse in HEPES-buffered NaCl **(A)**, NaCl plus amiloride **(B)**, and Cl-free Ringer's solution **(C)**. **(A)** The PC was treated with 30 mM $NH_4Cl$, which caused the characteristic increase of pHi. When the $NH_4$ was removed, pHi recovered rapidly. **(B)** If 1 mM amiloride was present during the time when $NH_4$ was washed out of the chamber, pHi recovery was blocked. **(C)** When the same experiment was performed in Cl-free solutions (cyclamate, CYC, replaced Cl, and $(NH_4)_2SO_4$ replaced $NH_4Cl$), the rate of pHi recovery was approximately the same as in NaCl. Note that pHi was initially 7.5 in the Cl-free solutions. (See also FIG. 4.) Experiments were performed at room temperature.

Thus, although the PC secretes large amounts of HCl across its apical membrane, pHi remains fairly constant, even in HCO$_3$- and CO$_2$-free solutions. However, as we will show, large changes occur in the activities of the various pHi regulatory mechanisms at the serosal membrane, and their coordinated activities give rise to the relative stability of pHi. We will now discuss the behavior of the Na/H, Cl/HCO$_3$, and Na/HCO$_3$ pathways during resting and stimulated conditions.

## Na/H EXCHANGE

The presence of Na/H exchange in PC has been shown from the fact that when cells are acid loaded, recovery of pHi back to the resting level is Na dependent, independent of the presence of Cl or HCO$_3$, and is blocked by 0.5–1.0 mM amiloride.[4-6] Experiments on isolated membrane vesicles have shown that the cation exchanger is located on the basolateral membrane.[6] Typical experiments on single PC in intact gastric glands are shown in FIGURE 3A, B, and C. In all cases the PC was acidified using the NH$_4$ prepulse technique. In FIGURE 3A, 30 mM NH$_4$Cl were added to the perfusion solution (HEPES-buffered, nominally HCO$_3$-free), and this caused pHi first to increase (presumably due to entry of the neutral weak base NH$_3$) and then to decrease (due to the combined activities of pHi regulatory mechanisms and NH$_4$ entry); when the NH$_4$ was removed, NH$_3$ quickly left the cell, and the cell acidified. Recovery of pHi occurs quite rapidly in these HEPES-buffered, NaCl Ringer's solutions (FIG. 3A), but recovery is completely blocked by the presence of 1.0 mM amiloride (FIG. 3B and TABLE 1). Similar experiments have been performed under conditions in which the Cl/HCO$_3$ exchanger has been blocked, that is, in Cl-free solutions (FIG. 3C) and in H$_2$DIDS-containing NaCl solutions, and pHi recovery appears normal. A comparison of rates of pHi recovery in Cl versus NaCl+H$_2$DIDS at pH 6.7 (to take account of the fact that most Na/H exchangers are pHi sensitive) shows essentially no difference between the two conditions. (See ref. 7 and the summary in TABLE 1.)

These experiments have demonstrated that resting PC in HEPES-buffered solutions recover from an acid load using only the Na/H exchanger. However, these experiments have not shown that the cation exchanger is operating in the

**TABLE 1.** Rates of pHi Recovery from an Acid Load in HCO$_3$-Free Solutions[a]

| Condition | R/S | Rate (pHi/min) |
|---|---|---|
| NaCl | R | 0.21 ± 0.03 (a) |
| NaCl | S | 0.12 ± 0.05 (a) |
| NaCl + 1 mM amiloride | R | 0 |
| NaCl + 1 mM amiloride | S | 0 |
| NaCl + 200 μM H$_2$DIDS | R | 0.21 ± 0.04 (b) |
| NaCl + 200 μM H$_2$DIDS | S | 0.13 ± 0.02 (b) |

[a] Parietal cells were acidified using the NH$_4$ prepulse technique.[7,10] Rates of pHi recovery (average ± SEM) were measured at pHi 6.7 during the first 45 seconds of pHi increase. All experiments were performed at 37°C in HEPES-buffered, nominally HCO$_3$-free, pH 7.45, NaCl Ringer's solutions.

"R" refers to PC that were treated with $10^{-4}$M cimetidine to insure the resting state. "S" refers to PC that were stimulated with $10^{-4}$M each histamine+IBMX. Data were taken from ref. 10. (a) and (b): Values were significantly different, $p < 0.05$, $t$ test for paired samples.

PC that has a normal pHi, nor do they tell how the antiporter might change its activity during histamine stimulation. Our approach to this problem has been to treat resting and stimulated PC with amiloride and look for changes in pHi immediately following addition of the inhibitor. Because Na/H exchange tends to alkalinize the cell, inhibition of this transporter by amiloride would be expected to acidify the cell. The rate of acidification reflects the activity of the exchanger. A comparison of amiloride's effects on the resting and stimulated PC then yields information about the relative activities of the antiporter in the two states.

As summarized in TABLE 2 and reported in detail elsewhere,[10] amiloride usually caused pHi to decrease more quickly in resting PC as compared to PC that had been stimulated with 100 $\mu$M histamine plus 100 $\mu$M IBMX. In 25 different experiments in which comparisons wre made in the same PC, amiloride caused pHi to decrease more rapidly in 19 resting cells, but there were 2 that decreased more rapidly in the stimulated state, and there were 4 others in which the rates of

TABLE 2. Rates of pHi Change during Treatment with Inhibitors

| Condition | R/S | Rate (pHi/min) |
|---|---|---|
| NaCl + 1 mM amiloride | R | $-0.04 \pm 0.001$ (a) |
| NaCl + 1 mM amiloride | S | $-0.01 \pm 0.004$ (a) |
| NaCl + 200 $\mu$M H$_2$DIDS | R | $+0.014 \pm 0.004$ (b) |
| NaCl + 200 $\mu$M H$_2$DIDS | S | $+0.070 \pm 0.020$ (b) |

Parietal cells were bathed in HEPES-buffered, nominally HCO$_3$-free, pH 7.45, NaCl Ringer's solutions at 37°C. When the pHi was in a steady state, either amiloride (to block Na/H exchange) or H$_2$DIDS (to block Cl/HCO$_3$ exchange) was added to the solution. Amiloride caused pHi to decrease (shown as a negative rate of change of pHi), whereas H$_2$DIDS caused pHi to increase (shown as a positive rate). Rates (averages $\pm$ SEM) were measured during the first 45 seconds of inhibitor treatment.
"R" refers to PC that were treated with 10$^{-4}$M cimetidine to insure the resting state.
"S" refers to PC that were stimulated with 10$^{-4}$M each histamine + IBMX.
Data were taken from ref. 10.
(a) and (b): Values were significantly different, $p < 0.05$, $t$ test for paired samples.

pHi decrease were the same. The average effect of amiloride on pHi in resting versus histamine+IBMX-stimulated PC is shown in TABLE 2. The implication from these studies was that the Na/H exchanger was *usually* becoming *inactivated* by these stimulants—the turnover rate of the Na/H exchanger was on average four times slower in histamine+IBMX versus resting PC.

Confirmation of these results has come from experiments in which rates of pHi recovery following an NH$_4$ prepulse were measured at pHi 6.7 in resting and stimulated PC in both NaCl and NaCl+H$_2$DIDS Ringer's solution. Under these conditions [Na]o≫Na]i and [H[i>[H]o, so this type of experiment yields information about the unidirectional rate of H efflux from the cells, that is, the maximal rate of transport of the Na/H exchanger at this pHi. In 38 different experiments in NaCl Ringer's solution the rate of pHi recovery was faster in 32 resting PC, but 6 PC showed no difference between resting and histamine+IBMX addition. Similar results were obtained with NaCl+H$_2$DIDS, which will block any contribution of the anion exchanger. As summarized in TABLE 1, the average rates of pHi recovery were about two times faster in resting as compared with stimulated PC, indicating that the inherent ability of the Na/H exchanger to transport H was usually decreased by histamine+IBMX.

Our findings are rather different from those of Muallem, Sachs, and their colleagues.[11] They showed that 100 $\mu$M histamine (alone) caused pHi of a suspension of purified PC to increase by up to 0.22 pH units, and this increase was nearly completely blocked by dimethylamiloride, the potent inhibitor of Na/H exchange. They interpreted their data as showing that the Na/H exchanger had been *activated* by histamine. One possible explanation for this discrepancy may be that Muallem *et al.*[11] recorded the activity of millions of PC, many of which have been stimulated, whereas we[10] have been unlucky enough to have chosen PC that have not really been stimulated to secrete HCl at maximal rates.

It should also be noted that a change from one steady-state pHi to another does not in itself yield unambiguous information. For example, let's assume that the resting PC has a Na/H exchanger and a Cl/HCO$_3$ exchanger, and they both turn over 5 times per second in a resting cell. Because under normal physiologic conditions the cation exchanger is a base loader and the anion exchanger is an acid loader, and assuming no other pHi-modifying processes are occurring in this resting cell, then pHi will remain constant (i.e., the cell is in a steady state). When the PC gets stimulated by histamine, the H/K-ATPase (another base loader) becomes activated. Let's assume for the sake of argument that its turnover rate is 10 times per second in the stimulated PC. If the anion exchanger also gets activated, and its turnover rate increases from 5 to 10 times per second while the *Na/H exchanger decreases* its activity from 5 to 2, then *pHi will still increase*, and this increase will also be blocked by amiloride. Thus, an amiloride-blockable increase of pHi alone does not necessarily indicate that the Na/H exchanger has been activated.

However, Muallem *et al.*[11] also showed that dimethylamiloride caused pHi to decrease more rapidly in histamine-stimulated PC as compared with resting PC; inspection of their FIGURE 8 shows that the rate of pHi decrease was about 3 times faster in the histamine-stimulated cells. It also should be noted that even in our experiments histamine+IBMX occasionally (2 of 25 PC) caused the Na/H exchanger to be stimulated (see above). Also, recent experiments[12] have shown that the weak secretagogue carbachol, which elicits many of its stimulatory effects by raising Cai,[13] causes a 2.5-fold activation of the Na/H exchanger, as judged by the criteria just mentioned. Because histamine and histamine+IBMX both cause increases in both cAMP[14–16] and Cai[17,18] while carbachol causes only increases in Cai,[16] it is possible that either or both of these second messengers are involved in controlling the activity of the antiporter.

Increases in Cai have been shown to stimulate the Na/H exchanger in fibroblasts,[19] and cAMP decreases the activity exchanger in the gallbladder.[20] In light of these facts, it might be proposed that the difference between the results of Muallem *et al.*[11] and our group[10] is that the Na/H exchanger gets activated when Cai is elevated (e.g., carbachol treatment) or when Cai and cAMP are both moderately increased (e.g., histamine alone). And when cAMP is raised to very high levels, the exchanger gets inhibited (e.g., histamine+IBMX). We feel that more experiments will be needed to determine whether histamine versus histamine+IBMX and cAMP versus Cai have different effects on the Na/H exchanger of the PC, or perhaps whether some subtle difference of experimental technique can explain the apparent contradictions between the data of Muallem *et al.*[11] and Paradiso *et al.*[10]

## Cl/HCO$_3$ EXCHANGE

The presence of the anion exchanger can be demonstrated by altering the transmembrane concentration gradients of Cl. A typical experiment is shown in FIGURE 4A, where it can be seen that Cl removal caused pHi to increase, presumably because Cl left and OH or HCO$_3$ entered the cell. Other studies[4-6] have shown that these Cl-dependent changes of pHi also occur in Na-free solutions, and as shown in FIGURE 4B, when PC were pretreated with the anion exchange inhibitor DIDS, Cl removal had no effect on pHi. The implication from these

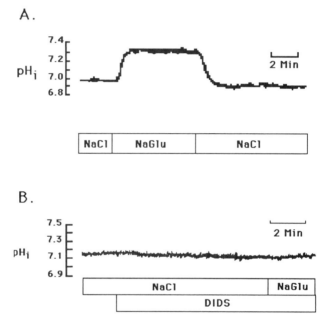

**FIGURE 4.** Effects of Cl removal on pHi in a single, resting PC. The perfusion solution was changed from a Cl-containing, HEPES-buffered solution to Cl-free solution as shown. In the control PC **(A)** this maneuver caused pHi to increase by 0.3 pH units, presumably due to the loss of cell Cl and the entry of OH or HCO$_3$. When PC were first treated with 200 $\mu$M DIDS for 10 minutes, the same Cl-free treatment elicited no change of pHi **(B)**. These experiments were conducted at room temperature.

studies is that the PC has an anion exchanger that is inhibited by disulfonic stilbenes. Experiments on isolated basolateral membrane vesicles have shown that this antiporter is located here.[6] Interestingly, this exchanger seems to operate at near-maximal rates in HEPES-buffered, nominally HCO$_3$-free solutions.[5] This offers the possibility of studying the anion exchanger in HCO$_3$-free solutions and subsequently adding HCO$_3$ to study the Na/HCO$_3$ cotransporter. (See below.)

An assessment of the relative rates of turnover of the Cl/HCO$_3$ exchanger during resting versus stimulated conditions has been made using the approach just described: pHi was measured immediately after the addition of H$_2$DIDS to resting and histamine+IBMX-stimulated PC.[10] The inhibitor always caused pHi to increase more rapidly in the stimulated cells. As summarized in TABLE 2, pHi increased on average five times faster in stimulated versus resting PC. (Also see ref. 10.) Thus, histamine+IBMX caused the anion exchanger to increase its rate of turnover by fivefold. This finding is in agreement with the results of Muallem, Sachs, *et al.*[11] who showed that during amiloride treatment of isolated PC pHi decreased about three times more quickly under histamine-stimulated versus resting conditions, and this decrease of pHi was blocked by DIDS. The implication of their experiment was that when the base-loading Na/H exchanger was blocked, the unopposed action of the anion exchanger caused the PC to acidify three times more rapidly in the stimulated state.

We have also analyzed the effect of Cl removal on the rate of pHi increase in resting versus stimulated PC. A typical experiment on a resting PC is shown in FIGURE 4A. Because there is an essentially infinite outward gradient of [Cl] under these conditions, this type of experiment yields information about the maximal rate of unidirectional OH or HCO$_3$ flux into the PC at pHi 7.1. When comparisons were made of resting versus stimulated PC that had pHi 7.1 and were incubated at room temperature, there was no difference in the rate of pHi increase during Cl removal: the rate of pHi increase was 0.34 ± 0.03 pH per minute in resting PC and 0.39 ± 0.10 pH per minute in histamine+IBMX-treated PC, $p > 0.20$ (ref. 10).

Thus, unlike the situation with the cation exchanger, activation of the anion exchanger appears *not* to be due to a change in its inherent ability to transport ions at pHi 7.1. It has been proposed that the Cl/HCO$_3$ exchanger becomes activated when pHi increases,[11] similar to the activation observed in Vero cells[21] and mesangial cells.[22] We recently showed that the anion exchanger of PC is indeed activated by increases of pHi, but there is only a 50–100% increase in activity between pHi 7.1 and 7.3,[23] so this effect cannot account for the three- to fivefold activation observed by us[10] and Muallem *et al.*[11] Also, many PC do not exhibit an increase of pHi during stimulation; therefore, this alkaline pH-induced activation of the anion exchanger is irrelevant for these cells. Rather, we believe that the histamine- or histamine+IBMX-induced activation of anion exchange occurs indirectly, perhaps through a change in intracellular [Cl]. This possibility will be discussed in more detail below. The specific second messengers involved in this activation have not been determined, but the situation is likely different from that of the gallbladder, in which the Cl/HCO$_3$ exchanger is inhibited by maneuvers that increase intracellular [cAMP].[24]

## Na/HCO$_3$ COTRANSPORT

The presence of this transporter in oxyntic cells of resting (i.e., not secreting HCl) frog gastric mucosa has been surmised by Curci and colleagues[25] from recent electrophysiologic studies. The technique of Demarest and Machen[26] was used to impale oxyntic cells across their basolateral membranes with microelectrodes. Curci *et al.*[25] found that changing the [Na] in the serosal solution caused the basolateral membrane potential (Vbl) to change in the direction opposite to that predicted for Na conductance, and these changes of Vbl occurred only if HCO$_3$ was also present in the serosal solution; changing the serosal [HCO$_3$] alone caused Vbl to change in the direction expected for HCO$_3$ conductance. In addition, the disulfonic stilbene SITS, which is known to inhibit Na/HCO$_3$ cotransport in other cell types,[27,28] caused Vbl to depolarize, and the inhibitor also blocked the

Na- and $HCO_3$-dependent changes in Vbl. It was concluded that resting frog oxyntic cells have an electrogenic $Na/HCO_3$ cotransporter that carries a net negative charge and therefore carries more $HCO_3$ than Na. Because SITS caused Vbl to depolarize, it seems likely that this negatively charged cotransporter normally operates as a *base loader* in the resting frog oxyntic cell. This is in contrast to the proximal tubule where the $Na/HCO_3$ cotransporter operates as an acid loader.[29,30]

With this as background, we have investigated whether this transporter is present in rabbit PC.[7] Because the amphibian oxyntic cell secretes both HCl and enzymes, it was possible that the $Na/HCO_3$ cotransporter was present in either (or both) the PC or the chief cell of the mammalian stomach. A typical experiment showing the presence of a $HCO_3$-dependent pHi regulator in the resting rabbit PC is shown in FIGURE 5. When this resting PC was acidified ($NH_4$ prepulse) in solutions containing 25 mM $HCO_3$ and gassed with 5% $CO_2$, pHi recovery occurred at a rapid rate even though the Na/H exchanger was blocked by the presence of 1 mM amiloride during the entire recovery phase (compare to FIG. 3A).

If this pHi regulatory mechanism were indeed a $Na/HCO_3$ cotransporter (i.e., as opposed to a Na-dependent $Cl/HCO_3$ exchanger), then it should be able to operate in the absence of Cl. We therefore investigated the ability of PC to regulate pHi after acid loading in Cl-free solutions, with and without $HCO_3$. Typical traces are shown in FIGURE 6A and B. In HEPES-buffered $HCO_3$- and Cl-free solutions, Na-dependent pHi recovery from an acid load was completely blocked by 1 mM amiloride. In contrast, in the presence of $HCO_3$ pHi recovery occurred as soon as Na was added to the perfusion solution, even though 1mM amiloride was present. These experiments have shown that PC have another (i.e., besides the Na/H exchanger) Na- and $HCO_3$-dependent mechanism for recovering from an acid challenge, and this transporter does not require Cl. As reported by Townsley and Machen[7] and summarized in TABLE 3, the rate of pHi recovery

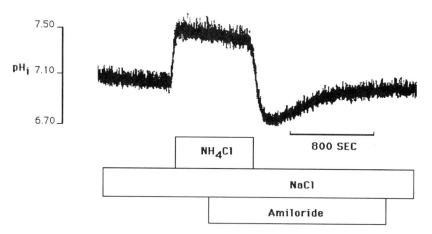

**FIGURE 5.** Recovery of pHi in resting and stimulated PC in $HCO_3$-containing solutions. The PC was perfused with a NaCl Ringer's solution that contained 25 mM $HCO_3$ and was gassed with 5% $CO_2$. The cell was treated with 55 mM $NH_4$. When the $NH_4$ was removed, the cell acidified and then recovered, even though 1 mM amiloride, which blocks the Na/H exchanger, was present throughout the recovery phase. The experiment was conducted at room temperature.

## A. HEPES

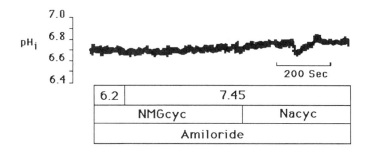

## B. HCO$_3$/CO$_2$

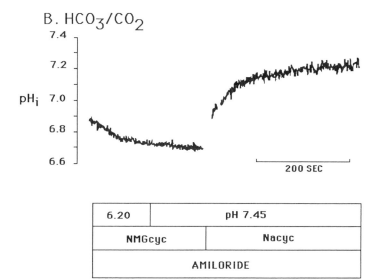

**FIGURE 6.** Na dependence of pHi recovery in Cl-free solutions buffered with either HEPES or HCO$_3$/CO$_2$. **(A)** The cell was first incubated in a Cl-free, HEPES-buffered solution for 20 minutes, and then it was acidified by exposure to a pH 6.2, Cl-free solution. This trace begins with the PC being incubated in a pH 7.45 solution in which all the Na was replaced with N-methyl glucamine (NMG) and the Cl with cyclamate (CYC), and there was also 1 mM amiloride present throughout. In this HCO$_3$-free, amiloride-containing solution, there was no effect on pHi of changing from Na-free to Na-containing solution. **(B)** A similar experiment was performed in HCO$_3$-containing solutions. The trace begins with the acidified PC bathed in a pH 7.45 Na- and Cl-free (NMG cyclamate) Ringer's solution that also contained 25 mM HCO$_3$ and 1 mM amiloride. When the NMG was replaced with Na, pHi recovered rapidly from 6.7 to 7.3, even though amiloride was present throughout the experiment. Experiments were conducted at room temperature.

following an acid load in HCO$_3$ plus amiloride-containing solutions (measured at pHi 6.7) was nearly identical in Cl-containing versus Cl-free solutions. And similar to the mechanism in the proximal tubule,[29,30] this pHi recovery was completely blocked by disulfonic stilbenes (in this case, 200 $\mu$M H$_2$DIDS). (See TABLE 3 and

**TABLE 3.** Rates of pHi Recovery from an Acid Load in Solutions Containing $HCO_3/CO_2$ and Amiloride

| Condition | Rate (pHi/min) |
|-----------|----------------|
| NaCl | 0.10 ± 0.03 (a) |
| Cl-free | 0.11 ± 0.004 (a) |
| Cl-free + 200 $\mu$M H$_2$DIDS | 0 |

Parietal cells (PC) were bathed in HEPES-buffered, nominally $HCO_3$-free, pH 7.45, NaCl or Cl-free (Cl replaced by cyclamate and $SO_4$) Ringer's solutions. PC were acidified using the $NH_4$ prepulse technique or lowering extracellular pH. Rates of pHi recovery (averages ± SEM) were then measured, *always in the presence of 1 mM amiloride*, at pHi 6.7, room temperature, during the first 45 seconds of pHi increase.
PC were treated with $10^{-4}$M cimetidine to insure the resting state.
Date were taken from ref. 7.
(a): Values were insignificantly different, $p > 0.20$.

ref. 7.) Thus, the PC seems to have a $Na/HCO_3$ cotransporter. Although our experiments have not addressed its membrane location (i.e., apical *vs* basolateral), the work of Curci *et al.*[25] and physiologic considerations have indicated that it must be basolateral.

An unusual aspect of the $Na/HCO_3$ cotransporter in the PC is that it seems to operate as a base loader (see above), which is inconsistent with the activity of the stimulated PC. Under these conditions the H/K-ATPase, Na/H exchanger, and the $Na/HCO_3$ cotransporter would all be alkalinizing the cell. We already mentioned that the Na/H exchanger seems to get inhibited (although only partially) during treatment with histamine+IBMX, and this would tend to decrease the base loading effect of this transporter. It appears from experiments on the intact frog gastric mucosa that the $Na/HCO_3$ cotransporter may also get inactivated by secretagogues. Pioneering experiments by Rehm and his colleagues showed that the transepithelial voltage (Vt) was absolutely unaffected by large changes in serosal pH and $[HCO_3]$ in tissues that were *spontaneously secreting H*.[31] With the advent of the histamine (H2) antagonists like metiamide and cimetidine,[32] it became possible to study the resting frog stomach. In these *resting* tissues, changes of serosal $[HCO_3]$ elicited large changes of Vt in the direction expected if there were a $HCO_3$ conductance in the serosal membrane.[33-35] Also, changes of serosal [Na] elicited changes of Vt that were in the direction opposite to that expected for a Na conductance.[36] And in $HCO_3$-free solutions there were essentially no effects of changing serosal [Na] (E. C. Manning and T. E. Machen, unpublished observations). It seems likely that these $HCO_3$- and Na-dependent changes of Vt were due to the operation of the $Na/HCO_3$ cotransporter in the basolateral membrane of the resting oxyntic cells and that the cotransporter was completely inactivated in the tissues that were secreting HCl spontaneously. It remains to be determined whether the $Na/HCO_3$ cotransporter of mammalian PC is inactivated by stimulation.

## A MODEL FOR pHi REGULATION IN THE PC

In the resting state the H/K-ATPase and its associated K and Cl conductances at the apical membrane are inactive, and pHi regulation is accomplished by the combined activities of the Na/H, $Cl/HCO_3$, and $Na/HCO_3$ mechanisms at the basolateral membrane. The Na/K-ATPase keeps intracellular [Na] low (about 8 mM; see ref. 12), and the two Na-dependent transporters appear to operate as base loaders, while the anion exchanger is an acid loader. Working together, these mechanisms keep pHi constant at 7.1 (FIG. 7).

Because the Na/H exchanger appears to be at least partially active at this pHi,[10,11] it will try to attain equilibrium, that is, [Na]i/[Na]o = [H]i/[H]o. For most situations, [Na]i/[Na]o ≪ [H]i/[H]o, and the cation exchanger will be constantly turning over, bringing Na into and removing H from the cell. Because the anion exchanger is also active at pHi 7.1,[23] it too will try to attain equilibrium, that is, [HCO$_3$]i/[HCO$_3$]o = [Cl]i/[Cl]o. Under physiologic conditions of 5% CO$_2$ gassing and pHo 7.4, [HCO$_3$]o = 25 mM, [HCO$_3$]i = 12.5 mM, and [Cl]o = 120 mM.

LUMINAL                                                     BASAL

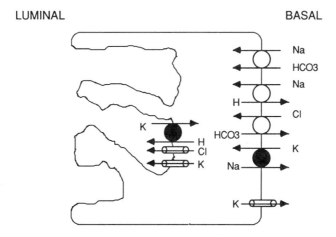

**FIGURE 7.** A model of the PC. Channels are shown by *cylinders*, ATP-utilizing pumps with *filled circles*, and exchangers and cotransporters that do not utilize ATP (at least directly as an energy source) as *open circles*. See text for details.

Therefore, the anion exchanger will transport Cl inward until [Cl]i = 60 mM. In the presence of other leak pathways for Cl (e.g., a Cl conductance at the apical membrane) [Cl]i will be lower than 60 mM, and the anion exchanger will be constantly exchanging cellular HCO$_3$ for extracellular Cl.

The activity of the Na/HCO$_3$ cotransporter under physiologic conditions is somewhat less clear. In the frog oxyntic cell the contransporter appears electrogenic and normally operates in the inward direction. In the rabbit parietal cell it is presently impossible to do accurate electrophysiologic testing to determine if the transporter carries a net charge and in which direction the transporter is normally operating.

Another possible approach to this problem is to add H$_2$DIDS to the resting PC and look for corresponding changes of pHi. However, both the anion exchanger and the cotransporter are inhibited by the stilbenes, so this approach will not work under physiologic conditions (i.e., when both Cl and $HCO_3$ are present). So, for the time being it is merely assumed that the PC works like the oxyntic cell and that the $Na/HCO_3$ cotransporter serves as a base loader in the resting state.

When the PC is treated with histamine+IBMX, the apical K and Cl conductances and the H/K-ATPase get activated. At the basolateral membrane the Na/H exchanger responds by reducing its turnover by a factor of 4, whereas the $Na/HCO_3$ cotransporter may also get inactivated. The $Cl/HCO_3$ exchanger increases its turnover by three- to fivefold. The inactivation of the Na/H exchanger seems to be due to changes in its inherent capability to transport H, but whether the $Na/HCO_3$ cotransporter of rabbit PC gets inactivated is still unresolved. Experiments on frog oxyntic cells indicate that the cotransporter is completely inactive in the stimulated state. Activation of the $Cl/HCO_3$ exchanger does not appear to be due to a change in its ability to transport ions. One possible explanation for its activation is that the apical Cl channels open suddenly, and [Cl]i decreases as Cl rapidly leaves the cell. This would increase the out>in gradient of [Cl] across the basolateral membrane and result in a more rapid uptake of Cl (and efflux of OH).

Because all these changes are occurring nearly simultaneously, pHi of PC often does not change very much, at least in $HCO_3$- and $CO_2$-free solutions. If the combined activities of the alkalinizing pathways are larger than the acidifying effect of the anion exchanger, then pHi will increase. However, if the anion exchanger gets turned on very quickly (perhaps because of the rapid opening of apical Cl channels and decrease of [Cl]i), then pHi will decrease. The variety of pHi changes seen on stimulation suggests that there may not be a fixed pattern or sequence of activation among the pHi regulatory mechanisms. However, the systems appear capable of balancing each other, because the pHi change is never very large. Because different stimulants and second messengers may affect the pHi-regulatory pathways differently, it will be important to see how the various mechanisms respond in the normal physiologic situation, that is, in $HCO_3/CO_2$-buffered solutions and during treatment with more than one secretagogue.

## ACKNOWLEDGMENT

We are grateful to Roger Tsien for the use of the SPEX spectrofluorimeter and for his valuable suggestions during much of this work.

## REFERENCES

1. ROOS, A. & W. F. BORON. 1981. Intracellular pH. Physiol. Rev. **61:** 296–434.
2. WOLOSIN, J. M. & J. F. FORTE. 1984. Stimulation of oxyntic cells triggers K and Cl conductances in apical H/K-ATPase membrane. Am. J. Physiol. **236:** C537–C545.
3. CUPPOLETTI, J. & G. SACHS. 1984. Regulation of gastric acid secretion via modulation of a chloride conductance. J. Biol. Chem. **259:** 14952–14959.
4. PARADISO, A. M., R. Y. TSIEN & T. E. MACHEN. 1987. Digital image processing of intracellular pH in gastric oxyntic and chief cells. Nature **325:** 447–450.
5. PARADISO, A. M., R. Y. TSIEN, J. R. DEMAREST & T. E. MACHEN. 1987. Na/H and Cl/OH(HCO₃) exchange in rabbit oxyntic cells using fluorescence microscopy. Am. J. Physiol. **253:** C30–C36.
6. MUALLEM, S., C. BURNHAM, D. BLISSARD, T. BERGLINDH & G. SACHS. 1985. Electrolyte transport across the basolateral membrane of the parietal cell. J. Biol. Chem. **260:** 6641–6653.

7. TOWNSLEY, M. C. & T. E. MACHEN. 1989. Na/HCO$_3$ cotransport in rabbit parietal cells. Am. J. Physiol. **257**: G350–G356.
8. RINK, T. J., R. Y. TSIEN & T. POZZAN. 1982. Cytoplasmic pH and free Mg$^{2+}$ in lymphocytes. J. Cell. Biol. **95**: 189–196.
9. PARADISO, A. M., P. A. NEGULESCU & T. E. MACHEN. 1986. Na$^+$-H$^+$ and Cl$^-$-OH$^-$(HCO$_3^-$) exchange in gastric glands. Am. J. Physiol. **250**: G524–G534.
10. PARADISO, A. M., M. C. TOWNSLEY, E. WENZL, & T. E. MACHEN. 1989. Regulation of intracellular pH in resting and stimulated parietal cells. Am. J. Physiol. **257**: C554–C562.
11. MUALLEM, S., D. BLISSARD, E. E. CRAGOE & G. SACHS. 1988. Activation of the Na/H and Cl/HCO$_3$ exchangers by stimulation of acid secretion in the parietal cell. J. Biol. Chem. **263**: 14705–14711.
12. NEGULESCU, P. A., A. HAROOTUNIAN, R. Y. TSIEN & T. E. MACHEN. Regulation of intracellular Na in gastric glands measured with a fluorescent sodium indicator. Proc. Natl. Acad. Sci., submitted.
13. NEGULESCU, P. A., W. W. REENSTRA & T. E. MACHEN. 1989. Intracellular Ca requirement for stimulus-secretion coupling in the parietal cell. Am. J. Physiol. **256**: C241–C251.
14. CHEW, C. S., S. J. HERSEY, G. SACHS & T. BERGLINDH. 1980. Histamine responsiveness of isolated rabbit gastric glands. Am. J. Physiol. **242**: G504–G512.
15. RUTTEN, M. J. & T. E. MACHEN. 1981. Histamine, cyclic AMP and activation events in piglet gastric mucosa. Gastroenterology **80**: 928–937.
16. SOLL, A. H. & A. WOLLIN. 1979. Histamine and cAMP in isolated canine parietal cells. Am. J. Physiol. **237**: E444–E450.
17. CHEW, C. S. & M. BROWN. 1986. Release of intracellular Ca and elevation of inosited trisphosphate by secretagogues. Biochim. Biophys. Acta **888**: 116–125.
18. NEGULESCU, P. A. & T. E. MACHEN. 1988. Intracellular Ca regulation during secretagogue stimulation of the parietal cell. Am. J. Physiol. **254**: C130–C140.
19. VILLEREAL, M. 1988. Regulation of Na/H exchange: Role of Ca. *In* Na/H Exchange, S. Grinstein, ed.: 243–253. CRC Press, Boca Raton.
20. REUSS, L. & K. U. PETERSEN. 1985. Cyclic AMP inhibits Na/H exchange at the apical membrane of Necturus gallbaldder epithelium. J. Gen. Physiol. **85**: 409.
21. OLSNES, S., T. I. TONNESSON & K. SANDVIG. 1986. pH-regulated anion transport in nucleated mammalian cells. J. Biol. Chem. **1986**: 967–971.
22. BOYARSKY, G., M. B. GANZ, R. B. STERNE & W. F. BORON. 1988. Intracellular pH regulation in single glomerular mesangial cells. II. Na-dependent and -independent Cl-HCL$_3$ exchangers. Am. J. Physiol. **255**: C857–C869.
23. WENZL, E. & T. E. MACHEN. 1988. Intracellular pH (pHi) regulates the anion exchanger in parietal cells. J. Gen. Physiol. **92**: 25a.
24. REUSS, L. 1987. Cyclic AMP inhibits Cl/HCO$_3$ exchange at the apical membrane of Necturnus gallbladder epithelium. J. Gen. Physiol. **90**: 173–196.
25. CURCI, S., L. DEBELLIS & E. FROMTER. 1987. Evidence for rheogenic Na/HCO$_3$ cotransport in the basolateral membrane of oxyntic cells of frog gastric fundus. Pflüg. Arch. **408**: 497–504.
26. DEMAREST, J. R. & T. E. MACHEN. 1985. Microelectrode measurements from oxyntic cells in intact Necturus gastric mucosa. Am. J. Physiol. **249**: C535–C540.
27. BORON, W. F. & E. L. BOULPAEP. 1983. Intracellular pH regulation in the renal proximal tubule of the salamander. Basolateral HCO$_3$ transport. J. Gen. Physiol. **81**: 53–94.
28. JENTSCH, T. J., M. MATTHES, S. K. KELLER & M. WIEDERHOLT. 1985. Anion dependence of electrical effects of bicarbonate and sodium on cultured bovine corneal endothelial cells. Pflug. Arch. **403**: 175–185.
29. LOPES, A. G., A. W. SIEBENS, G. GIEBISCH & W. F. BORON. 1987. Electrogenic Na/HCO$_3$ cotransport across the basolateral membrane of isolated, perfused Necturus proximal tubule. Am. J. Physiol. **253**: F340–F350.
30. ALPERN, R. J. 1983. Mechanisms of basolateral H/OH/HCO$_3$ transport in the rat convoluted tubule, a sodium-coupled electrogenic process. J. Gen. Physiol. **81**: 53–94.

31. SANDERS, S. S., J. O'CALLAGHAN, C. I. BUTLER & W. S. REHM. 1972. Conductance of submucosal-facing membrane of frog gastric mucosa. Am. J. Physiol. **222:** 1348–1354.
32. BLACK, J. W., W. A. M. DUNCAN, C. J. DURANT, C. R. GANELLIN & E. M. PARSONS. 1972. Definition and antagonism of histamine H2-receptors. Nature **236:** 385–390.
33. SCHLIESSEL, R., A. MERHAV, J. B. MATTHEWS, L. FLEISCHER, A. BARZILAI & W. SILEN. 1980. Role of nutrient $HCO_3$ in protection of amphibian gastric mucosa. Am. J. Physiol. **239:** G536–G542.
34. MANNING, E. C. & T. E. MACHEN. 1981. Effects of $HCO_3$ and pH on Cl transport by gastric mucosa. Am. J. Physiol. **243:** G60–G68. 1982.
35. SCHWARTZ, M., G. CARRASQUER & W. S. REHM. 1985. Evidence for $HCO_3$ conductance pathways in nutrient membrane of resting frog fundus. Biochim. Biophys. Acta **819:** 187–194.
36. MACHEN, T. E. & W. L. MCLENNAN. 1980. Na-dependent H and Cl secretion by in vitro frog gastric mucosa. Am. J. Physiol. **238:** G403–G413.

# Acid and Alkali Secretion by the Turtle Urinary Bladder

## A Model System for Neurohormonal Regulation of Acid-Base Homeostasis

WILLIAM A. BRODSKY

*Department of Physiology and Biophysics*
*Mt. Sinai School of Medicine*
*New York, New York 10029*

Isolated turtle urinary bladders, like isolated collecting tubules of the mammalian kidney, possess metabolically energized mechanisms for the active reabsorption of chloride[1,2] and sodium[3,4] and for the active excretion of alkali[5-10] and acid.[11-13] Recently it was shown that such bladders actively secrete calcium, and after ouabain-induced suppression of sodium reabsorption, actively reabsorb calcium.[14]

Sodium reabsorption is active and electrogenic in nature, not only at the transmural level,[3,4] but also at the transmembranal level.[15] At the transmembranal level in short-circuited turtle bladders, data obtained from microelectrodes kept within single Na-reabsorbing cells, together with macroelectrodes in the extracellular solutions, demonstrated that: (1) the first step in the sodium reabsorptive process is a passive, electrically conductive, lumen-to-cell flow across a Na-selective, amiloride-blockable path in the apical membrane; (2) the final step is an active, electrically conductive, cell-to-serosa flow driven by an ouabain-inhibitable, Na pump mechanism in the basolateral membrane; and (3) no change occurs in any parameter of these Na-reabsorbing cells during inhibition of acid excretion.[16]

These transmembranal findings are in harmony with the transmurally measured, Na-specific transport inhibitory effects of amiloride[17] and ouabain[18] in similarly short-circuited turtle bladders, and with the presence of a (Na + K)-stimulatable, ouabain-inhibitable ATPase activity in electrophoretically localized membrane fractions obtained from homogenates of isolated turtle bladder epithelial cells.[19]

Chloride reabsorption, acid excretion, and alkali excretion have been established as active electrogenic processes at the transmural, but not the transmembranal level, because no one has yet been able to insert a microelectrode into any identifiable non-Na-reabsorbing cell on the surface of an intact, isolated, functionally active turtle bladder. Nevertheless, much has been learned about these transport processes, including their contributions towards maintaining a constant state of water, electrolyte, and acid-base composition of the body fluids in the intact turtle. Consequently, the nature and regulation of acid-base and chloride transport processes across isolated turtle bladders will constitute the main content of this report.

## CHARACTERISTICS OF ACID EXCRETION IN TURTLE BLADDERS

### *An Active Electrogenic Process*

The active nature of acid excretion in the turtle urinary bladder was first demonstrated in isolated, open-circuited sac preparations filled with a relatively small volume (3.0 ml) of luminal fluid and incubated for 5–6 hours in a large volume (1,000 ml) of ($HCO_3$ + $CO_2$)-containing, NaCl-rich serosal Ringer's solutions, the composition of which remained constant. At the end of these incubation periods, the luminal $HCO_3$ concentration reached minimal levels of 0.005–0.01 mM, or less than 1/3500*th* of that in the serosal fluid; the concomitant luminal pH reached minimal levels of 3.85–4.15, or 3.0–3.2 pH units less than the serosal pH; and the maintained, serosa-positive electric potential averaged 54 mV.[11,12]

The magnitude of these chemical concentration gradients (equivalent to 212 mV for bicarbonate and 190 mV for proton activity), fourfold greater than that of the final Pd, not only are sufficient to stop the acid excretory process, but also require a source of metabolic energy (greater than that used for Na reabsorption) to account for the observed depletion of $HCO_3$ ions and accumulation of protons in the luminal fluid during the 5–6 hour incubation of these sac preparations.

These findings were subsequently confirmed in isolated turtle bladder sheet preparations incubating in NaCl-Ringer's media devoid of exogenously added $HCO_3$ or $CO_2$. (1) In the open-circuited state, such bladders spontaneously acidify the luminal fluids to minimal pH levels of 4.5–4.8, while the serosal pH is kept constant at pH 7.4.[13] (2) In the short-circuited state, the rate of luminal pH statting (maximal when luminal pH = serosal pH = 7.4) is nullified when the luminal fluid is reduced to and maintained at pH 4.7, while the serosal fluid is maintained at pH 7.4.[13,20] (3) The electrical analog of these acid excretory-nullifying chemical potential gradients was demonstrated in isolated short-circuited bladders incubating between identical Ringer's solutions devoid of exogenous $HCO_3$ or $CO_2$. Under these conditions, experimentally imposed electrical potential gradients in excess of 144mV (serosa negative) are required to nullify the rate of acid excretion.[21]

The electrogenic nature of this active acid excretory process has been demonstrated under the following incubation conditions.

1. During incubation of short-circuited bladders between identical NaCl Ringer's solutions devoid of exogenously added $HCO_3$ or $CO_2$, the rate of luminal pH statting is independent of the presence of ambient Na or K and partially independent of the presence of ambient Cl.[22]

2. During incubation between identical Cl-free, Na-free, ($HCO_3$ + $CO_2$)-containing, choline-rich Ringer's solutions, a steady-state mucosa (m) to serosal (s) flow of negative short-circuiting current, Isc (neg, ms), is carried exclusively by actively reabsorbed bicarbonate ions or secreted protons (as has been established by appropriate ion substitution) and is a direct saturation function of the luminal $HCO_3$ concentration.[23]

3. Using ouabain-treated, short-circuited turtle bladders incubating between identical NaCl-Ringer's solutions devoid of exogenously added $HCO_3$ or $CO_2$, Schwarts[24] found that the rate of luminal pH statting (at pH 7.4) approximates that of the concomitant m-to-s flow of negative short-circuiting current, Isc(neg, ms). The matching of these two parameters constitutes a direct demonstration of the active electrogenic nature of the acid excretory process.

4. Less direct, but equally compelling evidence for the electrogenicity of acid excretion has been obtained in short-circuited bladders incubating between identi-

cal ($HCO_3$ + $CO_2$)-supplemented choline Cl-rich Ringer's solutions.[2] This evidence will be given in the next major section of this report, because of its bearing on the nature of the Cl reabsorptive process.

## Sensitivity to Pharmacologic Agents

*Inhibitors*

*Acetazolamide.* Serosally added carbonic anhydrase inhibitors such as acetazolamide inhibit the rate of luminal pH statting by 80–90% in isolated, short-circuited turtle bladders incubating between identical NaCl Ringer's solutions, devoid of exogenously added $HCO_3$ or $CO_2$,[13] and decrease the m-to-s flow of negative short-circuiting current, Isc(neg, ms), to the same extent across bladders incubating between identical $HCO_3$ + $CO_2$-containing, Cl-free, Na-free, choline Ringer's solutions.[23]

*Disulfonic Stilbenes (SITS, DIDS).* Serosally added SITS or DIDS (at levels as low as $10^{-6}$M) induces near nullification of the acid excretory, short-circuiting current across bladders in Na-free, choline-rich Ringer's solutions with and without exogenous $HCO_3$ or $CO_2$.[5,6,25,26] This will be discussed in part III of this report.

*Other inhibitors* include mucosally added pseudomonas toxin A,[27] serosally added vanadate,[28] and furosemide.[29]

*Stimulators*

The stimulators include mucosally added amiloride,[30] mucosally added cholera and diphtheria toxins,[27] migration of turtles from a cooler to a warmer ambient temperature,[31,32] and serosally added aldosterone in isolated, substrate-depleted turtle bladders.[33]

## Subcellular Enzymes (Related to Acid Excretion)

*Carbonic Anhydrase*

In view of the acetazolamide-induced decrease of acid excretion in isolated intact turtle bladders, carbonic anhydrase activity was looked for and found in the cytosolic compartments of about 20% of epithelial cells. For example:

1. Directly determined (enzymatically) from the catalyzed (uninhibited) and uncatalyzed (acetazolamide-inhibited) rates of $CO_2$ hydration, carbonic anhydrase activity has been demonstrated almost exclusively in the cytosolic fraction of turtle bladder epithelial cells that had been separated from the submucosal layer by EDTA treatment.[34]

2. Determined histochemically with the Hansson technique in which cobalt and phosphate precipitates are found at carbonic anhydrase-containing regions of the cell, Rosen[35] first visualized these enzyme-selective reaction products in 20% of the epithelial cells in intact, glutaraldehyde-fixed turtle bladders.

3. Schwartz *et al.*[36] used a modification of the Ficoll density gradient technique, developed for epithelial cell systems by Scott *et al.*,[37,38] to separate the

carbonic anhydrase (CA)-rich from the granular (G)-rich cells and to obtain separate suspensions of each cell type. Under these conditions it was shown that acetazolamide induces a 30% inhibition of the $O_2$ consumption in isolated, low density CA cells, but no inhibition of that in isolated G cells, and that ouabain induces a 30% inhibition of $O_2$ consumption in isolated high-density G cells, but not in CA cells.

The implications of these findings are that the acetazolamide-sensitivity of the acid excretory process in the intact turtle bladder can be related to that of carbonic anhydrase activity in 20% of its epithelial lining cells. However, the free energy released from the CA-catalyzed hydration of cytosolic $CO_2$ is not sufficient to account for the observed rate or intensity of the acid excretory process. Conversely, the free energy released by the catalyzed hydrolysis ATP would be sufficient, and in this connection, the membrane-bound ouabain-resistant (Mg) ATPase enzyme complex has been found in subcellular fractions obtained from isolated turtle bladder epithelial cells.

*Ouabain-Resistant Mg-Dependent ATPase Activity.*

This enzyme activity has been demonstrated with an associated ouabain-sensitive, (NA + K)-stimulatable ATPase activity in isolated, mixed membrane fractions obtained from homogenates of turtle bladder epithelial cells.[18,39] In subsequent studies, the ouabain-resistant moiety was physically separated from the ouabain-sensitive moiety of ATPase and enriched in one of four electrophoretically separated subfractions of the isolated mixed membrane fraction.[19,26]

In the studies of Gluck et al.,[40] a DCCD-sensitive moiety of ouabain-resistant ATPase activity was found to comigrate with the ouabain-sensitive moiety of ATPase activity on a sucrose density gradient.

### Active Transport of Acid Into the Interior of Membrane Vesicles Obtained from Turtle Bladder Epithelial Cells

The Mg-dependent, ATP-induced quenching of acridine orange (AO) fluorescence in suspensions of isolated membrane vesicles has been used to estimate the initial rate and final magnitude of the transmembranal pH gradient that develops as a consequence of the active transport of acid from the external medium into the internal fluid compartments of such vesicles. It is generally accepted that the force required to drive this active transport process comes from the hydrolysis of ATP which is catalyzed by the membrane-bound, ouabain-resistant, Mg-dependent, ATPase complex. Isolated membrane vesicles containing this enzyme have been shown to transport protons under different *in vitro* conditions.

1. In the experiments of Gluck et al.,[40] membrane vesicles, maximally enriched in DCCD-inhibitable Mg ATPase activity, were found in one of the pooled fractions taken from a sucrose density gradient column through which a mixed membrane fraction had been passed. These vesicles were then suspended in AO-containing, buffered KCl solutions supplemented with oligomycin and valinomycin. Following the extravesicular addition of Mg ATP, the intravesicular pH rapidly decreases (half-time 30 seconds) to a minimal steady-state level, during which the *net* flow of acid across the vesicular membranes is near-zero.

2. In subsequent experiments,[41,42] mixed membrane vesicles were obtained from isolated dispersed turtle bladder epithelial cells. In these vesicles the ATP-

driven transport of *acid* (from the extra- to the intravesicular compartments) was shown to consist of two distinctly different, kinetically defined moieties. One moiety is maximally inhibited (25–30%) with nanomalor levels of vanadate in the external medium, but not inhibited at all by nanomolar levels of NEM. The other moiety, maximally inhibited (65–70%) with micromolar levels of NEM, is also maximally inhibited (25–30%) with micromolar levels of vanadate.

After these mixed membranes are electrophoretically separated in four pooled fractions (denoted E-I, E-II, E-III, and E-IV in order of decreasing electrophoretic mobility), the vanadate-sensitive acid transport process is localized only in vesicles of the E-II fraction, whereas the NEM-sensitive process is found not only in the vesicles of E-II, but also in those of E-III and, to a lesser extent, in those of the E-I fraction. There is no evokable acid transport function at all in vesicles of the E-IV fraction which is maximally enriched in Mg.ATPase activity. The level of this enzymatic activity in E-IV is 10-fold greater than that in any other electrophoretic fraction.

The implications of these findings are: (1) The NEM-sensitive proton transport function is physically separable from the vanadate-sensitive proton transport. (2) The maximal inhibitory capacity of NEM for its acid-transporting function is 2.5-fold greater than that of vanadate for its acid-transporting function. (3) But the concentration of vanadate required for evoking half of its maximal inhibition of proton transport (30 nanomolar) is less than 1/300th that of NEM for evoking half of its maximal inhibition of proton transport (10 micromolar), which means that the inhibitory affinity of vanadate is 2 exponential orders of magnitude greater than that of NEM. (4) The vanadate-sensitive acid excretory function is located in a single electrophoretic membrane fraction, whereas the NEM-sensitive acid transport function is found in at least two electrophoretic fractions. (5) Serosally added vanadate inhibits acid excretion in the intact isolated ouabain-treated short-circuited turtle bladder,[28] whereas we have found that NEM fails to change the acid excretory current, Isc(neg, ms), in these bladders (unpublished findings).

## CHARACTERISTICS OF Cl TRANSPORT ACROSS TURTLE BLADDERS

The active nature of Cl reabsorption has been demonstrated in isolated, open-circuited turtle bladder sacs incubating in Na-rich or Na-free, (Cl + $HCO_3$ + $CO_2$)-containing Ringer's solutions. Those incubating in Na-rich media are able to reabsorb net quantities of Cl, along with Na and water, until the luminal concentration of Cl is reduced to a minimal level of 0.18 mM, or to less than 1/500*th* of that in the serosal fluid.[1,3] Those incubating in Na-free (choline-rich) media reabsorb net quantities of Cl, along with choline and water, against the force of a spontaneously developed, unfavorably oriented (serosa-negative) transmural electrical potential gradient (PD) averaging 14 ± 3 mV.[3]

The transmural electrogenicity of Cl reabsorption has been demonstrated by Gonzalez *et al.*[2] in isolated short-circuited turtle bladders incubating between Na-free, choline Cl-rich, ($HCO_3$ + $CO_2$)-containing Ringer's solutions. When such bladders are incubated between identical solutions of this type, a steady-state m-to-s flow of negative short-circuiting current, Isc(neg, ms), develops, and the magnitude of this negative Isc is twice that of the concomitant net Cl reabsorptive flux, I(Cl,r) in microamps. This relationship can be represented formally by the equation,

$$\text{Isc(neg, ms)} = \text{I(Cl, r)} + \Delta\text{Isc(neg, ms)},$$

in which $\Delta$Isc(neg, ms), the *measured excess* of Isc(neg, ms) over I(Cl, r), is presumably carried by actively reabsorbed $HCO_3$ ions; for example, $\Delta$Isc(neg, ms) = I($HCO_3$, r).

When the luminal $HCO_3$ (but not the serosal $HCO_3$) is isohydrically (pH 7.6) replaced by $SO_4$, I(Cl, r) remains unchanged, whereas the magnitude of total Isc(neg, ms) decreases rapidly to approximate that of I(Cl, r); for example,

$$Isc(neg, ms) \geqslant I(Cl, r).$$

This means that I($HCO_3$, r) reaches very low, but nonetheless finite rates following the isohydric removal of luminal $HCO_3$, and that nearly all of the negative Isc is carried by actively reabsorbed Cl ions.

On the basis of these findings, one cannot deny that the active reabsorption of Cl is Na independent, electrogenic at the transmural level and independent of the concomitant rate of $HCO_3$ reabsorption. These claims have been verified experimentally as follows.

In short-circuited bladders incubating between identical *Cl-free*, ($HCO_3$ + $CO_2$)-containing, choline $SO_4$ Ringer's solutions, the total Isc(neg, ms), carried exclusively by actively reabsorbed $HCO_3$ ions, reaches and remains at a steady-state level, the magnitude of which is less than that reached in the presence of ambient Cl, but much greater than that found following the isohydric removal of luminal $HCO_3$.[23]

Symmetrical, extracellular addition of Cl is followed by a doubling of the transmural electrical conductance (Gt) in short-circuited bladders, initially incubating between identical *Cl-free*, ($HCO_3$ + $CO_2$)-containing choline-rich Ringer's solutions and then between identical *Cl-rich*, ($HCO_3$ + $CO_2$)-containing choline-rich Ringer's solutions. This increase in Gt is presumably due to an increase in the Cl- selective conductance paths in one or both plasma membranes.[43–45]

It has also been shown that the reabsorption of Cl is independent of the concomitant rate of Na reabsorption in similarly short-circuited bladders following the substitution of Na for choline in both bathing fluids[4] and following a ouabain-induced nullification of Na reabsorption across bladders incubating between identical Na-rich, Cl-rich, ($HCO_3$ + $CO_2$)-containing Ringer's solutions.[18]

All of the aforementioned findings can be replicated by a Thevinin type of lumped parameter model, used to represent the ion transporting cells of the turtle bladder. In such a model, the transmural flows of Na, Cl, and $HCO_3$ are driven across correspondingly ion-selective conductance paths in parallel with each other. This model has been extended to include a nonconductive,[2,44,45] anion-selective antiporter that mediates a passive Cl-for-$HCO_3$ exchange across the basolateral membranes of the Cl-reabsorbing cells. So extended, the model can be made to replicate the serosal $HCO_3$ dependence of Cl reabsorption as well as the transmural electrogenicity of that process.[44]

### Dependence of Cl Reabsorption on the Presence of Serosal $HCO_3$

This was first demonstrated by Leslie *et al.*[46] in ouabain-treated short-circuited turtle bladders, bathed on both surfaces by equimolar NaCl-rich Ringer's solutions. Before as well as after the incorporation of 20 mM $HCO_3$ into the serosal fluid, the mucosal fluid in these experiments was maintained at pH 4.6–4.8, and the serosal fluid at pH 7.4.

The findings were as follows: Before the serosal $HCO_3$ addition (with both bathing fluids devoid of exogenously added $HCO_3$ and $CO_2$), the rate of Cl reabsorption, along with the m-to-s flow of a positive short-circuiting current, Isc(pos, ms), remained at near-zero levels, and there was little or no titrimetrically detectable alkali flow into the acidic luminal fluid. After the isohydric (pH 7.4) substitution of $HCO_3$ for $SO_4$ in the serosal fluid, the rate of active Cl reabsorption, along with an equimolar, s-to-m flow of $HCO_3$, increased to sustained finite levels, and each of these transport rates was over fivefold greater than the concomitant increment of Isc(pos, ms). In parallel experiments[47] carried out in 1975, it was shown that the magnitude of the serosal $HCO_3$ induced increments of Cl and $HCO_3$ transport were reduced by 50–60% in anoxic bladders.

In summary, given the presence of $HCO_3$ in the serosal fluid, the reabsorption of Cl is maintained with a concomitant equimolar, downhill flow of $HCO_3$ in the reverse (secretory) direction. This observation extends that of Gonzalez et al.[2] who showed that the reabsorption of Cl is equally well maintained with a concomitant level flow of $HCO_3$ in the reabsorptive direction. Therefore, the active reabsorption of Cl is independent of the direction of the concomitant transmural flow of $HCO_3$ in the presence of serosal $HCO_3$. The discovery of a serosal $HCO_3$ requirement for the maintenance of finite rates of Cl reabsorption,[46] a significant advance in our understanding of the steps involved during the active reabsorptive process of Cl reabsorption, has nevertheless led to different interpretations.

*The original interpretation*, by the Leslie and Oliver groups, suggested that the aforementioned characteristic of Cl reabsorption could be due to a metabolically energized, electroneutral antiporter, the action of which drives an obligatorily coupled (one-for-one) exchange of Cl reabsorption for $HCO_3$ secretion across the apical membrane.

*An alternative interpretation* is that although the findings of Leslie et al.,[46] as well as those of Oliver et al.,[47] are indisputable, their interpretation is open to the following criticism. The experimentally induced flow of $HCO_3$ along the direction of a favorable transmural concentration gradient in excess of 16,000 to 1 is no evidence for the assumed active nature of that flow. Neither is the metabolically induced reduction of such a downhill flow sufficient to establish the active nature of the observed s-to-m flow of $HCO_3$. Instead, this could be attributed to the inhibition of a discrete electrogenic Cl pump in the apical membrane followed by a reduced delivery of Cl ions to a $Cl:HCO_3$ antiporter in the basolateral membrane.

The unique serosal sidedness of the $HCO_3$-induced reabsorption of Cl, together with the equimolar s-to-m flow of $HCO_3$, is no evidence for locating the $Cl:HCO_3$ antiporter in the apical membrane (be it a mediator of active or passive exchange of these anions across that membrane), especially on the basis of an ad hoc assumption on the unknown relative permeability of the basolateral membrane to $HCO_3$ and Cl.

Over and above these theoretic considerations, it has been shown experimentally[5-8] that a primary active electrogenic secretion of $HCO_3$, in the absence of exogenuously added Cl, is evoked in isolated short-circuited turtle bladders bathed on both surfaces by identical ($HCO_3$ + $CO_2$)-containing, Na-free, choline-rich Ringer's solution. This observation effectively excludes any obligatory coupling of active $HCO_3$ secretion to the Cl reabsorptive process. As shown in the next section, the discovery of the Cl-independent, active electrogenic secretion of $HCO_3$ in no way confirms the previously observed, s-to-m downhill flow of $HCO_3$,[46] as was recently implied by Stetson et al.[10]

## USE OF DISULFONIC STILBENES (SITS AND DIDS) AS SURFACE-BINDING BLOCKERS OF ANION-SELECTIVE FLOW PATHS

### Early Studies on Erythrocytes and Turtle Bladders

Using erythrocytes suspended in Na Ringer's solution, Cabantichik and Rothstein[48,49] showed that the extracellular addition of SITS or DIDS reduces the membrane permeability to $SO_4$ as well as to Cl, but not to that of Na or K, and that the kinetics of radioactive Cl and $SO_4$ fluxes out of preloaded erythrocytes are *not* those of carrier-mediated processes. Moreover, neither SITS nor DIDS penetrates the plasma membranes, because the isothiocyano groups of each agent form covalent thiocarbamate complexes with free amino groups on the external surface of the plasma membrane. Finally, a 93-kD, $^3$H-DIDS-occupied protein (the presumed anion-selective carrier or conductance path) was isolated from membranes of erythrocytes that had been exposed to $^3$H-DIDS.

To determine the conductive or nonconductive nature of the DSS-sensitive anion transport processes, it was advantageous to determine if DSS compounds can induce changes in the ion transport-related electrical parameters across an epithelial cell system, namely, the turtle urinary bladder.

Using isolated, short-circuited turtle bladders, incubating between *identical* ($HCO_3$ + $CO_2$)-containing, choline Cl-rich Ringer's solutions (or NaCl-rich Ringer's solutions in some cases), Ehrenspeck and Brodsky[25,26] found that the serosal addition of SITS or DIDS (and explicitly not the mucosal addition), at levels of $10^{-6}$ M, nullifies the m-to-s flow of negative short-circuiting current, Isc(neg, ms). But even at levels up to $10^{-3}$ M in both the serosal and mucosal fluid, these compounds do not change the rate of Na reabsorption.

Clearly, the electrical nature of this induced change cannot readily be ascribed to the blockade of any electroneutral anion-exchanging process, even if such a process is operative (or operating) in the turtle bladder.

Additional evidence for the irreversibility of DSS binding to the basolateral membrane of turtle bladders, obtained in subsequent studies,[19] is based on several observations. Repeated substitutions of albumin-supplemented solutions for SITS-containing solutions bathing the serosal surfaces of isolated turtle bladder fail to reverse the SITS-induced suppression of Isc(neg, ms), once that suppression has been initiated. Moreover, after the serosal surfaces of intact turtle bladders have been exposed to $^3$H-DIDS, some of the electrophoretically localized membrane fractions (isolated from such bladders) are enriched in $^3$H-labeled proteins. No such enrichment is found in similarly isolated membranes from bladders exposed to $^3$H-DIDS on the mucosal surface.

### Differential Effects of Disulfonic Stilbenes on Cl Reabsorption and Acid Excretion across Isolated Turtle Bladders

In principle, the necessary and sufficient conditions required for a rigorous demonstration of the active, electrogenic nature of any specified ion flow across epithelial tissue are those prevailing when the transmural gradients of chemical concentration are zero (i.e., when the tissue is incubating between identical mucosal and serosal solutions), and when the transmural gradients of electrical potential are clamped at zero by an externally delivered short-circuiting current. This principle was applied to resolve some questions arising from previously reported data on Cl and $HCO_3$ reabsorption across the turtle bladder. Therefore, in all of the experiments on disulfonic stilbenes (to be described), the transmural electrochemical potential gradients were continuously kept at zero.

Under such symmetrical incubation conditions, disulfonic stilbene-induced changes of anion transport[5,6] have been evoked in three separate groups of isolated, short-circuited turtle bladders incubating between identical Na-free, choline-rich Ringer's solutions. The anion content of these *identical pairs* of incubation solutions, set specifically for bladders in each group, was as follows: group A, Cl-rich (25 mM), $HCO_3$-poor; group B, Cl-rich (25 mM), $HCO_3$-rich (20 mM); and group C, $HCO_3$-rich (20 mM), Cl-free. Under these incubation conditions, Isc, PD, Gt, and unidirectional fluxes (m-to-s and s-to m) of Cl were determined.

*Group A. Bladders in Cl-Rich, $HCO_3$-Poor Media: A Condition under which There is No Net Transport of Cl.* Before SITS, the net reabsorptive flux of Cl is near-zero, whereas Isc (neg,ms) reaches low, finite levels of 3.0 $\mu$amps/cm². The concomitant serosa-negative PD remains at 20 mV and the transmural conductance (Gt) at 0.14 ms/cm². After SITS, net Cl flux remains at near-zero levels, Isc and PD decrease rapidly to zero levels, whereas Gt remains low and unchanged.

*Group B. Bladders in $HCO_3$-Rich, Cl-Rich Media: Demonstration of an Active $HCO_3$ Secretion.* Before SITS, the level of Isc (neg,ms), 24 $\mu$a/cm², is twice the rate of net Cl reabsorption, 12 $\mu$a/cm² or 0.45 $\mu$eq/h/cm², whereas the intermittently measured, serosa-negative PD remains at 36 mV and the transmural electrical conductance (Gt) remains at 0.66 ms/cm². This level of Gt, fivefold greater than that in Group A, can be attributed to the increased concentration of ambient $HCO_3$ (25 mM) bathing the bladders in Group B relative to that (0.1 mM) in Group A. After SITS, the Cl reabsorptive rate decreases by 62% to reach a new steady-state level of 4.6 $\mu$a/cm² or 0.17 $\mu$eq/h/cm², whereas Isc and PD decrease rapidly to zero. The net transfer of Cl ions across electrically silent bladders requires an equal countertransport of anions (or cotransport of cations). Ion substitution tests showed that the charge-balancing ion flow must have been that of $HCO_3$ secretion or proton reabsorption.

In summary, serosally added SITS (or DIDS) nullifies the $HCO_3$ reabsorptive function, thereby unmasking an active process of $HCO_3$ secretion, the magnitude of which is equal to that of the partially inhibited process of Cl reabsorption.

*Group C. Bladders in Cl-Free, $HCO_3$-Rich Media: Evidence for the Primary Active and Electrogenic Nature of $HCO_3$ Secretion.* Before SITS, Isc(neg, ms), PD, and Gt reach steady-state levels of 15 $\mu$a/cm², 39 mV, and 0.35 ms/cm², respectively. This level of transmural Gt in the absence of ambient Cl, half that across bladders in group B, cannot be reconciled with the claim that Cl reabsorption is a nonconductive (electroneutral) transport process.

After SITS, Isc and PD decrease rapidly and reverse in directional orientation, so that the resulting Isc(*pos*, ms) and serosa-positive PD reach steady-state levels of 2.7 ± 0.3 $\mu$a/cm² and 7.3 ± 0.3 mV, whereas Gt remains unchanged. Moreover, the serosal addition of theophylline to such SITS-treated bladders induces a doubling in magnitude of the $HCO_3$ secretory current, Isc(pos,ms), as well as of the serosa-positive PD.

The implications of these findings are that although the maintenance of a finite rate of Cl reabsorption requires the presence of serosal $HCO_3$, the active electrogenic secretion of $HCO_3$, like the active electrogenic reabsorption of $HCO_3$, is a Cl-independent, Na-independent transport process. Moreover, the accelerating effect of theophylline (a well-known phosphodiesterase inhibitor) on the $HCO_3$ secretory rate in SITS-treated bladders implies that intracellularly accumulated cyclic AMP acts like a second messenger to upregulate alkali secretion by the turtle bladder. This was subsequently confirmed experimentally.

## ENVIRONMENTAL CHANGES AND NEUROHORMONAL FACTORS IN THE REGULATION OF ION EXCRETION BY THE TURTLE URINARY BLADDER

Data to be reviewed in this section show that the *in vitro* rate of acid or alkali excretion in an isolated turtle bladder is reproducibly dependent upon the prior *in vitro* temperature to which its donor turtle had become adapted, and/or the prior *in vivo* state of acid-base balance in its donor turtle. These data also include the effects of specific neurohormonal agonists (of the first and second messenger types) on the *in vitro* rates of acid or alkali excretion as well as on that of chloride reabsorption in isolated urinary bladders from acidotic, alkalotic, and euhydric donor turtles.

### Adaptation to Changes in the In Vivo Ambient Temperature

*General (Nonexcretory) Effects.* In their review of this field, White and Somero[50] have discussed the following phenomena. During the adaptation of turtles and other ectothermic vertebrates to a sustained increase in ambient temperature, the concomitant increase in body temperature induces nearly instantaneous increases in the dissociation constants of buffer systems, the kinetic constants of enzymatically catalyzed reactions, and the activity of thermosensitive neurones in the brain stem and spinal cord. During the ensuing 1-2 weeks of exposure to the elevated ambient temperature, the *in vivo* rates of $O_2$ consumption, physical activity, food intake, and gastrointestinal secretion all increase gradually to new and higher steady levels. Throughout the 1-2 weeks of this adaptation period, neurogenically stimulated changes in the respiratory and cardiovascular functions contribute to the maintenance of the systemic acid-base balance.

*Excretory Effects.* Initial evidence for an *in vivo*, thermally regulated transport process in any urine-producing epithelium was that obtained by Schilb[31] on the rate of acid excretion generated by the turtle bladder. These and subsequently extended experiments[32] indicated that the *in vitro* rate of acid excretion (at 25°C) in isolated bladders from 32°C-adapted donor turtles is twice that in isolated bladders (at the same *in vitro* temperature) from 21°C- or 26°C-adapted donor turtles. This correlation parallels the *in vivo*, time-dependent increases in the acidity of bladder urine following the transfer of a 21°C-adapted turtle to a 32°C-statted environment.

### Neurohormonally Regulated Excretory Responses to Changes in the In Vivo Acid-Base Balance

It has been shown in several nonurinary epithelia of mammals and birds that vasoactive intestinal peptide (VIP), a neuroparacrine hormone, acts like a first messenger to upregulate ion-secreting pump mechanisms, presumably by stimulating the adenylate cyclase (AC)-catalyzed reaction cascade.[51] It has also been shown that porcine-derived vasoactive intestinal peptide (pVIP) or cyclic AMP (a well-known intracellular second messenger) can upregulate alkali excretion in the isolated turtle urinary bladder (see below) and that enrichments of adenylate cyclase as well as cAMP-stimulated protein kinase activities have been demonstrated in some of the electrophoretically isolated membrane fractions obtained from turtle bladder epithelial cells.[19]

To determine whether the *in vitro* excretion of alkali or acid in isolated bladders can be correlated with changes of the prior *in vivo* state of acid-base balance in 32°C-adapted donor turtles, the following procedures were used. Bladders removed from alkalotic, acidic, postprandial, and postabsorptive turtles were short circuited and incubated at 25°C between identical ($HCO_3$ + $CO_2$)-containing, Na-free, choline-rich Ringer's solutions with and without exogenously added Cl. Under these conditions, the short-circuiting current, Isc(pos,ms) or Isc(neg,ms) and the rate of Cl reabsorption were monitored before and after the serosal addition of pVIP or cyclic AMP in the presence or absence of the exogenously added phosphodiesterase inhibitor IBMX.

### In Vitro *Changes of Alkali and Acid Excretion*

Alkali excretion cannot be evoked in isolated bladders from acidotic turtles. Conversely, acid excretion cannot be evoked in bladders from alkalotic turtles.[8,44] Therefore, the singly operating acid-base transport mechanism in alkalotic bladders is that responsible for alkali excretion; the singly operating acid-base transport mechanism in acidotic bladders is that responsible for acid excretion. Both mechanisms are operative or potentially operative in bladders from postprandial or postabsorptive turtles.

*Before VIP or cAMP.* In *alkalotic* bladders, alkali excretion, at maximal levels during the early phase (5–10 minutes) of incubation, decreases to near-zero in the ensuing 1–2 hours. In *acidotic* bladders, acid excretion remains at near-maximal levels throughout this 1–2 hour period. In *postprandial* bladders, alkali excretion, Isc(pos,ms), at near-maximal levels during the early phase (5–10 minutes) of incubation, decreases to zero, reverses its directional orientation, and consequently becomes an acid excretory process, Isc(neg, ms), during the ensuing 1–2 hours of incubation. In *postabsorptive* bladders (as in acidotic bladders), acid excretion, (Isc(neg, ms), remains at constant levels during the entire 1–2 hours of incubation.

*After pVIP or cAMP.* In *alkalotic* bladders, under steady-state conditions during which Isc reaches zero levels, serosal additions of cyclic AMP[8,44] or pVIP[52] are followed by a rapid restoration of alkali secretion, Isc(pos, ms), to its initial maximal level, if a phosphodiesterase inhibitor, such as isobutyl methyl xanthine (IBMX), is added to the bathing fluids before, during, or after cAMP or VIP. In acidotic bladders, serosal additions of pVIP or cAMP together with IBMX fail to alter the steady-state rate of acid excretion of Isc(neg,ms). In postprandial bladders, serosal additions of pVIP or cAMP together with IBMX first nullify the steady-state level of acid excretion or Isc(neg,ms), then reverse its directional orientation to that of an Isc(pos,ms) carried by alkali excretion which is restored to its initial maximal level. These data show that both mechanisms of acid-base excretion operate or can be made to operate in bladders of postprandial turtles. In postabsorptive bladders, the steady-state rate of acid excretion, Isc(neg, ms), is nullified but not reversed in orientation after the aforementioned serosal additions. However, after data from acidotic turtle bladders forces us to conclude that this apparent decrease in Isc(neg, ms) is really an increase in Isc(pos,ms).

*In Vitro Change of Cl Reabsorption*

The active reabsorption of chloride, demonstrable in postabsorptive, post-prandial, and alkalotic bladders, is nearly nullified while alkali excretion is enhanced following the addition of cAMP or VIP in the presence of IBMX.[8,10,44] Of further interest with respect to *in vivo* regulatory phenomena is that the net reabsorptive flux of chloride remains at near-zero levels in isolated short-circuited bladders obtained from $NH_4Cl$-loaded (acidotic) turtles.[44]

### Native VIP in Turtle Epithelia

An endogenously formed, reptilian type of vasoactive intestinal peptide (rVIP), located in submucosal neuronal structures of turtle bladder and small intestine, has been identified on the basis of its determined immunochemical characteristics under different conditions as follows.

*Immunoreactivity.* Radioimmunoassays performed on acid extracts of turtle bladder and small intestine in collaboration with Drs. Sudir Paul and Sami Said have shown that the pattern of rVIP binding to an anti-porcine VIP antibody is similar but not identical to that of pVIP binding to that same antibody. For example, serially diluted aliquots of rVIP-containing acid extracts displace more radioactive pVIP from the pVIP-anti-VIP antibody complex than do correspondingly diluted aliquots of nonradioactive pVIP.[53]

*Molecular Size.* Polyacrylamide gel electrophoresis demonstrated that the estimated molecular weight of rVIP is close to that of pVIP.

*Storage Sites.* In collaboration with Dr. Richard Dey, we found that the storage sites of rVIP are localized in submucosal neurones of turtle bladder epithelia. This has been demonstrated from the histologically localized, immunospecific fluorescence obtained after having added a fluorescein-conjugated secondary rabbit serum antibody against the primary pVIP antibody which had been added to sections of whole bladder wall.

*Effect of Systemic Alkalosis on Tissue Content of rVIP.* The concentration of rVIP in the bladders of normal turtles (59 mg/g dry tissue, on the average) is almost fivefold greater than that in alkalotic bladders (12mg/g dry tissue weight, on the average). This finding led us to postulate that adrenergic or cholinergic agonists (released during the development of alkalotic states) could act directly on the VIP-storing neurones, which respond by ejecting rVIP into the surrounding submucosal compartment in or near VIP receptor sites on the serosal surface of nearby alkali-excreting cells of the bladder. However, the IBMX independence of the carbachol-induced alkali excretion suggests that cholinergic agonists cannot be involved as VIP-releasing factors.[54]

### Release of VIP from Its Submucosal Storage Cells

*Adrenergically Initiated Alkali Excretion.* We[55] recently found that the serosal addition of norepinephrine, an alpha-2-adrenergic agonist, induces an IBMX-dependent, phentolamine-inhibitable excretion of alkali in isolated urinary bladders from alkalotic or postprandial turtles, but not from acidotic turtles. Moreover, the electrophysiologic characteristics of this induced alkali transport are indistinguishable from those of the IBMX-dependent pVIP or cAMP-induced alkali transport.

It is now postulated that norepinephrine, acting like a 'zeroth' messenger, stimulates the release of endogenously formed VIP from submucosal neuronal cells of the turtle bladder, and that this VIP (and not the norepinephrine) acts directly in the manner of a first messenger on serosal surface receptors of the alkali-excreting cells to induce a cyclic AMP: protein kinase A-mediated, PDE-

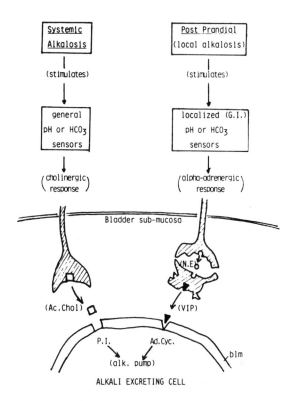

**FIGURE 1.** Tentative working hypothesis for the physiologic role of separate but convergent pathways for the up-regulation of alklai secretion in the turtle. Because *in vitro* acetylcholine (Ac.Chol) or VIP can initiate alkali secretion, it is postulated that the VIP-induced cascade is initiated during gastrointestinal activity, therefore accounting for the finding that postprandial bladders spontaneously secrete alkali, and that during systemic metabolic alkalosis, cholinergic responses initiate the acetycholine-induced alkali secretion, therefore accounting for the finding that alkalotic bladders spontaneously secrete alkali. P.I. = phosphoinositol reaction cascades; Ad.Cyc. = cAMP reaction casacades; alk. pump = components of the alkali secretory pathway stimulated by the two intracellular messenger systems; N.E. = norepinephrine; blm = baso-lateral membrane of the alkali-secreting cell.

modulated activation of the pump mechanism that generates the active excretion of alkali.

*VIP-Independent Cholinergically Initiated Alkali Excretion.* The serosal addition of the muscarinic cholinergic agonist carbachol induces an atropine-blockable stimulation of alkali excretion in the absence as well as the presence of any PDE inhibitor, and this response is evoked in bladders from alkalotic or postprandial turtles, but not from acidotic turtles.

Although the active electrogenic characteristics of carbachol-induced alkali excretion are the same as those of pVIP or cAMP-induced alkali excretion, the IBMX independence of the carbachol-induced effect suggests that it is mediated by an intracellular reaction cascade other than that initiated by adenylate cyclase. In this connection, it has been shown that quasi-second messengers of the phosphoinositide reaction system such as oleyl acryl glycerol (OAG), A-23187, and phorbol myristate acetate also induce an upregulation of alkali excretion in nonacidotic but not in acidotic turtle bladders.[54]

## *Currently Held Tentative Hypotheses*

Two separate, apparently independent intracellular reaction cascades (adenylate cyclase-catalyzed and phopholipase C-catalyzed) are needed for the upregulation of a single transport mechanism, that for alkali excretion. One could postulate that under conditions associated with the development of systemic alkalosis, alkali-sensitive neurones of the central nervous system might be stimulated to release acetylcholine which acts directly on specific muscarinic receptors on the alkali-excreting cells of the turtle bladder. Another hypothesis is that under conditions associated with the development of the postprandial alkaline urinary tide, alkali-sensitive neurones in local interstitial regions of the gastrointestinal tract might transmit signals that release norepinephrine into the interstitium of the urinary bladder where it acts on submucosal neurosecretory cells to induce the release of endogenous VIP from these cells into the surrounding interstitium at or near the VIP receptors on the alkali-excreting cells (Fig. 1).

## REFERENCES

1. BRODSKY, W. A. & T. P. SCHILB 1965. Osmotic properties of isolated turtle bladder. Am. J. Physiol. **208:** 46–57.
2. GONZALEZ, C. F., Y. E. SHAMOO & W. A. BRODSKY. 1967. Electrical nature of active chloride transport across short-circuited turtle bladders. Am. J. Physiol. **212:** 641–650.
3. BRODSKY, W. A. & T. P. SCHILB. 1966. Ionic mechanisms for sodium and chloride transport across turtle bladders. Am. J. Physiol. **210:** 987–995.
4. GONZALEZ, C. F., Y. E. SHAMOO, H. R. WYSSBROD, R. E. SOLINGER & W. A. BRODSKY. 1967. Electrical natural of active sodium transport across short-circuited turtle bladders. Am. J. Physiol. **213:** 333–340.
5. BRODSKY, W. A., G. EHRENSPECK & J. H. DURHAM. 1978. Some new aspects of anion pumping by the turtle bladder. Acta Physiol. Scand. (Suppl.): 341–351.
6. BRODSKY, W. A., J. H. DURHAM & G. EHRENSPECK. 1979. The effects of a disulphonic stilbene on chloride and bicarbonate transport in the turtle bladder. J. Physiol. **287:** 559–573.
7. SATAKE, N., J. H. DURHAM & BRODSKY W. A. 1981. Reversal of the luminal acidification current by a phosphodiesterase inhibitor in the turtle bladder: Evidence for active electrogenic bicarbonate secretion. *In* Structure and Function in Epithelia and Membrane Biophysics. M. Dinno, ed.: 13–24. Alan R. Liss, New York, NY.
8. SATAKE, N., J. H. DURHAM, G. EHRENSPECK & W. A. BRODSKY. 1983. Active electrogenic mechanisms for alkali and acid transport in turtle bladders. Am. J. Physiol. **244**(Cell Physiol. 13): C259–C269.
9. EHRENSPECK, G. 1982. Effect of 3-isobutylmethylxanthine on $HCO_3$ secretion. Biochim. Biophys. Acta **684:** 219–227.

10. STETSON, D. L., R. BEAUWENS, J. PALMISANO, P. P. MITCHELL & P. R. STEINMETZ. 1985. A double-membrane model for urinary bicarbonate secretion. Am. J. Physiol. **249:** F546–F552.

11. SCHILB, T. P. & W. A. BRODSKY. 1963. The acidification mechanism of isolated turtle bladder. Proceedings of the IInd International Congress of Nephrology. Excerpta Medica International Congress Series No. **78:** 103–105. Prague.

12. SCHILB, T. P. & W. A. BRODSKY. 1966. Acidification of mucosal fluid by transport of bicarbonate ion in turtle bladders. Am. J. Physiol. **210:** 997–1008.

13. STEINMETZ, P. R. 1967. Characteristics of hydrogen ion transport in urinary bladder of water turtle. J. Clin. Invest. **46:** 1531–1540.

14. SABATINI, S. & N. A. KURTZMAN. 1987. Calcium transport in turtle bladder. Am. J. Physiol. **253:** R917–R931.

15. NAGEL, W., J. H. DURHAM & W. A. BRODSKY. 1981. Electrical characteristics of the apical and basal-lateral membranes in the turtle bladder epithelial cell layer. Biochim. Biophys. Acta **645:** 77–87.

16. DURHAM, J. H. & W. NAGEL. 1986. Evidence for separate cellular origins of Na and acid-base transport in the turtle bladder. Am. J. Physiol. **250:** C609–C616.

17. WILCZEWSKI, T. & W. A. BRODSKY. 1975. Effects of ouabain and amiloride on the Na pathways in turtle bladder. Am. J. Physiol. **228:** 781–790.

18. SOLINGER, R. E., C. F. GONZALEZ, Y. E. SHAMOO, H. R. WYSSBROD & W. A. BRODSKY. 1968. Effect of ouabain on ion transport mechanisms in the isolated turtle bladders. Am. J. Physiol. **217:** 652–660.

19. BRODSKY, W. A., Z. I. CABANTCHIK, N. DAVIDSON, G. EHRENSPECK, E. KINNE-SAFFRON & R. KINNE. 1979. Localization and characterization of transport related elements in the plasma membrane of turtle bladder epithelial cells. Biochim. Biophys. Acta **556:** 490–508.

20. STEINMETZ, P. R. & L. R. LAWSON. 1971. Effect of luminal pH on ion permeability and flows of Na and H in turtle bladder. Am. J. Physiol. **220:** 1573–1580.

21. AL-AWQATI, Q., A. MUELLER & P. R. STEINMETZ. 1977. Transport of $H^+$ against electrochemical gradients in turtle urinary bladder. Am. J. Physiol. **233:** (Renal Fluid Electrolyte Physiol. 2): F502–F508.

22. STEINMETZ, P. R., R. S. OMACHI & H. S. FRAZIER. 1967. Independence of hydrogen ion secretion and transport of other electrolytes in turtle bladder. J. Clin. Invest. **46:** 1541–1548.

23. GONZALEZ, C. F. & T. P. SCHILB. 1969. Acetazolamide-sensitive short-circuiting current versus mucosal $HCO_3$-concentration in turtle bladders. Biochim. Biophys. Acta **193:** 419–429.

24. SCHWARTZ, J. H. 1976. $H^+$ current response to $CO_2$ and carbonic anhydrase inhibition in turtle bladder. Am. J. Physiol. **231:** 565–572.

25. EHRENSPECK, G. & W. A. BRODSKY. 1975. Effects of 4-acetamido-4-isothiocyano-2, 2-disulfonic stilbene on ion transport in turtle bladders. Biochim. Biophys. Acta **419:** 555–558.

26. BRODSKY, W. A. & G. EHRENSPECK. 1977. The localization of ion-selective pumps and paths in the plasma membranes of turtle bladders. *In* Membrane Toxicity. M. W. Miller & A. E. Shamoo, eds.: 41–50 Plenum Press, New York, NY.

27. BRODSKY, W. A., J. C. SADOFF, J. H. DURHAM, G. EHRENSPECK, M. SCHACHNER & B. M. IGLEWSKI. 1979. Effects of pseudomonas toxin A, diphtheria toxin and cholera toxin on electrical parameters of the turtle bladder. Proc. Natl. Acad. Sci. USA **76:** 3562–3566.

28. EHRENSPECK, G. 1980. Vanadate-induced inhibition of sodium transport and of sodium-independent anion transport in turtle bladder. Biochim. Biophys. Acta **601:** 427–432.

29. EHRENSPECK, G. & C. VONER. 1985. Effect of furosemide on ion transport in the turtle bladder: Evidence for direct inhibition of active acid-base transport. Biochim. Biophys. Acta **817:** 318–326.

30. EHRENSPECK, G., J. H. DURHAM & W. A. BRODSKY. 1978. Amiloride-induced stimulation of $HCO_3$ reabsorption in turtle bladder. Biochim. Biophys. Acta **509:** 390–394.

31. SCHILB, T. P. 1978. Bicarbonate ion transport: Mechanisms for the acidification of urine in the turtle. Science **200:** 208–209.

32. SCHILB, T. P., J. H. DURHAM & W. A. BRODSKY. 1988. Evidence for HCO₃ ion reabsorption: In-vivo environmental temperature and the in-vitro pattern of luminal acidification in turtle bladders. J. Gen. Physiol. **92**(11): 613–642.

33. AL-AWQATI, Q. 1978. H⁺ transport in urinary epithelia. Am. J. Physiol. **35:** 77–88.

34. SCOTT, W. N., Y. E. SHAMOO & W. A. BRODSKY. 1970. Carbonic anhydrase in turtle urinary bladder mucosal cells. Biochim. Biophys. Acta **219:** 248–250.

35. ROSEN, S. 1972. Localization of carbonic anhydrase activity in turtle and toad urinary bladder mucosa. J. Histochem. Cytochem. **20:** 696–702.

36. SCHWARTZ, J. H., D. BETHENCOURT & S. ROSEN. 1982. Specialized function of carbonic anhydrase-rich and granular cells of turtle bladder. Am. J. Physio. **242:** F627–F633.

37. SCOTT, W. N., V. S. SAPIRSTEIN & M. J. YODER. 1974. Partition of tissue functions in epithelia: Localization of enzymes in "mitochondria-rich" cells of toad urinary bladder. Science **184:** 794–799.

38. SCOTT, W. N., M. J. YODER & J. F. GENNARO, JR. 1978. Isolation of highly enriched preparations of two types of mucosal cells of the turtle urinary bladder. Proc. Soc. Exp. Biol. Med. **158:** 565–571.

39. SHAMOO, Y. E. & W. A. BRODSKY. 1970. The (Na + K)-dependent adenosine triphosphatase in the isolated mucosal cells of turtle bladder. Biochim. Biophys. Acta **203:** 111–123.

40. GLUCK, S., S. KELLY & Q. AL-AWQATI. 1982. The proton translocating ATPase responsible for urinary acidification. J. Biol. Chem. **257:** 9230–9233.

41. YOUMANS, S. J., H. WORMAN & W. A. BRODSKY. 1983. ATPase activity and ATP-dependent proton translocation in plasma membrane vesicles of turtle bladder epithelial cells. Biochim. Biophys. Acta **730:** 173–177.

42. YOUMANS, S. J. & W. A. BRODSKY. 1987. Vanadate inhibition of ATP-dependent H transport in membrane vesicles from turtle bladder epithelial cells. Biochim. Biophys. Acta **900:** 88–102.

43. DURHAM, J. H., C. MATONS, G. EHRENSPECK & W. A. BRODSKY. 1982. Electrophysiological parameters associated with chloride reabsorption and bicarbonate secretion in the turtle bladder. *In* Chloride Transport in Biological Membranes. J. Zadunaisky, ed.: 243–259. Academic Press. New York, NY.

44. DURHAM, J. H. & C. MATONS. 1984. Cl-induced increment in short-circuiting current of the turtle bladder: Effects of in-vivo acid-base state. Biochim. Biophys. Acta **769:** 297–310.

45. DURHAM, J. H. & W. A. BRODSKY. 1984. Chloride reabsorption by the reptilian (turtle) urinary bladder. *In* Chloride Transport Coupling in Biological Membranes and Epithelia. G. Gerencser, ed.: 249–270. Elsevier Press, The Netherlands.

46. LESLIE, B. R., J. H. SCHWARTZ & P. R. STEINMETZ. 1973. Coupling between Cl-absorption and HCO₃-secretion in turtle urinary bladder. Am. J. Physiol. **225:** 610–617.

47. OLIVER, J. A., S. HIMMELSTEIN & P. R. STEINMETZ. 1975. Energy dependence of urinary bicarbonate secretion in turtle bladder. J. Clin. Invest. **55:** 1003–1008.

48. CABANTCHIK, Z. I. & A. ROTHSTEIN. 1972. The nature of the membrane sites controlling anion permeability of human red blood cells as determined by studies with disulfonic stilbene derivatives. J. Membr. Biol. **10:** 311–330.

49. CABANTCHIK, Z. I. & A. ROTHSTEIN. 1974. Membrane proteins related to anion permeability of human red blood cells: Localization of disulfonic stilbene binding sites in proteins involved in permeation. J. Membr. Biol. **15:** 207–226.

50. WHITE, F. N. & G. SOMERO. 1982. Acid-base regulation and phospholipid adaptations to temperature: Time courses and physiological significance of modifying the milieu for protein function. Physiol. Rev. **62:** 40–90.

51. SAID, S. I. 1984. Vasoactive intestinal polypeptides. Peptides **5:** 143–150.

52. DURHAM, J. H., C. MATONS & W. A. BRODSKY. 1987. Vasoactive intestinal peptide stimulates alkali excretion in the turtle urinary bladder. Am. J. Physiol. **252:** C428–C435.

53.  BRODSKY, W. A., C. MATONS & J. H. DURHAM. 1989. Endogenous VIP: A paracrine regulator of alkali excretion. Kid. Int. **35:** 453.
54.  SCHNEIDER, E., J. H. DURHAM, C. MATONS & W. A. BRODSKY. 1986. Alkali secretion in the turtle bladder: Up-regulation by the phosphoinositol cascade and inhibition by diphenyl-amino carboxylate (DPC). Membr. Biophys. II: Biol. Transport: 81–92. Alan R. Liss, New York, NY.
55.  BRODSKY W. A., J. H. DURHAM & C. MATONS. 1989. Adrenergic release of VIP during up-regulation of alkali excretion in turtle bladders. Fed. Proc. **3:** 1143.

# Effect of Loop Diuretics on Bullfrog Cornea Epithelium

WOLFRAM NAGEL AND GASPAR CARRASQUER

*Department of Physiology*
*University of Munich*
*Federal Republic of Germany*
*and*
*Department of Medicine (Nephrology)*
*University of Louisville*
*Louisville, Kentucky 40292*

The frog cornea epithelium actively transports Cl from stroma to tear.[1] It has been postulated that Cl moves from stroma to cell via a Na-Cl symport located in the basolateral membrane[2] and from cell to tear via a simple Cl pathway located in the apical membrane. The active transport of Cl takes place across the basolateral membrane, and it is secondary to the Na-K ATPase pump which maintains a low intracellular concentration of Na. With a high intracellular electrochemical potential, Cl can move passively across the apical membrane into the tear side. Loop diuretics inhibit the transport of Cl by the cornea[3] as they do in other epithelia.[4] Whereas inhibition in absorptive epithelia has been found at the level of the Na-Cl symport,[4] it was proposed that in the bullfrog cornea, which has a Cl secreting epithelium, inhibition takes place at the level of the simple Cl conductance located in the epical membrane.[3] Present experiments were performed to reassess the site of action of loop diuretics in the cornea. Our data support the concept that loop diuretics inhibit the entrance of Cl via the Na-Cl symport.

Experiments were performed on an *in vitro* preparation of the bullfrog cornea. Bumetanide ($10^{-4}$ M) or furosemide ($10^{-3}$ M) added to the stromal solution (TABLES 1 and 2) decreased short-circuit current $I_{sc}$ to near zero values, decreased transepithelial conductance ($g_t$), hyperpolarized the intracellular potential ($V_o$), increased the transpical membrane fractional resistance ($fR_o$), decreased the intracellular Cl activity ($a_c^{Cl}$), did not affect the apical membrane permeability ($p_o^{Cl}$), and decreased the electrochemical potential difference between the cell and the bathing media ($\Delta\mu^{Cl}/F$) to near zero.

If bumetanide exerted its effect on the Cl exit site, that is, on the Cl conductance located in the apical membrane, an increase in $a_c^{Cl}$ and especially an increase in $\Delta\mu^{Cl}/F$ should be observed. If the effect were exerted purely on the entry site, that is, on the Na-Cl symport, $a_c^{Cl}$ should decrease and $\Delta\mu^{Cl}/F$ should approach zero because the intracellular and the tear Cl electrochemical potential should approach equilibrium across the apical membrane. Results just presented on $a_c^{Cl}$ and on $\Delta\mu^{Cl}/F$ support the concept that bumetanide affects the Na-Cl symport and not the apical membrane Cl conductance. Further support of this concept is the lack of effect of bumetanide on $p_o^{Cl}$. The decrease in $g_t$, increase in $fR_o$, and increase in $V_o$ are probably a result of a decrease in the intracellular Cl concentration.

**TABLE 1.** Effect of Furosemide ($10^{-3}$ M) or Bumetanide ($10^{-4}$ M) on Transport Parameters of Bullfrog Cornea ($n = 19$)[a]

|  | $I_{sc}$ ($\mu A/cm^2$) | $g_t$ ($mS/cm^2$) | $V_o$ (mV) | $fR_o$ |
|---|---|---|---|---|
| Control | 16.9 ± 7.1 | 1.01 ± 0.37 | 50.9 ± 13.3 | 0.34 ± 0.11 |
| Inhibitor | 1.6 ± 1.3 | 0.70 ± 0.4 | 57.0 ± 18.5 | 0.48 ± 0.15 |

[a] "Control" represents readings immediately before addition of the respective inhibitor. "Inhibitor" refers to the steady state in the presence of furosemide ($n = 8$) or bumetanide ($n = 11$). All values after inhibition were significantly different from the controls ($p < 0.01$). (Reproduced, with permission, from Nagel et al.[5])

In conclusion, data have been presented that support the concept that loop diuretics inhibit the Na-Cl symport located in the basolateral membrane of the bullfrog cornea epithelium.

## REFERENCES

1. ZADUNAISKY, J. A. 1966. Active transport of chloride in frog cornea. Am. J. Physiol. **211:** 506–512.
2. CANDIA, O. A. 1982. The active translocation of Cl and Na by the frog corneal epithelium: Cotransport or separate pumps? In Chloride Transport in Biological Membranes. J. A. Zadunaisky, ed.: 223–242. Academic Press. New York, NY.
3. PATARCA, R., O. A. CANDIA & P. S. REINACH. 1983. Mode of inhibition of active chloride transport in the frog cornea by furosemide. Am. J. Physiol. **245:** F660–F669.
4. FRIZZELL, R. A., M. FIELD & S. C. SCHULTZ. 1979. Sodium-coupled chloride transport by epithelial tissues. Am. J. Physiol. **231:** F1–F8.
5. NAGEL et al. 1989. Am. J. Physiol. **256:** C750–755.

**TABLE 2.** Effect of Bumetanide ($10^{-4}$ M) in the Stromal Solution on Transepithelial Transport Parameters ($I_{sc}$ and $g_t$) and on the Magnitude of $Dl^{Cl}/F$ and $a_c^{Cl}$ of Bullfrog Cornea Epithelium[a]

|  | $I_{sc}$ ($\mu A/cm^2$) | $g_t$ ($mS/cm^2$) | $Dl^{Cl}/F$ (mV) | $V_o$ (mV) | $a_c^{Cl}$ (mM | $p_o^{Cl}$ $10^{-6}$ cm/s) |
|---|---|---|---|---|---|---|
| Control | 19.8 | 0.87 | −16.9 | −52.9 | 19.6 | 9.3 |
| ± SEM | 8.4 | 0.24 | 4.1 | 4.5 | 4.0 | 4.3 |
| Bumetanide | 2.7 | 0.54 | −4.1 | −61.5 | 8.5 | 8.1[b] |
| ± SEM | 1.3 | 0.27 | 2.0 | 4.8 | 1.8 | 4.5 |

[a] Also listed are intracellular potential ($V_o$) and the apparent Cl permeability of the apical membrane ($p_o^{Cl}$). The data are mean values ± SEM from 8 tissues.

[b] $P > 0.05$; all other values are significantly different from the corresponding control values ($P < 0.01$). (Reproduced, with permission, from Nagel et al.[5])

# Transport Compartments in the Turtle Urinary Bladder

## An Electron Microprobe Analysis

ADOLF DÖRGE, PETER BUCHINGER, FRANZ BECK,
ROGER RICK, AND KLAUS THURAU

*Department of Physiology*
*University of Munich*
*D-8000 Munich 2, FRG*

There is compelling evidence that the turtle urinary bladder transports H, $HCO_3$, and Na and that these transports may be accomplished by different cell types.[1] To elucidate the functional organization of the turtle urinary bladder (Pseudemys scripta elegans) further, element concentrations in individual cells were determined by electron microprobe analysis. Blocking the transepithelial Na transport was expected to lead to characteristic electrolyte changes by which the putative cell types, with respect to the different transport compartments, could be identified.

After the short-circuit current had been determined on isolated bladder pieces in Ussing-type chambers, the epithelium was covered on the luminal side with a thin layer of an albumin standard solution, and freeze-dried cryosections were prepared for electron microprobe analysis. Energy dispersive X-ray spectra were obtained from the cells and the albumin standard layer in a scanning electron microscope (Cambridge S150) with a LINK detector. Quantification of the cellular element concentrations in millimoles per kilogram wet weight was achieved by direct comparison of the element characteristic X-ray signals of cellular and standard spectra.

The finding that blocking the transepithelial Na transport reversed the short-circuit current indicated that the bladders of these postabsorptive turtles secreted H. After ouabain the short-circuit current decreased from 26.5 to $-3.9$, and after amiloride from 25.4 to $-8.0$ $\mu A/cm^2$. The freeze-dried cryosections revealed in about 70% of the cases a two-layered and in about 30% a three-layered epithelium. The surface cells could be classified morphologically as granular or granule-free cells. TABLE 1 shows the electrolyte concentrations obtained under the different experimental conditions. Whereas under control conditions the cellular Na and K concentrations were similar in all cells and the same was true for Cl in most of the cells, some cells exhibited very low Cl concentrations. The Cl-poor cells mainly represented granule-free surface cells and, to a smaller extent, cells of deeper layers. After ouabain the Na concentration increased in most of the surface and basal cells by about 90 mmol/kg wet weight and the K concentration decreased by the same amount (ouabain-sensitive). Almost no alteration in electrolytes was observed in the remainder of the cells (ouabain-insensitive). According to their Cl concentrations the ouabain-insensitive cells were subdivided into a Cl-rich and a Cl-poor population. Whereas the ouabain-sensitive cells could be identified as

**TABLE 1.** Cellular Electrolyte Concentrations[a]

|  |  | Na | K | Cl |
|---|---|---|---|---|
| Cell Types | *n* | (mmol/kg wet weight) | | |
|  |  | Control | | |
| *Cl-rich* |  |  |  |  |
| Surface cells | 330 | 10.1 ± 7.4 | 117.0 ± 13.3 | 31.2 ± 9.2 |
| Basal cells | 112 | 9.7 ± 7.8 | 119.6 ± 12.4 | 29.6 ± 7.4 |
| *Cl-poor* | 39 | 8.9 ± 6.5 | 115.4 ± 12.1 | 13.1 ± 3.5 |
|  |  | Ouabain | | |
| *Ouabain-sensitive* |  |  |  |  |
| Surface cells | 302 | 101.2 ± 17.3 | 26.7 ± 12.6 | 37.2 ± 9.9 |
| Basal cells | 113 | 102.6 ± 13.8 | 30.1 ± 12.1 | 32.1 ± 5.5 |
| *Ouabain-insensitive* |  |  |  |  |
| Cl-rich | 49 | 25.2 ± 12.1 | 99.5 ± 14.7 | 34.6 ± 7.6 |
| Cl-poor | 22 | 17.3 ± 10.7 | 112.3 ± 14.0 | 12.0 ± 5.6 |
|  |  | Amiloride and Ouabain | | |
| *Cl-rich* |  |  |  |  |
| Surface cells | 333 | 22.2 ± 11.1 | 99.3 ± 12.9 | 33.1 ± 7.2 |
| Basal cells | 128 | 21.3 ± 9.2 | 105.1 ± 11.5 | 29.8 ± 5.0 |
| *Cl-poor* | 25 | 8.1 ± 5.4 | 115.6 ± 9.4 | 11.5 ± 6.2 |

[a] Mean values ± SD, 12–13 experiments under each condition. Ouabain ($10^{-4}$ M) was applied for 90 minutes to the basal and amiloride ($10^{-4}$ M) for 100 minutes, 10 minutes before ouabain, to the apical side.

granular and basal cells, the ouabain-insensitive cells were granule free and more than 70% were located in the surface layer. After the successive application of amiloride and ouabain the changes in electrolytes of Cl-rich surface cells (mainly granular cells) and basal cells were much less pronounced than after ouabain

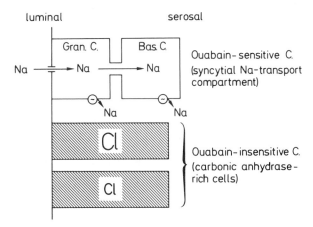

**FIGURE 1.** Cellular organization of turtle urinary bladder. The upper part illustrates the syncytial Na transport compartment, constituted by ouabain-sensitive granular and basal cells. The lower part represents the Cl-rich and Cl-poor populations of ouabain-insensitive cells.

alone. The Cl-poor cells found under this experimental condition had similar electrolyte concentrations to control. The observation that blocking the Na entrance on the apical side with amiloride prevented the typical electrolyte changes observed after ouabain in the granular and basal cells suggests that both cell types constitute a syncytial Na transport compartment as depicted in the upper part of FIGURE 1. Na enters the granular cells via amiloride-sensitive Na channels in their apical membranes, diffuses via intercellular bridges into the basal cells, and is extruded by Na/K pumps into the intercellular spaces. The ouabain-insensitive cells were regarded as carbonic anhydrase-rich cells, as depicted in the lower part of FIGURE 1, because they did not contain large mucin granules, when located in the surface layer. The relation of Cl-rich to Cl-poor ouabain-insensitive cells agrees well with that found for $\alpha$ and $\beta$ types of carbonic anhydrase-rich cells which are thought to be involved in the H and $HCO_3$ secretion.[1]

## REFERENCE

1. STEINMETZ, P. R. 1986. Cellular organization of urinary acidification. Am. J. Physiol. **251:** F173–F187.

# Vibrating Probe Method for Measuring Local Transepithelial Current

CARL SCHEFFEY

*Department of Biological Sciences*
*Columbia University*
*New York, New York 10027*

The vibrating probe has now been used successfully in combination with Ussing chamber to localize ionic transport pathways in five epithelia: teleost opercular membrane, amphibian skin, gastric mucosa, turtle urinary bladder, and chicken coprodeum.[1-6] I summarize here the method and recent technical improvements.

The vibrating probe itself is a way to measure the electrical current density outside a cell with a spatial resolution as good as 5 $\mu$m. It uses a metal electrode moving back and forth (vibrating) over a short distance.

When the electrogenic transport of an epithelium is restricted to a minority of its cells, then those cells may be located by scanning the epithelial surface with the vibrating probe and seeking out peaks of current density. The tissue is viewed with a microscope to allow identification of cells and simultaneously voltage clamped in Ussing chamber to control transepithelial currents.

The effective spatial resolution of the probe has been improved by the recent development of a version that simultaneously measures both the horizontal and vertical components of the current density.[7] Even so, peaks of current density corresponding to individual transporting cells cannot be resolved when the transporting cells cover more than 30% of the epithelial surface.

## REFERENCES

1. Scheffey, C., J. K. Foskett & T. E. Machen. 1983. J. Membr. Biol. **75:** 193–203.
2. Foskett, J. K. & T. E. Machen. 1985. J. Membr. Biol. **85:** 25–35.
3. Katz, U. & C. Scheffey. 1986. Biochim. Biophys. Acta **861:** 480–482.
4. Demarest, J. R., C. Scheffey & T. E. Machen. 1986. Am. J. Physiol. **251:** C643–C648.
5. Durham J. H., A. M. Shipley & C. Scheffey. 1989. This volume.
6. Holtug, K., E. Skadhauge & A. Shipley. Personal communication.
7. Scheffey, C. 1988. Rev. Sci. Instrum. **59:** 787–792.

# Vibrating Probe Localization of Acidification Current to Minority Cells of the Turtle Bladder

JOHN H. DURHAM,[a] ALAN SHIPELY, AND
CARL SCHEFFEY

*Department of Physiology and Biophysics*
*Mt. Sinai School of Medicine*
*New York, New York 10029–6574*

*Marine Biological Laboratories*
*Woods Hole, Massachusetts*
*and*
*Department of Biology*
*Columbia University*
*New York, New York*

There has been considerable evidence that the electrogenic Na-independent acidification process of the turtle urinary bladder is produced by the carbonic anhydrase rich (CA) cells which comprise about 20% of the bladder epithelial cell population. To date, the data which indicate that these cell types are the source of the electrogenic acidification process have arisen from morphologic,[1,2] endocytosed dye,[3] and cell pH[4] studies as well as transmembrane electrophysiologic studies of the majority granular (G-cell) population.[5] In the present study we applied the cell current localization technique of the 2-D vibrating probe as developed by Scheffey[6] to determine if the acidification current of the bladder can be directly localized to a minority cell population of the bladder epithelial cell layer.

Bladders were isolated, mounted, short-circuited, and perfused in a glass-bottomed Ussing chamber and the apical surface was viewed with an upright microscope using a 40× water immersion objective. The luminal and serosal perfusates were initially NaCl (80 mM), NaHCO$_3$ (20 mM) containing media equilibrated with 5% CO$_2$ in O$_2$, pH 7.3. After inhibition of Na reabsorption by adding 0.1 mM amiloride to the luminal perfusate, the short-circuit current was used as a measure of the luminal acidification current.[5]

The epithelial cell surface was brought into focus and scanned until a field of view which contained both G cells (denoted by their granulated appearance) and minority cells (presumably CA cells) was located (FIG. 1). The vibrating probe (5 $\mu$m diameter tip, platinized stainless steel probe) was then brought to within 20 $\mu$m of the cell surface by remotely controlled manipulators and the surface scanned for current sources. The probe was vibrated in planes parallel and perpendicular to the tissue surface, and the magnitude and direction of detected currents were calculated and plotted (by on-line computer analysis[6]) as vectors placed at the probe position in the digitized image of the field of view. The vector

---

[a] To whom correspondence should be addressed.

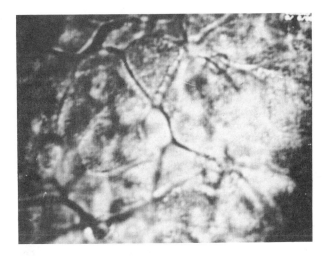

**FIGURE 1.** Photograph of a video image of the apical surface of the perfused, short-circuited turtle bladder observed with DIC optics. The numerous G-cells are identified by their granular appearance. The central minority cell is probably a CA cell (× 600).

component convention used is: a flow of positive charges from the cells into the luminal fluid (the direction of acidification current) which is perpendicular to the cell surface would be plotted in the upward direction (or along the Y-axis); a flow of positive charges which is parallel and flowing to the right (tissue viewed from

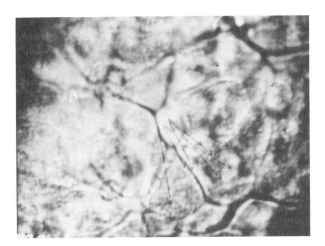

**FIGURE 2.** Photograph of the cells shown in FIGURE 1 following a scan of the surface with the 2-D vibrating probe. The tail of each vector is placed at the pixel location of the probe tip. Large current densities in the direction of acidification are observed over the central minority cell, whereas negligible currents are observed over the G-cells. (Note vectors in the lower left quadrant of the photograph.) (× 600).

above) would be plotted to the right (or along the X-axis). The addition of these components then provides the detection of both the magnitude and the direction of currents in two dimensions with respect to the tissue surface.

FIGURE 2 shows the result of such a scan over the cells shown in FIGURE 1 during the production of an acidification short-circuit current of about 10 $\mu$A. Note the grouping of positive mostly Y-component current vectors over the minority cell. Current measurements over the G-cells, seen in the lower left, show little or no current. Also note that the current vectors become closer to 90 degrees as the center of the minority cell is approached.

This successful demonstration of a localized current production in a bladder producing an acidification short-circuit current provides the first direct evidence that: (1) acidification current is indeed localized to a minority cell population of the bladder, and (2) the G-cells do not detectably contribute to electrogenic acidification.

## REFERENCES

1. HUSTED, R. F., A. K. MUELLER, R. G. KESSEL & P. R. STEINMETZ. 1981. Kid. Int. **19:** 491–502.
2. STETSON, D. L. & P. R. STEINMETZ. 1985. Am J. Physiol. **249:** F553–F565.
3. GLUCK, S., C. CANNON & Q. AL-AWQATI. 1982. Proc. Natl. Acad. Sci. USA **79:** 4327–4331.
4. GRABER, M. L., T. E. DIXON, D. COACHMAN, K. HERRING, A. REIMES, T. GARDNER & E. P. MUNOZ. 1986. Am. J. Physiol. **250:** F159–F168.
5. DURHAM, J. H. & W. NAGEL. 1986. Am. J. Physiol. **250:** C609–C616.
6. SCHEFFEY, C. 1988. Rev. Sci. Instrum. **59:** 787–792.

# Phorbol Myristate Acetate Controls the Mode of Acid/Base Transport in the Turtle Urinary Bladder

MARK GRABER, PHILIP DEVINE, DENISE COACHMAN,
AND TROY DIXON

*VAMC*
*Northport, New York 11768*
*and*
*SUNY at Stony Brook*
*Stony Brook, New York 11794*

Resembling the human collecting duct, the turtle bladder can secrete either $H^+$ or $HCO_3$ into the urine, depending on the environment. Phorbol myristate acetate (PMA) was recently shown to stimulate $HCO_3$ secretion by the turtle bladder.[1] To test for comparable effects of PMA on acid secretion, we measured $H^+$ transport as the reverse short-circuit current in hemibladders where $Na^+$ transport was inhibited by ouabain.

As seen in FIGURE 1, $HCO_3/CO_2$ stimulated acid secretion in control bladders, whereas 0.2 $\mu$M mucosal PMA inhibited baseline levels of acid secretion and abolished this stimulation, producing a paradoxical inhibition of acid secretion in most bladders. Mucosal PMA was found to inhibit acid secretion in a dose-dependent manner with a $K_i$ of $5 \times 10^{-8}$ M and with maximal inhibition rates approaching 100% on large doses of PMA. The PMA inhibition was not reversed by 1 mM azide, which restores transport that has been inhibited by acetazolamide. As measured from the fluorescence of intracellular 4-methylumbelliferone (4MU), the bladder's CA cells were significantly acidified from 0.2 $\mu$M mucosal PMA by $0.14 \pm 0.04$ pH units versus a nonsignificant change of $0.04 \pm 0.04$ in time-control bladders. The appearance of the 4MU-fluorescing cells changed noticeably after treatment with PMA, with a retraction of subapical processes and a "rounding-up" of the cellular appearance. To quantitate such a morphologic change, the planar apical surface area was defined by staining cell borders with Di-O-C5. This area was reduced significantly by 18% in response to mucosal PMA.

We conclude that PMA blocks $CO_2$-stimulated and basal acid secretion by an azide-insensitive mechanism distinct from the inhibition of ACZL. The morphologic changes induced by PMA fit the general pattern noticed earlier by ourselves and others that acid secretion by the turtle bladder correlates directly with the size/shape of the CA cell population.[2] Because PMA stimulates $HCO_3$ secretion by this tissue and inhibits proton transport, our results suggest that protein kinase C-mediated phosphorylation may be a central event in the transition from the secretion of acid to the secretion of base.

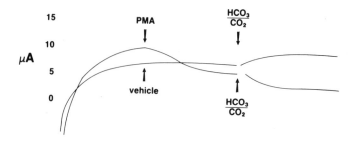

**FIGURE 1.** Effects of PMA on proton transport in response to $HCO_3/CO_2$. The reverse short-circuit current immediately after mounting is typically negative. As sodium transport is progressively inhibited by ouabain, the current becomes positive, representing proton secretion into the mucosal buffer. In the control hemibladder $HCO_3/CO_2$ stimulated proton transport. Mucosal PMA at 0.2 $\mu$M inhibited the baseline transport rate before $HCO_3/CO_2$, and $HCO_3/CO_2$ now produced a paradoxical inhibition of transport. The abscissa represents time over the total observation period of 90 minutes.

## REFERENCES

1. DURHAM, J. H., E. SCHNEIDER, C. MATONS & W. A. BRODSKY. 1988. Two independent, neurohormonally-induced reaction cascades for up-regulation of alkali secretion into the urine (abstr.). Kidney Int. **33:** 399.
2. GRABER, M. L., T. E. DIXON, D. COACHMAN, K. HERRING, A. RUENES, T. GARDNER & E. PASTORIZA-MUNOZ. 1986. Fluorescence identifies an alkaline cell in turtle urinary bladder. Am. J. Physiol. **250:** F159–F168.

# Chloride-Transport Inhibitors *in Vitro* Do Not Inhibit Intestinal Fluid Secretion *in Vivo*

HÅKAN LARSSON,[a] MARIE-LOUISE BERGLUND, AND
JAN FRYKLUND

*Gastrointestinal Research*
*Department of Biology*
*AB Hässle*
*Mölndal, Sweden*

Chloride transport is generally believed to constitute the "driving force" for fluid secretion from intestinal crypts,[1] although direct evidence for this seems to be lacking. We have tried to verify the hypothesis by investigating the effects of $Cl^-$-transport inhibitors on both $Cl^-$ secretion *in vitro* and on fluid secretion *in vivo*. A putative chloride-channel blocker, 5-nitro-2-(3-phenylpropylamino)-benzoate (NPPB),[2] and an inhibitor of the $Na^+/K^+/2Cl^-$ cotransporter, bumetanide (a "loop diuretic;" a gift from P. Feit, Leo Pharm., Ballerup, Denmark), were used to inhibit $Cl^-$ secretion either directly by blocking the apical $Cl^-$ channel or indirectly by blocking basolateral $Cl^-$ uptake and thus the cellular source of $Cl^-$.

## METHODS

$Cl^-$ secretion *in vitro*, stimulated by VIP, dibuturyl-cyclic AMP (db-cAMP), carbachol, or cholera toxin (CT), was measured indirectly by monitoring the short-circuit current ($I_{SC}$) over $T_{84}$ monolayers mounted in a Ussing chamber.[3] $^{36}Cl^-$ uptake in $T_{84}$ cells was measured under $Na^+$-free conditions, representing $Cl^-$ transport through the apical $Cl^-$ channel. The $^{36}Cl^-$-uptake experiments were performed essentially as described by Mandel *et al.*[4]

To measure fluid secretion *in vivo* we used an anesthetized rat model, in which fluid secretion could be followed for several hours by a gravimetric method.[5] CT (20 $\mu$g) was added into the lumen of a 10-cm intestinal segment 1 hour before beginning to record net intestinal fluid movement.

## RESULTS

*In vitro:* VIP-stimulated ($10^{-8}$ M, serosal side) $I_{SC}$ in $T_{84}$ monolayers *in vitro* was inhibited by both NPPB and bumetanide. The $EC_{50}$ values were calculated to be 100–200 $\mu$M for NPPB added to either side of the monolayer and 2–4 $\mu$M and about 1 mM for bumetanide added to the basolateral and apical side, respectively

---

[a] Address for correspondence: Håkan Larsson, Ph.D., Dept. of Biology, Gastrointestinal Research, AB Hässle, S-431 83 Mölndal, Sweden.

(FIG. 1). NPPB also inhibited CT-, db-cAMP-, and carbachol-induced increases in $I_{SC}$ with the same potency (not shown). Furthermore, both VIP- and db-cAMP-stimulated cellular $^{36}Cl^-$ uptake under $Na^+$-free conditions was concentration dependently inhibited by NPPB (FIG. 2). Db-cAMP-stimulated uptake was not affected by bumetanide ($10^{-4}$ M), ouabain ($10^{-3}$ M), or $BaCl_2$ ($2 \times 10^{-2}$ M), suggesting that the $^{36}Cl^-$ was taken up specifically through $Cl^-$ channels.

*In vivo:* In the *in vivo* rat model CT induced a net fluid secretion which, after about 2–3 hours, usually amounted to 100–200 $\mu l$/hour per $cm^2$ intestine. Neither bumetanide, given close intraarterially directly to the secreting intestinal segment at infusion rates of up to 180 $\mu mol/kg$ per hour (i.e., corresponding to concentrations of up to 60 mM given at infusion rates of 6–30 $\mu l$/min) (this was enough to systemically result in a massive diuresis) nor NPPB, added either luminally at concentrations from 0.1 up to 50 mM or infused close intraarterially at rates up to 40 $\mu mol/kg$ per hour, or even given simultaneously by both routes of administration, did affect net fluid transport.

## DISCUSSION

Clearly, both NPPB and bumetanide inhibit $Cl^-$ transport *in vitro*. Also, NPPB blocked the uptake of $^{36}Cl^-$ into $T_{84}$ cells, verifying an action on the $Cl^-$ channel as shown previously in kidney preparations.[2] Despite their actions on $Cl^-$ transport

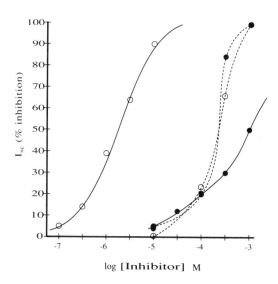

**FIGURE 1.** Effect of bumetanide (——) and NPPB (5-nitro-2-(3-phenylpropylamino)-benzoate) (– – –) on VIP-stimulated $I_{SC}$ in $T_{84}$ cell monolayers mounted in Ussing chambers. VIP was given at the serosal side at a concentration of $10^{-8}$ M, and bumetanide or NPPB was given either at the serosal (○—○) or the mucosal side (●—●) at the time when the maximal VIP-response level ($I_{SC}$ approximately 50 $\mu A/cm^2$) had been attained, that is, approximately 30 minutes after the addition of VIP.

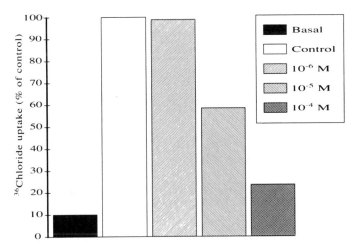

**FIGURE 2.** Effect of NPPB (5-nitro-2-(3-phenylpropylamino)-benzoate) on db-cAMP-stimulated (3 mM added directly into the culture dish) $^{36}Cl^-$ uptake in $T_{84}$ monolayers grown in small Petri dishes ($n = 2-6$). Uptake was monitored during 45 seconds essentially as described by Mandel *et al.*[4]

*in vitro*, neither compound affected net fluid secretion in the rat *in vivo* even though very high concentrations were given. In contrast, the local anesthetic lidocaine, given luminally (2 mg/ml) as a positive control in this rat model, was able to inhibit net fluid transport (not shown). Therefore, our results indicate that the mechanism for CT-induced intestinal fluid secretion *in vivo* may not solely be explained by the secretion of $Cl^-$. An alternative explanation may be that the compounds did not reach the crypt cells. However, bumetanide was shown to be biologically active in this rat model because it caused massive diuresis (not shown). Future experiments have to prove whether or not intestinal $Cl^-$ transport is affected by these compounds *in vivo*.

## REFERENCES

1.  DONOWITZ, M. & M. J. WELSH. 1987. Regulation of mammalian small intestinal electrolyte secretion. *In* Physiology of the Gastrointestinal Tract, 2nd Edition. L. R. Johnson, ed. :1351–1388. Raven Press, New York.
2.  WANGEMANN, P., M. WITTNER, A. DI STEFANO *et al.* 1986. $Cl^-$-channel blockers in the thick ascending limb of the loop of Henle. Structure activity relationship. Pflügers Arch. **407**(suppl 2): S128–S141.
3.  DHARMSATHAPHORN, K., K. G. MANDEL, J. A. MCROBERTS, L. D. TISDALE & H. MASUI. 1984. A human colonic tumor cell line that maintains vectorial electrolyte transport. Am. J. Physiol. **246**: G204–G208.
4.  MANDEL, K. G., K. DHARMSATHAPHORN & J. A. MCROBERTS. 1986. Characterization of a cyclic AMP-activated $Cl^-$ transport pathway in the apical membrane of a human colonic epithelial cell line. J. Biol. Chem. **261**: 704–712.
5.  CASSUTO, J., M. JODAL & O. LUNDGREN. 1982. The effect of nicotinic and muscarinic receptor blockade on cholera toxin induced intestinal secretion in rats and cats. Acta Physiol. Scand. **114**: 573–577.

# Integrated Response of Na-HCO₃ Cotransporter and Na-H Antiporter in Chronic Respiratory Acidosis and Alkalosis

ZVI TALOR,[a] OFELIA S. RUIZ,
AND JOSE A. L. ARRUDA

*Section of Nephrology*
*University of Illinois at Chicago*
*and West Side Veterans Administration Medical Center*
*Chicago, Illinois 60612*

Renal acidification in the renal proximal tubule is thought to be mediated by luminal Na-H antiporter, and the $HCO_3^-$ generated by this antiporter is removed from the cell by a basolateral Na-HCO₃ cotransporter. To study the effect of respiratory acid-base disorders on these transport systems, we measured the Na-HCO₃ cotransport in basolateral membranes and Na-H antiporter in luminal membranes in control rabbits, rabbits exposed to 10% $CO_2$ (chronic hypercapnia), and rabbits exposed to 10% $O_2$ and 90% $N_2$ (chronic hypocapnia).

Renal luminal and basolateral membranes were prepared from cortical homogenates by a sequence of differential and gradient centrifugations with ionic precipitation as described by Kinsella *et al.*[1] with slight modifications.[2] Na-HCO₃ cotransport activity was measured by ²²Na uptake using the rapid filtration technique as described by Akiba *et al.*[3] $HCO_3^-$-dependent Na uptake by basolateral membrane vesicles was measured with an uptake medium containing either gluconate (control group) or bicarbonate (experimental group) at 3 seconds. There were no significant differences among membranes prepared from control, hypercapnic, or hypocapnic rabbits in enzyme markers for both luminal and basolateral membranes. In all three groups, alkaline phosphatase and gamma-glutamyl transverse were enriched approximately 10 times in the luminal membranes, whereas Na-K-ATPase was enriched 14-fold in the basolateral membranes.

FIGURE 1 (upper panel) summarizes the $V_{max}$ of the Na-HCO₃ cotransport in experiments in hypercapnic rabbits, control rabbits, and hypocapnic rabbits. The $V_{max}$ of the Na-HCO₃ cotransport was significantly higher in hypercapnic rabbits than in controls ($2.54 \pm 0.03$ *vs* $1.18 \pm 0.21$ nmol/mg protein/3 seconds, $p$ <0.001). In contrast, the $V_{max}$ was significantly lower in hypocapnic rabbits than

[a] Address for correspondence and reprint requests: Zvi Talor, M.D., Section of Nephrology (m/c 793), University of Illinois Hospital, 840 South Wood Street, Chicago, IL 60612.

controls (0.72 ± 0.11 *vs* 1.18 ± 0.21 nmol/mg protein/3 seconds, $p$ <0.05). FIG-URE 1 (lower panel) summarizes the V$_{max}$ of the Na-H antiporter in the three experimental groups. In agreement with our previous study[4] the V$_{max}$ of the Na-H antiporter was significantly higher in hypercapnic rabbits than in controls (924.9 ± 42.1 *vs* 549.1 ± 62.8 FU/300 $\mu$g protein/minute, $p$ <0.001). There was

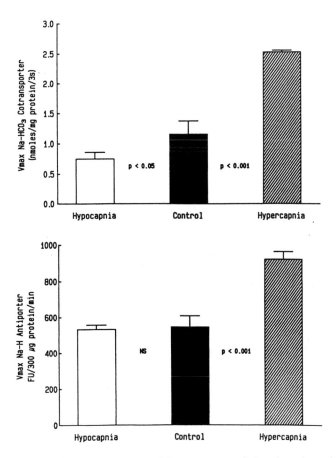

**FIGURE 1.** (*Upper panel*) V$_{max}$ of Na-HCO$_3$ cotransporter in basolateral membranes pre-pared from hypocapnic, control, and hypercapnic rabbits. $n = 5$ in each group. (*Lower panel*) V$_{max}$ of Na-H antiporter in luminal membranes prepared from the three experimental groups. $n = 5$ in each group.

no difference, however, in the V$_{max}$ of Na-H antiporter between hypocapnic rab-bits and controls (524.2 ± 24.3 *vs* 549.1 ± 62.8 FU/300 $\mu$g protein/minute, re-spectively). The V$_{max}$ of Na-HCO$_3$ cotransport was linearly related to the V$_{max}$ of the Na-H antiporter (Y = −0.65 + 0.003X, $r = 0.81$, $p$ <0.001). Thus, there is a good correlation between the V$_{max}$'s of the two transport systems in hypercapnic, control, and hypocapnic rabbits, suggesting a simultaneous adaptation of the two systems in respiratory acid-base disorders.

FIGURE 2 (upper panel) shows the $V_{max}$ of Na-HCO$_3$ cotransport plotted against plasma HCO$_3^-$ concentration for the three groups of animals. It can be clearly seen that the $V_{max}$ of Na-HCO$_3$ cotransport is linearly related to plasma HCO$_3^-$ concentration ($r = 0.84$, $p < 0.001$). Likewise the $V_{max}$ of Na-H antiporter (lower panel) is also linearly related to plasma HCO$_3^-$ ($r = 0.76$, $p < 0.001$). These results do not agree with the conclusion of Akiba et al.[3] that the $V_{max}$'s of both

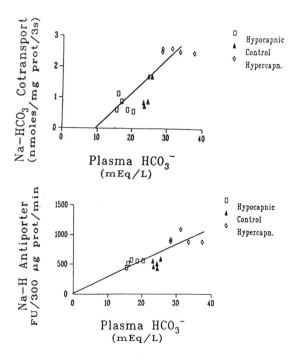

**FIGURE 2.** (*Upper panel*) $V_{max}$ of Na-HCO$_3$ cotransporter plotted against plasma HCO$_3^-$ in hypocapnic, control, and hypercapnic rabbits ($r = 0.84$, $p < 0.001$). (*Lower panel*) $V_{max}$ of Na-H antiporter plotted against plasma HCO$_3^-$ in the three experimental groups ($r = 0.76$, $p < 0.001$).

transport systems correlate inversely with plasma bicarbonate concentration. This apparent discrepancy can be reconciled if the relationship of the transport systems to plasma H$^+$ concentration is analyzed. In our study the $V_{max}$'s of both Na-H antiporter and Na-HCO$_3$ cotransporter were linearly related to plasma H$^+$ concentration (not shown), and although Akiba et al.[3] did not report this analysis, it is likely that such correlation was also present. Thus, our results support the hypothesis that high extracellular H$^+$ concentration stimulates, whereas a low extracellular H$^+$ concentration suppresses both transport systems.

## REFERENCES

1. KINSELLA, J. L., P. D. HOLOHAN, N. I. PESSAH & C. R. ROSS. 1979. Isolation of luminal and antiluminal membranes from dog kidney cortex. Biochim. Biophys. Acta **552:** 468–477.
2. TALOR, Z., D. S. EMMANOUEL & A. I. KATZ. 1982. Insulin binding and degradation by luminal and basolateral tubular membranes from rabbit kidney. J. Clin. Invest. **69:** 1136–1146.
3. AKIBA, T., V. K. ROCCO & D. G. WARNOCK. 1987. Parallel adaptation of the rabbit renal cortical sodium/proton antiporter and sodium/bicarbonate cotransporter in metabolic acidosis and alkalosis. J. Clin. Invest. **80:** 308–315.
4. TALOR, Z., W. C. YANG, J. SHUFFIELD, E. SACK & J. A. L. ARRUDA. 1987. Chronic hypercapnia enhances $V_{max}$ of Na-H antiporter of renal brush-border membranes. Am. J. Physiol. **253:** F394–F400.

# Evidence for Luminal KCl Exit As Mechanism for Active Chloride Secretion by Rabbit Cortical Collecting Duct

C. S. WINGO

*Division of Nephrology*
*University of Florida*
*and*
*Veterans Administration Medical Center*
*Gainesville, Florida 32610–0224*

The cortical collecting duct (CCD) represents a major site of K secretion by the distal nephron.[1] When perfused *in vitro*, this segment typically exhibits net K secretion and net Cl absorption, yet the magnitude of net Cl absorption is small compared to unidirectional tracer Cl flux. In part, this is because the majority of Cl flux occurs by exchange diffusion.[2,3] However, direct measurement of paracellular Cl conductance and Cl permeability in the absence of exchange diffusion demonstrates that the paracellular pathway possesses a large Cl permeability.[3,4] This permeability is sufficient to drive substantial passive Cl absorption for voltages that are typically present in the CCD. Furthermore, our recent studies provide unequivocal evidence that active Cl secretion is present in the CCD when animals are fed a $KHCO_3$-rich NaCl-restricted diet.[5] Hence, this study examines net Cl flux ($J_{Cl}^N$), active Cl flux ($J_{Cl}^A$), and passive Cl flux ($J_{Cl}^P$) by the CCD when rabbits were conditioned to a low K diet or one of two separate methods of K loading.

## METHODS AND MATERIALS

Female New Zealand White rabbits were maintained on one of three diets for 3–10 days before experimentation. The diets were identical except for the addition of 241 mmol $NaHCO_3$ ($NaHCO_3$ diet), 241 mmol $KHCO_3$ ($KHCO_3$ diet), or 241 mmol KCl (KCl diet) per kilogram of a basic diet. All tubules were studied under identical conditions using *in vitro* microperfusion. The bath and dissection solution contained (in mmol): Na 145, K 5, Cl 112, $HCO_3$ 25, Ca 1.5, phosphate 2.3, Mg 1, sulfate 1, acetate 10, glucose 8, alanine 5, and 5% vol/vol fetal calf serum (FCS). Perfusate was identical except that FCS was omitted and 50 $\mu$Ci of $^3$H-inulin was added. All solutions were gased to a pH of 7.4 with 95% oxygen and 5% $CO_2$. Tubules with leaks were discarded. After measurement of volume flux at 37°C at least three collections were obtained for [Cl]. [Cl] was measured in triplicate by electrotitration on six samples of perfused fluid and three samples of collected fluid. Net Cl flux was calculated as $J_{Cl}^N = V_0 ([Cl]_i-[Cl]_0)/L$, where $V_0$ is the rate of fluid collection, and L is the length. Subscripts for [Cl] denote perfused ($[Cl]_i$) and collected ($[Cl]_0$) fluid.

Net Cl transport represents the sum of active Cl transport ($J_{Cl}^A$, strictly speaking, nonpassive Cl transport) and passive Cl transport ($J_{Cl}^P$): $J_{Cl}^N = J_{Cl}^A + J_{Cl}^P$. $J_{Cl}^P$ was derived from the Goldman equation:

$$J_{Cl}^P = \frac{zF}{RT} V_T P_{Cl} \frac{[Cl]_b - [Cl]_l exp[zFV_T/(RT)]}{1 - exp[zFV_T/(RT)]}$$

where $[Cl_l]$ is the mean luminal [Cl] and $[Cl]_b$ is the bath [Cl]; $P_{Cl}$ is the paracellular Cl permeability, $V_T$ is the transepithelial voltage, and z, F, R, and T have their usual meaning.

Statistics were performed by analysis of variance with posthoc comparison by the REGWF test. The null hypothesis was rejected at the 0.05 level of significance.

## RESULTS

Twenty-nine tubules were perfused *in vitro* (13 in the NaHCO₃ group, 8 in the KHCO₃ group, and 8 in the KCl group). There was no significant difference in

**TABLE 1.** Diet Group

|  | Control | High K, KHCO$_3$ | High K, KCl |
|---|---|---|---|
| $V_T$ (mV) | 0.1 ± 1.5 | −14.8 ± 5.9[a] | −35.4 ± 5.4[a] |
| $J_{Cl}^N$ (pmol · mm$^{-1}$ · min$^{-1}$) | −3.4 ± 6.5 | −26.3 ± 9.3[a] | 17.2 ± 5.1[a] |
| $J_{Cl}^A$ (pmol · mm$^{-1}$ · min$^{-1}$) | −5.3 ± 6.6 | −43.8 ± 8.9[a] | −20.0 ± 4.4[a] |
| $J_{Cl}^P$ (pmol · mm$^{-1}$ · min$^{-1}$) | 1.9 ± 1.8 | 17.6 ± 6.4[a] | 37.2 ± 6.2[a] |

[a] Significantly different from zero.

tubule length or flow rate for the three groups. TABLE 1 lists $V_T$ (in mV), $J_{Cl}^N$, $J_{Cl}^A$, and $J_{Cl}^P$ (all in pmol · mm$^{-1}$ · mm$^{-1}$) for each group. $V_T$ for the control group was close to zero (0.1 ± 1.5) and net Cl transport was not significantly different from zero ($J_{Cl}^N = -3.4 \pm 6.5$). Likewise, neither $J_{Cl}^A$ (−5.3 ± 6.6) nor $J_{Cl}^P$ (1.9 ± 1.8) was significantly different from zero. In the KHCO₃ group, $V_T$ (−14.8 ± 5.9) was more lumen negative than in the NaHCO₃ group and significant net Cl secretion was present ($J_{Cl}^N = -26.3 \pm 9.3$). Moreover, there was significant active Cl secretion ($J_{Cl}^A = -43.8 \pm 8.9$) and passive Cl absorption ($J_{Cl}^P = 17.6 \pm 6.4$). In contrast to the KHCO₃ group, the KCl group exhibited significant net Cl absorption ($J_{Cl}^N = 17.2 \pm 5.1$). However, this group displayed the most lumen-negative voltage (−35.4 ± 5.4). Hence, active Cl secretion ($J_{Cl}^A = -20.0 \pm 4.4$) and passive Cl absorption ($J_{Cl}^P = 37.2 \pm 6.2$) were stimulated also.

To examine whether direct coupling of K and Cl secretion was present, tubules were perfused in random order with a luminal [Cl] of either 112 or 5 mM (gluconate substitution). In these studies, rabbits were adapted to the KCl diet. Reducing luminal Cl concentration did not significantly alter $V_T$ (−10.6 ± 2.7 high Cl *vs* −16.8 ± 6.7 low Cl, $p$ = ns) or net Na absorption (24.3 ± 6.0 high Cl *vs* 12.0 ± 9.3 low Cl, $p$ = ns). However, K secretion consistently increased in each of six tubules from 13.6 ± 3.1 high Cl to 20.1 ± 3.2 low Cl, $p < 0.01$ (FIG. 1).

## DISCUSSION

Previous studies have demonstrated that Cl absorption occurs in the CCD by voltage-driven paracellular transport Cl, Cl-HCO$_3$ exchange, and NaCl absorption. The present observations demonstrate that Cl secretory flux (bath-to-lumen) exceeds absorptive flux (lumen-to-bath) under conditions of KHCO$_3$ loading. Under these conditions, we have unequivocal evidence for active (and hence transcellular) Cl secretion, because Cl entry was against its electrochemical gradient. However, equally convincing is the evidence for significant bidirectional Cl transport in the KCl group. Using a value for P$_{Cl}$ of $6.9 \times 10^{-6}$ cm/s[3] that is less than the directly measured P$_{Cl}$,[4] there was significant paracellular Cl absorptive flux ($37.2 \pm 6.2$) and significant transcellular Cl secretion ($20.0 \pm 4.4$). In fact, all 16 tubules in the KCl and the KHCO$_3$ groups exhibited active Cl secretion. Thus, both methods of K loading enhanced active transcellular Cl secretion. Moreover,

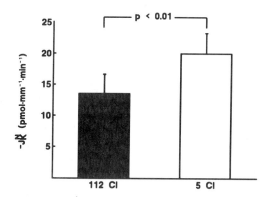

**FIGURE 1.** Effect of reducing luminal [Cl] from 112 mM to 5 mM on net Cl secretion ($-J_K^N$) by the CCD. Reducing luminal [Cl] stimulated K secretion without significantly changing V$_T$ or net Na absorption.

reducing luminal [Cl] consistently stimulated K secretion without affecting V$_T$ or net Na flux. These facts suggest a coupling of a component of Cl and K secretory flux at the luminal membrane of the CCD.

## REFERENCES

1. STOKES, J. B., M. J. INGRAM, A. D. WILLIAMS & D. INGRAM. 1981. Heterogeneity of the rabbit collecting tubules: Localization of mineralocorticoid hormone action to the cortical portion. Kidney Int. **20:** 340–347.
2. STONER, L. C., M. B. BURG & J. ORLOFF. 1974. Ion transport in cortical collecting tubule: Effect of amiloride. Am. J. Physiol. **227:** 453–459.
3. HANLEY, M. D. & J. P. KOKKO. 1978. Study of chloride transport across the rabbit cortical collecting tubule. J. Clin. Invest. **62:** 39–44.

4.  SANSOM, S. C., E. J. WEINMAN & R. G. O'NEIL. 1984. Microelectrode assessment of chloride-conductive properties of cortical collecting duct. Am. J. Physiol. **247** (Renal Fluid Electrolyte Physiol. **16**): F291–F302.
5.  WINGO, C. S. 1989. Potassium secretion by the cortical collecting tubule: Effect of Cl gradients and ouabain. Am. J. Physiol **256** (Renal Fluid Electrolyte Physiol. **25**): F306–F313.

# Effect of pH on Cl⁻ and K⁺ Permeabilities of the Frog Corneal Epithelium[a]

ALDO ZAMUDIO,[b] OSCAR A. CANDIA,[b,c,d] AND
LAWRENCE ALVAREZ[b]

*Departments of Ophthalmology,[b] and of Physiology
and Biophysics[c]
Mount Sinai School of Medicine of the City University
of New York
New York, New York 10029-6574*

Previous work established that lowering the pH of the tear (apical) bathing solution from 8.6 to 7.2 by increasing $pCO_2$ from 0.3 to 5% produced a reversible 40% reduction of the Cl⁻-dependent short-circuit current ($I_{SC}$) and net Cl⁻ transport across the isolated frog corneal epithelium.[1] These effects could have resulted from either direct reduction of apical Cl⁻ permeability or depolarization of the cell compartment produced by closing the basolateral K⁺ channels. Thus, the effects of pH on intracellular electrical parameters were investigated with conventional microelectrodes, as used previously.[2] Increasing $pCO_2$ to 5% hyperpolarized the intracellular potential difference ($PD_{SC}$) by 5 mV, increased the apical/basolateral resistance ratio ($R_a/R_b$), and reduced $I_{SC}$, effects consistent with a decrease in apical Cl⁻ permeability by cellular acidification. The addition of 20 mM $NH_4^+$ to the apical solution (to induce cellular alkalinization due to the permselectivity of the apical membrane to $NH_3$) elicited an immediate and large stimulation of $I_{SC}$, a 5–12 mV depolarization of $PD_{SC}$ and a decrease in $R_a/R_b$ from 0.49 to 0.43. Within 90 seconds $PD_{SC}$ reversed towards hyperpolarization, and $R_a/R_b$ began to increase, usually surpassing their control values, while $I_{SC}$ remained stimulated (FIG. 1). The biphasic effects of $NH_4^+$ are consistent with an increase in apical Cl⁻ permeability, followed by a comparable increase in basolateral K⁺ permeability.

Additional experiments with Cl⁻-sensitive microelectrodes determined that acidification produced an increase of about 2 mV in the cellular-to-bath Cl⁻ electrochemical gradient ($\Delta\bar{\mu}_{Cl}$), whereas alkalinization reduced $\Delta\bar{\mu}_{Cl}$ by about 4 mV (FIG. 2).

To confirm that the changes in electrical parameters were actually due to changes in the intracellular pH ($pH_i$) of the corneal epithelium a method that uses the pH-sensitive, cytoplasmic-entrapped, fluorescent dye 2′,7′-biscarboxyethyl-

[a] This work was supported by grants EY00160 and EY01867 from NIH and a Research to Prevent Blindness Award to O.A.C.

[d] Corresponding author.

5(6)-carboxyfluorescein (BCECF) was implemented. The pH-sensitive probe was essentially confined to the epithelial cells with little or no diffusion to the stroma. The details for the tissue mount, dye loading, superfusion, and dye excitation were identical to those described for the crystalline lens.[3] Because the apical side of the corneal epithelium was unresponsive to the "K⁺-nigericin" method,[4] an

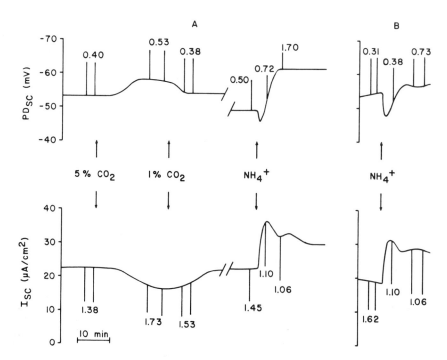

**FIGURE 1.** Effects of increased $pCO_2$ and of $NH_4^+$ on the intracellular potential difference and short-circuit current of the isolated frog corneal epithelium. Chart recordings from two experiments (**A** and **B**) were traced. The NaCl Ringer's solution that bathed each side of the cornea contained 25 mM $HCO_3^-$ and 75 mM Cl⁻, and was continuously bubbled with 1% $CO_2$. *Arrows* indicate the points at which $pCO_2$ was increased and reduced, and when 10 mM $(NH_4)_2SO_4$ was added to the apical-side solution. (*Upper traces*) Recording of intracellular potential difference with respect to the bathing solution. Spikes are the current-induced deflections that are used to calculate $R_a/R_b$, the value of which is indicated by numbers next to the spikes. (*Lower traces*) Short-circuit current (indicating transepithelial net Cl⁻ transport). Downward spikes represent resistance measurement elicited by a 14-mV offset (14 $mV/\Delta I_{SC}$). Numbers indicate transepithelial resistance in $K\Omega \cdot cm^2$.

external calibration curve was used to obtain a minimal estimate of changes in $pH_i$. It was found that the addition of 20 mM $NH_4^+$ to Ringer's solution (1% $CO_2$ bubbling, pH 8.1) increased $pH_i$ by $0.21 \pm 0.06$ units (mean $\pm$ SD, $n = 6$ corneas) and that increasing $pCO_2$ by bubbling the Ringer's solution with 5% $CO_2$ (pH 7.2) decreased $pH_i$ by $0.47 \pm 0.04$ units ($n = 7$).

This study suggests that cellular pH modifies, in a manner similar to $Cl^-$ secretagogues,[2] both apical $Cl^-$ and basolateral $K^+$ permeabilities so that substantial changes in $Cl^-$ secretion ensue.

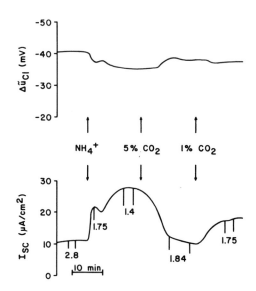

**FIGURE 2.** Effects of increased $pCO_2$ and of $NH_4^+$ on the intracellular electrochemical potential of $Cl^-$ and short-circuit current of the isolated frog corneal epithelium. Chart recordings from a typical experiment were traced. With the exception that apical-side $[Cl^-]$ was reduced to 10 mM, experimental details were as in FIGURE 1. (*Upper trace*) Recording of intracellular electrochemical potential of $Cl^-$ with respect to the apical-side bathing solution. (*Lower trace*) Short-circuit current (indicating transepithelial net $Cl^-$ transport). Downward spikes represent resistance measurements (14 $mV/\Delta I_{SC}$). Numbers indicate transepithelial resistance in $K\Omega \cdot cm^2$. In other experiments $pCO_2$ changes were done before the addition of $NH_4^+$, and the results were symmetrical.

## REFERENCES

1. CANDIA, O. A. 1973. Effect of pH on chloride transport across the isolated bullfrog cornea. Exp. Eye Res. **15:** 375–382.
2. WOLOSIN, J. M. & O. A. CANDIA. 1987. $Cl^-$ secretagogues increase basolateral $K^+$ conductance of frog corneal epithelium. Am. J. Physiol. **253:** C555–C560.
3. WOLOSIN, J. M., L. J. ALVAREZ & O. A. CANDIA. 1988. Cellular pH and $Na^+$-$H^+$ exchange activity in lens epithelium of *Bufo marinus* toad. Am. J. Physiol. **255:** C595–C602.
4. THOMAS, J. A., R. N. BUCHSBAUM, A. ZIMNIAK & E. RACKER. 1979. Intracellular pH measurements in Ehrlich ascites tumor cells utilizing spectroscopic probes generated *in situ*. Biochemistry **18:** 2210–2218.

# Mechanism of pH-Independent Transport of $NH_4^+$ by Mouse Renal Medullary Thick Limbs

D. KIKERI, A. SUN, M. L. ZEIDEL, AND S. C. HEBERT

*Harvard Medical School*
*Boston, Massachusetts 02115*

The medullary thick ascending limb of Henle (MAL) actively absorbs ammonium in the absence of a favorable transepithelial pH gradient.[1-4] To define the mechanisms involved in pH-independent transport of $NH_4^+$, intracellular pH ($pH_i$) was measured using BCECF in both the isolated perfused mouse MAL (IP-MAL) and suspensions of mouse MAL tubules (S-MAL). In nonbicarbonate HEPES-buffered Ringer's solution pH 7.4, the steady-state $pH_i$ was $7.26 \pm 0.04$ in IP-MAL ($n = 10$) and $7.43 \pm 0.01$ in S-MAL ($n = 12$). In IP-MAL, the addition of 20 mM $NH_4Cl$ to the medium bathing the luminal surface resulted in a striking acidification of $pH_i$ to $6.43 \pm 0.08$, whereas the addition of 20 mM $NH_4Cl$ to the medium bathing the basolateral side of the tubule led to rapid alkalinization of $pH_i$ from $7.09 \pm 0.08$ to $7.4 \pm 0.16$. Thus the basolateral membranes of MAL cells are highly $NH_3$ permeable, while apical membranes are highly $NH_4^+$ permeable. The addition of $NH_4Cl$ in concentrations greater than 0.5 mM to the media bathing S-MAL or to both apical and basolateral media bathing IP-MAL led to significant acidification of $pH_i$, indicating that overall cell membrane $NH_4^+$ entry was greater than entry of $NH_3$. $NH_4Cl$-induced acidification was also observed in S-MAL in the presence of $CO_2/HCO_3^-$-buffered media, and in HEPES-buffered media, the stilbene DIDS (0.2–0.5 mM) did not inhibit or attenuate the $NH_4Cl$-induced acidification response. Thus, bicarbonate exit pathways do not participate in the acidification response.

In IP-MAL, the acidification response to luminal $NH_4Cl$ was completely inhibited by the combination of 10 mM $BaCl_2$ and 0.1 mM furosemide, whereas in S-MAL either $Ba^{2+}$ or bumetanide (0.1 mM) partially inhibited ammonium-induced acidification, indicating that both apical $K^+$ channels and $Na:K:2Cl$ cotransporters mediate luminal $NH_4^+$ entry into MAL cells; however, the major fraction of $NH_4^+$ entry was via the $Ba^{2+}$-sensitive pathway. The decrement in $pH_i$ observed with the luminal 20 mM $NH_4^+$ addition in IP-MAL was approximately two times greater than that in S-MAL at the same $NH_4Cl$ concentration. This was the result of a very low permeability of the apical cell membrane to $NH_3$ as demonstrated by the lack of cell alkalinization in IP-MAL with luminal addition of $NH_4Cl$ when $NH_4^+$ entry pathways were completely blocked. Recovery of $pH_i$ in S-MAL following $NH_4^+$ withdrawal was inhibited more than 85% by either sodium removal or amiloride (0.1 mM), but not by $Ba^{2+}$ or furosemide. Thus, $NH_4^+$ exit from MAL cells requires $Na:H$ exchange. Moreover, because this cation cannot exit cells by $K^+$ channels or $Na:K:2Cl$ cotransport, $NH_4^+$ transport by these pathways is apparently rectified (i.e., they can mediate the entry, but not exit, of $NH_4^+$). The $Na:H$ exchanger required for $NH_4^+$ exit was localized to the apical plasma membrane in the IP-MAL; apical but not basolateral amiloride inhibited $pH_i$ recovery after $NH_4^+$ withdrawal.

A model for pH-independent absorption of $NH_4^+$ is presented in FIGURE 1. $NH_4^+$ entry from the lumen occurs via an apical $Ba^{2+}$-sensitive pathway, likely $K^+$ channels, and furosemide (or bumetanide)-sensitive Na : K : 2Cl cotransporters. $NH_4^+$ exit involves proton exit into the lumen via Na : H exchange coupled to $NH_3$ diffusion across the highly $NH_3$-permeable basolateral membranes. Protons could be transported from the lumen to the medullary interstitium via the cation-selective paracellular pathway. $NH_4^+$ backleak from the interstitium to the lumen would

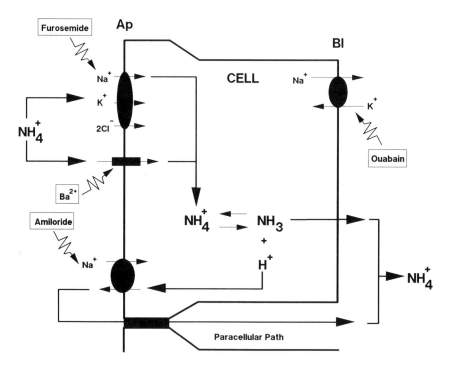

**FIGURE 1.** Model for the pH-independent transport of ammonium. *Arrows* represent the modes for absorptive transport of ammonium. $NH_4^+$ enters cells via the apical (Ap) $Ba^{2+}$-inhibitable transporter, probably a $K^+$ channel and the apical furosemide (bumetanide)-inhibitable $Na^+ : K^+ : 2Cl^-$ cotransporter. $NH_4^+$ exit from cells depends on $NH_3$ exit across basolateral (Bl) membrane, presumably mediated by a solubility-diffusion process, coupled with $H^+$ exit across the apical amiloride-inhibitable $Na^+ : H^+$ exchanger. $H^+$ may recycle to the basolateral solution by diffusion through the cation-selective paracellular pathway.

be minimized, because apical membranes are poorly $NH_3$ permeable, and because apical $NH_4^+$ transport pathways do not mediate the exit of $NH_4^+$ from MAL cells.

## REFERENCES

1. GOOD, D. W. & M. A. KNEPPER. 1985. Am. J. Physiol. **248:** F459–F471.
2. GOOD, D. W., M. A. KNEPPER & M. B. BURG. 1984. Am. J. Physiol. **247:** F35–F44.
3. GOOD, D. W. 1987. J. Clin. Invest. **80:** 1358–1365.
4. GOOD, D. W. 1988. Am. J. Physiol. **255:** F78–F87.

# Index of Contributors